Springer

Berlin
Heidelberg
New York
Hong Kong
London
Milan
Paris
Tokyo

Flow Modulation and Fluid–Structure Interaction at Airplane Wings

Research Results of the Collaborative Research Center SFB 401 at RWTH Aachen, University of Technology, Aachen, Germany

Josef Ballmann (Editor)

Springer

Professor Dr. Josef Ballmann
RWTH Aachen
Lehr- und Forschungsgebiet Mechanik
Templergraben 64
D - 52062 Aachen
Germany

ISBN 978-3-642-53613-7 ISBN 978-3-540-44866-2 (eBook)
DOI 10.1007/978-3-540-44866-2

Library of Congress Cataloging-in-Publication-Data

Flow modulation and fluid-structure interaction at airplane wings : research results of the collaborative research center SFB 401 at RWTH Aachen, University of Technology, Aachen, Germany / Josef Ballmann (editor).
p. cm. -- (Notes on numerical fluid mechanics and multidisciplinary design ; v. 84)
Includes bibliographical references.

1. Air flow. 2. Airplanes -- Wings. 3. Aerodynamics. I.Ballmann, Josef. II. Rheinisch-Westfälische Technische Hochschule Aachen. III. Notes on numerical fluid mechanics and multidisciplinary design ; v. 84)
TL574.F5F58 2003
629.132'32--dc21

Springer-Verlag Berlin Heidelberg New York
a member of BertelsmannSpringer Science+Business Media GmbH
http://www.springer.de

Softcover reprint of the hardcover 1st edition 2003

Typesetting: camera-ready by editor
Cover design: deblik Berlin
Printed on acid free paper 62/3020/M - 5 4 3 2 1 0

NNFM Editor Addresses

Prof. Dr. Ernst Heinrich Hirschel
(General editor)
Herzog-Heinrich-Weg 6
D-85604 Zorneding
Germany
E-mail: e.h.hirschel@t-online.de

Prof. Dr. Kozo Fujii
Space Transportation Research
Division
The Institute of Space
and Astronautical Science
3-1-1, Yoshinodai, Sagamihara,
Kanagawa, 229-8510
Japan
E-mail: fujii@flab.eng.isas.ac.jp

Dr. Werner Haase
Höhenkirchener Str. 19d
D-85662 Hohenbrunn
Germany
E-mail: wug.haase@t-online.de

Prof. Dr. Bram van Leer
Department of Aerospace Engineering
The University of Michigan
Ann Arbor, MI 48109-2140
USA
E-mail: bram@engin.umich.edu

Prof. Dr. Michael A. Leschziner
Imperial College of Science,
Technology and Medicine
Aeronautics Department
Prince Consort Road
London SW7 2BY
U. K.
E-mail: mike.leschziner@ic.ac.uk

Prof. Dr. Maurizio Pandolfi
Politecnico di Torino
Dipartimento di Ingegneria
Aeronautica e Spaziale
Corso Duca degli Abruzzi, 24
I - 10129 Torino
Italy
E-mail: pandolfi@polito.it

Prof. Dr. Jacques Periaux
Dassault Aviation
78, Quai Marcel Dassault
F-92552 St. Cloud Cedex
France
E-mail: periaux@rascasse.inria.fr

Prof. Dr. Arthur Rizzi
Department of Aeronautics
KTH Royal Institute of Technology
Teknikringen 8
S-10044 Stockholm
Sweden
E-mail: rizzi@aero.kth.se

Dr. Bernard Roux
L3M – IMT La Jetée
Technopole de Chateau-Gombert
F-13451 Marseille Cedex 20
France
E-mail: broux@l3m.univ-mrs.fr

PREFACE

The research work of the collaborative research center SFB401 *Flow Modulation and Fluid-Structure Interaction at Airplane Wings* at the Rheinisch-Westfälische Technische Hochschule (RWTH) Aachen, which is reported in this book, was possible due to the financial support of the Deutsche Forschungsgemeinschaft (DFG). The proposal has been approved after evaluation by the referees of DFG selected from other universities and industry, which is gratefully acknowledged. The work is still in progress and now approved to continue until the end of year 2005.

More than 50 scientists from universities of the United States, Russia, France, Italy, Japan, Great Britain, Sweden, Netherlands, Switzerland, Austria and research organizations NASA, ONERA, NLR, DLR could be invited and have visited the research center, gave seminars on their research on related topics and some of them stayed longer for joined work.

Besides its scientific value, also the importance of the program for scientific education becomes evident by looking at the numbers of completed theses, which are up to now about 15 doctoral theses, 40 diploma theses and 70 study theses.

The authors of this book acknowledge the valuable support coming from all these persons and institutions. They are especially grateful to the referees having reviewed this work, A. Cohen (Université Pierre et Marie Curie), J. Cooper (Manchester School of Engineering), W. Devenport (Virginia Tech.), M. Drela (MIT), F. Gern (Avionics Specialties Inc.), A. Griewank (TU Dresden), H. Hönlinger (DLR), P. Hovland (Argonne National Laboratory), F. Menter (AEA Technology), M. Pandolfi (Politecnico di Torino), V.J. Rossow (NASA Ames), G. Schewe (DLR), F. Seiler (ISL), T. Sonar (TU Braunschweig), S.P. Spekreijse (NLR), I.W. Tjatra (NLR), H. Tijdeman (University of Twente), R. Voss (DLR), P. Wernert (ISL), for the valuable time they spent to improve the papers collected in this volume.

Furthermore, it is the desire of the authors to express their deep gratitude to Dr. Dieter Funk from Deutsche Forschungsgemeinschaft for all his valuable suggestions and the competent guidance during all processes related to the funding.

Finally, the editor gives thanks to his students Frank Bramkamp, Carsten Braun and Michael Hesse who did the technical editing work and last but not least to Professor E.H. Hirschel and Springer Verlag who made this edition possible in the series Notes on Numerical Fluid Mechanics and Multidisciplinary Design.

Aachen, January 2003
Josef Ballmann
(Co-ordinator of SFB401)

CONTENTS

1. Introduction

The Collaborative Research Center SFB 401: *Flow Modulation and Fluid-Structure Interaction at Airplane Wings* is concerned with fundamental problems of very high capacity aircrafts with large elastic wings designated for transonic transport. This volume presents a survey of the current research of the center from the first up to the fifths year, with emphasis on the results achieved so far.

The content of the three chapters following this introduction correspond to the three main topics of large transport aircrafts, which the center concentrates on: (i) Model flow, wakes and vortices of airplanes in high lift configuration, (ii) Numerical tools for large scale adaptive flow simulation based on multiscale analysis and a parametric mapping concept for grid generation, and (iii) Validated computational design tools based on direct aeroelastic simulation with reduced structural models.

On these topics fundamentals and methods for flow modulation, the physical and mathematical modelling, the mathematical well-posedness and the numerical solution are elaborated in altogether 14 projects in cooperation of the departments for Aircraft Design and Aeronautics (ILR), Aerodynamics (AIA), Lightweight Structures (IfL), Flight Dynamics (FD), Mathematics and Numerical Mathematics (IGPM), Scientific Computing (SC), and Mechanics (LFM), all at RWTH Aachen, University of Technology. All mathematical/numerical efforts are joined by experimental analysis in the full range of Mach and Reynolds numbers. For experimental analysis water tunnels, low speed wind tunnels and transonic wind and shock tunnels are employed.

The Collaborative Research Center has defined a reference configuration with a supercritical profile, which is based on the three-element high-lift airfoil system BAC3-11/RES/30/21 (described in AGARD-AR 303).

The first two contributions of Chapter 2 are concerned with design features minimizing the hazard by the vortex system which is generated by very large aircrafts during take-off and landing for following aircrafts. This purpose requires the knowledge of the vortical flow field within a distance up to 100 spans behind the airplane. Presently, neither large scale wind tunnel measurements nor numerical Navier-Stokes solvers can provide the necessary data without an extreme effort. Therefore, it is useful to choose a water tunnel as experimental facility which is now available in the center for small scale wake investigations in a distance of up to 60 spans. In this context it seems also advisable to apply available vortex transport methods in reduced dimensions for layout of test configurations and measurement interpretation. Nevertheless, at least in the near wake wind tunnel measurements together with the most sophisticated available Navier-Stokes solver have to be applied. The third contribution of Chapter 2 is concerned therewith. Furthermore, it provides improved turbulence modelling and novel boundary conditions and simulation techniques for the analysis of the near wake and wake-jet interaction behind high-lift configurations. The fourth and fifth contributions of this chapter consider transonic flow experimentally and numerically. The first one is concerned with the optimum design of an adaptive airfoil with respect to viscous drag reduction by downstream displacement of transition in the transonic regime. The other describes a new transonic shock tunnel for high Reynolds number flow corresponding to real flight conditions. The facility has been developed and built in the center and is now operating up to $Re_c = 40 \cdot 10^6$ at transonic speeds. First results are presented.

Chapter 3 faces the challenge to essentially improve the quality of numerical solutions and to include later on the aeroelastic quality into the design processes of the aerodynamic shape and the supporting structure. For these tasks large scale simulations of compressible fluid flow and fluid-structure interaction are required. The first paper of Chapter 3 reports the present state in the development of the new solver QUADFLOW for this purpose. The method is developed by a group of mathematicians and engineers. In order to keep the size of the discrete problems at every stage as small as possible for any given target accuracy, a multi-resolution local adaptation strategy is employed. The aim is a solution method which automatically detects all relevant flow features such as shocks and shear layers in flows about complex configurations even when starting the computation on a very coarse grid and without any knowledge about the expected flow field. A mesh generation concept has been introduced that is to support the adaptive strategy as well as possible. One of its key ideas is to understand meshes as parametric mappings determined by possibly few control points, as opposed to storing every mesh cell separately. A mapping based on B-Splines with low number of control points allows for a surface conforming grid evaluation. An unstructured finite volume discretization is employed which is capable to operate on meshes with hanging nodes, occurring due to the local mesh adaptation. In the presented state of development, the method has proven to apply excellently for shock configurations, as e.g. the famous fish-tail shock formation at the NACA0012 airfoil, and for shear and boundary layers as well. The second contribution of the chapter reports on automatic differentiation, which is used to calculate sensitivities of flow solutions with respect to parameters as e.g. Mach number, angle of attack or surface shape parameters. It can also be used to evaluate Jacobi matrices arising in implicit flow solvers. A key feature is its ability to generate accurate derivatives rather than approximations obtained from numerical differentiation by use of divided differences.

Chapter 4 is devoted to computational and experimental aeroelasticity. In the first contribution, which is fundamental for computational aeroelasticity, well-posedness and modelling of nonlinear aeroelasticity is considered. An elastic panel in transonic flow was chosen as a model problem for mathematical analysis and computational experiments. Based on energy methods a uniqueness proof for small times has been established. In that context the energy transfer between structure and fluid became evident as an essential feature. As a consequence, independently of possibly different discretizations in the fluid and in the structural domain, it is very important to simulate the energy transfer between the different phases in the discretized system correctly. On the one hand, it will be essential for the further development of QUADFLOW, and on the other hand, it is to be observed in the already established aeroelastic codes which are based on algorithmic code coupling of single purpose solvers, i.e. existing finite volume methods for the Euler or Navier-Stokes equations and finite element codes for structural dynamics.

In order to predict the aeroelastic quality of large span airplane wings already in the design phase and to exploit afterwards the respective knowledge for the detailed design of the supporting wing structure, reduced structural models are preferred firstly. Correspondingly, the second contribution of Chapter 4 is concerned with concepts for those reduced models. A membrane panel model is presented as an example. It extends the multi-axial Timoshenko beam model presently employed for aeroelastic simulations in the third contribution of the chapter. Additional effects are incorporated, such as taper, restrained warping and the effect of ribs as well as different material properties within the wing cross section. The fluid flow in the already applied aeroelasticity code is modelled optionally by the conservation laws according to Navier-Stokes or Euler. Hamiltons principle applied to the Timoshenko beam forms the base for the coupling and the structural dynamics code. Results are presented for the transonic dip of an airfoil and for fluid-structure interaction of a highly flexible rectangular wing in subsonic flow, which validate the code in this flow regime.

A method for approximated stability analysis of nonlinear dynamic systems and its application to an aircraft with a flexible wing has been developed in the fourth contribution. It extends existing averaging methods imitating the linearized eigenvalue analysis, but defines the eigenvalue and eigenvector as functions of the deflection from the actually examined equilibrium state.

The fifth contribution analyzes in a basic unsteady airfoil study at low speeds the influence of fundamental unsteady parameters on the practical transition limit from a quasi-steady to a truly unsteady flow field. The problem is investigated experimentally in a water tunnel and numerically employing a finite element solver for incompressible fluids. The sixth contribution of this chapter presents the experimental setup for the measurement of unsteady pressure and forces on a 2–D rectangular wing section oscillating in pitch at transonic flow conditions. First, steady flow fields with free and forced transition in the wind tunnel with straight and adapted walls are analyzed and on this basis detailed analyzes of unsteady flow fields have been performed to provide validation data in the transonic regime for the aeroelastic code development. Extended aeroelastic experimemts in the subsonic regime are reported in the last contribution. It describes the design and development of a highly flexible wind tunnel model of a straight wing and its behaviour under wind off and wind on conditions in steady subsonic flow. The shapes of the deformed wing model with zero lift have been determined experimentally and numerically as well for various flow conditions. Unsteady pressure, forces and displacement measurements have been conducted in pluck tests. All measurements compare well with numerical results of the method described in the third contribution of this chapter.

2. Model Flow, Wakes and Vortices in High Lift Configuration

Experimental Analysis and Modulation of Vortices

Erol Özger, Ingmar Schell, Dieter Jacob
Lehrstuhl für Luft- und Raumfahrt, RWTH Aachen
Wüllnerstraße 7, D-52062 Aachen, Germany

Summary

This paper presents the results of an experimental investigation of the wake vortex structure behind different wings with and without flaps. The wind-tunnel test is part of a project funded to investigate the flow structure in the wake of different wings. The purpose of this investigation is to study the main features of lift generated vortices in order to find ways to alleviate hazardous wake vortex encounters for following airplanes during take-off and approach such that an increase in airport capacity can be achieved. Therefore, the wakes of different wings without flaps are investigated to examine the influence of different wing planforms on the structure of the wing tip vortex and the induced rolling moment. Then, the wake of a swept and tapered wing is investigated for different flap settings. Finally, a wing fin is mounted on the wings with and without flaps to examine its alleviating influence on the wake structure. The results show that the wing planform and the flap deflection have a considerable influence on the wake structure and the induced rolling moment on a following aircraft and that a wing fin can lead to an alleviation of the wake.

1 Introduction

The hazard posed to following aircraft during take-off and approach is mainly caused by the wake vortices generated by a preceding aircraft. Because of the low altitude of the aircraft during take-off and approach an induced rolling moment caused by the vortices may lead to a dangerous flight attitude. Therefore minimum separation distances between aircraft during take-off and approach are to be kept, which have been introduced in the 1960s and which are conservative in order to avoid dangerous situations. However the growth of commercial aviation has placed such a demand on the air traffic system that many major airports are capacity limited and are experiencing significant traffic delays so that a reduction of the separation distances is desirable. The wake vortex problem and the associated safety issues become even more important for heavy future high capacity transport aircraft (e.g. A380, B-747X) because the strength of the vortices is approximately proportional to the gross weight of the wake generating aircraft. Important issues to be tackled are the structure of the wake and methods to influence the far field wake to alleviate the effects of a wake vortex encounter. Therefore this paper investigates experimentally the wake structure in the near field of different wings without flaps and of a swept and tapered wing for different flap settings as well as the effect of a wing fin on the wake structure and the achievable alleviation. Moreover the far field of the different wing configurations is investigated numerically in [24].

In the past the wake structure of unflapped wings has been investigated both experimentally [4, 7, 8] and numerically [1, 7, 9, 11, 14, 20], starting from simple Betz methods [1, 9, 20] to elaborated Navier-Stokes computations [7, 11]. But these investigations did not lead to a definite understanding of the influence of the geometry on the wake structure, mainly because of the lack of appropriate measurements and computations as well as non-dimensional wake vortex parameters. Experimental and numerical investigations on the wake of flapped wings are even fewer in number [2, 3, 5, 12, 13, 7, 16, 21, 22] due to the higher complexity of the wake structure. In [2, 3] a detailed survey of the wake structure was carried out for different flaps settings but a quantitative characterization of the hazard posed to a following aircraft was not performed. In contrast Rossow [18] quantified the hazard posed to following aircraft but did not investigate the detailed structure of the wake.

The attenuation of the vortex wake is still a challenging issue although many investigations have been performed since the 1960s [5, 6, 10, 17, 18, 19]. The prime goal is the large scale distribution of vorticity in the wake. Basically, this can be accomplished by influencing the lift distribution or by increasing the dispersion of vorticity. As an example, the investigations of Rossow [18, 19] show that the induced rolling moment on a following aircraft can be reduced to one third by means of wing fins without great loss in lift. Here, a wing fin is defined as a small wing mounted perpendicular to the wing surface with an angle of incidence to the flow.

In the current investigation the detailed wake structure of different wings without flaps and slats as well as of a swept and tapered wing with flaps was experimentally determined. Vortex parameters such as core radius, maximum tangential velocity, circulation and vorticity distribution as well as the induced rolling moment on a following aircraft were evaluated and non-dimensionalized in an appropriate way. The alleviating effect of wing fins, as used in Rossow's investigations [18, 19], on the wake structure of the wings with and without flaps was investigated in detail by comparing the vortex parameters and the induced rolling moment on a following aircraft for the configurations with and without fin.

The emphasis of the whole investigation is on the influence of configuration changes (wing planform, flap settings, fins) on the wake and vortex parameters and particularly on the induced rolling moment on a following aircraft. Correspondingly, the discussion is mainly concerned with the differences in the above mentioned parameters and not their absolute values. As the effects of vortex meandering, minor probe interferences or other sources of errors are assumed to be nearly the same for all investigated configurations and, moreover, negligible for the hazard posed to following aircraft this approach seems justified.

2 Experimental Setup

The experiments were conducted in the Göttinger type low speed wind-tunnel of the Institut für Luft- und Raumfahrt (ILR) (see figure 1). The test section has a length of $L = 3m$ and a nozzle diameter of $D = 1.5m$. The maximum freestream velocity is $U_\infty = 75m/s$. In these experiments a half-model mounted on a test rack was used (see figure 2). Suction is applied at the nose of the rack to reduce the boundary layer growth on the half model symmetry plane. In order to adjust the angle of attack α of the wing models an electrically powered turn table is integrated within the rack. Aerodynamic forces and moments are measured by a six-component strain gauge balance. Velocities (u, v, w) in the wake of the wing are measured by means of an automatically traversed probe. The probe can be traversed in all three axis directions (x, y, z) so that velocities can be recorded in arbitrary planes perpendicular to the freestream velocity.

Figures 3 and 4 show the wing models used in this investigation. To investigate the influence of different wing planforms on the vortex wake a rectangular and a tapered wing without flaps were used. Theses wings are segmented in spanwise direction so that various configurations can be realized. Figure 3 shows the rectangular and tapered wing model (no. 1 and 4) and three other configurations with different sweep angles and aspect ratios that were used in this investigation (no. 2, 3 and 5). The wing geometry data (span b, mean cord c, sweep angle φ, taper ratio λ and aspect ratio Λ) are specified in the table of figure 3. The wing profile is a BAC 3-11/RES/30/21 profile with a typical shape for transonic speeds.

To investigate the effect of different flap settings on the vortex wake a swept and tapered wing model with slats and flaps was used as displayed in figure 4. The wing has a span of $b = 1.35m$ and a mean chord of $c = 0.153m$. The sweep angle of the leading edge is $\varphi = 34°$, the aspect ratio amounts to $\Lambda = 8.8$ and the wing profile is also a BAC 3-11/RES/30/21 profile. The flap section consists of two segments, an inboard and an outboard flap, that can be deployed independently at $\delta_i, \delta_o = 0°, 10°, 20°$. For the configuration with retracted flaps the slats were also retracted while for all configurations with deployed flaps they were fully deployed at $\delta_s = 25°$.

The experiments were performed at a freestream velocity of $U = 40m/s$ without fixed transition and a Reynolds number based on the mean wing chord of about $Re = 3 \cdot 10^5$ for the wings without flaps and $Re = 4 \cdot 10^5$ for the swept and tapered wing with flaps. At first, force measurements and oil painting visualizations were conducted for each configuration to assure that no major separation occurs and to analyze the flow structure on the suction side of the wing. Moreover smoke was injected at the wing tip and the flap edge to gain a three dimensional view of the vortices in the wake. Then velocity measurements were performed by means of hotwire anemometry. The 3D-hotwire probes have a diameter of the sensors of $d_P = 5\mu m$ and an active length of $l_P = 1mm$ so that the measurement volume of the hotwire probe is about $V_P = 1mm^3$. The probe size and the Reynoldsnumber correspond approximately to the values in [8] so that a probe interference is deemed to be minor. The velocity data were recorded at a frequency of $f = 1kHz$ for a measurement period of $t = 1sec$. Each of the hotwire sensors is operated separately using a Dantec Frame 90N10 with three anemometer units CTA 90C10. For calibration of the probes the Dantec calibration system 90H10 is used.

The velocities downstream of the wing were measured in planes perpendicular to the freestream velocity at four distances x based on the span b of the complete wing: $x/b = 0.0, 0.2, 0.4$ and 0.8 respectively 1.0. First, planes with a coarse step size ($\Delta y = \Delta z = 7.5mm$) were measured to evaluate the position of the vortices and to get an overall impression of the wake structure. Then, grids with a fine step size ($\Delta y = \Delta z = 2.0mm$) were used to resolve finer structures of the vortices. In figure 5 the coordinate system and the investigated areas are shown.

3 Evaluation of data

Force measurements were evaluated to compute the lift coefficient:

$$c_L = \frac{L}{q \cdot S} \quad . \tag{1}$$

where q is the dynamic pressure and S the wing surface. The vorticity ω was computed from the velocity data:

$$\omega = \frac{\partial w}{\partial y} - \frac{\partial v}{\partial z} \quad . \tag{2}$$

From the vorticity distribution in the wake of the wings the position of the vortex centers was determined as the locus of the local maximum of vorticity in the wake. The local distribution of the radial and circumferential velocity (v_r, v_θ) of the vortices was computed on beams emanating from the vortex center. The core radius R_c and the maximum tangential velocity $V_{\theta,max}$ were evaluated as an average value of the local core radius and the local maximum tangential velocity as shown in figure 6. The circulation Γ in the wake of the wing was computed according to

$$\Gamma = \oint \vec{v} \cdot d\vec{s} \quad . \tag{3}$$

along the outer contour of the coarse grid. To measure the hazard of the vortex wake posed to a horizonatally flying, following aircraft the induced rolling moment coefficient c_l based on the halfspan b_f of the following aircraft was computed from the velocity data in the wake of the wing:

$$c_l = \frac{2}{b_f^2} \int_{-b_f/2}^{b_f/2} c_{L\alpha} \cdot \arctan\left(\frac{w}{u}\right) \cdot \xi \cdot d\xi \quad . \tag{4}$$

The evaluation of the induced rolling moment coefficient c_l is based on the strip theory where the following wing is divided into a series of cordwise strips (see figure 7). Each strip is treated as a two-dimensional airfoil and the lift induced by the generator aircraft is computed as a function of the local flow angle, which is obtained from the velocity components u and w. Together with the lift gradient $c_{L\alpha}$ and the lever arm ξ the incremental rolling moment is integrated on the whole span b_f. Here, the span ratio of $b_f/b = 0.2$ is chosen. To determine the maximum induced rolling moment coefficient $c_{l,max,rel}$ the induced rolling moment $c_{l,rel}$ is computed at each point of the measured planes in the wake of the wing by moving the following aircraft.

To present the results the evaluated wake vortex parameters were non-dimensionalized in an appropriate way to ensure a reasonable comparison among the investigated configurations. Thus, it is possible to compare the wake and vortex parameters of wings with different lift coefficients and wing spans. A detailed derivation of the non-dimensional wake vortex parameters can be found in [23]. The circulation Γ_{rel} was non-dimensionalized by means of the average circulation $\bar{\Gamma}$ so that the circulation of wings with different lift coefficients can be compared:

$$\Gamma_{rel} = \frac{\Gamma}{\bar{\Gamma}} = \frac{\Gamma}{1/2 \cdot U_\infty \cdot c_L \cdot c} \quad . \tag{5}$$

The non-dimensionalized average core radius R_c was defined according to:

$$R_{c,rel} = \frac{R_c}{b} \tag{6}$$

where the wing span b and not the mean chord c is taken as a reference value. Thus, the effect of different wing spans can be compensated so that only the influence of the wing planform or the flap setting remain. This approach is advantageous when for a required wing span the vortex geometry is to be determined. The average maximum tangential velocity V_θ was non-dimensionalized by means of the average circulation $\bar{\Gamma}$ and the core radius R_c that were used to compute a reference tangential velocity $\bar{V_\theta}$:

$$V_{\theta,rel} = \frac{V_\theta}{\bar{V_\theta}} = \frac{V_\theta}{\bar{\Gamma}/R_c} \cdot const. = \frac{V_\theta \cdot \Lambda}{U_\infty \cdot c_L} \quad . \tag{7}$$

The non-dimensional vorticity ω_{rel} was determined accordingly by means of the average circulation $\bar{\Gamma}$ and the core radius R_c that were used to compute a reference vorticity $\bar{\omega}$:

$$\omega_{rel} = \frac{\omega}{\bar{\omega}} = \frac{\omega}{\bar{\Gamma}^2/R_c} \cdot const. = \frac{\omega \cdot \Lambda \cdot b}{U_\infty \cdot c_L} \quad . \tag{8}$$

The induced rolling moment c_l was non-dimensionalized by means of the reference tangential velocity $\bar{V}_\theta$ and the freestream velocity U_∞:

$$c_{l,rel} = \frac{c_l}{\bar{V}_\theta / U_\infty} = \frac{c_l \cdot \Lambda}{c_L} \quad . \tag{9}$$

Thus, the effect of different lift coefficients and wing spans are compensated for the maximum tangential velocity, the vorticity and the induced rolling moment so that just the influence of the flap settings, the wing planform or fins remain.

To quantify the alteration of the vortex structure and the induced rolling moment in the wake of the different wings due to the influence of the wing fin the change in the core radius

$$\Delta R_c / R_c = \frac{R_{c,withfin} - R_{c,withoutfin}}{R_{c,withoutfin}} \quad , \tag{10}$$

in the maximum tangential velocity

$$\Delta V_{\theta,max} / V_{\theta,max} = \frac{V_{\theta,max,withfin} - V_{\theta,max,withoutfin}}{V_{\theta,max,withoutfin}} \quad , \tag{11}$$

and in the maximum induced rolling moment

$$\Delta c_{l,max} / c_{l,max} = \frac{c_{l,max,withfin} - c_{l,max,withoutfin}}{c_{l,max,withoutfin}} \tag{12}$$

was calculated.

4 Results

The presentation of the results is divided into three parts. First, the wake structure and the induced rolling moment on a following aircraft in the wake of wings without flaps are analyzed. Then, the results obtained with the swept and tapered wing with slats and flaps are presented. Finally, alleviation with a wing fin is discussed.

4.1 Wings without flaps

The vorticity distribution ω_{rel} in the wake of the rectangular wing computed from the velocity components v, w perpendicular to the freestream is shown in figure 8.

High values of vorticity ω_{rel} are encountered in the tip region of the wake. The wing tip vortex can be clearly seen as the locus of maximum vorticity. Due to the influence of the wing tip vortex the wake wraps around the vortex core.
Detailed distributions of the vorticity ω_{rel} in the region of the wing tip vortex are shown in figure 9. The vortex remains concentrated within the investigated wake region $x/b \leq 1.0$ so that the maximum values of vorticity do not decrease significantly. Due to the influence of the wake the vortex moves on a curved path in negative y-direction.

The distribution of circulation Γ_{rel} in the wake of the different wings without flaps is shown in figure 10 and 11. There is almost no difference in the circulation Γ_{rel} at the trailing edge of the various untapered wings. The circulation Γ_{rel} increases steeply in the region of the wing tip and is nearly constant in the middle section of the wing. In contrast, the tapered wing has a circulation distribution Γ_{rel} that increases constantly along the wing span. Thus the circulation is more evenly distributed in the wake and the overall circulation in the wake is higher for the tapered wing. At the wing root a decrease in circulation can be discerned for all configurations due to the influence of the boundary layer of the wall.

For wings without flaps the hazard posed to a following aircraft depends largely on the distribution of circulation in the wing tip region of the wake generating aircraft. It determines the tangential velocities and the induced rolling moment on a following aircraft. As can be seen in figure 11 the amount of circulation Γ_{rel} in the wing tip region is lower for the tapered wing than for the configurations of the rectangular wing. Moreover the small rectangular wing with the lower aspect ratio of $\Lambda = 5.0$ has the lowest gradient of circulation Γ_{rel} in the region of the wing tip vortex.

The structure of the wing tip vortex of the various wings without flaps is described by means of the core radius $R_{c,rel}$, the maximum tangential velocity $V_{\theta,max,rel}$ and the maximum vorticity $\omega_{max,rel}$ in the figures 12, 13 and 14. Due to the viscosity in the flow the core radius $R_{c,rel}$ slightly increases while the maximum tangential velocity $V_{\theta,max,rel}$ and the maximum vorticity $\omega_{max,rel}$ decrease. The differences in the core radius $R_{c,rel}$ are mainly caused by the different aspect ratios Λ of the wing. Thus, the wing tip vortex of the small rectangular wing ($\Lambda = 5.0$) has the greatest core radius $R_{c,rel}$ that is twice as large as the core radius of the wing tip vortex of the rectangular wing ($\Lambda = 10.0$). Therefore a high core radius $R_{c,rel}$ of the wing tip vortex corresponds to a wing with a low aspect ratio Λ. As the core radius $R_{c,rel}$ decreases with increasing aspect ratio Λ the maximum tangential velocity $V_{\theta,max,rel}$ and the maximum vorticity $\omega_{max,rel}$ increase correspondingly.

In figure 15 the maximum induced rolling moment coefficient $c_{l,max,rel}$ in the wake of the wings without flaps is shown. At the trailing edge the value of the maximum induced rolling moment $c_{l,max,rel}$ is high caused by the downwash of the wing. Further downstream of the wing the maximum induced rolling moment $c_{l,max,rel}$ decreases and is approximately constant downstream from $x/b = 0.4$. The highest value of the maximum induced rolling moment $c_{l,max,rel}$ is encountered in the wake of the rectangular wing. A decreasing aspect ratio Λ of the wing leads to a lower maximum induced rolling moment $c_{l,max,rel}$ which can be seen for the small rectangular wing and the highly swept wing. The lowest value of the maximum induced rolling moment $c_{l,max,rel}$ occurs in the wake of the tapered wing.

The reduction of the induced rolling moment $c_{l,max,rel}$ caused by the lower aspect ratio is due to the structure of the wing tip vortex that is less concentrated which can be clearly seen in the higher value of the core radius $R_{c,rel}$ and the lower value of the maximum tangential velocity $V_{\theta,max,rel}$ and vorticity $\omega_{max,rel}$ in figure 12, 13 and 14. In contrast to this the lower taper ratio of the wing leads to a reduction of circulation Γ_{rel} (see figure 10 and 11) within the wing tip vortex that effects a decrease of the maximum induced rolling moment $c_{l,max,rel}$ while the vortex structure is nearly unchanged (see figure 12, 13 and 14).

4.2 Wing with flaps

A typical vorticity distribution ω_{rel} in several planes measured in the wake of the swept and tapered wing with a flap setting of $\delta_i = \delta_o = 20°$ is shown in figure 16. The wake of the wing rolls up into two strong and stable vortices, the wing tip vortex and the outboard flap vortex. The wing tip vortex shows a single branched structure while the outboard flap vortex shows a double branched structure. The value of the maximum vorticity slightly decreases downstream for both vortices due to the friction in the wake flow while the change in size is negligible. Moreover the flap vortex is larger in size and shows higher values of vorticity ω_{rel} compared to the wing tip vortex. Additionally, a weak inboard flap vortex is generated caused by the discontinuity of the geometry between the inboard and midboard wing segment.

Figure 17 shows the circulation distribution Γ_{rel} at the trailing edge of the various flap configurations. By deploying the flaps the circulation Γ_{rel} is redistributed from the wing tip to the inner part of the wing so that the amount of circulation Γ_{rel} in the wing tip vortex decreases while the amount of circulation Γ_{rel} in the flap vortices increases correspondingly.

Figure 18 to 23 show the structure of the wing tip and outboard flap vortex by means of the core radius $R_{c,rel}$, the maximum tangential velocity $V_{\theta,max,rel}$ and the maximum vorticity $\omega_{max,rel}$. The size of the wing tip vortex is not influenced by the flap deflection as can be seen from the core radius $R_{c,rel}$ of the wing tip vortex in figure 18. The maximum tangential velocity $V_{\theta,max,rel}$ and the maximum vorticity $\omega_{max,rel}$ decrease with increasing flap angle δ_i, δ_o (see figure 19 and 20) caused by the decrease of the circulation Γ_{rel} in the wing tip vortex (see figure 17).

In contrast to the size of the wing tip vortex the size of the outboard flap vortex is influenced by the flap deflection. This can be seen by the core radius $R_{c,rel}$ in figure 21 that decreases with increasing flap angle δ_i, δ_o. This concentration as well as the increase of the circulation Γ_{rel} of the outboard flap vortex (see figure 17) leads to an increase of the maximum tangential velocity $V_{\theta,max,rel}$ and the maximum vorticity $\omega_{max,rel}$ of the outboard flap vortex as shown in figure 22 and 23.

In figure 24 the maximum induced rolling moment $c_{l,max,rel}$ is shown. The highest value of the maximum induced rolling moment $c_{l,max,rel}$ occurs in the wake of the wing when the flaps are retracted. This may be explained by the fact that for this configuration the circulation Γ_{rel} in the wake rolls up into one wing tip vortex whereas for the other configurations with deployed flaps the circulation in the wake is divided into two vortices (three vortices for $\delta_i = 20°, \delta_o = 10°$). This division leads to a reduction of the maximum induced rolling moment $c_{l,max,rel}$, at least in the investigated wake field up to $x/b = 0.8$, where no merger of the wing tip and the outboard flap vortex occurs.

If the circulation in the wake is distributed more equally on the vortices the maximum induced rolling moment $c_{l,max,rel}$ shows the smallest values. This is true for the configurations with moderate flap deflection ($\delta_i = \delta_o = 10°$, two vortices, and $\delta_i = 20°, \delta_o = 10°$, three vortices). Higher flap deflections, e.g. flaps angles of $\delta_i = \delta_o = 20°$, produce an outboard flap vortex that is much stronger than the wing tip vortex so that the value of the maximum induced rolling moment $c_{l,max,rel}$ is much higher.

4.3 Alleviation by means of wing fins

4.3.1 Rectangular wing without flaps

In figure 25 a sketch of the rectangular wing with a wing fin at the tip is shown. The wing fin was fastened in between the spanwise segments of the rectangular wing, both on the suction (positive sign of ε_F) and the pressure side (negative sign of ε_F), so that four configurations with a wing fin were realized, $\varepsilon_F = 8.53°, 34.0°$, $-8.53°$, $-34.0°$.

The smoke visualization in figure 26 shows that without fin the wing tip vortex is concentrated and stable within the investigated wake area. Due to the wing fin the wing tip vortex is disturbed which leads to a widening and a fluctuation of the position of the wing tip vortex (see figure 27). The lift curves in figure 28 show that the lift c_L of the wing is slightly influenced by the wing fin. The small loss in lift is caused by the separation in the wake of the wing fin that leads to a change in the circulation distribution Γ_{rel} at the wing tip (see figure 29). Therefore the angle of attack of the configurations with a wing fin was adjusted such that data at a constant lift coefficient can be compared.

The wake of the wing fin consists of a fin tip vortex whose sense of rotation depends on the angle of incidence ε_F (same sense of rotation as the wing tip vortex for positive ε_F), and of a necklace vortex at the base of the fin. The vorticity distribution ω_{rel} in the wake of the rectangular wing with a wing fin at $\varepsilon_F = 34°$ is shown in figure 30. The necklace vortex that wraps around the root of the wing fin could not be resolved sufficiently in the measurements. But in comparison with the vorticity distribution ω_{rel} for the rectangular wing without fin in figure 8 it can be seen that the wing fin generates additional vorticity in its wake. Moreover the wing fin produces additional turbulence u'/U_∞ that interferes with the wing tip vortex which can be seen by comparing figure 31 and 32. The amount of vorticity and turbulence that is produced in the wake of the fin depends on its angle of incidence ε_F. At a high angle of incidence ε_F the fin is stalled so that the flow separates around the wing fin and the level of turbulence in the wake of the fin is high. In contrast to this a lower angle of incidence ε_F leads to a higher production of vorticity in the wake of the fin.

The amount of wake diffusion can be defined by determining the influence of the wing fin on the core radius $R_{c,rel}$ and on the maximum tangential velocity $V_{\theta,max,rel}$ of the vortices in the wake. The alteration of the structure of the wing tip vortex is shown in figure 33 and 34 by the change of the core radius $\Delta R_c/R_c$ and the maximum tangential velocity $\Delta V_{\theta,max}/V_{\theta,max}$. As can be seen in figure 33 the wing tip vortex is widened due to the influence of the wing fin. Correspondingly, the maximum tangential velocity $V_{\theta,max,rel}$ of the wing tip vortex decreases. This can be attributed to the additional turbulence u'/U_∞ and vorticity ω_{rel} that is generated by the wing fin.

The circulation distribution Γ_{rel} in the wing tip vortex at a distance of one span behind the wing ($x/b = 1.0$) is shown in figure 35. The wing fin effects a reduction of circulation in the wing tip vortex that is especially distinct for the wing fin with an angle of incidence of $\varepsilon_F \pm 34°$. This is caused by the large production of turbulence in the wake of the stalled fin at high angles of incidence ε_F that leads to a widening of the wing tip vortex.

The alteration of the structure and the circulation distribution of the wing tip vortex due to the wing fin influences the maximum induced rolling moment $c_{l,max,rel}$ in the wake of the wing. The change of the maximum induced rolling moment $\Delta c_{l,max}/ c_{l,max}$ compared to the wing without wing fin is shown in figure 36. The strongest decrease of maximum induced rolling moment $c_{l,max,rel}$ (ca. 25 percent) corresponds to the lowest value of circulation Γ_{rel} in the wing tip vortex that is caused by the wing fin with an angle of incidence of $\varepsilon_F = \pm 34°$.

4.3.2 Swept and tapered wing with slats and flaps

In figure 37 a sketch of the swept and tapered wing with a wing fin (fin height h_F, fin cord c_F and angle of incidence ε_F) and the testing program is shown. The wing fin ($c_F/b = h_F/b = 0.04$) was mounted at two different angles of incidence ($\varepsilon_F = \pm 20°$) at a spanwise position of the swept and tapered wing with fully deployed flaps ($\delta_i = \delta_o = 20°$) that corresponds to the outer flap edge.

The influence of the wing fin with positive angle of incidence on the wake flow can be seen by means of smoke visualization. Figure 38 displays the wing tip and the outboard flap vortex of the wing without fin. Both the wing tip and outboard flap vortex are concentrated and do not decay within the test section. Figure 39 displays the wing tip and outboard flap vortex of the wing with fin ($\varepsilon_F = 20°$). Due to the wake of the wing fin the outboard flap vortex diffuses rapidly downstream while the wing tip vortex is hardly influenced by the fin.

The influence of the wing fin on the overall lift coefficient c_L is shown in figure 40. The curves show that a large lift decrease of 7 percent occurs for the fin with positive angle of incidence. Therefore the angle of attack of the configurations with wing fin was adjusted to compensate for the loss in lift due to the flow separation induced by the wing fin. The effect of the wing fin on the circulation distribution Γ_{rel} as shown in figure 41 is correspondingly small.

In figure 42 the vorticity distribution ω_{rel} in planes perpendicular to the freestream velocity behind the wing with fin is shown for $\varepsilon_F = 20°$. The additional vorticity in the outboard flap vortex region produced by the wing fin can be seen at the first station $x/b = 0.0$ by comparing figure 42 and 16.

The increased diffusion of vorticity in the outboard flap region is mainly caused by the higher level of turbulence generated by the partially separated flow behind the wing fin. The difference in the turbulence level between the cases without and with wing fin can be seen in figure 43 and 44 which show the turbulence distribution u'/U_∞ in freestream direction.

As the outboard flap vortex dominates the wake flow of the swept and tapered wing only the alteration of the structure of the outboard flap vortex is discussed. For this, the change of the core radius and of the maximum tangential velocity ($\Delta R_c/R_c$, $\Delta V_{\theta,max}/\ V_{\theta,max}$) of the outboard flap vortex is shown in figure 45 and 46. A fin angle of $\varepsilon_F = 20°$ causes an increase of the core radius $\Delta R_c/R_c$ of the outboard flap vortex by more than 200 percent whereas for $\varepsilon_F = -20°$ almost no change of the core radius occurs. The widening of the outboard flap vortex leads to a corresponding decrease in the maximum tangential velocity as shown in figure 46. Therefore the configuration with a wing fin at an angle of incidence of $\varepsilon_F = -20°$ causes almost no decrease of maximum tangential velocity at $x/b = 0.8$ while a decrease of maximum tangential velocity of around 40 percent is achieved for the configuration with the wing fin at an angle of $\varepsilon_F = 20°$.

In figure 47 the change in maximum induced rolling moment $\Delta c_{l,max}/c_{l,max}$ is presented in terms of the distance to the trailing edge x/b. Only the configuration with the fin at a negative angle of incidence shows a decrease of 24 percent in maximum induced rolling moment in the last measured plane $x/b = 0.8$ compared to the reference configuration without fin. The configuration with a fin at a positive angle of incidence does not show a change in the maximum induced rolling moment at $x/b = 0.8$ compared to the reference configuration without fin.

The influence of the angle of incidence of the fin is presented in figure 48 and 49 which show the vorticity distribution ω_{rel} in the region of the outboard flaps vortex at $x/b = 0.8$. At a positive angle of incidence ε_F the flow at the wing fin is separated because of the high local angle of attack due to the inward directed flow on the upper side of the wing. This leads to a highly turbulent wake behind the wing fin which interacts with the flap vortex diffusing the vortex core. At a negative angle of incidence ε_F the flow at the wing fin leads to strong clockwise rotating fin tip vortex.

Therefore the level of turbulence in the wake of the fin is much lower compared to the wake of the fin with a positive angle of incidence. The fin tip vortex wraps around parts of the vorticity of the wing boundary layer so that the circulation in the boundary layer is rolled up into both the outboard flap vortex and the fin tip vortex. The smaller amount of circulation in the outboard flap vortex compared to the circulation in the outboard flap vortex for the configuration without fin causes the reduction of the maximum induced rolling moment.

Figure 50 shows the circulation distribution Γ_{rel} in the outboard flap vortex. Although the fin with positive angle of incidence ε_F diffuses the circulation within the inner part of the vortex it fails to distribute the circulation in the outer part (from $r/c = 0.3$). In contrast to this the fin with negative angle of incidence lowers the amount of circulation in the outer part (from $r/c = 0.3$) of the outboard flap vortex which leads to an appreciable decrease of the maximum induced rolling moment.

5 Conclusion

An experimental investigation of the wake structure of various wings without flaps and a swept and tapered wing with flaps and slats was conducted to examine the influence of the wing geometry on the wake structure and the induced rolling moment on a following aircraft. Velocity distributions in the wake of different wings without flaps and a swept and tapered wing with flaps and slats were measured by means of hotwire anemometry. Vortex parameters such as core radius, maximum tangential velocity as well as vorticity and circulation distributions were investigated. To present and compare the results the vortex parameters were non-dimensionalized in an appropriate way.

The investigation of the wake of the different wings without flaps showed that the structure of the wing tip vortex depends considerably on the aspect ratio of the wing. With increasing aspect ratio the wing tip vortex is more concentrated leading to a lower core radius and to higher values of the maximum tangential velocity and maximum vorticity. Correspondingly, the maximum induced rolling moment in the wake of the wing diminishes with decreasing aspect ratio. The taper ratio of a wing influences the circulation distribution in its wake, whereby the circulation is more evenly distributed in the wake of the wing for decreasing taper ratio. By this, the circulation of the wing tip vortex decreases which leads to a reduction of the induced rolling moment on a following aircraft though the structure of the wing tip vortex remains nearly unchanged. The effect of the sweep angle on the structure of the wing tip vortex and the induced rolling moment could not be clarified in this investigation because the sweep angle of the wings could not be adjusted independent from the aspect ratio. This topic will be dealt among others in the numerical investigation in [24].

For the swept and tapered wing with flaps and slats the structure of the vortices shed at the wing tip and the flap edge is highly dependent on the flap angle. For retracted flaps the circulation in the wake is rolled up into one tip vortex whereas deploying the flaps leads to a division of circulation into the wing tip and the flap vortex. This leads, especially for moderate flap angles, to a reduction of the maximum induced rolling moment within the investigated wake field where no merger of the wing tip and the flap vortex occurred.

By means of a wing fin mounted at the wing tip of the rectangular wing without flaps an alleviation of the wake of the wing was achieved. The wing fin produces a fin tip vortex, whose sense of rotation depends on the angle of incidence, and of a highly turbulent wake. The interaction of the fin wake with the wing tip vortex leads to a widening of the vortex and a dispersion of circulation especially for a high angle of incidence of the wing fin. As a result the core radius of the wing tip vortex increases and the maximum tangential velocity decreases correspondingly so that the maximum induced rolling moment diminishes.

For the swept and tapered wing with slats and flaps the wake of the wing fin which was mounted upstream of the outboard flap edge interacts with the flap vortex leading to a widening of the flap vortex and therefore to a decrease of maximum tangential velocity. A positive angle of incidence of the wing fin generates a co-rotating fin tip vortex which merges with the flap vortex leading to a large region of vorticity and turbulence. Correspondingly, a negative angle of incidence of the wing fin produces a counter-rotating fin tip vortex that does not merge with the flap vortex and influences the flap vortex by disturbing the roll-up process. This leads to a reduction of the maximum induced rolling moment at approximately one span behind the wing.

Due to the limitation of the test section length this experimental investigation was restricted to the near field of the different wings. Additionally, a numerical investigation of the far field of the wings was performed that is presented in [24].

References

[1] Betz, A., "Verhalten von Wirbelsystemen", ZAMM, Bd. XII, No. 3, pp. 164-174, 1932

[2] de Bruin, A. C., Hegen, S. H., Rohne, P. B., Spalart, P. R.,"Flowfield survey in trailing vortex system behind a civil aircraft model at high lift", Paper presented at the AGARD FDP Symposium on "The Characterization and Modification of Wakes from Lifting Vehicles in Fluids" held in Trondheim, Norway from 20-23 May 1996, and published in CP-584

[3] Chen, A. L., Jacob, J. D., Savas, Ö., "Dynamics of corotating vortex pairs in the wake of flapped airfoils", Journal of Fluid Mechanics, Vol. 382, pp. 155-193, Mar. 1999

[4] Chow, J. S., Zilliac, G. G., Bradshaw, P., "Near-Field Formation of a Turbulent Wing Tip Vortex", AIAA Paper 93-0551, Jan. 1993

[5] Corsiglia, V. R., Dunham, R. E., "Aircraft Wake Vortex Minimization by Use of Flaps", NASA SP-409, Symposium held at Washington D.C., Feb. 1976, pp. 305-338, 1977

[6] Croom, D. R., "The Development and Use of Spoilers as Vortex Attenuators", NASA SP-409, Symposium held at Washington D.C., Feb. 1976, pp. 339-368, 1977

[7] Dacles-Mariani, J., Zilliac, G. G., Chow, J. S., Bradshaw, P.,"Numerical/Experimental Study of a Wingtip Vortex in the Near Field", AIAA Journal, Vol. 33, No. 9, pp. 1561-1568, Sep. 1995

[8] Devenport, W. J., "The structure and development of a wing-tip vortex", Journal of Fluid Mechanics, Vol. 312, pp. 67-106, 1996

[9] Donaldson, C. DuP., Snedeker, R. S., Sullivan, R. D.,"Calculation of Aircraft Wake Velocity Profiles and Comparison with Experimental Measurements", Journal of Aircraft, Vol. 11, No. 9, pp. 547-555, 1974

[10] Dunham, R. E., "Unsuccessful Concepts for Aircraft Wake Vortex Minimization", NASA SP-409, Symposium held at Washington D.C., Feb. 1976, pp. 221-249, 1977

[11] Eaton, J. A., O' Flaherty, M. P., "Flowfield prediction of three-dimensional wing trailing vortices using advanced turbulence models", Paper presented at the AGARD FDP Symposium on "The Characterization and Modification of Wakes from Lifting Vehicles in Fluids" held in Trondheim, Norway from 20-23 May 1996, and published in CP-584

[12] Holbrook, G. T., Dunham, D. M., Greene, G. C. "Vortex Wake Alleviation Studies with a Variable Twist Wing", NASA Technical Paper-2442, 1985

[13] Hueneke, K., "Structure of a transport aircraft-type near field wake", Paper presented at the AGARD FDP Symposium on "The Characterization and Modification of Wakes from Lifting Vehicles in Fluids" held in Trondheim, Norway from 20-23 May 1996, and published in CP-584

[14] Krasny, R., "Computation of Vortex Sheet Roll-Up in the Trefftz Plane", Journal of Fluid Mechanics, Vol. 184, pp. 123-155, 1987

[15] Özger, E., Schell, I., Jacob, D., On the Structure and Attenuation of an Aircraft Wake, AIAA Journal of Aircraft, Vol. 38, No. 5, pp. 878-887, 2001

[16] Özger, E., Schell, I., Jacob, D., Wirbelbeeinflussung an Tragflügeln mit Hilfe von Finnen, DGLR-Jahrestagung 2000-053, Sep. 2000

[17] Rossow, V. J., "Inviscid Modelling of Aircraft Trailing Vortices", NASA SP-409, Symposium held at Washington D.C. Feb. 1976, pp. 9-59, 1977

[18] Rossow, V. J., "Experimental Investigation of Wing Fin Configurations for Alleviation of Vortex Wakes of Aircraft", NASA TM 78250, Nov. 1978

[19] Rossow, V. J., Fong, R. K., Wright, N. S., Bisbee, L. S., "Vortex Wakes of Two Transports in 80 by 120 Foot Wind Tunel", Journal of Aircraft, Vol. 33, No. 2, pp. 399-406, 1996

[20] Rossow, V. J., "Extended-Betz Methods for Roll-Up of Vortex Sheets", Journal of Aircraft, Vol. 34, No. 5, pp. 592-599, 1997

[21] Schell, I., Özger, E., Jacob, D., "Influence of Different Flap Settings on the Wake-Vortex Structure of a Rectangular Wing with Flaps and Means of Alleviation with Wing Fins", Aerospace Science and Technology, No. 4 (2000), pp. 79-90, 2000

[22] Schell, I., Özger, E., Jacob, D., "Influence of Different Flap Settings on the Hazard Posed to Following Aircraft", Notes on Numerical Fluid Mechanics, Vol. 77, pp. 66-73, 2002

[23] Schell, I., "Untersuchung zum Einfluss der Geometrie eines Flügels auf seinen Nachlauf", Ph.D. thesis, Institut für Luft- und Raumfahrt, RWTH Aachen, 2001

[24] Schell, I., Özger, E., Jacob, D., "Simulation of Vortex Sheets at Take-Off and Landing", chapter 2.2 of this volume

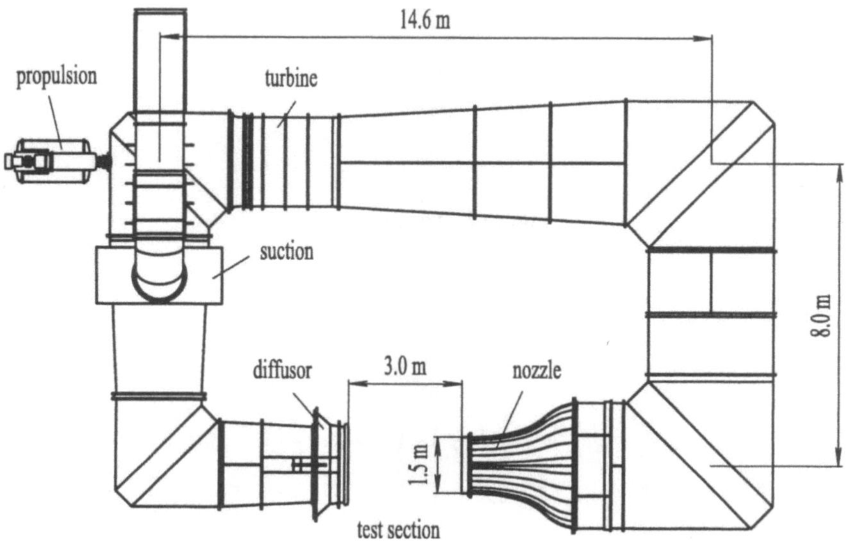

Figure 1 Göttinger type wind tunnel of the Institut für Luft- und Raumfahrt (ILR)

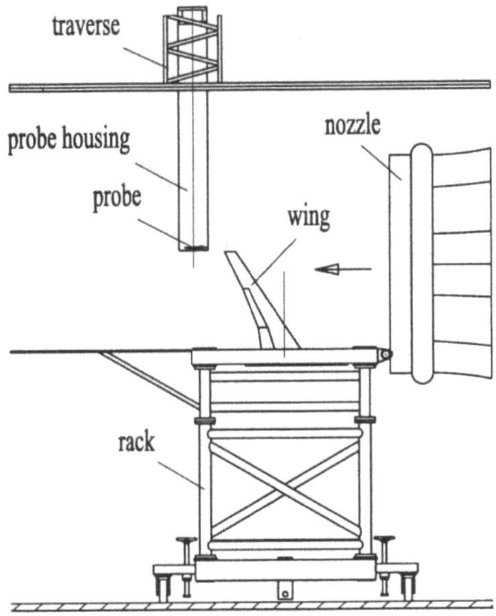

Figure 2 Test rack

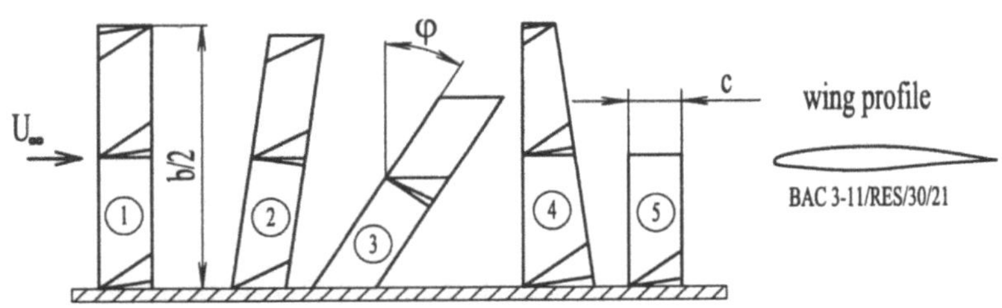

#	wing model	b [m]	c [m]	φ [°]	λ	Λ	c_L
①	rectangular	1.0	0.1	0.0	1.0	10.0	0.551
②	weakly swept	0.947	0.101	8.53	1.0	9.369	0.529
③	highly swept	0.697	0.121	34.0	1.0	5.781	0.421
④	tapered	1.0	0.1	0.0	0.455	10.0	0.549
⑤	small rectangular	0.5	0.1	0.0	1.0	5.0	0.450

Figure 3 Wings without flaps

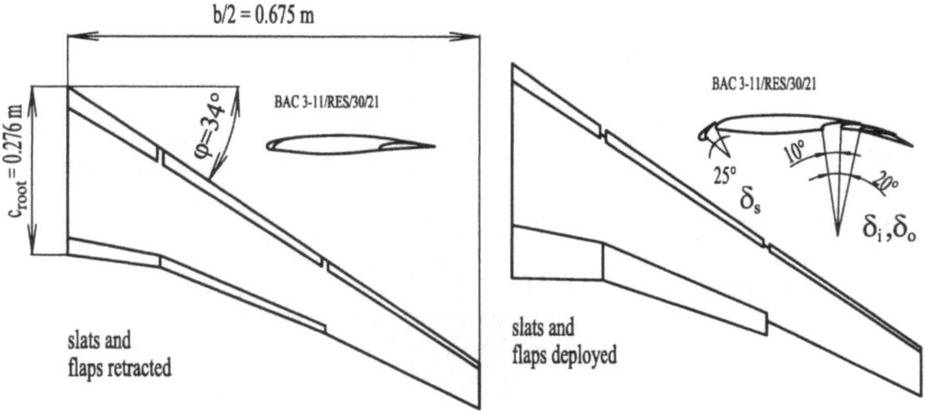

swept and tapered wing with slats and flaps	δ_i [°]	δ_o[°]	δ_s[°]	c_L
	0	0	0	0.652
	10	10	25	0.889
	20	20	25	1.337
	20	10	25	1.095

Figure 4 Swept and tapered wing with flaps and slats

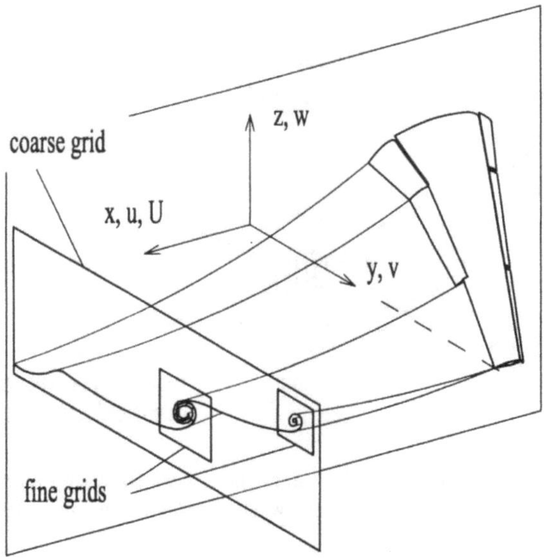

Figure 5 Wing with coordinate system and arrangement of measured planes

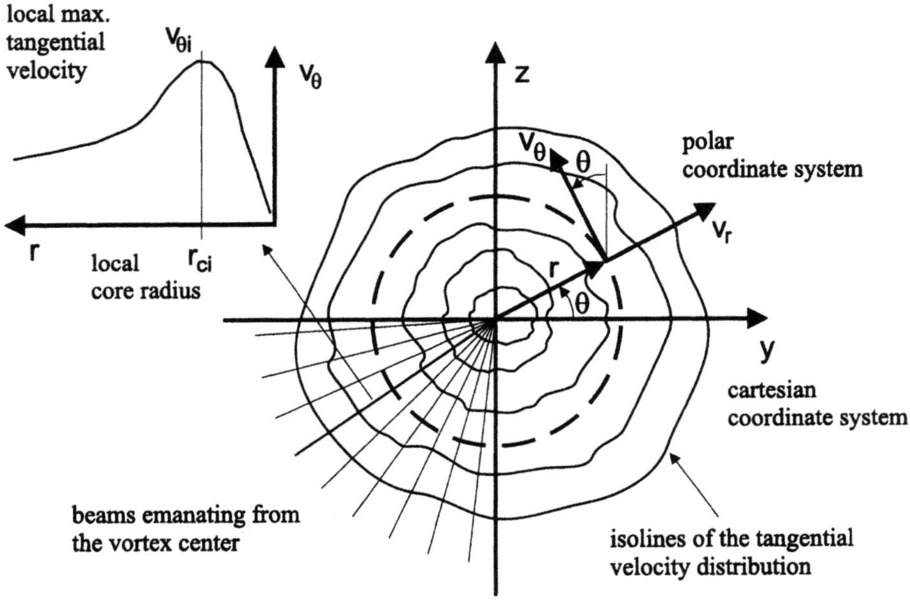

Figure 6 Evaluation of core radius R_c and maximum tangential velocity $V_{\theta,max,rel}$ in a vortex

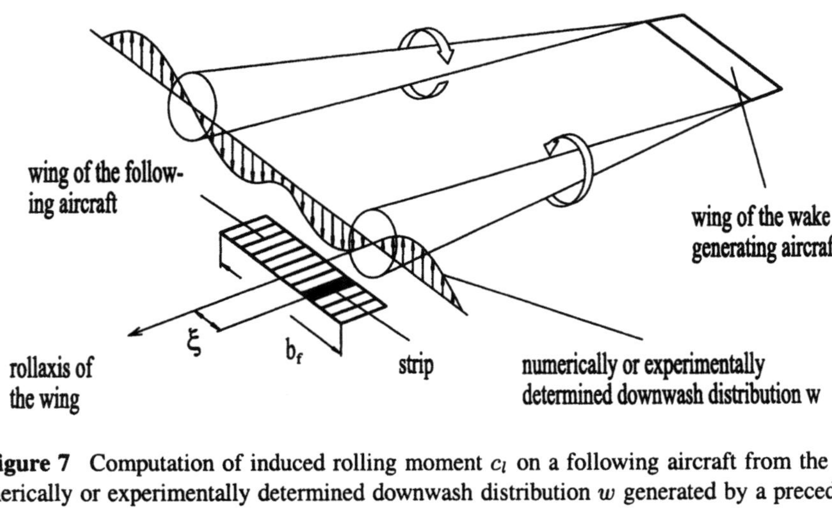

Figure 7 Computation of induced rolling moment c_l on a following aircraft from the numerically or experimentally determined downwash distribution w generated by a preceding aircraft

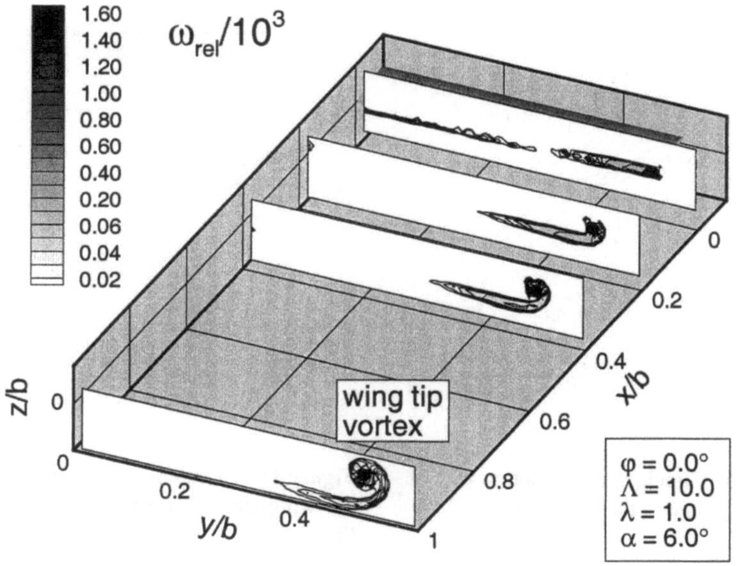

Figure 8 Distribution of vorticity ω_{rel} in the wake of the rectangular wing without flaps

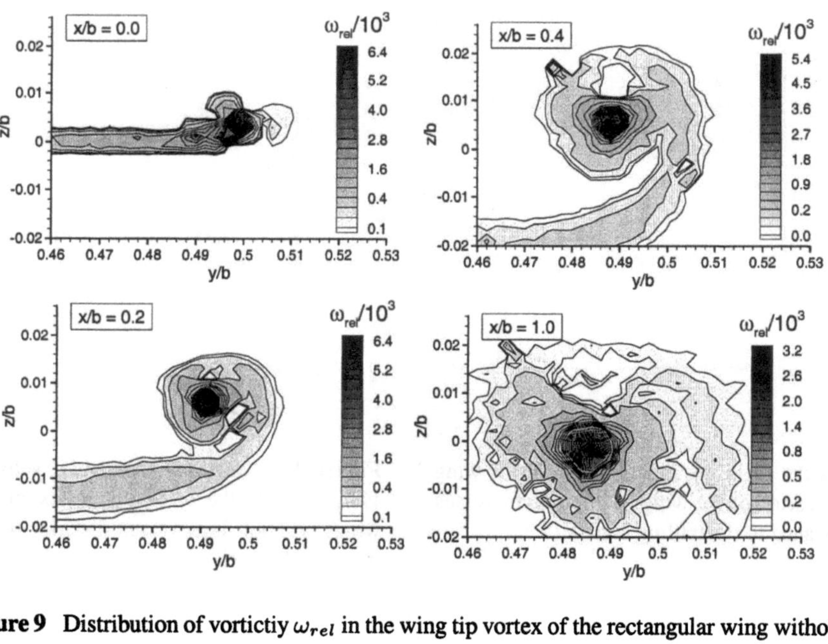

Figure 9 Distribution of vorticity ω_{rel} in the wing tip vortex of the rectangular wing without flaps

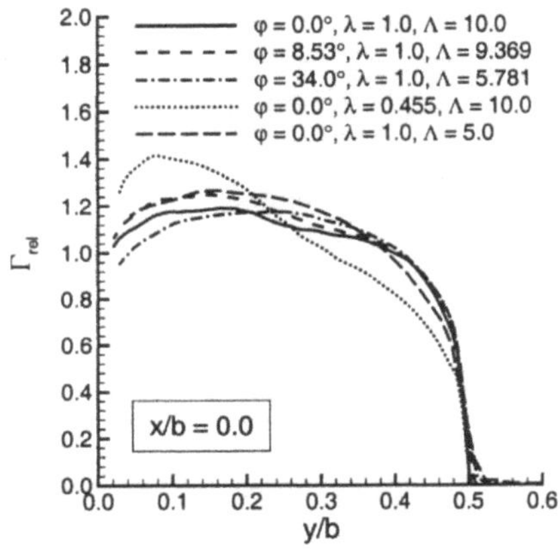

Figure 10 Circulation distribution Γ_{rel} at the trailing edge of the wings without flaps

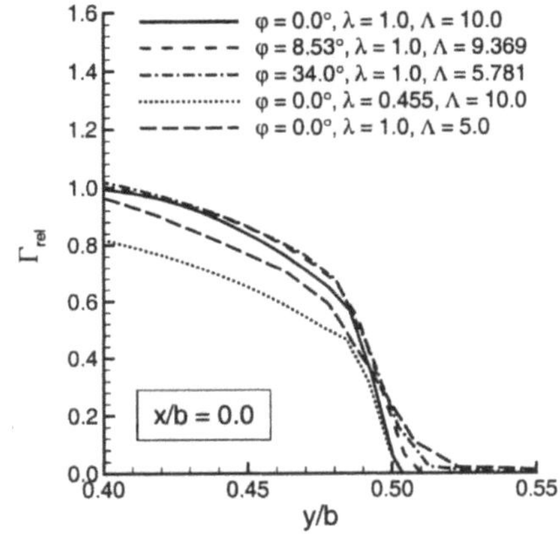

Figure 11 Circulation distribution Γ_{rel} in the wing tip region of the wings without flaps

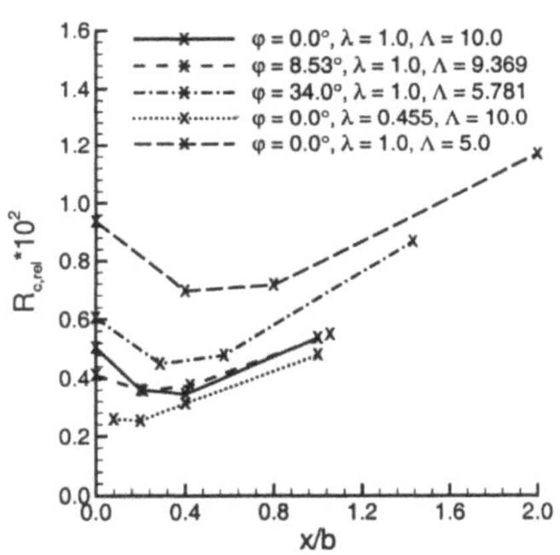

Figure 12 Core radius $R_{c,rel}$ of the wing tip vortex of the wings without flaps versus distance to the trailing edge x/b

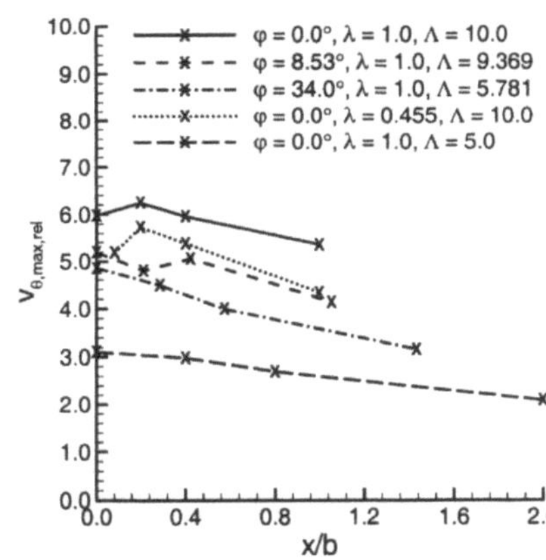

Figure 13 Maximum tangential velocity $V_{\theta,max,rel}$ of the wing tip vortex of the wings without flaps versus distance to the trailing edge x/b

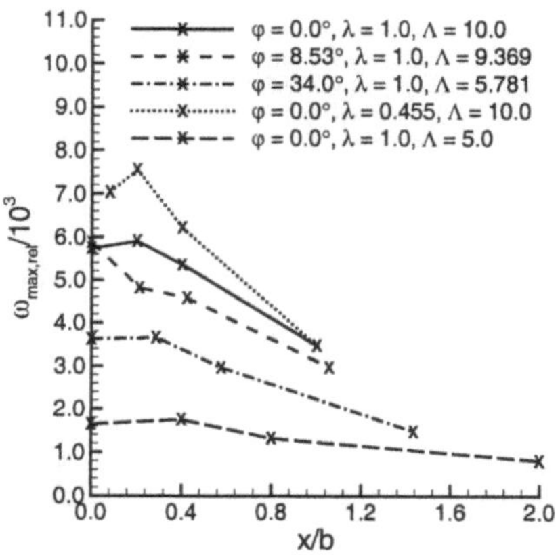

Figure 14 Maximum vorticity $\omega_{max,rel}$ of the wing tip vortex of the wings without flaps versus distance to the trailing edge x/b

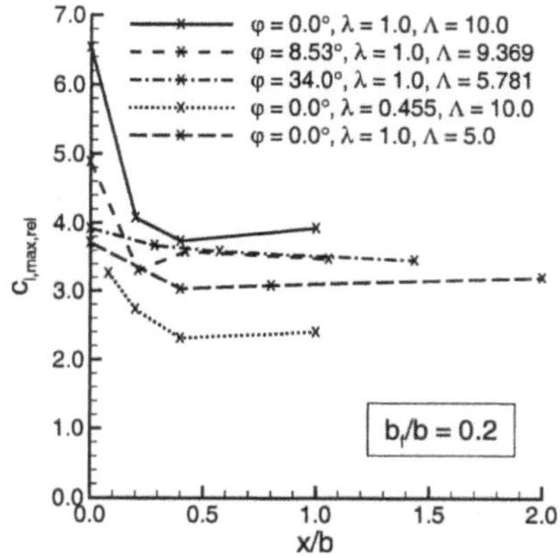

Figure 15 Maximum induced rolling moment $c_{l,max,rel}$ in the wake of the wings without flaps versus distance to the trailing edge x/b

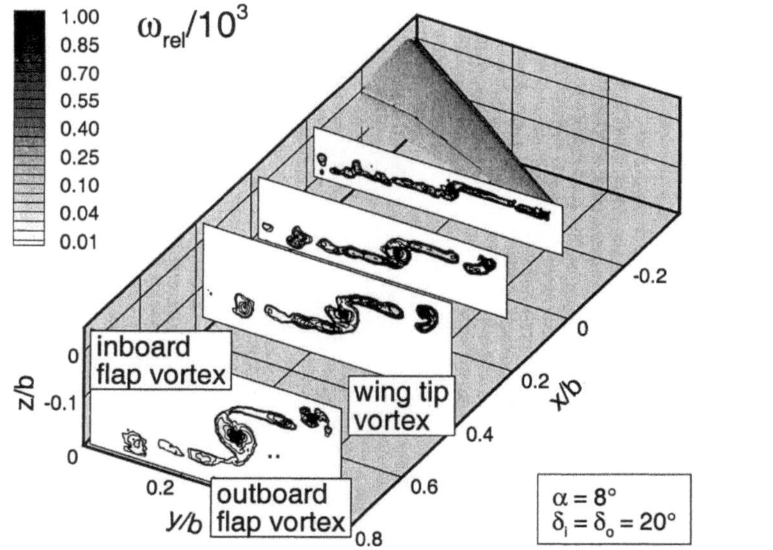

Figure 16 Distribution of vortictiy ω_{rel} in the wake of the swept and tapered wing with flaps and slats ($\delta_i = \delta_o = 20°$)

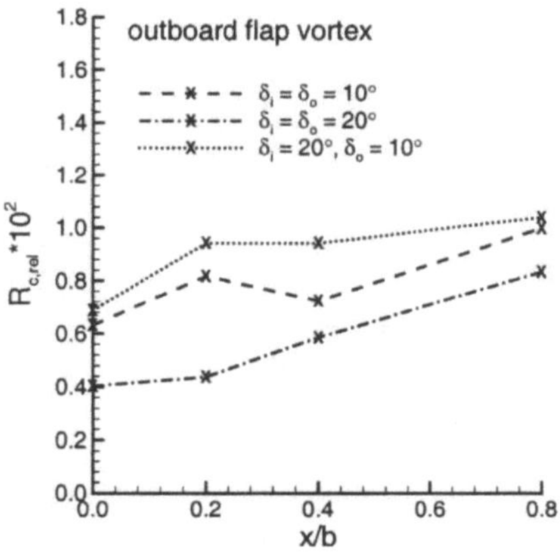

Figure 21 Core radius $R_{c,rel}$ of the outboard flap vortex versus distance to the trailing edge x/b for the swept and tapered wing with flaps and slats

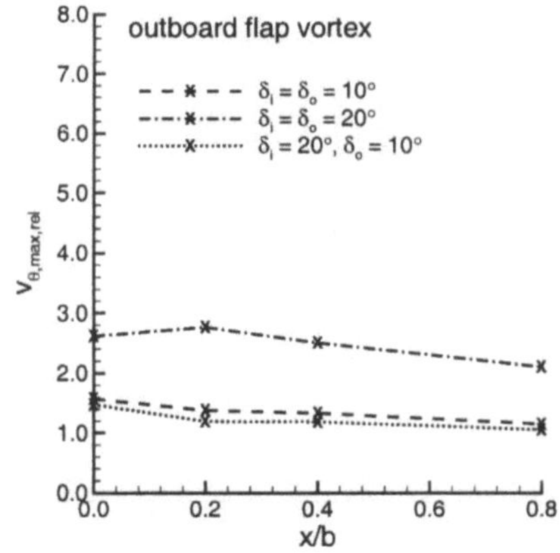

Figure 22 Maximum tangential velocity $V_{\theta,max,rel}$ of the outboard flap vortex versus distance to the trailing edge x/b for the swept and tapered wing with flaps and slats

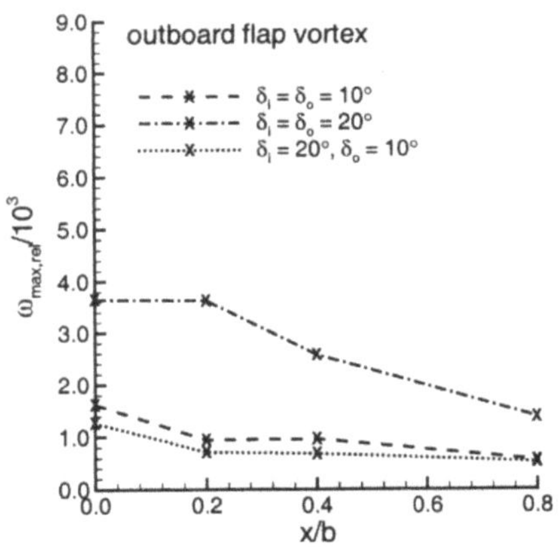

Figure 23 Maximum vorticity $\omega_{max,rel}$ of the outboard flap vortex versus distance to the trailing edge x/b for the swept and tapered wing with flaps and slats

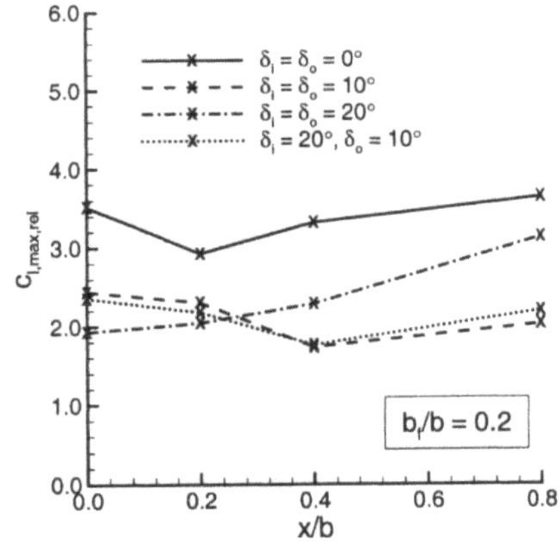

Figure 24 Maximum induced rolling moment $c_{l,max,rel}$ in the wake of the swept and tapered wing with flaps and slats

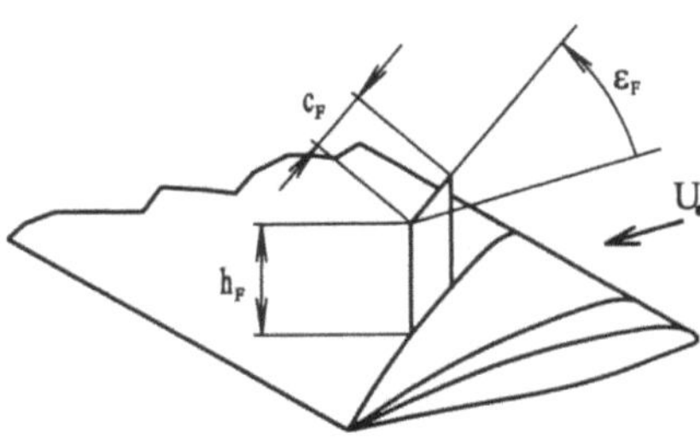

rectangular wing with a wing fin at the tip	c_F/b	h_F/b	ε_F [°]
	0.03	0.025	8.53
	0.03	0.025	34.0
	0.03	0.025	-8.53
	0.03	0.025	-34.0

Figure 25 Rectangular wing with a wing fin at the wing tip (height h_F, cord c_F and angle of incidence ε_F)

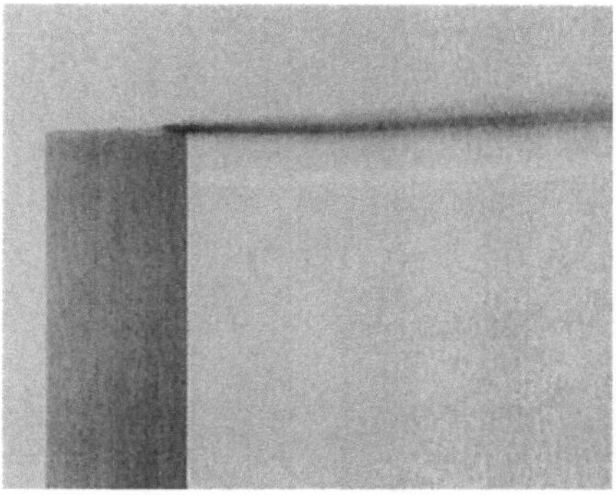

Figure 26 Smoke visualization of the wing tip vortex in the wake of the rectangular wing without wing fin

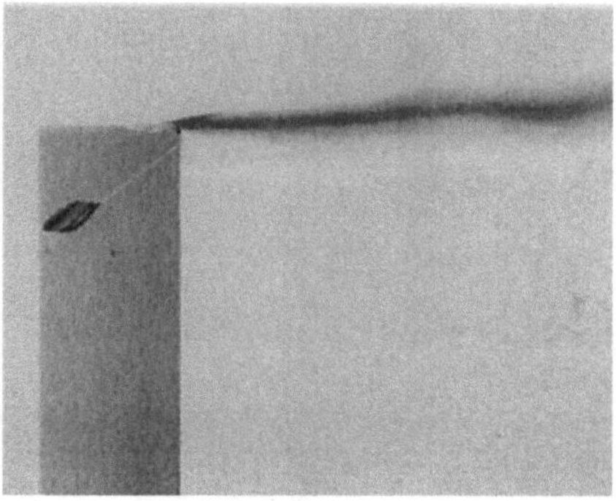

Figure 27 Smoke visualization of the wing tip vortex in the wake of the rectangular wing with a wing fin at the wing tip ($\varepsilon_F = 34°$)

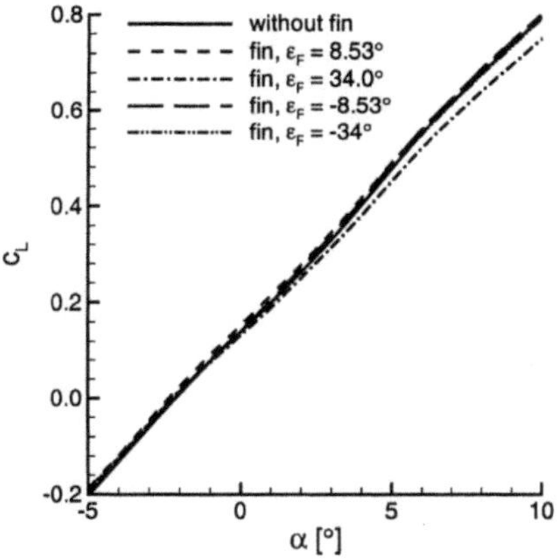

Figure 28 Lift coefficient c_L versus angle of attack α of the rectangular wing without and with a wing fin

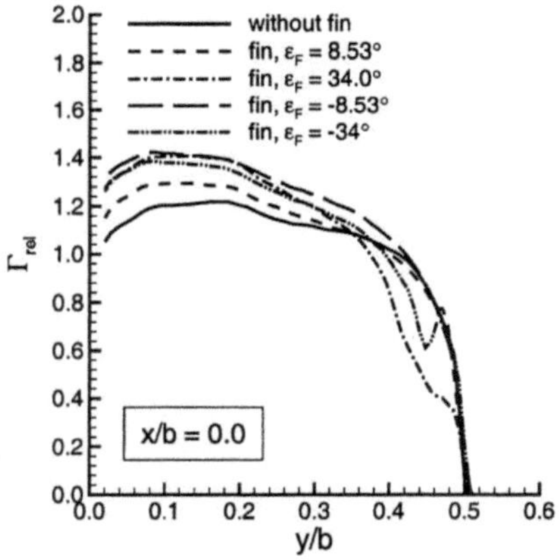

Figure 29 Circulation distribution Γ_{rel} at the trailing edge $x/b = 0.0$ of the rectangular wing without and with a wing fin

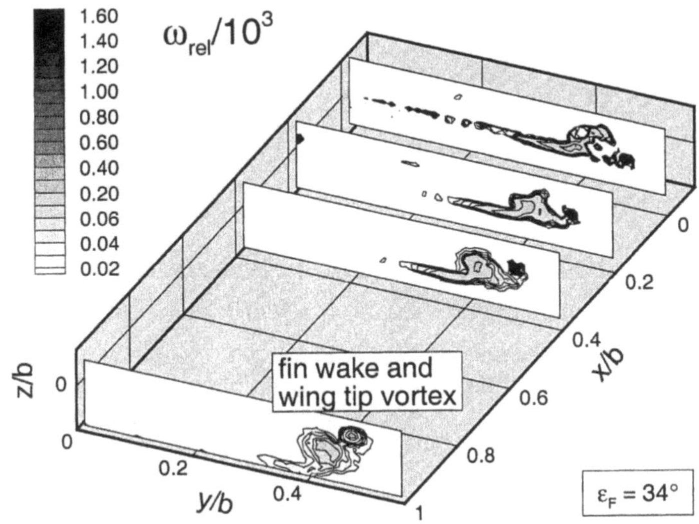

Figure 30 Distribution of vortictiy ω_{rel} in the wake of the rectangular wing with a wing fin at wing tip ($\varepsilon_F = 34°$)

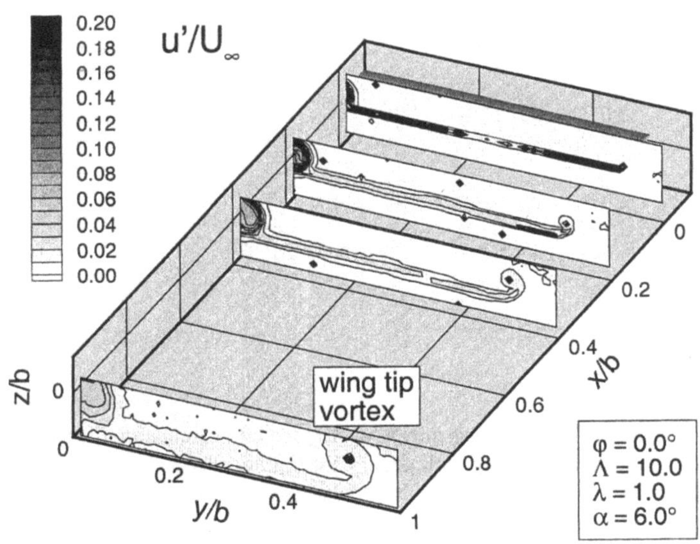

Figure 31 Distribution of turbulence u'/U_∞ in the wake of the rectangular wing without wing fin

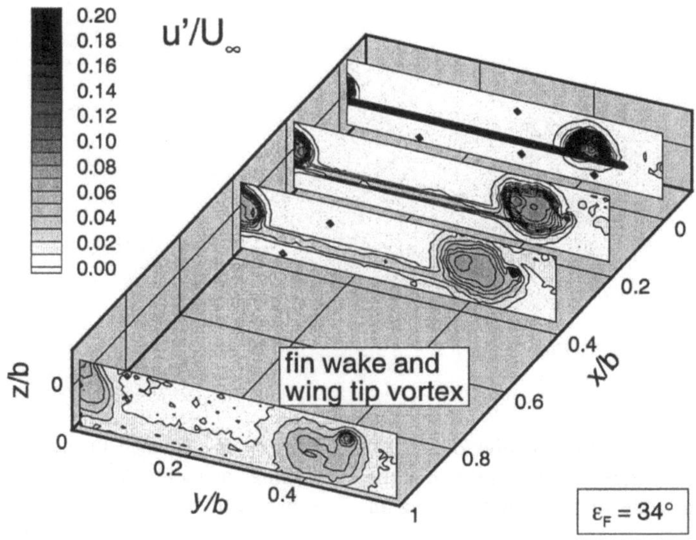

Figure 32 Distribution of turbulence u'/U_∞ in the wake of the rectangular wing with a wing fin at the wing tip ($\varepsilon_F = 34°$)

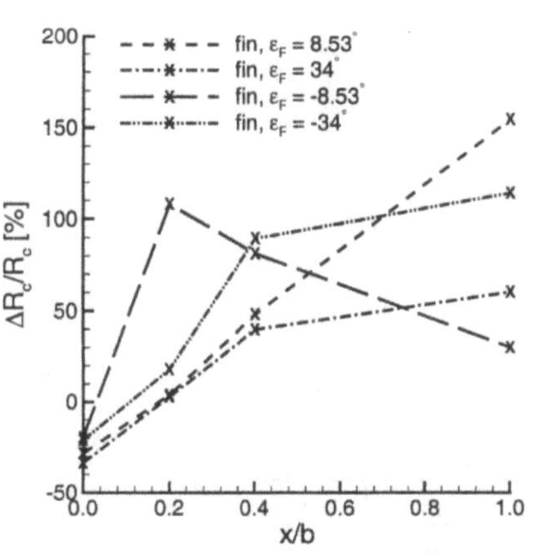

Figure 33 Change in core radius $\Delta R_c/R_c$ of the wing tip vortex of the rectangular wing without flaps due to the influence of the wing fin versus distance to the trailing edge x/b

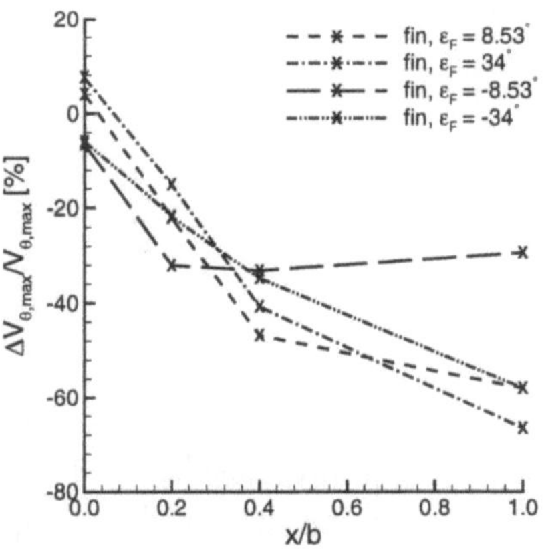

Figure 34 Change in maximum tangential velocity $\Delta V_{\theta,max}$ of the wing tip vortex of the rectangular wing without flaps due to the influence of the wing fin versus distance to the trailing edge x/b

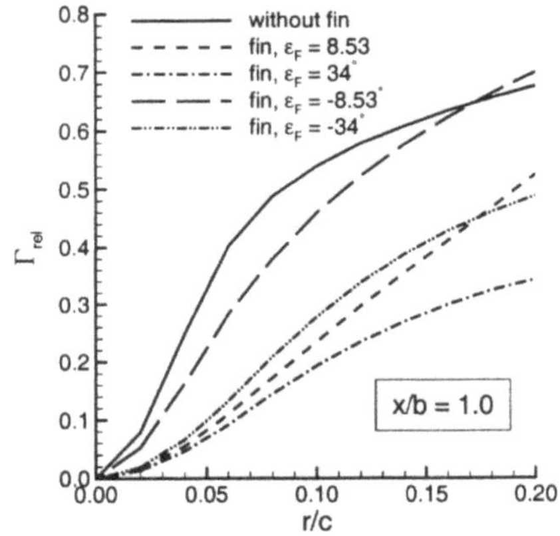

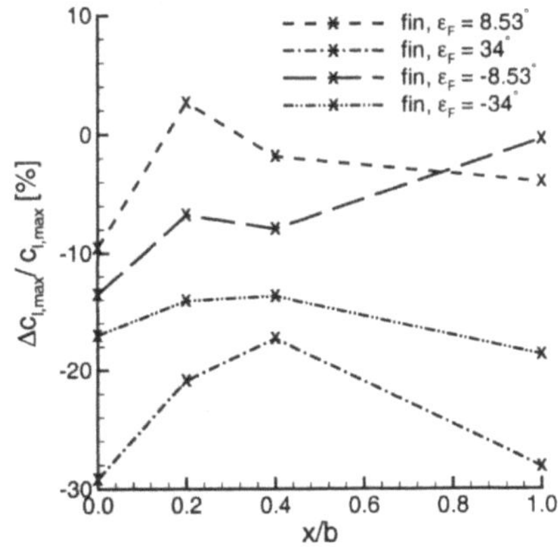

Figure 35 Circulation distribution Γ_{rel} in the wing tip vortex versus radius r/c of the rectangular wing without and with wing fin

Figure 36 Change in maximum induced rolling moment $\Delta c_{l,max}/c_{l,max}$ in the wake of the rectangular wing without flaps due to the influence of the wing fin versus distance to the trailing edge x/b

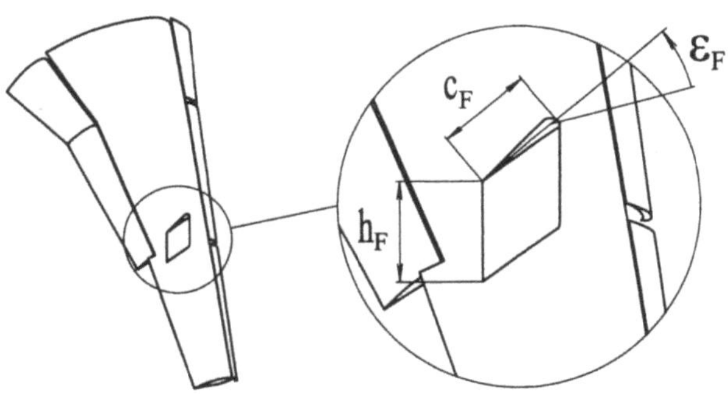

swept and tapered wing with a wing fin	c_F/b	h_F/b	$\varepsilon_F[°]$
	0.04	0.04	20
	0.04	0.04	-20

Figure 37 Swept and tapered wing with a wing fin at the outboard flap edge (height h_F, cord c_F and angle of incidence ε_F)

Figure 38 Smoke visualization of wing tip vortex in the wake of swept and tapered wing without wing fin

Figure 39 Smoke visualization of wing tip vortex in the wake of swept and tapered wing with a wing fin at the wing tip $\varepsilon_F = 20°$

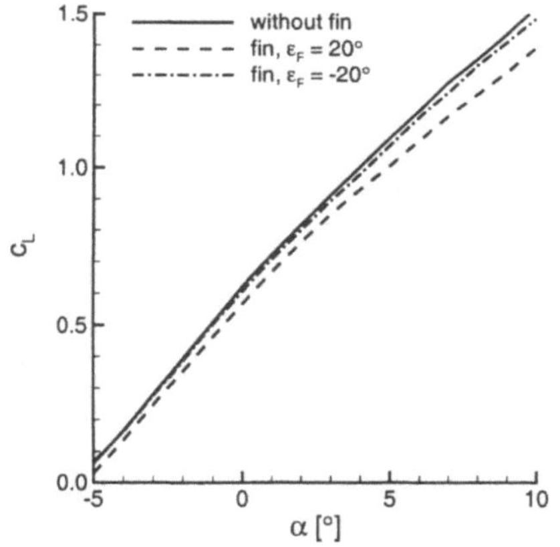

Figure 40 Lift coefficient c_L versus angle of attack α of the swept and tapered wing without and with wing fin

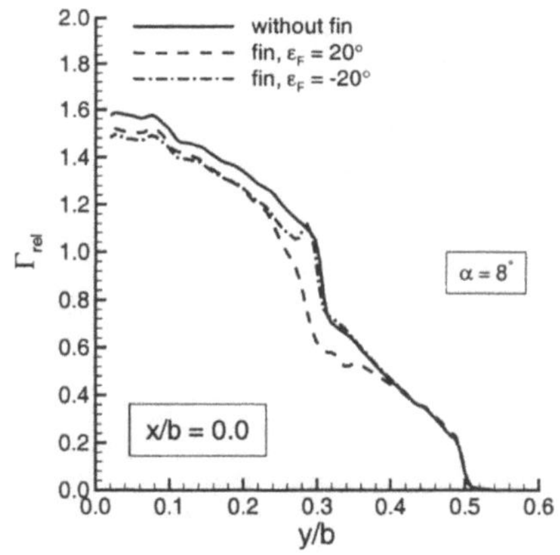

Figure 41 Circulation distribution Γ_{rel} at the trailing edge $x/b = 0.0$ of the swept and tapered wing without and with wing fin

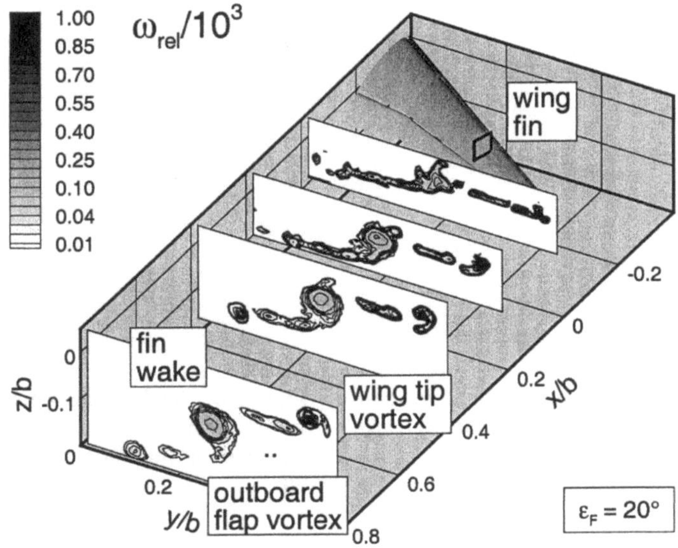

Figure 42 Distribution of vorticity ω_{rel} in the wake of the swept and tapered wing with a wing fin ($\varepsilon_F = 20°$)

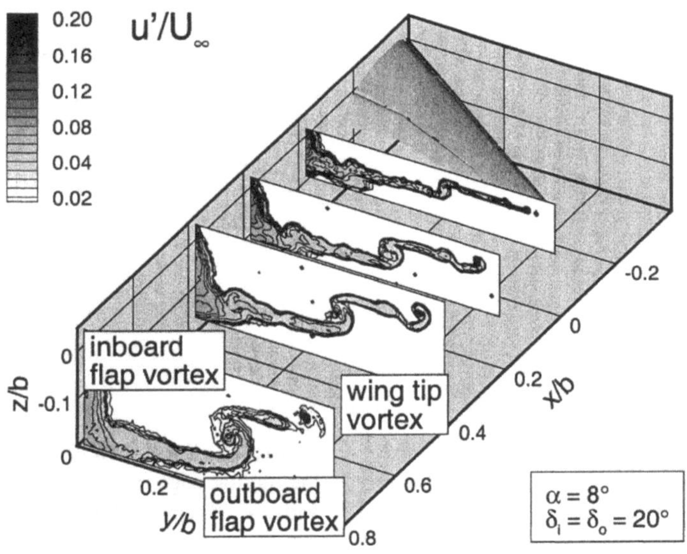

Figure 43 Distribution of turbulence u'/U_∞ in the wake of the swept and tapered wing with flaps and slats ($\delta_i = \delta_o = 20°$)

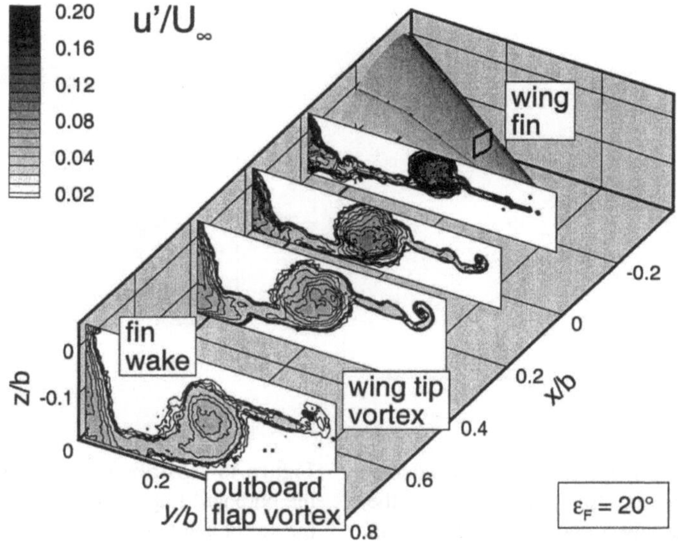

Figure 44 Distribution of turbulence u'/U_∞ in the wake of the swept and tapered wing with a wing fin ($\varepsilon_F = 20°$)

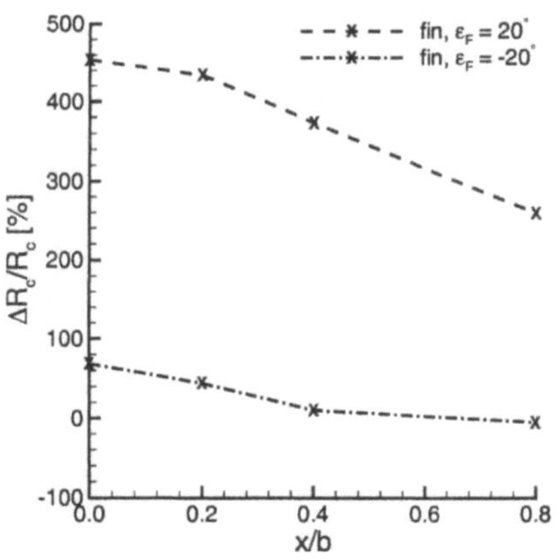

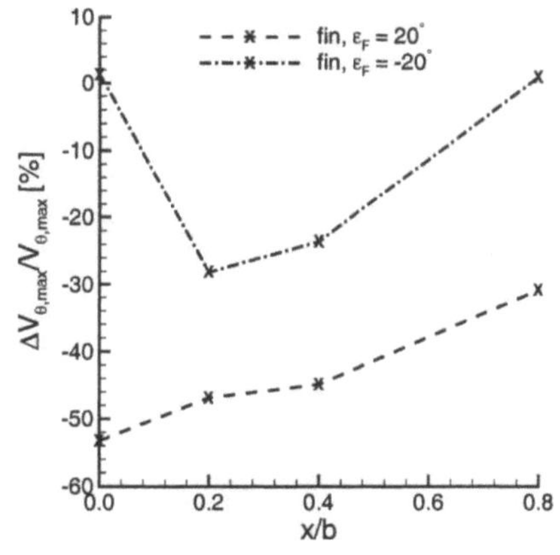

Figure 45 Change in core radius $\Delta R_c/R_c$ of the outboard flap vortex of the swept and tapered wing with flaps and slats due to the influence of the wing fin versus distance to the trailing edge x/b

Figure 46 Change in maximum tangential velocity $\Delta V_{\theta,max}/V_{\theta,max}$ of the outboard flap vortex of the swept and tapered wing with flaps and slats due to the influence of the wing fin versus distance to the trailing edge x/b

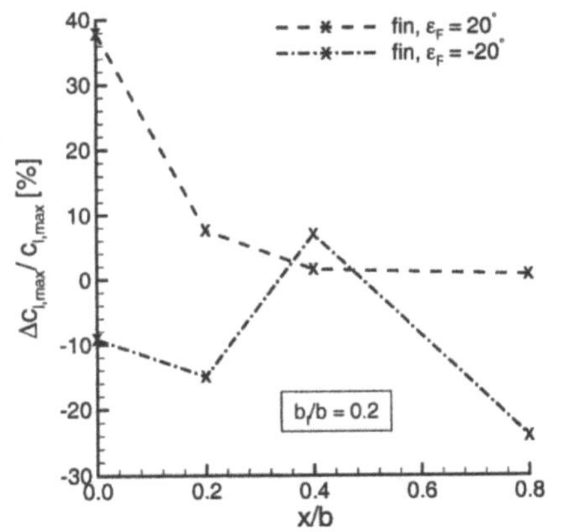

Figure 47 Change in maximum induced rolling moment $\Delta c_{l,max}/c_{l,max}$ in the wake of the swept and tapered wing with flaps and slats due to the influence of the wing fin versus distance to the trailing edge x/b

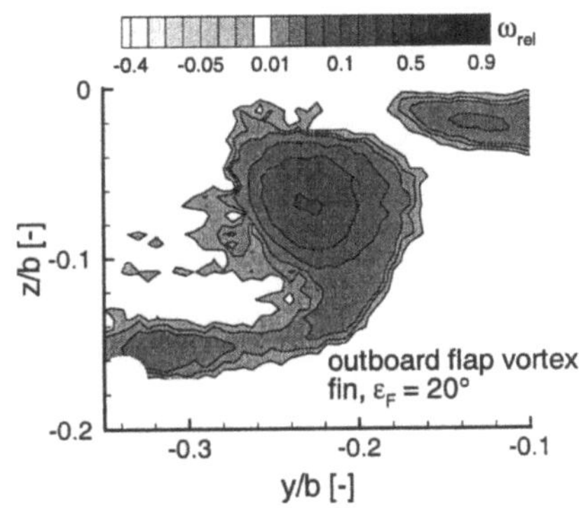

Figure 48 Distribution of vorticity ω_{rel} at the outboard flap vortex at $x/b = 0.8$ for the swept and tapered wing with a wing fin ($\varepsilon_F = 20°$)

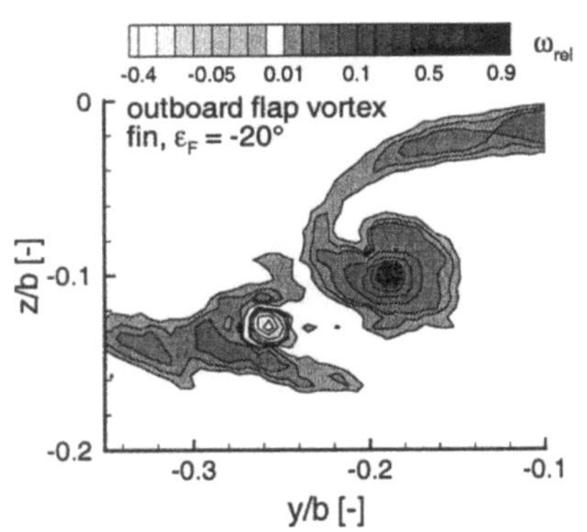

Figure 49 Distribution of vorticity ω_{rel} at the outboard flap vortex at $x/b = 0.8$ for the swept and tapered wing with a wing fin ($\varepsilon_F = -20°$)

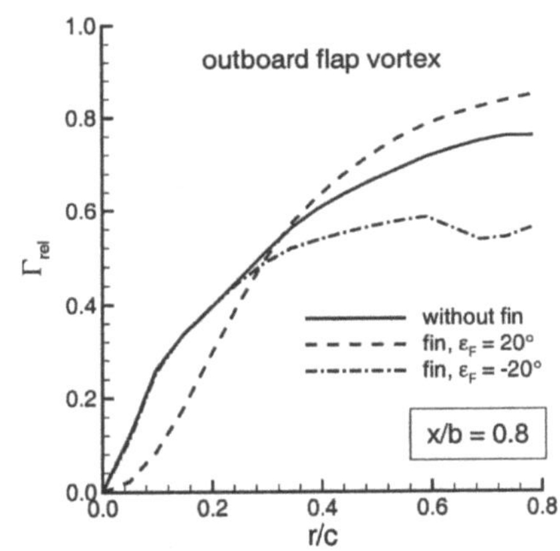

Figure 50 Circulation distribution Γ_{rel} in the outboard flap vortex versus radius r/c of the swept and tapered wing without and with wing fin

Simulation of Vortex Sheets at Take-Off and Landing

Ingmar Schell, Erol Özger, Dieter Jacob
Lehrstuhl für Luft- und Raumfahrt, RWTH Aachen
Wüllnerstraße 7, D-52062 Aachen, Germany

Summary

In this paper a numerical investigation of the wake structure of wings with different planforms and flap settings as well as of wings with wing fins is presented. The near wake region is computed by means of a 3D panel method while the far region is simulated by a 2D Navier-Stokes scheme. The hazard posed to a following aircraft is quantified by the value of the maximum induced rolling moment which is calculated by means of 2D strip theory. The results of the numerical investigation confirm the tendencies that were obtained experimentally in the near field of the wings. Additionally, the development of the wake structure in the far field could be determined. As a result, the alleviating effect of a suitable wing geometry as well as of an additional wing fin that was obtained in the near field is decreased in the far field of the wings.

1 Introduction

In previous experimental investigations the near field wake structure of various wings with and without flaps was examined. The experiments [8, ?, 11, 12, 13] showed the influence of different wing geometries and flap settings on characteristic parameters in the wake. Because of the restricted length of the test section the experiments could only be conducted for the near wake region up to one span behind the trailing edge of the wings. Up to this point in the wake the roll-up and merging process of the vortices is not completed so that a further numerical investigation of the farer wake is necessary.

Numerical investigations of the wake of wings have been conducted for many years. A first simple method was derived by Betz [2] in 1932. He calculated the rolled-up structure of the wing tip vortex of an elliptical wing from the circulation distribution with the invariance of the two dimensional inviscid vortex flow. This method is simple and efficient and was used in many following investigations [4, 10] but it is not reasonably applicable to compute the wake of wings with flaps. In contrast to this, 3D-Navier-Stokes methods [3, 5] model the physics of the wake flow accurately but are time consuming with regard to the generation of grids and the final computation. Because of this, an investigation of the wake flow of various wings cannot be conducted efficiently with 3D-Navier-Stokes methods.

In this investigation the near wake region of the different wings with and without flaps was computed by means of a 3D-panel method [1]. Subsequently, the far wake region was simulated by a 2D-Navier-Stokes scheme in vorticity form [6] with the results of the panel calculation serving as an initial condition. Additionally, the influence of a wing fin on the far wake structure was investigated by means of a 2D-Navier-Stokes computation [6]. In this case the experimentally determined vorticity distribution one span behind the different wings served as an initial condition. The hazard posed to following aircraft was quantified by calculating the maximum induced rolling moment. From the computed data the non-dimensional vortex parameters already defined in [8] were evaluated.

It should be emphasized that the influences of the configuration changes (wing planform, flap setting and fin configurations) on the wake parameters, and particularly, on the induced rolling moment on a following aircraft are of main interest. In contrast to [8] the impact of the above mentioned parameters is investigated not for the near wake but for the far wake. Therefore, the choice of simpler but more robust and faster methods such as the above mentioned 2D- and 3D-Navier-Stokes method seems justified. Only these methods make it possible to determine the far wake-structure approximately. Of course, three dimensional instabilities, vortex aging and other effects of atmospheric turbulences are not accounted for. But these methods are sufficient to determine the vortex positions and to estimate the influence of vortex merger on the induced rolling moment of a following aicraft. Furthermore, neglecting above mentioned effects (vortex aging, 3D-instabilities etc.) leads to a more conservative estimation of the wake physics with respect to the maximum induced rolling moment.

2 Numerical methods

For the computation of the flow behind the wing the wake is divided into two distinct parts, a near region extending from the trailing edge of the wing where the flow is considered to be three dimensional (one span in length) and a far region where the flow is assumed to be time dependent two dimensional. To evaluate the three dimensional wake roll-up in the near region behind the different wing configurations a low order panel method [1] is used. The vorticity distribution evaluated from the panel computation one span behind the wing serves as an initial condition for the calculation of the far region with a two dimensional unsteady Navier-Stokes solver [6]. For wings with fin (small wing mounted perpendicular to the surface of the wing) the experimentally determined vorticity distribution at one span behind the trailing edge (see [8]) serves as an initial condition for the computation of the far wake. Both methods are validated in former investigations (see [1] for the panel method and [15] for the 2D-Navier-Stokes method) so that this is not repeated here again. In the following both numerical methods are described.

2.1 Panel method

Three dimensional irrotational incompressible flows around arbitrary configurations are characterized by a perturbation velocity potential φ [1, 14, 17]. This potential satisfies the Laplace equation

$$\nabla^2 (\varphi_\infty + \varphi) = 0 \quad , \quad \varphi = \underbrace{\frac{1}{4\pi} \int_S \sigma \cdot \left(\frac{1}{r}\right) \cdot dS}_{sources} + \underbrace{\frac{1}{4\pi} \int_{S,W} \mu \vec{n} \cdot \vec{\nabla} \left(\frac{1}{r}\right) \cdot dS}_{doublets} \tag{1}$$

in a region surrounding the body with a surface S, whereat φ_∞ is the potential of the freestream. In order to obtain a unique solution in lifting case, a potential discontinuity surface, the so-called wake with a surface W, leaving the sharp trailing edge and extending to infinity, has to be introduced. According to Green's third identity, a solution of 1 may be expressed at any point p as a potential introduced by a combination of source singularities of strength σ and doublet singularities of strength μ, disturbed on the surfaces S and W, where r is a distance from the field point p to the surface. A required solution is found by imposing the Neumann condition on S and W. The free wake surface is adjusted iteratively in order to guarantee tangential velocity along the wake surface and zero pressure jump across the wake. The computation of the free wake of the different wings is performed for a wake length of approximately two spans to obtain an unperturbed wake geometry at $x/b = 1$ (see figure 1).

2.2 Two dimensional Navier-Stokes method

The far region is calculated by treating the three dimensional steady flow in the wake as two dimensional and unsteady based on the work of Westwater [16], Rosenhead [9] and Fell [6]. The method assumes a flow with weakly bend vortex filaments and a dominating vorticity component $\omega_x \approx \omega$. The three dimensional effects caused by vortex stretching and bending and by the axial jets within the vortex are neglected. Another simplification is achieved by assuming that the vortices are dominated by laminar viscosity rather than turbulence, so that the calculated vortex decay rates are smaller than those found in real wakes. The laminar treatment of the flow is therefore a worst case scenario. The 3D-Navier-Stokes equations are transformed in the 2D-vorticity transport equation 2

$$\frac{\partial \omega}{\partial t} + v \frac{\partial \omega}{\partial y} + w \frac{\partial \omega}{\partial z} = \nu \left(\frac{\partial^2 \omega}{\partial y^2} + \frac{\partial^2 \omega}{\partial z^2} \right) \quad , \tag{2}$$

where $t = x/U_\infty$. The velocities v, w are obtained by solving the Poisson equation 3 for the streamfunction Ψ.

$$\frac{\partial^2 \Psi}{\partial y^2} + \frac{\partial^2 \Psi}{\partial z^2} = -\omega \quad with \quad v = \frac{\partial \Psi}{\partial z}, \quad w = -\frac{\partial \Psi}{\partial y} \,. \tag{3}$$

As initial condition a vorticity distribution is necessary which is obtained from the panel calculation or from the experimental data for the case with wing fin as discussed in the following chapter.

2.3 Generation of an initial condition for the 2D-Navier-Stokes computation

To obtain the initial condition for the 2D-Navier-Stokes computation of the far region the geometry of the wake and the circulation distribution are interpolated at one span behind the wing ($x/b = 1$) as shown in figure 2a for an elliptical wing. This distance proved well in former investigations [13] as the downwash of the bound circulation around the wing can be neglected and the bending of the vortex filaments is small ($\omega_x \approx \omega$). Then, the wake at $x/b = 1$ is divided into the wing tip vortex (and possible flap vortices), where the circulation is mainly concentrated, and the vortex sheet. The circulation of the wing tip vortex (and possible flap vortices) is conserved and the position is evaluated according to the invariance of the first moment of circulation (see Betz [2]). More line vortices are introduced into the vortex sheet in order to obtain a more continuous distribution of circulation (see figure 2b). Finally, the circulation Γ_i of the singular line vortices is distributed by use of Gauss vortices to obtain a continuous vorticity distribution $\omega(y, z)$ according to

$$\omega(y,z) = \sum_{i=1}^{n} \frac{\Gamma_i}{2\pi \cdot k^2 \cdot R_{c,i}^2} \cdot e^{-\frac{(y-y_i)^2+(z-z_i)^2}{2k^2 R_{c,i}^2}} \quad , \quad k = 0.6308348... \tag{4}$$

where y_i, z_i are the positions of the line vortices, k is a constant and $R_{c,i}$ is the core radius of the Gauss vortices. The core radius of the wing tip vortex (and possible flap vortices) is evaluated according to the invariance of the second moment of circulation (see Betz [2]). The core radius $R_{c,i}$ of the vortices in the vortex sheet hardly influences the wake dynamics and is therefore assumed to be constant and set to $R_{c,i} = 0.1 \cdot c$ (figure 2c) according to experimental results.

3 Results

The presentation of the results is divided into three parts. First, the calculated wake structure and induced rolling moment in the wake of the wings without flaps is presented up to 20 spans behind the wing. Then, the numerical results obtained with the swept and tapered wing with slats and flaps are discussed. Finally, the influence of a wing fin on the far field of the wake of the rectangular wing without flaps and the swept and tapered wing with fully deployed slats and flaps is investigated. The evaluation of the computed data corresponds to the evaluation of the experimental data in [8] and the results are also presented by means of the non-dimensional vortex parameters defined in [8].

3.1 Wings without flaps

Figure 3 shows the surface grids of the wings without flaps. Starting from an unswept and untapered rectangular wing with an aspect ratio of $\Lambda = 10$ the sweep angle φ, taper ratio λ and aspect ratio Λ were varied. As a reference an elliptical wing was also used for the calculation. The testing program is summarized in figure 4. The wake of all wings was calculated for approximately constant lift (the maximum deviation of lift is about 2.5 percent). Angles of attack with equal lift were obtained from the lift curves of the wings, that were calculated with the panel method using a rigid wake.

Figure 5,6 and 7 show the circulation distribution Γ_{rel} of the different wings without flaps at $x/b = 1.0$. The influence of the sweep angle φ on the circulation distribution Γ_{rel} is shown in figure 5. With increasing sweep angle φ the circulation Γ_{rel} in the region of the wing tip vortex also increases whereas the circulation Γ_{rel} in the inner part of the wake decreases correspondingly. The influence of the taper ratio λ is opposite as can be seen in figure 6. Consequently, a lower taper ratio (lower value of λ) leads to a shift of circulation Γ_{rel} from the wing tip vortex to the inner part of the wake. Compared to the influence of the sweep angle φ and the taper ratio λ the effect of the aspect ratio Λ on the circulation distribution Γ_{rel} is smaller. An increasing aspect ratio Λ leads to a slight concentration of circulation in the wing tip vortex region (see figure 7). In contrast to the circulation of the rectangular wing the circulation of the elliptical wing is less concentrated in the wing tip vortex region and more pronounced in the inner part of the wake.

The wake geometry at one span behind the various wings is presented in figure 8,9 and 10. For increasing sweep angle φ the wake of the wings moves further downward in the region of the wing tip vortex and less downward in the inner region so that the sense of rotation of the whole wake is opposite in comparison with the wing tip vortex (see figure 8). Therefore for constant lift the circulation of the wing tip vortex increases with increasing sweep angle φ. In contrast to this a decreasing taper ratio λ causes a further downward movement of the wake in the inner region and a less pronounced downward movement in the wing tip vortex region so that the whole wake rotates with the same sense of rotation as the wing tip vortex (see figure 9). Accordingly, for constant lift the circulation of the wing tip vortex decreases with decreasing taper ratio λ. The effect of the aspect ratio Λ on the wake geometry is small and only discernible in the wing tip vortex region (see figure 10). Here, the wake rolls further up with increasing aspect ratio Λ caused by the higher amount of circulation in the wing tip vortex. The wake geometry of the elliptical wing is similar to that of the tapered wing with $\lambda = 1/3$.

In figure 11,12 and 13 the maximum rolling moment coefficient $c_{l,max,rel}$ of the different wings induced on a following aircraft with $b_f/b = 0.2$ is shown. Up to five spans behind the wing the maximum induced rolling $c_{l,max,rel}$ is slightly increasing due to the roll-up process in the wake of the wings. Further downstream the maximum induced rolling moment $c_{l,max,rel}$ is approximately constant and decreases slightly due to the influence of the viscosity in the wake. An increase of the sweep angle φ leads to higher values of $c_{l,max,rel}$ in the wake because of the higher circulation of the wing tip vortex. A decrease of the taper ratio λ causes a lower maximum induced rolling moment $c_{l,max,rel}$ due to the lower amount of circulation in the wing tip vortex. Finally, increasing the aspect ratio Λ leads to slightly higher values of $c_{l,max,rel}$ caused by the higher circulation of the wing tip vortex which structure is also more concentrated.

3.2 Wing with flaps

The testing program for the swept and tapered wing with slats and flaps (for the dimensions of the wing see [8]) is shown in figure 14. The computation of the wake for various flap settings was performed at constant lift. The inboard and outboard flaps were deployed with both equal flap deflection ($\delta_i = \delta_o$) and unequal flap deflection where the outboard flap angle was chosen to be half the inboard flap angle ($\delta_i = 2 \cdot \delta_o$).

In figure 15 and 16 the circulation distribution Γ_{rel} in the wake of the swept and tapered wing with equal and unequal flap deflection is shown for $x/b = 1.0$. For increasing flap angles δ_i, δ_o the circulation Γ_{rel} in the wake of the wing is redistributed from the wing tip vortex to the flap vortices. An approximately uniform distribution of circulation on the vortices is accomplished for moderate flap angles ($\delta_i = \delta_o = 10°$ for equal flap deflection and $\delta_i = 10°, \delta_o = 5°$ for unequal flap deflection). For fully deployed flaps ($\delta_i = \delta_o = 20°$) the outboard flap vortex has the largest amount of circulation Γ_{rel} in the wake of the wing and therefore dominates the wake flow. In comparison to this, a maximum unequal flap deflection ($\delta_i = 20°, \delta_o = 10°$) causes a very strong inboard flap vortex that comprises most of the circulation in the wake of the wing.

The wake geometry at one span behind the swept and tapered wing with flaps and slats is presented in figure 17 and 18. For equal flap deflection $\delta_i = \delta_o$ the wake of the flaps moves further downward with increasing flap angle whereas the wake of the outer part of the wing moves less downward (see figure 17). Although the inboard and outboard flaps are equally deployed a weak inboard flap vortex is generated because of the discontinuity of the geometry between the inboard and midboard wing segment. For unequal flap deflection this inboard flap vortex is much stronger and dominates the wake flow (see figure 18). For increasing flap angles the wake of the inboard flap moves further downward and the wake of the outer part of the wing moves less downward whereas the movement of the wake of the outboard flap is not affected.

In figure 19 and 20 the development of the vortices in the wake of the swept and tapered wing with flaps and slats is shown starting from the initial condition at $x/b = 1$ calculated with the panel method up to $x/b = 61$ for two different flap settings. For $\delta_i = \delta_o = 20°$ the outboard flap vortex dominates the flow in the wake of the wing whereby the wing tip vortex and the inboard flap vortex rotate around the outboard flap vortex (see figure 19). This leads to a simultaneous merger of the wing tip vortex with the outboard and inboard flap vortex 21 spans behind the wing and finally to the merger of the inboard and outboard flap vortex 59 spans behind the wing. For $\delta_i = 20°, \delta_o = 10°$ the inboard flap vortex dominates the flow in the wake of the wing whereby the outboard flap vortex rotates around the inboard flap vortex and merges with it 39 spans behind the wing (see figure 20). The weaker wing tip vortex is not influenced by this interaction of the vortices and rotates slowly around them without merging.

The maximum induced rolling moment coefficient $c_{l,max,rel}$ in the wake of the swept and tapered wing with flaps and slats is shown in figure 21 and 22. For the configuration with stowed flaps the value of $c_{l,max,rel}$ does not change significantly and decreases slightly downstream due to the viscosity. The maximum induced rolling moment $c_{l,max,rel}$ of the configurations with deployed flaps oscillates due to the rotation of the wing tip and the flap vortices around each other. With merging of the vortices in the wake of the wing this oscillation diminishes.

When comparing the maximum induced rolling moment $c_{l,max,rel}$ for the configurations with equal and unequal flap deflection at a distance of $x/b = 1.0$ it can be seen that for increasing flap angles until $\delta_i = \delta_o = 10°$ respectively $\delta_i = 20°, \delta_o = 10°$ the maximum induced rolling moment $c_{l,max,rel}$ is decreasing which was also observed in the experimental results presented in [8]. A further increase of the flap angle leads to an increase of the maximum induced rolling moment $c_{l,max,rel}$ to values that can be even higher than for the configuration with no flap deflection. This can be attributed to the increasing circulation of the flap vortices which become dominant in the wake flow. Downstream at a distance of $x/b = 61$ where the merging process is almost completed the effect of wake alleviation by a moderate flap deflection compared to the configuration with stowed flaps is not discernible anymore. However, the maximum induced rolling moment $c_{l,max,rel}$ in the wake of the configuration with moderate flap deflection is still much lower than that of the configuration with fully deployed flaps.

3.3 Alleviation by means of wing fins

In order to investigate the influence of a wing fin on the wake flow in the far field computations by means of the 2D-Navier-Stokes method were performed. For this, the experimentally determined vorticity distribution at approximately one span behind the wings was used as an initial condition.

3.3.1 Rectangular wing without flaps

Figure 23 shows the change of the maximum induced rolling moment coefficient $\Delta c_{l,max}/c_{l,max}$ in the wake of the rectangular wing with a wing fin compared to the case without fin up to $x/b = 11$. The roll-up process is nearly completed at $x/b = 5.0$ behind the wing so that further downstream the value of $\Delta c_{l,max}/c_{l,max}$ remains almost constant. Corresponding to the experiments the strongest alleviation in the wake of approximately 20 percent is achieved with the wing fin at $\varepsilon_F = 34°$.

3.3.2 Swept and tapered wing with slats and flaps

The trajectory of the vortices in the wake of the swept and tapered wing with flaps and slats up to $x/b = 70$ is presented in figure 24 for the configurations without and with a wing fin. It can be seen that the wing tip and the outboard flap vortex rotate around each other, descend due to the velocity induction of their counterparts beyond the symmetry plane and merge with each other. The wing fin leads to a merging of the wing tip and outboard flap vortex that occurs at a smaller distance to the trailing edge of the wing. Moreover the wake of the wing descends further downward due to the influence of the wing fin. Figure 25 shows the change of the maximum induced rolling moment coefficient $\Delta c_{l,max}/c_{l,max}$ due to the influence of the wing fin. Caused by the merging process of the wing tip and outboard flap vortex the alleviation of the wake of the wing that was discernible near the trailing edge is considerably diminished in the far wake.

4 Conclusion

A numerical investigation of the wake structure of various wings without flaps and a swept and tapered wing with flaps and slats was conducted to examine the influence of the wing geometry on the wake structure and the induced rolling moment in the far wake. The numerical simulations were performed for constant lift by means of a 3D-panel method for the calculation of the near region and an unsteady 2D-Navier-Stokes method for the calculation of the far region.

The planform of a wing has a great influence on its wake structure and the induced rolling moment on a following aircraft. In comparison with the unswept and untapered rectangular wing a wing sweep leads to a concentration of the circulation in the wing tip vortex and thus to an increase of the maximum induced rolling moment. The reduction of the taper ratio and aspect ratio causes a weaker wing tip vortex and a lower maximum induced rolling moment.

At constant lift moderate flap angles lead to a large reduction of the maximum induced rolling moment in the near wake of the swept and tapered wing because of the nearly equal distribution of the circulation on the wing tip and the flap vortices. High flap angles lead to dominant flap vortices and to higher induced rolling moments because of the unequal distribution of circulation on the wing tip and flap vortices. The results also show that the merging of the wing tip and the flap vortices further downstream reduces the alleviating effect of moderate flap deflection. Nevertheless, for a moderate flap deflection the maximum induced rolling moment in the wake of the swept and tapered wing is much lower than for fully deployed flaps.

The calculation of the wake behind the rectangular wing without flaps and with a wing fin shows that the alleviating effect of the wing fin is still discernible after completion of the roll-up process of the wake. In contrast to this the reconcentration of the circulation due to the merging process of the wing tip and flap vortex leads to a considerable decrease of the alleviating effect of the wing fin in the wake of the swept and tapered wing with fully deployed flaps and slats.

References

[1] Behr, R., Wagner, S., "Ein nichtlineares Wirbelgitterverfahren zur Berechnung von Interferenzeffekten zwischen freien Wirbelschichten und Tragflächen", DGLR-Bericht 88-05, pp. 303-316,1988

[2] Betz, A., "Verhalten von Wirbelsystemen", ZAMM, Bd. XII,No. 3, pp. 164-174, 1932

[3] Dacles-Mariani, J., Zilliac, G. G., Chow, J. S., Bradshaw, P.,"Numerical/Experimental Study of a Wingtip Vortex in the Near Field", AIAA Journal, Vol. 33, No. 9, pp. 1561-1568, Sep. 1995

[4] Donaldson, C. DuP., Snedeker, R. S., Sullivan, R. D.,"Calculation of Aircraft Wake Velocity Profiles and Comparison with Experimental Measurements", Journal of Aircraft, Vol. 11, No. 9, pp. 547-555, 1974

[5] Eaton, J. A., O' Flaherty, M. P., "Flowfield prediction of three-dimensional wing trailing vortices using advanced turbulence models", Paper presented at the AGARD FDP Symposium on "The Characterization and Modification of Wakes from Lifting Vehicles in Fluids" held in Trondheim, Norway from 20-23 May 1996, and published in CP-584

[6] Fell, S., Staufenbiel, R., "Formation and Structure of Vortex Systems Generated by Unflapped and Flapped Wing Configurations", Z. Flugwiss. Weltraumforschung, Vol. 19, No. 6, pp. 366-379, December 1995

[7] Özger, E., Schell, I., Jacob, D., On the Structure and Attenuation of an Aircraft Wake, AIAA Journal of Aircraft, Vol. 38, No. 5, pp. 878-887, 2001

[8] Özger, E., Schell, I., Jacob, D., "Experimental Analysis and Modulation of Vortices", chapter 2.1 of this volume

[9] Rosenhead, L., "The Formation of Vortices from a Surface of Discontinuity", Proc. Roy. Soc. London, A134, pp. 170-192, 1931

[10] Rossow, V. J., "Extended-Betz Methods for Roll-Up of Vortex Sheets", Journal of Aircraft, Vol. 34, No. 5, pp. 592-599, 1997

[11] Schell, I., Özger, E., Jacob, D., "Influence of Different Flap Settings on the Wake-Vortex Structure of a Rectangular Wing with Flaps and Means of Alleviation with Wing Fins", Aerospace Science and Technology, No. 2, pp. 79-90, April 2000

[12] Schell, I., Özger, E., Jacob, D., "Influence of Different Flap Settings on the Hazard Posed to Following Aircraft", Notes on Numerical Fluid Mechanics, Vol. 77, pp. 66-73, 2002

[13] Schell, I., "Untersuchung zum Einfluss der Geometrie eines Flügels auf seinen Nachlauf", Ph.D. thesis, Institut für Luft- und Raumfahrt, RWTH Aachen, 2001

[14] Sytsma, H. S., Hewitt, B. L., Rubbert, P. E., "A Comparison of Panel Methods for Subsonic Flow Computation", AGARDograph, No. 241, February 1979

[15] Türk, L., Jacob, D.,"Das Verschmelzen von Wirbeln", Schlussbericht für Airbus im Rahmen des BMFT-Forschungsprogramms, 1999

[16] Westwater, F. L., "Rolling Up of a Surface of Discontinuity behind an Airfoil of Finite Span", Aeronautical Research Council Reports and Memoranda, No. 1962, pp. 116-131, 1935

[17] Yeh, D. T., Plotkin, A., "Vortex Panel Calculation of Wake Rollup behind a Large Aspect Ratio Wing", AIAA Journal, Vol. 24, No. 9, pp. 1417-1423, September 1986

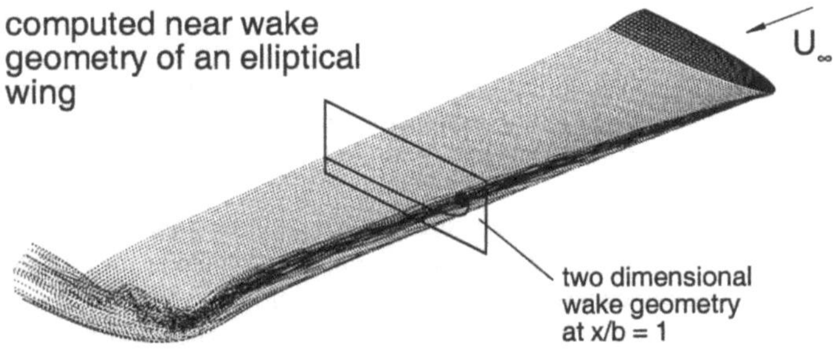

Figure 1 Calculated geometry of the near wake of the elliptical wing

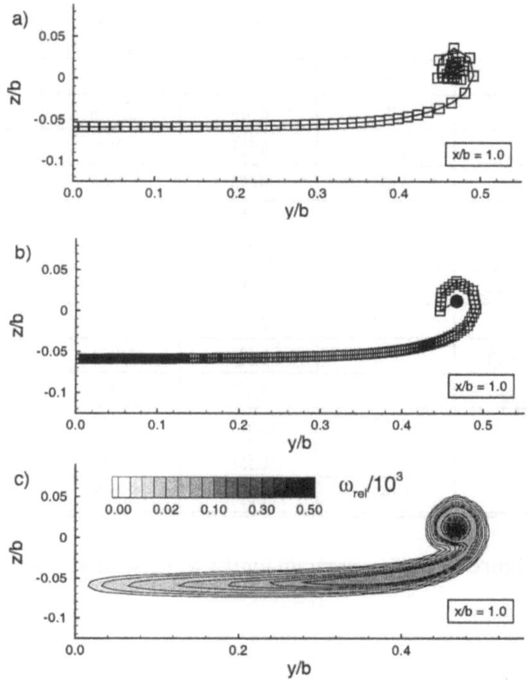

Figure 2 Evaluation of an initial condition for the 2D-Navier-Stokes computation of the wake of the elliptical wing

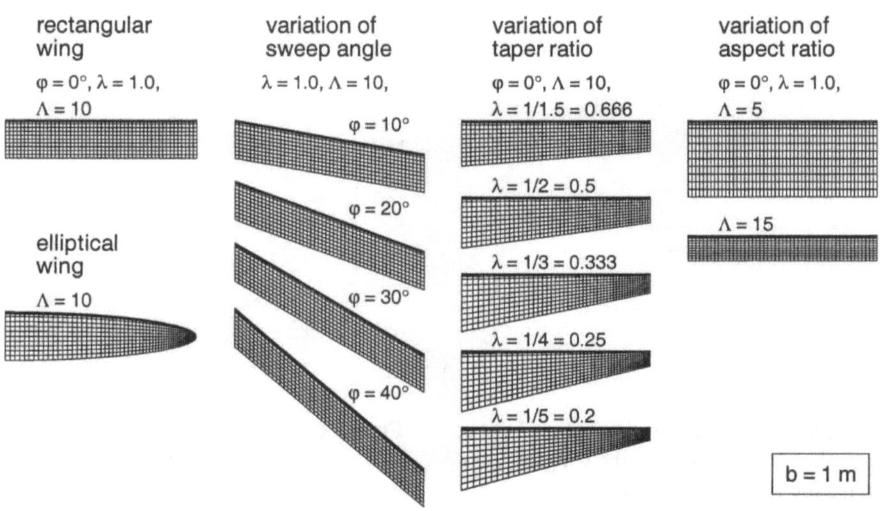

Figure 3 Surface grids of the wings without flaps

	α	c_A	c_AS	$\Delta(c_AS)$ [%]
rectangular wing	6.000	0.684	0.0684	0.0
elliptical wing	5.721	0.676	0.0676	-1.183
$\varphi = 10°$	6.223	0.681	0.0681	-0.527
$\varphi = 20°$	6.631	0.676	0.0676	-1.129
$\varphi = 30°$	7.272	0.672	0.0672	-1.776
$\varphi = 40°$	8.245	0.668	0.0668	-2.422
$\lambda = 1/1.5 = 0.666$	5.826	0.683	0.0683	-0.111
$\lambda = 1/2 = 0.5$	5.771	0.682	0.0682	-0.293
$\lambda = 1/3 = 0.333$	5.751	0.681	0.0681	-0.514
$\lambda = 1/4 = 0.25$	5.768	0.679	0.0679	-0.704
$\lambda = 1/5 = 0.2$	5.792	0.678	0.0678	-0.849
$\Lambda = 5$	2.907	0.351	0.0702	2.576
$\Lambda = 15$	9.067	1.014	0.0676	-1.231

Figure 4 Testing program for the wings without flaps

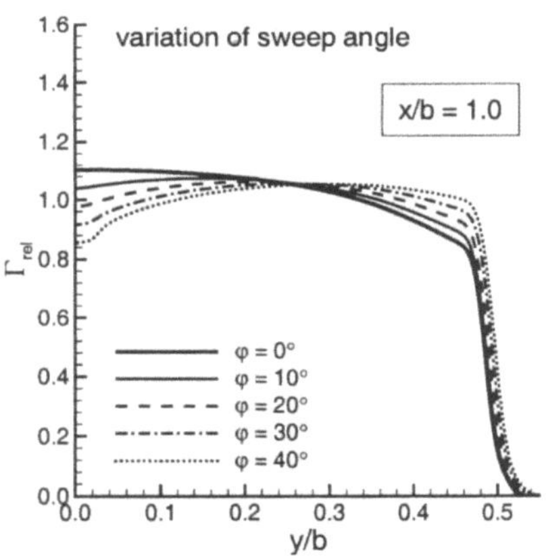

Figure 5 Circulation distribution Γ_{rel} at $x/b = 1.0$ in the wake of the wings without flaps for different sweep angles φ

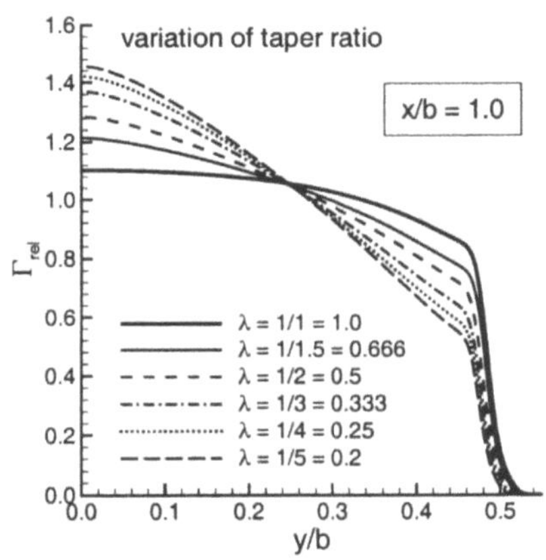

Figure 6 Circulation distribution Γ_{rel} at $x/b = 1.0$ in the wake of the wings without flaps for different taper ratios λ

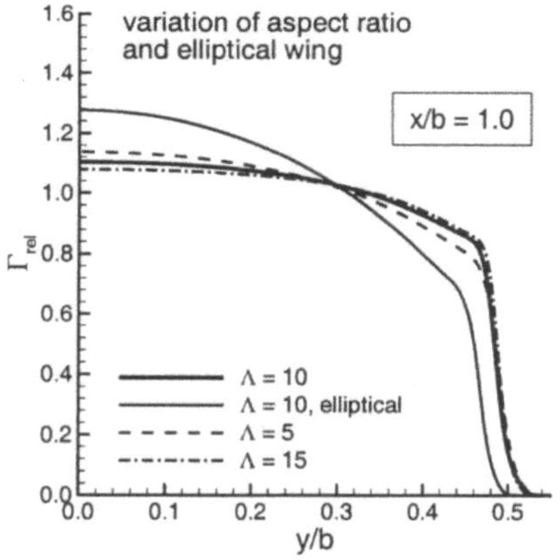

Figure 7 Circulation distribution Γ_{rel} at $x/b = 1.0$ in the wake of the wings without flaps for different aspect ratios Λ and for the elliptical wing

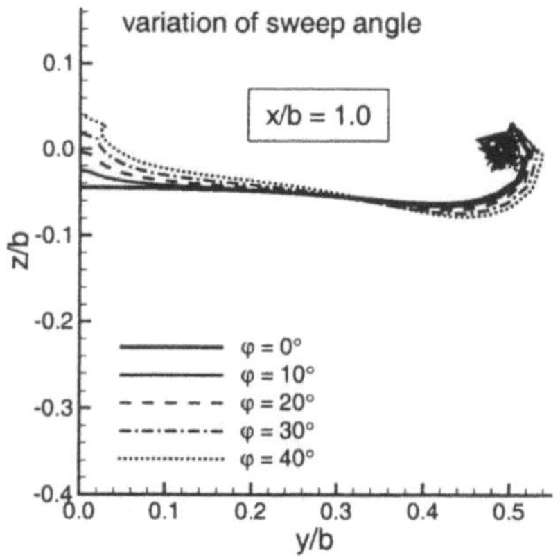

Figure 8 Wake geometry at $x/b = 1.0$ in the wake of the wings without flaps for different sweep angles φ

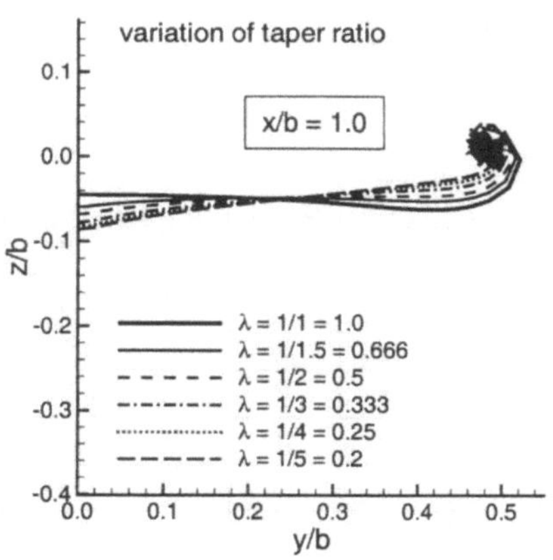

Figure 9 Wake geometry at $x/b = 1.0$ in the wake of the wings without flaps for different taper ratios λ

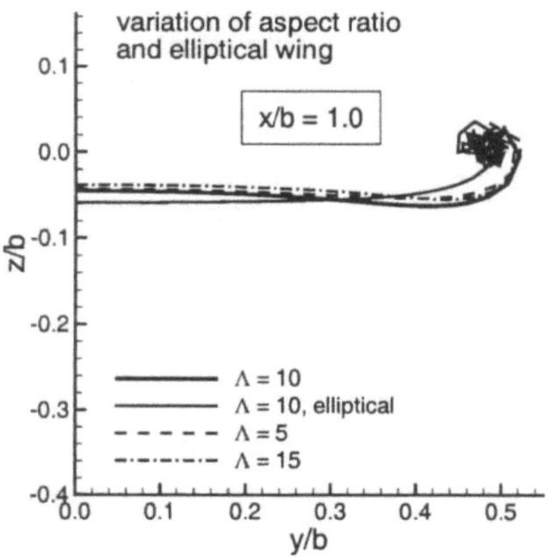

Figure 10 Wake geometry at $x/b = 1.0$ in the wake of the wings without flaps for different aspect ratios Λ and of the elliptical wing

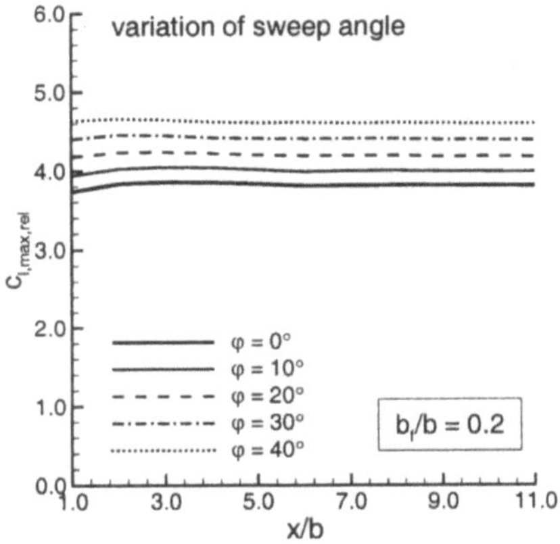

Figure 11 Maximum induced rolling moment coefficient $c_{l,max,rel}$ in the wake of the wings without flaps for different sweep angles φ

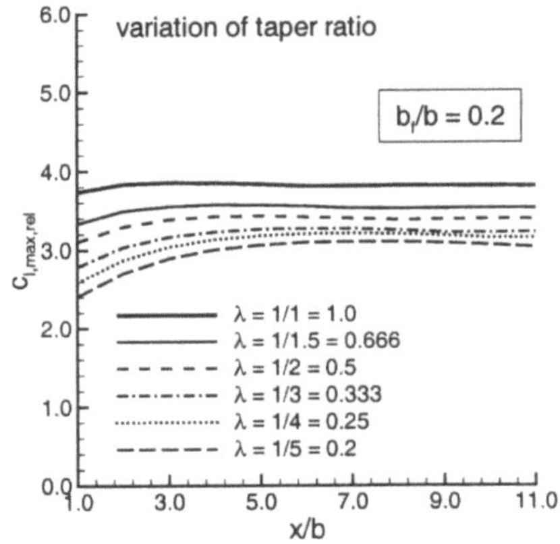

Figure 12 Maximum induced rolling moment coefficient $c_{l,max,rel}$ in the wake of the wings without flaps for different taper ratios λ

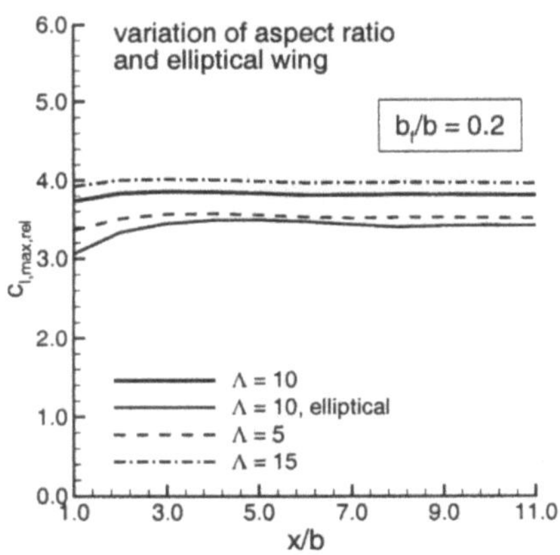

Figure 13 Maximum induced rolling moment coefficient $c_{l,max,rel}$ in the wake of the wings without flaps for different aspect ratios Λ and of the elliptical wing

swept and tapered wing with flaps and slats

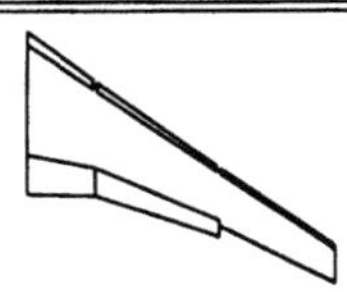

α [°]	δ_i [°]	δ_o [°]	δ_s [°]	c_A
10.000	0	0	0	0.915
7.025	5	5	25	0.916
4.160	10	10	25	0.921
1.589	15	15	25	0.928
-1.369	20	20	25	0.932
5.728	10	5	25	0.919
1.333	20	10	25	0.904

Figure 14 Testing program for the swept and tapered wing with flaps and slats

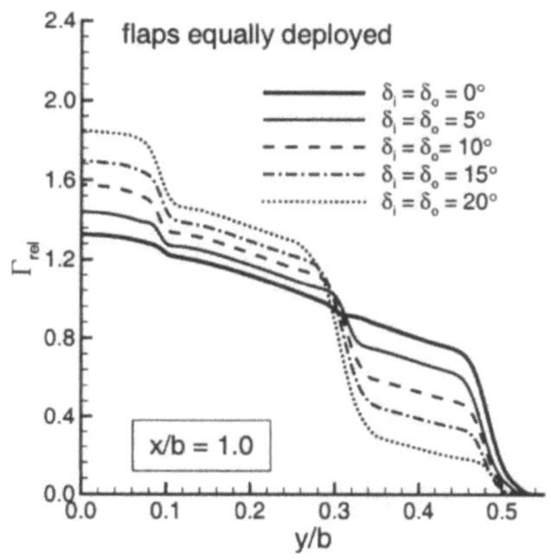

Figure 15 Circulation distribution Γ_{rel} at $x/b = 1.0$ in the wake of the swept and tapered wing with flaps and slats for equal flap deflection $\delta_i = \delta_o$

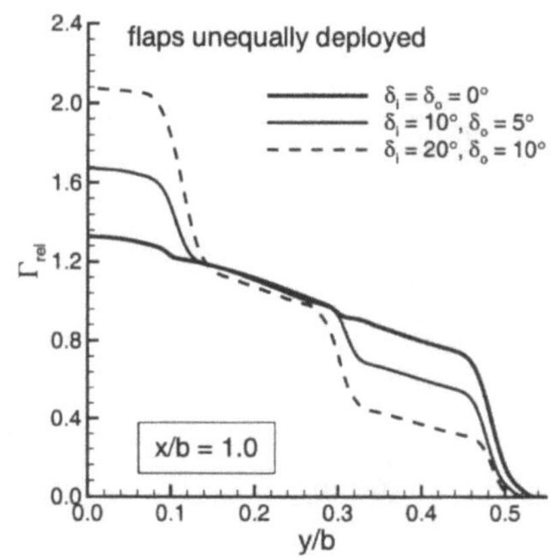

Figure 16 Circulation distribution Γ_{rel} at $x/b = 1.0$ in the wake of the swept and tapered wing with flaps and slats for unequal flap deflection $\delta_i \neq \delta_o$

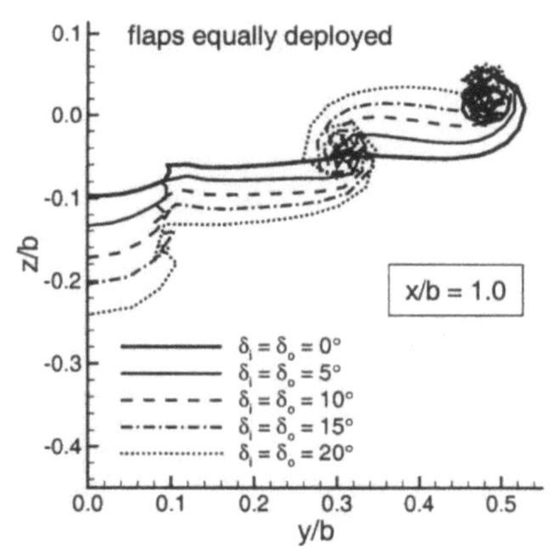

Figure 17 Wake geometry at $x/b = 1.0$ in the wake of the swept and tapered wing with flaps and slats for equal flap deflection $\delta_i = \delta_o$

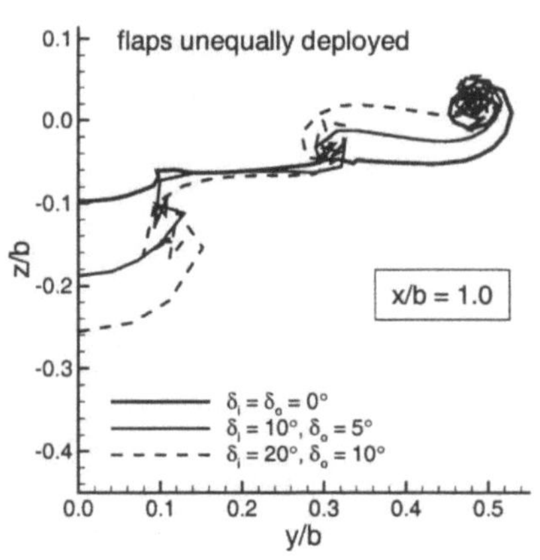

Figure 18 Wake geometry at $x/b = 1.0$ in the wake of the swept and tapered wing with flaps and slats for unequal flap deflection $\delta_i \neq \delta_o$

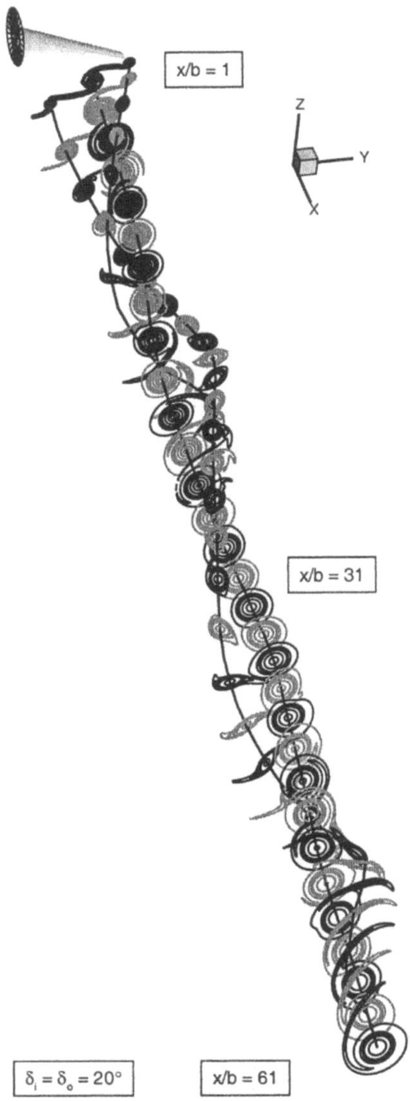

Figure 19 Vortex trajectory in the wake of the swept and tapered wing with flaps and slats for equal flap deflection $\delta_i = \delta_o$

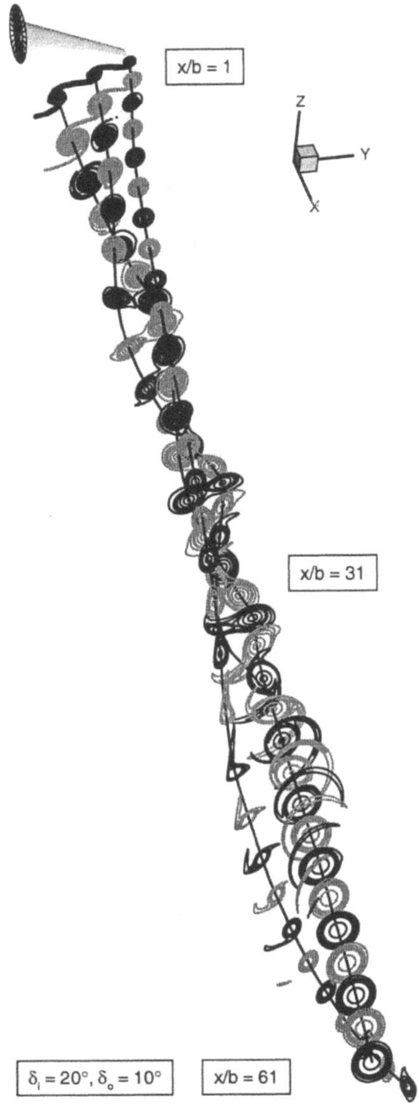

Figure 20 Vortex trajectory in the wake of the swept and tapered wing with flaps and slats for unequal flap deflection $\delta_i \neq \delta_o$

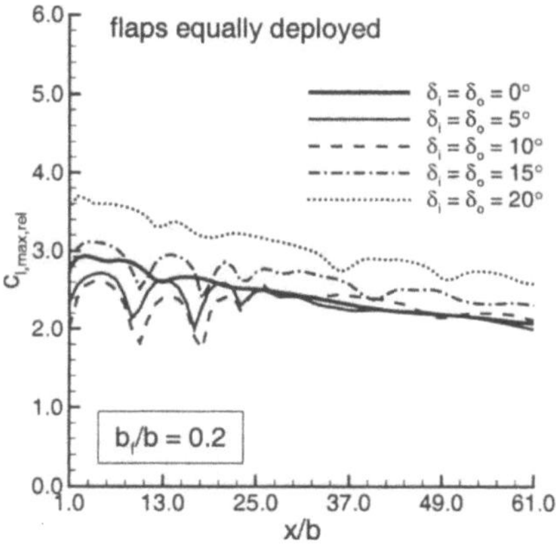

Figure 21 Maximum induced rolling moment coefficient $c_{l,max,rel}$ in the wake of the swept and tapered wing with flaps and slats for equal flap deflection $\delta_i = \delta_o$

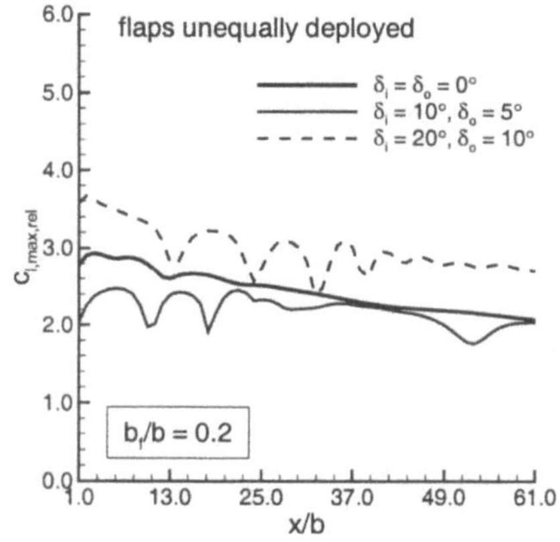

Figure 22 Maximum induced rolling moment coefficient $c_{l,max,rel}$ in the wake of the swept and tapered wing with flaps and slats for unequal flap deflection $\delta_i \neq \delta_o$

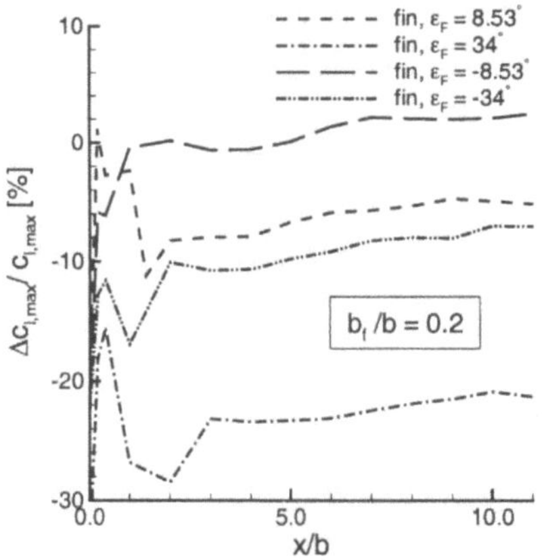

Figure 23 Change in the maximum induced rolling moment coefficient $\Delta c_{l,max}/c_{l,max}$ in the wake of the rectangular wing without flaps and slats and with a wing fin at the wing tip

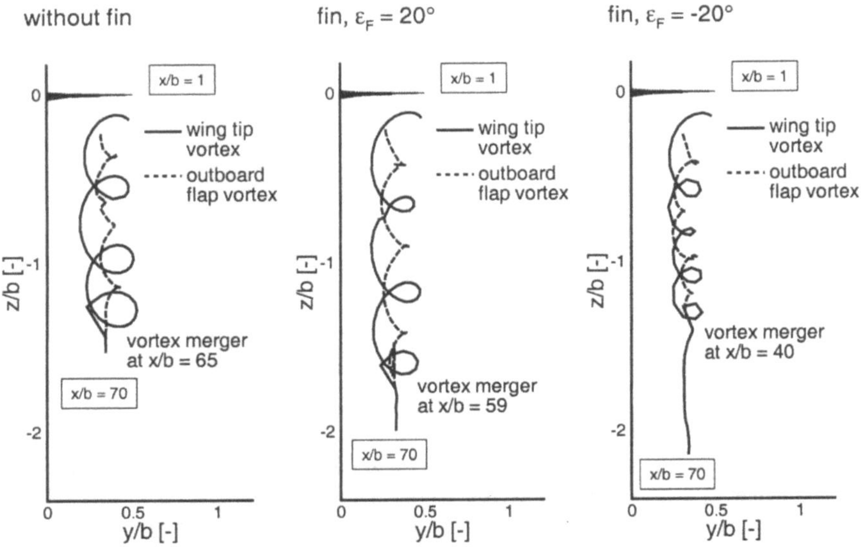

Figure 24 Influence of the wing fin on the vortex trajectory in the wake of the swept and tapered wing with flaps and slats, view from the back

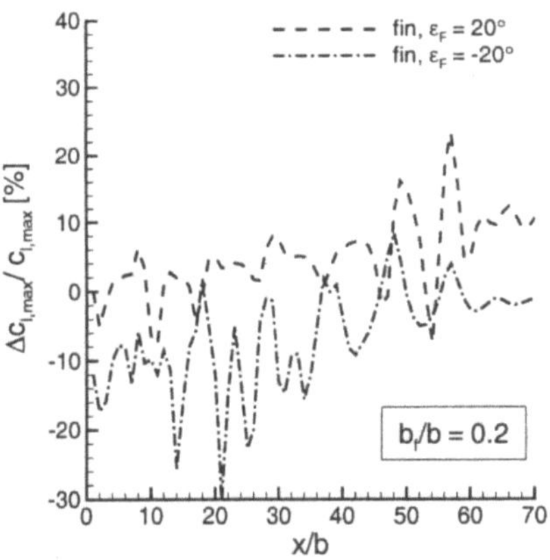

Figure 25 Change in the maximum induced rolling moment coefficient $\Delta c_{l,max}/c_{l,max}$ in the wake of the swept and tapered wing with flaps and slats and with a wing fin

Analysis of Wakes and Wake-Jet Interaction

E. Fares, W. Schröder
Aerodynamisches Institut der RWTH Aachen
Wüllnerstraße zw. 5 und 7, D-52062 Aachen, Germany

Summary

The near wakes of a wing/engine configuration and a three-element wing segment are investigated. The analysis is focused on the development of the flap and wingtip vortices and their interaction with the engine jets in the near field. The essential details of the solution technique such as grid generation, turbulence modeling, boundary conditions and their implementation are presented. A modification of the Spalart-Allmaras one-equation model is proposed including a determination of the wall distance using a partial differential equation.
The methods are validated with solutions for an isolated analytical vortex and the near wake of a flapped wing based on experimentally determined inflow boundary conditions. Furthermore, comparisons with measurements of the 3 element high lift airfoil are presented. Finally, physical aspects of the flows in the near wake of the wing segment for the high lift configuration and the wing/engine configuration with and without the engine jets are discussed.

1 Introduction

The increasing aircraft traffic requires optimized take-off and landing frequencies. The minimum distance of landing or starting aircrafts is usually based on the assumption that the circulation of the trailing vortices decreases by dissipative mechanisms. It is known, however, that the trailing vortices can break down suddenly, see e.g. the review of the dynamics of trailing aircraft vortices by Spalart [3]. The exact reason for the destruction of the vortices is still unknown. It is assumed that the interaction of the engine jets, flap side-edge and trailing-edge vortices in the near field is one possible mechanism. Another aspect of the engine-jet wake-interaction is the entrainment and detrainment of the engine exhaust into the vortex cores, which is of importance in the study of aircraft emissions in the atmosphere.

Recent measurements [4] and numerical studies of the isolated turbulent wingtip vortex [5] were successfully carried out for a rectangular wing with small aspect ratio. It was shown in the numerical investigations that good agreement with experimental results could only be achieved with appropriate grid resolution, turbulence models and boundary conditions. Other authors [6, 7, 8] investigated the dynamics of the engine jets and vortices in the far field using simplified models to predict the inflow conditions from the near wake. These investigations included a wide range of parameters that influence the wake flow, such as stratification, wind shear, buoyancy and geometrical parameters.

Since the dynamics of vortices is largely dependent on the axial and circumferential velocity distribution, see e.g. [9], a detailed prediction of the velocity field would require the simulation of the wake flow field including all vortices and also the engine jets. Due to the high computational costs, a numerical simulation of the flow field around the wing including the far wake is not possible today. The method used here was developed in [1] and consists of a two step approach: first, the flow field around the wing is determined, subsequently, the data of this simulation is used to provide the inflow condition for a separate wake simulation.

2 Method of Solution

2.1 Flow Solver

The method of solution used here has been developed at the Aerodynamische Institut of the RWTH Aachen and was successfully used for predicting a wide variety of internal and external flow problems [10, 11]. The version used in this paper solves the Favre-averaged steady Navier-Stokes equations of a compressible ideal gas in three space dimensions on the basis of a finite-volume discretization on block-structured grids. The spatial discretization of the convective terms follows the $AUSM$ scheme variant $AUSM^+$ proposed by Liou [12, 13]. A $MUSCL$-interpolation [14] is used to achieve second-order accuracy.It is combined with a van Albada [15] limiter which switches to first-order approximation in the vicinity of extrema. The viscous stresses are centrally discretized to second-order accuracy.
Two different schemes were considered for the explicit integration in time: An explicit 5-step Runge-Kutta time stepping scheme [16] using the coefficients $\alpha_i = (0.059, 0.145, 0.273, 0.5, 1.)$ and the two-stage rational Runge-Kutta scheme described in [17, 18] for steady flows with the coefficients $\{b, c\} = \{-1., 0.5\}$. Both schemes are second-order accurate and allow time steps at a maximum Courant number of approx. 3.5. The rational Runge-Kutta scheme was found to be more efficient only in the beginning of the calculations. Convergence up to machine accuracy could only be achieved for the 5-step Runge-Kutta scheme, since the rational Runge-Kutta scheme uses averaging factors depending on the vanishing residuals. To further accelerate the solution process to steady state local time stepping, multigrid [19] and residual smoothing [20] were used. The algorithm was developed on a PC cluster using MPI and PVM and run on vector/parallel architectures.

2.2 Multiblock Topology

Small grid cells are necessary in regions with strong gradients, i.e., in boundary layers, wake regions and of transonic flows also in the vicinity of shocks. One disadvantage of simple structured grids is an unnecessary clustering of grid lines in the far field which can be avoided by a complex "clamp" grid topology as proposed in [22]. The basic idea behind this technique is to limit the regions with small spatial steps e.g. near rigid walls by redirecting the grid lines such that they begin and end at the surface, see e.g. Fig. 1. Such grids were used throughout this paper to improve the grid resolution without increasing the number of grid points.

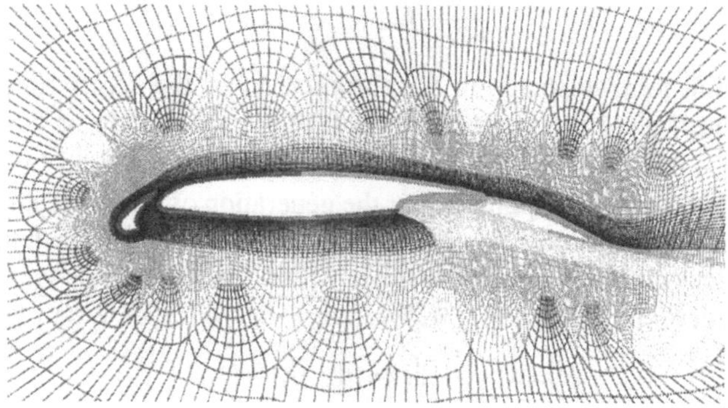

Figure 1 Grid for multi-element airfoil, 41 blocks, 43279 nodes

2.3 Turbulence Modeling

The flow investigated in this work can be divided into two regions. The boundary layer dominated flow around the wing and the vortex and jet dominated wake. At the moment a direct numerical simulation of high Reynolds number flows is not possible, since the computational costs typically scale according to Re_L^3 [23]. Attempts have been made to perform large-eddy simulations of temporally developing wakes [7, 24, 25] using spatially periodic boundary conditions, but only at high computational costs. Therefore, in this study the turbulent flow is modeled based on the standard Reynolds-averaging concept. Several simple algebraic turbulence models as well as transport equations of varying complexity have been developed and investigated for aerodynamic wall-bounded flows. For an overview see [26, 27]. The performance and the predictive capabilities of many models, such as the Baldwin-Lomax [28], Baldwin-Barth [29], Spalart-Allmaras [30], Jones and Launder's $k-\varepsilon$ [31], Wilcox's $k-\omega$ [32] and Menter's SST [33] are documented in studies works [27, 34, 35, 37, 36, 38]. Simple algebraic models have severe limitations, since they do not take transport and diffusion effects into account. Two- or higher-equation models on the other hand require much more computing time compared to one-equation models. However, often hardly any improvement is achieved.

The Spalart-Allmaras one-equation model has proven to be very efficient and robust for the challenging aerodynamic flow and was used in similar calculations [39]. This model was slightly modified, since in its original formulation it is not suitable for free-shear flows. The original transport equation was reformulated in a "conservative" manner for the determination of the turbulent viscosity ν_t:

$$\begin{aligned}\frac{D(\varrho\tilde{\nu})}{Dt} &= \frac{\varrho}{\sigma}\left[\nabla\cdot((\nu_l+(1+c_{b2})\tilde{\nu})\nabla\tilde{\nu})-c_{b2}\tilde{\nu}\nabla^2\tilde{\nu}\right] \\ &\quad +\varrho c_{b1}\tilde{S}\tilde{\nu}-\varrho c_{w1}f_w\left(\frac{\tilde{\nu}}{d}\right)^2\end{aligned}$$

with

$$\tilde{S} = S+\frac{\tilde{\nu}}{\kappa^2 d^2}f_{v2} \qquad \nu_t=\tilde{\nu}\cdot f_{v1}\ .$$

The production term depends on the formulation of the scalar norm of the deformation tensor S. In the original form [30] the magnitude of vorticity $\Omega_{ij} = \sqrt{(\partial U_i/\partial x_j - \partial U_j/\partial x_i)^2}$ was incorporated, which is motivated by the fact, that the near-wall vorticity is responsible for the generation of turbulence. However, this is not true for the vortex core where circumferential effects diminish turbulence resulting in a decay of the overall turbulence within the vortex [40, 41, 4]. In [5] it is therefore suggested to use $s_{ij} = \sqrt{(\partial U_i/\partial x_j + \partial U_j/\partial x_i)\partial U_i/\partial x_j}$ to limit S:

$$S = \Omega_{ij} + 2min(0., s_{ij} - \Omega_{ij}) \quad .$$

It is worth noting that this modification has no influence on the near-wall region, where Ω_{ij} and s_{ij} are of the same order of magnitude. Furthermore, it could be considered a drawback of the original model that the destruction term disappears completely in the far-wall region ($d \gg 1$), which is not true for turbulent free-shear flows like jets, for example. The absence of this term leads to an unphysical growth of $\tilde{\nu}$ when a jet flow occurs far from a wall and, consequently, results in an overprediction of the spreading rate as reported in [27].
By investigating the ratio between production and dissipation of the standard high Reynolds number Jones/Launder $k-\varepsilon$ model [31] and assuming that the term $P_k = \frac{\partial U_i}{\partial x_j}\left(\frac{\partial U_i}{\partial x_j}+\frac{\partial U_j}{\partial x_i}-\frac{2}{3}\delta_{ij}\frac{\partial U_k}{\partial x_k}\right)$ is the main mechanism for production and destruction of the k and ε equations this term was modified. As a first step towards an improvement of the original model, the destruction term is reformulated only for the far-wall regions as follows:

$$\begin{aligned}dest_1 &= \varrho c_{w1}f_w\left(\frac{\tilde{\nu}}{d}\right)^2 \\ dest_2 &= \frac{c_\mu}{c_1c_2c_{b1}}\varrho\tilde{\nu}\cdot\frac{1}{2}\sqrt{P_k} \\ dest &= max(dest_1, dest_2)\end{aligned}$$

where c_μ, c_1, c_2 are taken from the original $k-\varepsilon$ model. In the vicinity of the wall the first term $dest_1$ prevails due to its dependence on $1/d^2$. In free shear layers far away from the wall, the term $dest_1$ will tend to zero, so that $dest_2$ dominates the destruction term [1].

The flows considered in this paper can be assumed fully turbulent, since in the experimental investigation the boundary layers were tripped at the leading edge. For this reason, no tripping corrections were incorporated into the turbulence model.

It was found that the initial value of the variable $\tilde{\nu}$ had to be larger than the physical free stream value $\tilde{\nu} \approx 0.0$ to provide a sufficiently strong trigger for the production term and to avoid convergence to the trivial solution of the equation for the turbulent viscosity ($\tilde{\nu} \equiv 0$). At the inflow boundary $\tilde{\nu} = 0.1\nu_l$ was specified, where ν_l is the laminar viscosity. At the outflow boundaries $\tilde{\nu}$ was extrapolated.

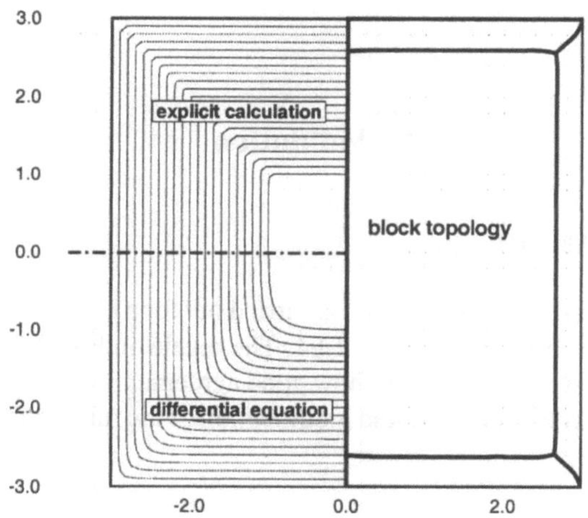

Figure 2 Lines of constant distance from the walls. Comparison between exact distance and the solution of the PDE

In 3D problems the computation of the distance *d*, especially when a massively parallel method is used, usually involves a complex search algorithm. To avoid the search procedure *d* was determined by solving a partial differential equation for the inverse of the distance:

$$\frac{\partial g}{\partial t} = c_r \left(\frac{1}{g} \frac{\partial}{\partial x_j} \left(g \frac{g}{\partial x_j} \right) + (\sigma - 1) \frac{\partial^2 g}{\partial x_j^2} - \frac{2\sigma + 1}{(\sigma + 1)^2} g^3 \right)$$
$$g = \left(\frac{1}{d + d_0} \right) \cdot (\sigma + 1) \tag{1}$$

[1]The exact influence of this modification on the turbulent boundary layer still has to be investigated [49]. In this study the modification was used only for the wake flow that didn't include any walls.

with the following initial and boundary conditions:

$$g(t=0) \approx 0.0 \quad g_{wall} = \frac{\sigma+1}{d_0} \quad \left.\frac{\partial g}{\partial n}\right|_{other} = 0.0 \quad .$$

The constant c_r represents a relaxation factor which can be chosen according to the integration scheme. The quantity σ introduces an additional elliptic character into the equation to account for intersecting walls, say at corners. The offset d_0 is used to avoid large values of g near rigid walls. Since d grows linearly with the wall distance, this offset has no effect on the solution. The values chosen here are $\sigma = 0.1$ and $d_0 = 1.0$. An example for a converged solution of a multi-block domain is shown in Fig. 2. This equation is solved iteratively in the beginning of the simulation.

3 Validation

3.1 Plane Turbulent Jet

A test of the proposed turbulence model is carried out for a plane turbulent jet flow at $Ma_0 = 0.2$, $Re_d = 10^4$ and $u_0/u_1 = 1.84$. The results in Fig. 3 for the half width and in Fig. 4 for the velocity profiles show good agreement with the similar solution according to [53] , when the suggested modification of the turbulence model is used.

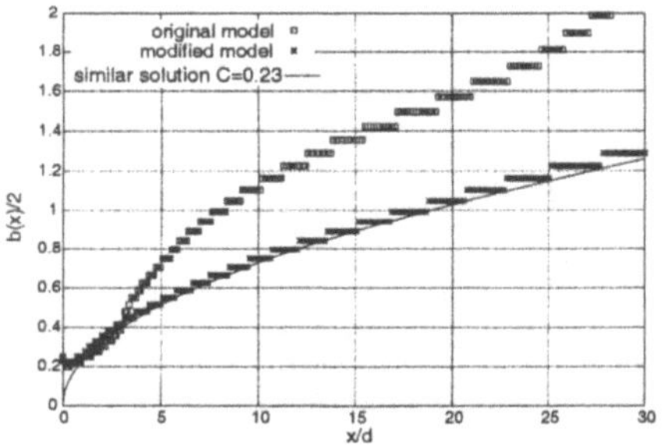

Figure 3 Analytical and simulated half width $b(x)/2$ of the turbulent plane jet

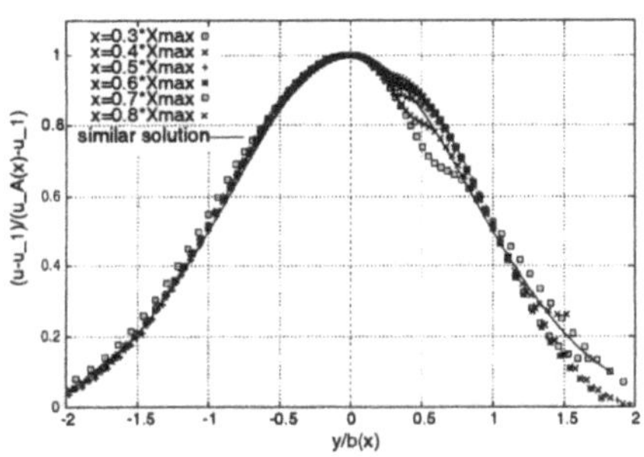

Figure 4 Similar velocity profile of turbulent plane jet **left:** modified model **right:** original model

3.2 Analytical Laminar Vortex

Different outflow boundary conditions and grid resolutions were investigated by studying an analytical laminar vortex [43, 44] :

$$
\begin{aligned}
v_t &= \frac{\Gamma}{2\pi r}\left(1 - e^{-\frac{U_\infty r^2}{4\nu z}}\right) \\
p &= -\frac{\Gamma^2 U_\infty}{32\pi^2 \nu z} P(\zeta) \\
w &= -\frac{\Gamma^2 U_\infty}{32\pi^2 \nu z} e^{-\zeta}\left[ln\frac{uz}{\nu} \right. \\
&\qquad \left. - \int_0^\zeta (1 - \xi e^\xi + P(\zeta)e^\xi) d\xi + K \right]
\end{aligned}
$$

with

$$
P(\zeta) = \int_\zeta^\infty \frac{(1 - e^{-\xi})^2}{\xi^2} d\xi \quad \text{and} \quad \zeta = \frac{U_\infty r^2}{4\nu z} \; .
$$

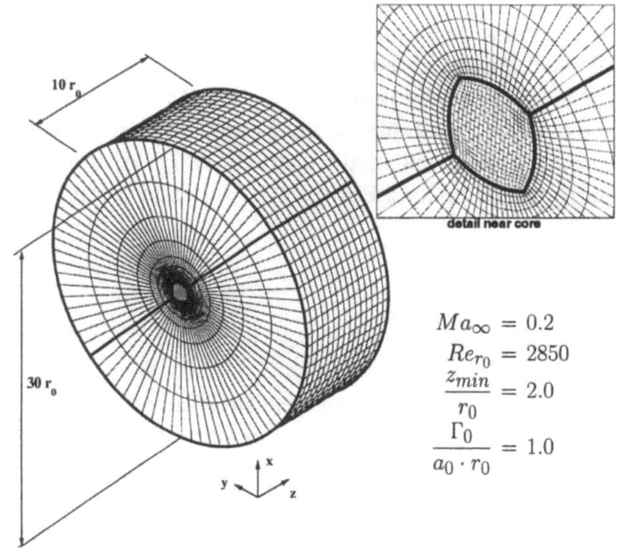

Figure 5 3-block grid for the analytical laminar vortex

The grid used and the extents of the boundary are shown in Fig. 5. The velocity field was explicitly imposed at the inflow boundary. Additionally, the temperature was chosen to be constant and the pressure was evaluated using one-dimensional Riemann invariants on the inflow boundary. A boundary condition based on the Euler equations was used at the cross flow boundary. At the outflow boundary the pressure level was prescribed from the analytical solution allowing pressure fluctuations to cross the boundary and the remaining conservative variables were evaluated by two different approaches, the method of linearized characteristics [45] and a non-reflecting boundary condition [46] . In Fig. 6 results obtained with the two outflow boundary conditions are compared for a fine and a coarse grid. The comparison is done for the analytical and the simulated values of the core radius and the maximum tangential velocity of the vortex as a function of the main flow direction. It should be noted that the core radius location is determined at discrete grid nodes without interpolation which is why a step-like curve is obtained. Compared to the boundary condition based on an extrapolation of linearized characteristics the non-reflecting condition minimizes the influence of the far field boundary condition on the solution. It is apparent, that this boundary condition type is superior to the other at negligible computational costs.

The grid resolution in the vortex core plays a major role for the accuracy in the prediction of vortical flows. Different grid resolutions ranging from $16-64$ grid points in a crossflow section (x-y plane) within the vortex core were investigated. Different streamwise (z-direction) grid resolutions were also investigated (Fig. 7). Two main factors have shown to be relevant within this study. The crossflow resolution, that is the resolution of the vortex core in both crosswise directions, and the aspect ratio of the cells. The latter is the ratio of the grid spacing between the streamwise and crosswise direction. A coarse crosswise resolution could only be adequately realized, if the aspect ratio is kept small. A large aspect ratio resulted in a poor solution especially near the outflow boundary. As a compromise between the computational costs and a reasonable accuracy a resolution of about 32 points in the vortex core was chosen with a maximal aspect ratio of 10. A similar result was also found in another investigation for a Burgers type vortex in compressible flows [47]

3.3 3 Element High Lift Airfoil

The two-dimensional flow of the multi-element airfoil still constitutes a challenging task for the computational fluid dynamics community since many details like grid quality and turbulence modeling may play a significant role in the correct reproduction of the flow. Some effort has been invested in the grid generation of the 2 dimensional problem to ensure the quality of the grid for the desired flow and to minimize the number of grid points necessary since the real flow to be investigated in this work is the 3D extension of this flow including the engine-jet. In other words the 2D grid has the same topology as the 3D grid used later.

Figs. 8 and 9 depict the pressure distribution on the surface compared to experimental data from Moir [50] and the calculated Mach number distribution at $\alpha = 4°$. A

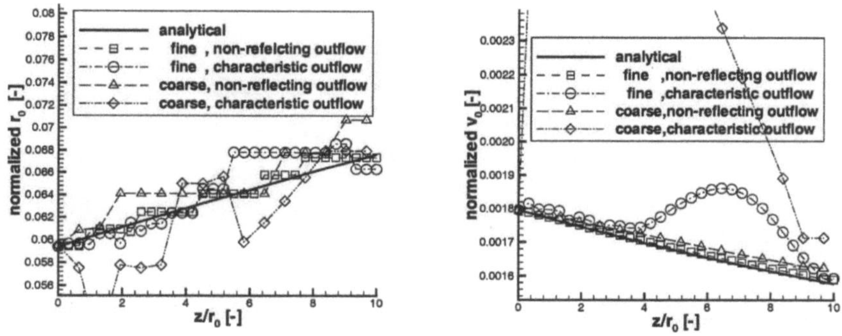

Figure 6 Impact of the outflow boundary condition on the development of the core radius r_c and the maximum tangential velocity v_0

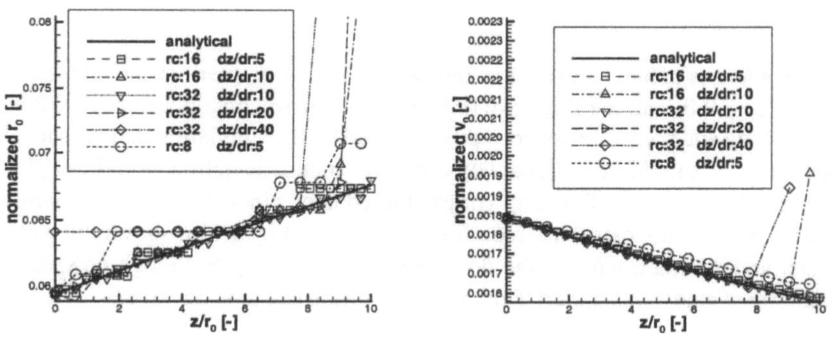

Figure 7 Influence of the grid spacing on the development of the core radius r_0 and the maximum tangential velocity v_0

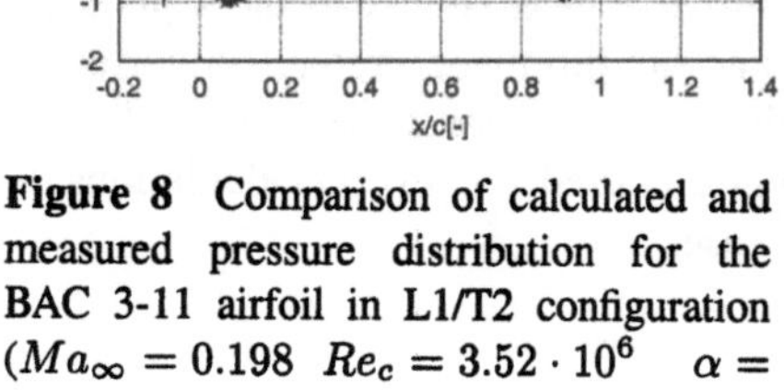

Figure 8 Comparison of calculated and measured pressure distribution for the BAC 3-11 airfoil in L1/T2 configuration ($Ma_\infty = 0.198$ $Re_c = 3.52 \cdot 10^6$ $\alpha = 4°$)

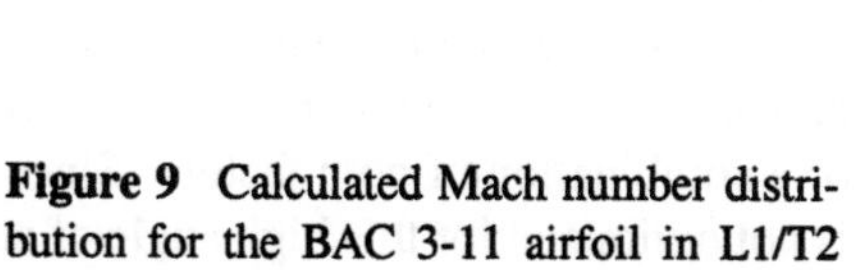

Figure 9 Calculated Mach number distribution for the BAC 3-11 airfoil in L1/T2 configuration ($Ma_\infty = 0.198$ $Re_c = 3.52 \cdot 10^6$ $\alpha = 4°$)

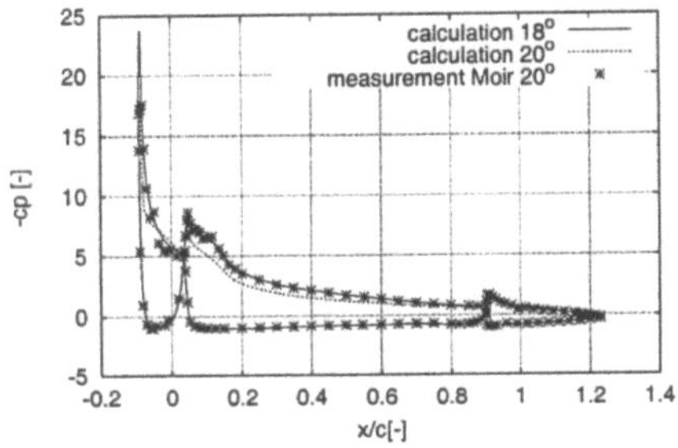

Figure 10 Comparison of calculated and measured pressure distribution for the BAC 3-11 airfoil in L1/T2 configuration ($Ma_\infty = 0.198$ $Re_c = 3.52 \cdot 10^6$ $\alpha = 18°/20°$)

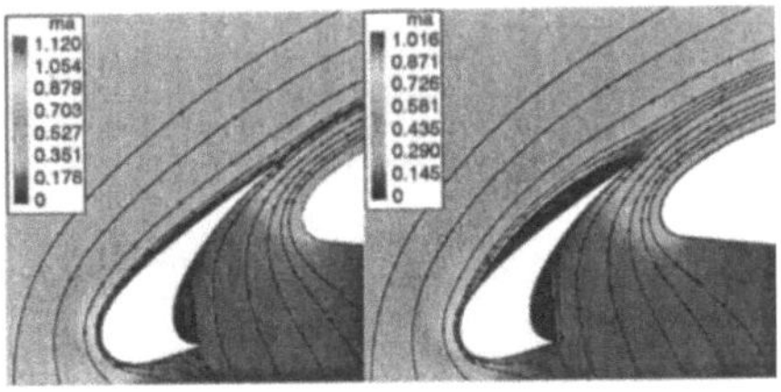

Figure 11 Calculated Mach number distribution and streamlines showing separation near the slat ($Ma_\infty = 0.198$ $Re_c = 3.52 \cdot 10^6$ $\alpha = 18°$(left)/20°(right)

good agreement is achieved except in the two recirculation areas at the slat lower surface ($x/c = 0.0$) and the flap cove ($x/c = 0.8$). These contain a separated flow that is unsteady and therefore not correctly captured by the simulation that uses local time stepping. It should be noticed, however, that such a deviation would not play a role in the development of the flap vortex studied later, since this vortex is mainly created by the pressure distribution on the flap surface. Fig. 9 also shows the wake of the slat and the main wing in the flow field downstream even though no special grid refinement study was undertaken in the resolution of the wake area.

The solution at $\alpha = 20$ is shown in Figs. 10 and 11. The comparison between the calculated and the measured pressure distribution evidences some differences. According to the literature [51] we correct the angle of attack by 2° in the calculation to achieve a much better agreement for the main wing and the slat distribution. The reason for this can be understood by investigating the flow field on the slat upper surface as shown in Fig. 10. The streamlines show a separation on the upper surface at $\alpha = 20°$ that influences the flow on the slat and on the main wing suction side. By reducing the angle of attack by two degrees this effect disappears and a much better agreement between the calculation and the measurements is achieved. It is also remarkable that the flow reaches supersonic speeds at the slat upper side leading to a very strong peak in the pressure distribution. The maximum Mach number is almost six times as large as the free stream Mach number. The existence of the separation and its size can be caused by the weakness of the turbulence model to correctly predict such a flow or a grid dependence of the simulation for these flow conditions or a combination of both factors. Since one grid is used for several calculations, it is not optimized for this high angle of attack case.

3.4 Wake of a Wing with Flaps

The wake of a swept wing with flaps and slats was experimentally investigated by Özger et al. [48] with hotwire probes in the near wake. The measurement points were located on a cartesian grid in 4 planes perpendicular to the free stream direction located at $x/b = 0.0, 0.2, 0.4$ and 0.8, where b is the wingspan, which is $b = 4.9c$ and c is the root chord length. Special care was undertaken by local refinement of the measurement grid to enhance the resolution of the vortex dominated regions. This wake flow was studied numerically for further validation of the proposed turbulence model in vortical flows. A structured almost cartesian grid with 2.5 million grid points was used for this investigation.

The velocity distribution in the first measured plane at $x/b = 0.0$ was imposed explicitly at the inflow boundary. Since the numerical grid contains much more grid points than the experiment data measurement points, a second-order interpolation was used for grid points within the experimental domain and a third-order extrapolation for the points outside the experimental domain. The outflow boundary conditions were the same as for the analytic vortex. It was sufficient for the other boundaries to use a first order extrapolation of the velocity field to permit inflow and outflow as induced by the vortices. This was found to be accurate as long as the vortices are well captured within the integration domain. The pressure was evaluated by solving the Euler equations for the pressure gradient normal to the boundary using the prescribed velocity field, which takes centrifugal forces into account. A specification of the turbulent viscosity based on the fluctuation and correlation measurements was not possible as a consequence of the missing information in the main flow direction and the coarse grid used for measurements, which would lead to an inaccurate calculation of the flow gradients. Therefore, as a crude assumption, a value of $\tilde{\nu} = 10.0\nu_l$ was specified for the eddy viscosity at the inflow boundary. At the other boundaries the value of $\tilde{\nu}$ was extrapolated. In the experimental investigations the wing was mounted on a flat plate, in the numerical simulation no effort was undertaken to approximate this condition at the wing root.

The result of the calculation is presented in Fig. 12. The rolling up of the wing tip vortex and the outboard flap vortex could be captured and the decrease of maximum vorticity could be simulated. A deviation of the location of the flap vortex compared to the experimental data was found, due to the different boundary conditions at the wing root.

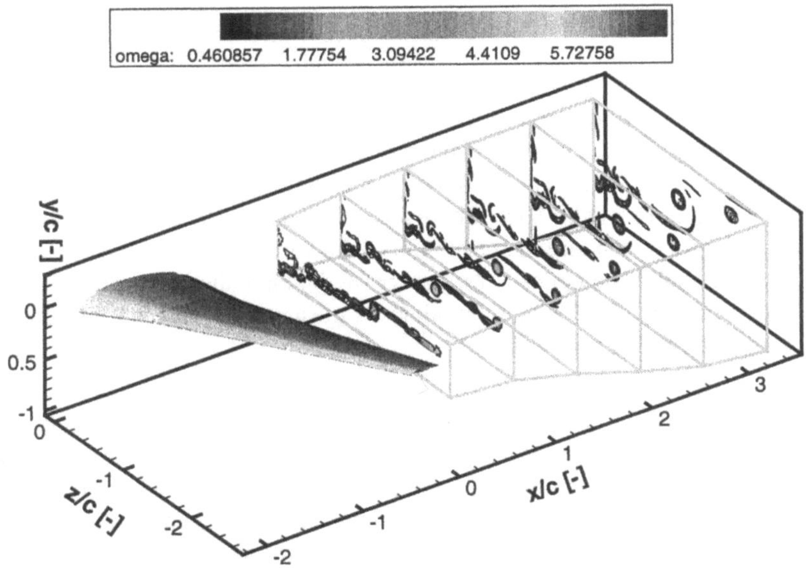

Figure 12 Vorticity distribution in the near wake of a swept wing with a flap based on experimental inflow data at $Ma_\infty = 0.17$ $Re = 800000$ $\alpha = 8°$

In Fig. 13 the turbulent viscosity obtained with the modified Spalart-Allmaras and the original version is juxtaposed. The comparison shows the strong difference between the proposed and the original turbulence model. The original model predicts unphysical high values of the turbulent viscosity in the vortex cores due to the missing destruction mechanism and the dependency of the production term on the vorticity.

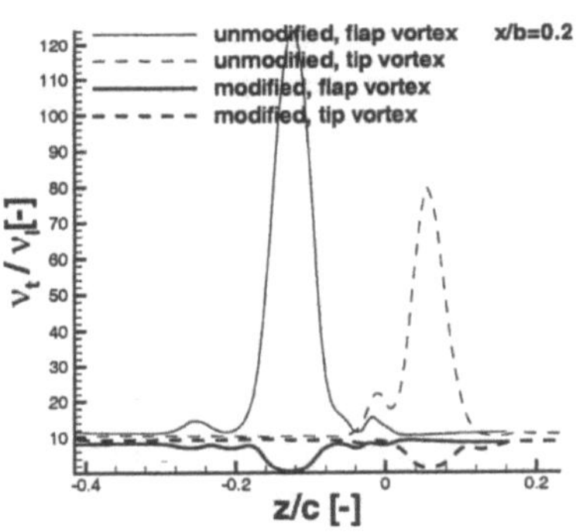

Figure 13 Turbulent viscosity distributions across vortex cores for the original and modified turbulence model

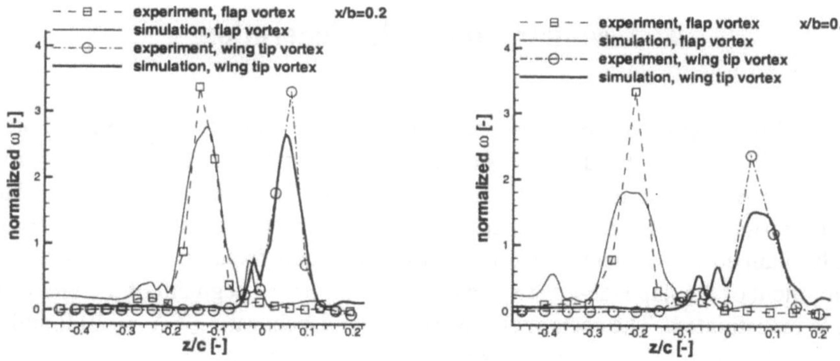

Figure 14 Calculated and measured vorticity across vortex cores at different locations in the wake

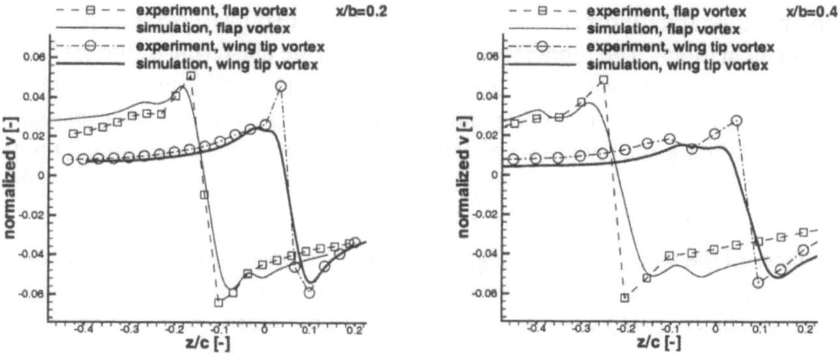

Figure 15 Calculated and measured velocity profiles across vortex cores at different locations in the wake

The profiles of the vorticity ω and the tangential velocity v through the vortex cores at distances $x/b = 0.2$ and $x/b = 0.4$ are given in Figs. 14 and 15. It can be seen from the experimental results that the velocities reach a maximum in the wing tip and the flap vortex. The simulation predicts a lower value of the vorticity compared to the measurements and a stronger decrease in the streamwise direction especially for the flap vortex (Fig. 14). The velocity profiles in the vortex cores (Fig. 15) show good agreement for $x/b = 0.2$ and moderate agreement for $x/b = 0.4$ with the exception of the peak velocities. Although the calculations were carried out on an almost cartesian grid based with a resolution determined according to the analytic case, the resolution in the streamwise direction has still to be improved. In addition, the interpolated experimental data prescribed at the inflow boundary has an influence. It contains measurement and interpolation errors that are convected through the integration domain. Another strong bearing was found in the turbulence modeling which motivated the modification discussed in the previous section. Even though the modification improved the solution in the vortex regions, it still cannot be considered completely satisfactory and requires further investigations.

4 Geometry and Grid Generation

The wing/engine configuration has been defined in the collaborative research programme (SFB 401) at the RWTH Aachen and is typical for future large aircrafts like e.g. the A380. Limitations of the geometry, such as engine cross section size or pylon cross section were due to constraints in the model construction for additionally planned experimental investigations. The main focus in this study is the development of the flap side-edge vortex. Therefore, only a segment of the unswept wing is investigated. The rectangular wing has an aspect ratio of $\Lambda = 10.0$ and a BAC 3-11 [50] cross section profile. The engine geometry was chosen to be similar to the Trent 900, which is expected to be utilized in similar configurations.
A commercial grid generator was used to determine the topology of the multi-block grid. To shorten the interactive grid generation process and to stay within the available computational resources it was necessary to generate only a coarse grid suitable for Euler calculations. The refinement of the grid and the clustering of the grid lines normal to the solid walls were done according to the work in [52]. First the grid was transformed from physical into computational space using cubic splines as mapping functions. Then a global and a local refinement was preformed in computational space. The clustering was done using a tan^{-1}-similar function for the redistribution of the grid points normal to rigid walls.

The surfaces of the wing/engine configuration were wrapped with C-grid type topology to correctly capture the shear flow at the sharp trailing edges of the wing, pylons and nacelle (Fig. 16) whereas the outer part of the mesh possessed an O-type topology. Special care was invested in an adequate resolution of the engine and its exhaust plume, see details in Fig. 17.

The grids for the wing section configuration are shown in Figs. 18, 19 and 20. An acceptable resolution of the flow was achieved even though the trailing edges were wrapped with an O-grid type topology. Especially the clamp topology has been extensively used to reduce the number of grid points. The minimum distance from the wall was chosen $y^+ = 1.5$. The stretching factor of the grid lines normal to the wall was less than 1.5 in the boundary layer.

In the beginning of the simulation there was no prior experimental knowledge of the wake flow and the behaviour of the flap vortex. Therefore, this region in the wake was not especially resolved. Keeping in mind that the ratio of the vortex core diameter to the root chord length of the wing might be as small as 1/60, a classical simple structured C-O or C-H type grid will lead to a number of grid points in the order of 10^8 to adequately resolve the boundary layer and the wake. Since a solution on such a number of grid points cannot be computed in an acceptable time, a two step approach is used for the wake simulation. First, the flow field around the wing is simulated without trying to adequately resolve the wake flow field. Then, the solution at about $0.2c$ downstream from the trailing edge is interpolated and prescribed at the inflow boundary of a cartesian grid to compute the wake.

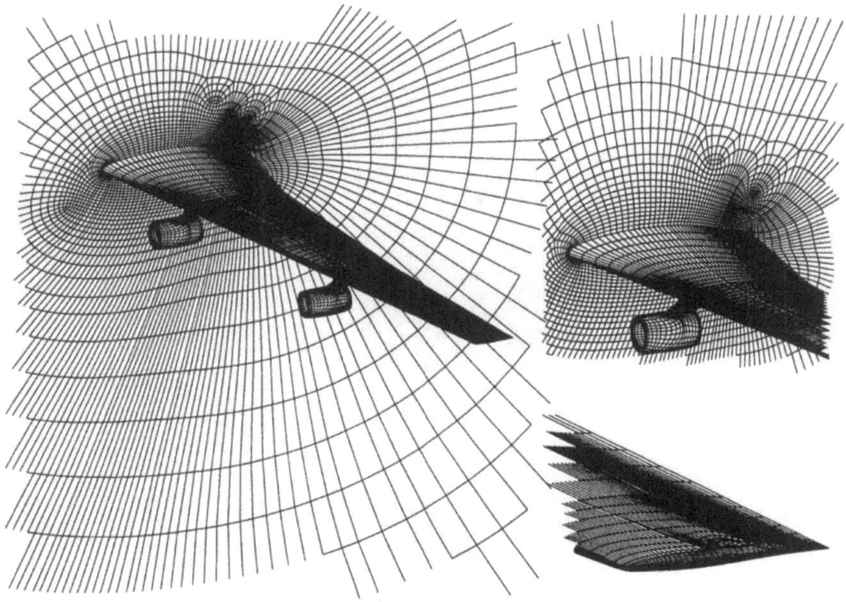

Figure 16 Surface grids of the wing with details at the wing root and wingtip (used for Euler computations)

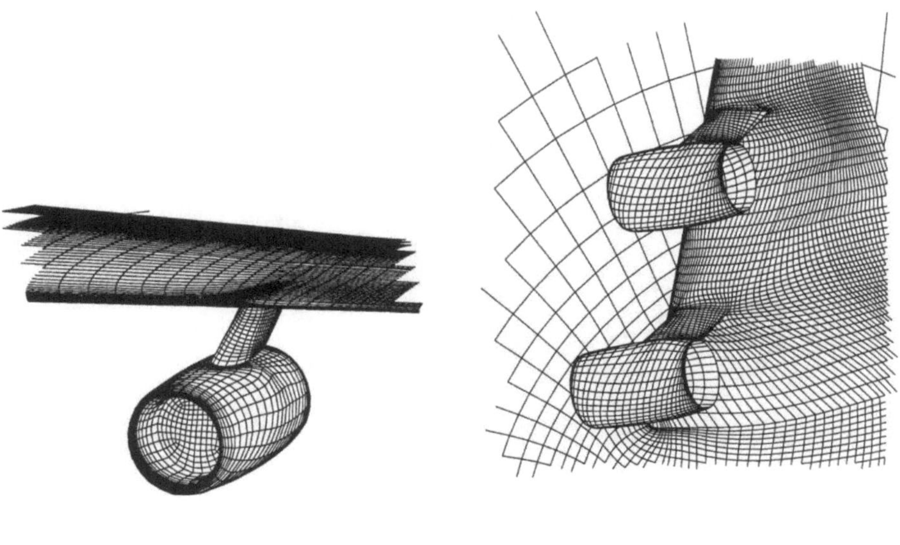

Figure 17 Surface grid of the engines (used for Euler computations)

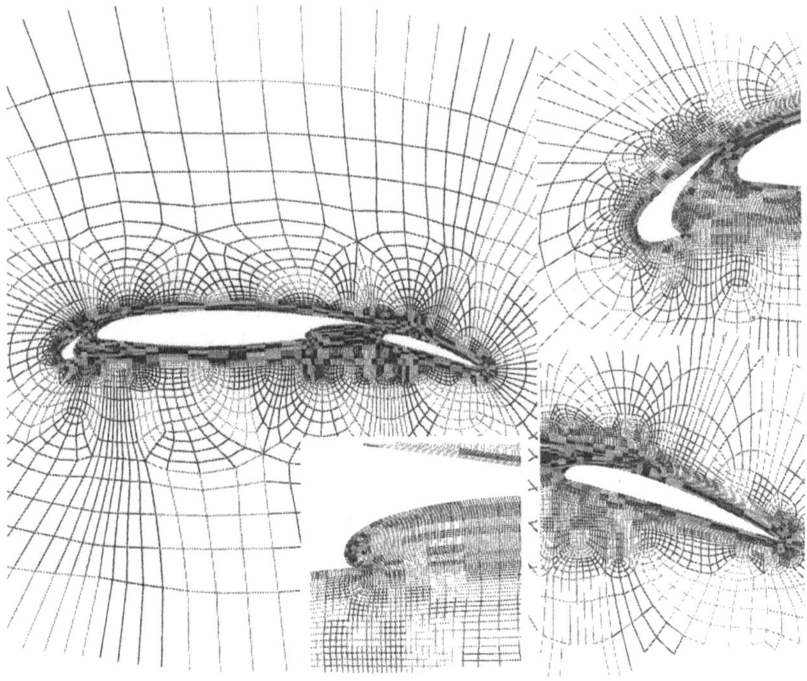

Figure 18 2d grid of the wing segment including enlargements of the slat and flap region

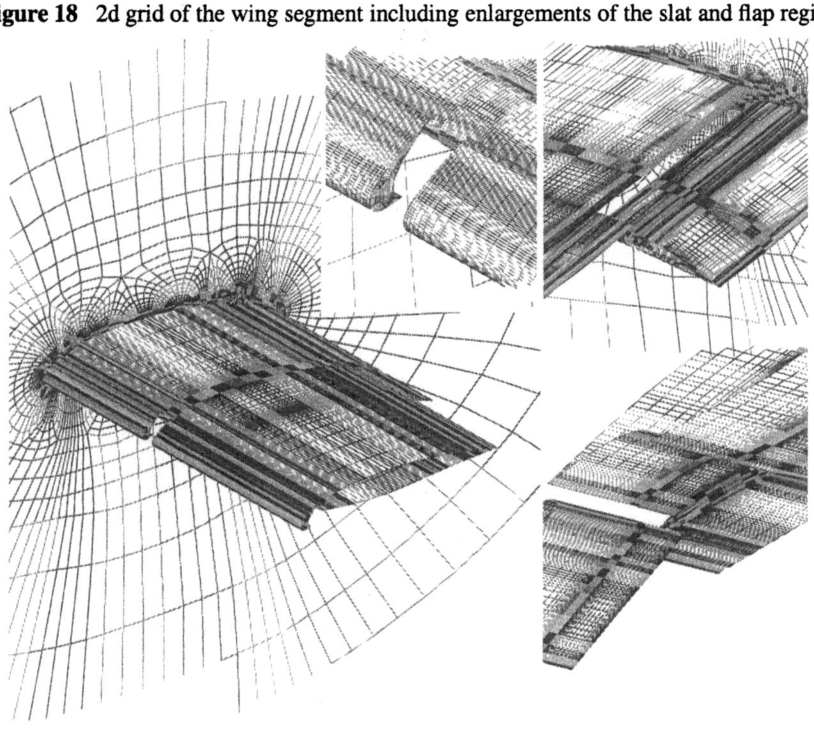

Figure 19 3d surface grid and zooms at slat and flap 90 blocks, 1.5 million nodes

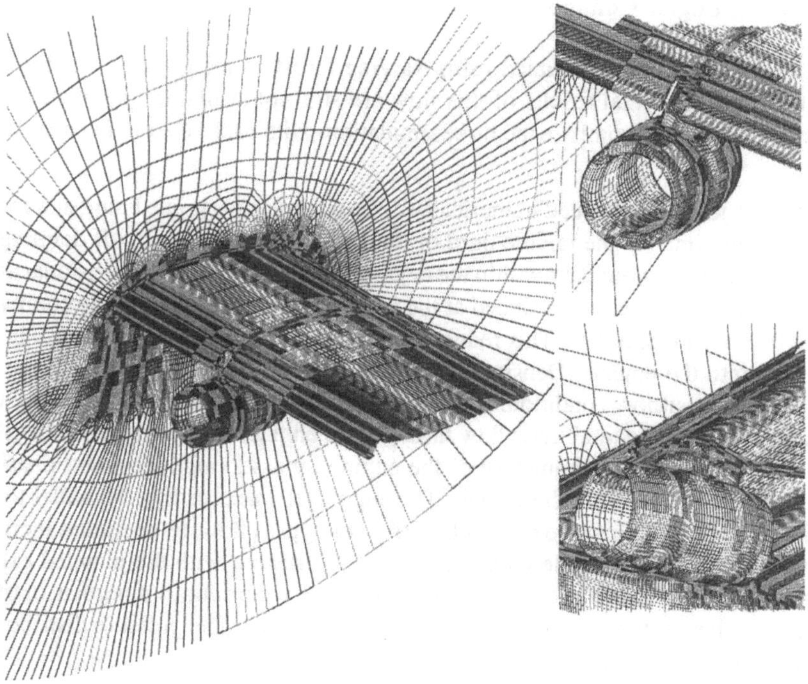

Figure 20 3d surface grid with details at pylon and nacelle 221 blocks, 1.8 million nodes

5 Results and Discussion

The numerical simulation was done as mentioned above in 2 steps. First, the flow of the wing and the engine is simulated until lift and drag converged to a constant value then the variables were interpolated using a weighted inverse distance algorithm into the first plane of a rectangular cartesian wake grid with 3.5 million (wing/engines configuration) and 9 million (wingsegment) grid points. Subsequently, the solution was continued on the cartesian grid similar to the wake case considered for the validation of the algorithm. Considering the amount of computational resources needed for such calculations the rectangular grid could be extended only from $x = 0.2c$ to $x = 10c$ in the main flow direction, where c is the unflapped wing chord length. It should be noted that there is no feedback from the wake grid to the original grid, which is motivated by the approximation that the wake flow can be considered parabolic [6].

5.1 Wing/Engine Configuration

Three simulations were initiated using varying thrust of the engines. At the moment, the experimental results of this flow are still not available, therefore it was considered useful to compare the three simulations with each other in the near wake field.

The engine jet was simulated numerically by imposing appropriate boundary conditions. Approximately the middle third of the engine nacelle was cut off defining an inflow and outflow boundary within the engine. According to one-dimensional analysis, the pressure level was reduced at the inflow boundary of the engine compared to the free stream pressure. This reduction in pressure resulted in an increase in the flow rate into the engine. Density and velocities were calculated in the same way as in a regular outflow condition. In addition, the flow rate at the engine inflow was integrated to give the total mass flow through one engine. This was then prescribed explicitly at the outflow boundary of the engine by assuming a constant distribution of the velocity within the engine. The engine jet was prescribed parallel to the undisturbed flow at an angle of attack of 0°. No bypass flow, temperature increase or swirl of the jet is considered here, which is consistent with the planned experimental setup. Pressure and density are calculated according to a regular outflow condition whereas the eddy viscosity is set to $\tilde{\nu} = 10.0\nu_l$. The pressure distribution on the wing and the outboard engine nacelle for the case of $u_{engine}/u_\infty = 1.6$ is shown in Fig. 21.

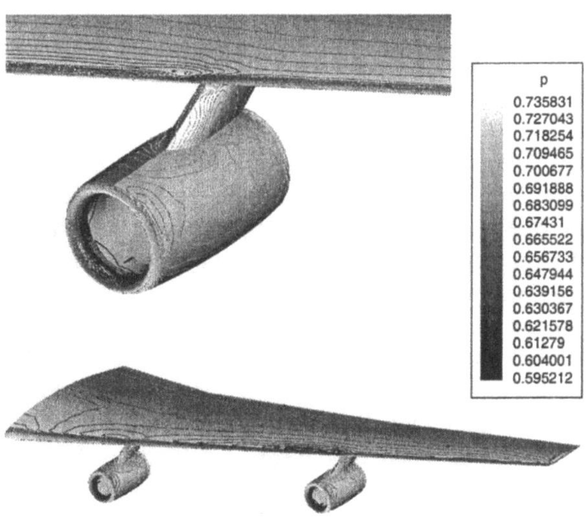

Figure 21 Pressure distribution for $Ma_\infty = 0.27$, $Re_c = 8 \cdot 10^5$, $\alpha = 4.0°$ and $u_{engine}/u_\infty = 1.6$

The engine jet plays a minor role on the flow around the wing since the distance between the engine and the main wing is relatively large compared to current designs of modern civil aircrafts. This is a consequence of the choice of the pylon geometry dictated by the experimental model. The location of the stagnation line on the engine from the inflow side moved as expected with varying thrust. The cross flow component w in the spanwise direction at the location of the pylons is also influenced by the engine suction.

The behaviour of the wingtip vortex as well as the engine jet for three different flow rates of the engine is illustrated in Figs. 22, 23 and 24. These figures include a perspective view of the vorticity distribution in planes perpendicular to the main flow direction and a top view of isosurfaces calculated for a constant vorticity and a constant flow velocity u in the main flow direction. The wingtip vortex is in the roll-up regime and is not influenced in its path by the jet flow within the simulated domain for the investigated cases. The jet plume on the other hand experiences an outward movement towards the wingtip vortex, which is a consequence of the vortex induced velocities. This mechanism is stronger for the outboard jet as it is nearer to the vortex center. The vortex influence also leads to a deformation of the jet as documented in Figures 25 and 26 for the outboard jet region, which could be considered the starting stage of the entrainment of the plume. It is interesting to note, that the jet flow with the higher exhaust velocity seems to deform more rapidly than the weaker one. Within the limited domain of integration downstream of the wing, the region of decay or entrainment was not reached. This would require an extent of 5 to 10 spans behind the wing. Due to the extreme computational effort we performed such a calculation just for the high-lift configuration, the results of which are discussed in the following subsection.

5.2 Wing Segment of the High Lift Configuration

The simulations were performed for the flapped wing segment. Classical boundary conditions were used for the inflow/outflow boundary and the adiabatic wall. The velocities were linearly extrapolated on both planes normal to the spanwise direction thus permitting non-zero spanwise velocity components on these boundaries. This boundary condition was tested previously [1] and has shown to be accurate enough as long as the vortices are well captured within the integration domain. Non-reflecting outflow boundary conditions were prescribed in the streamwise direction [46], since they have less influence on the upstream flow than a characteristic boundary condition.

The near-wall streamlines on the upper surface of the 3-dimensional flow of the wing segment (Fig. 27) show a discrepancy to the 2-dimensional counterpart. A separation region is found on the part of the segment without the flap. This zone extends up to the flapped segment. The structure of the step-like separation line (Fig. 27) evidences the impact of the slat gap generated vortices (Fig. 28) and the flap side edge vortex on the pressure distribution of the wing. The development of

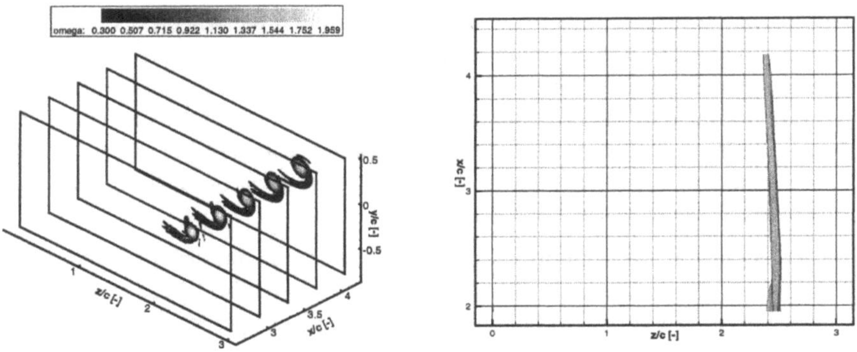

Figure 22 Distribution of the magnitude of vorticity (left:perspective view) and surfaces of constant vorticity and constant velocity u (right:top view) for $u_{engine}/u_{\infty} \approx 1.0$

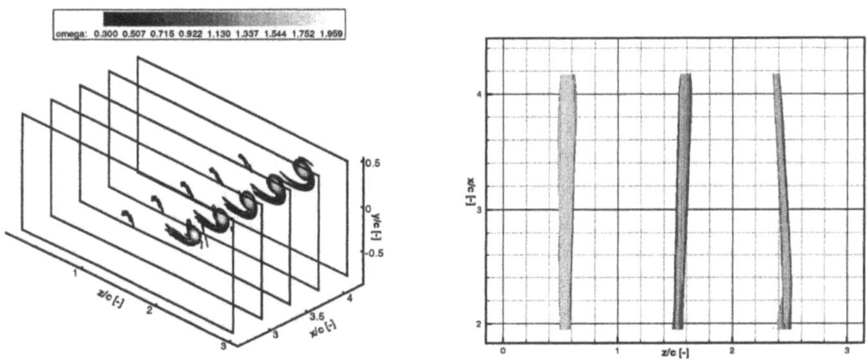

Figure 23 Distribution of the magnitude of vorticity (left:perspective view) and surfaces of constant vorticity and constant velocity u (right:top view) for $u_{engine}/u_{\infty} = 1.6$

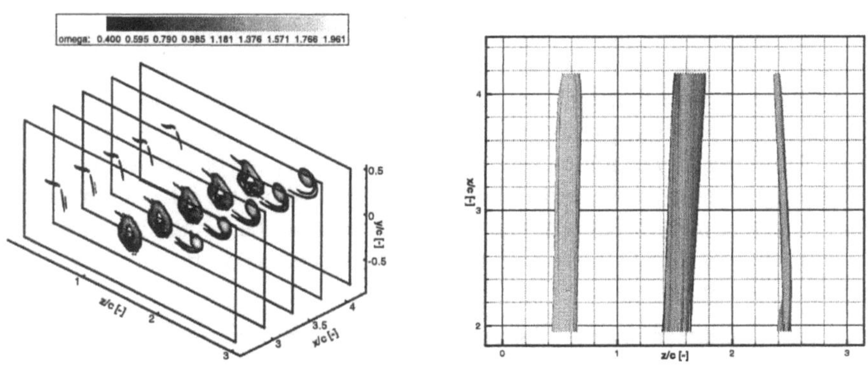

Figure 24 Distribution of the magnitude of vorticity (left:perspective view) and surfaces of constant vorticity and constant velocity u (right:top view) for $u_{engine}/u_{\infty} = 2.5$

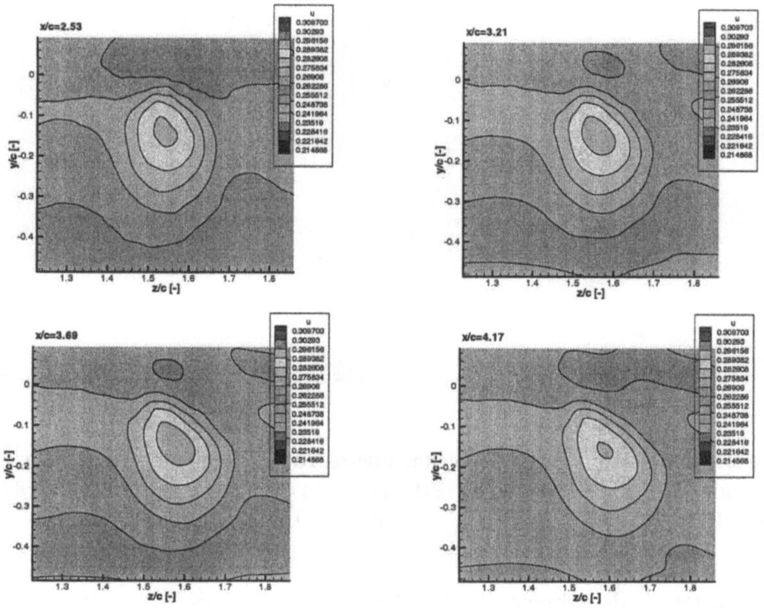

Figure 25 Distribution of the velocity component in the main flow direction u in the wake of the outboard jet for $u_{engine}/u_{\infty} = 1.6$ at four locations downstream of the wing

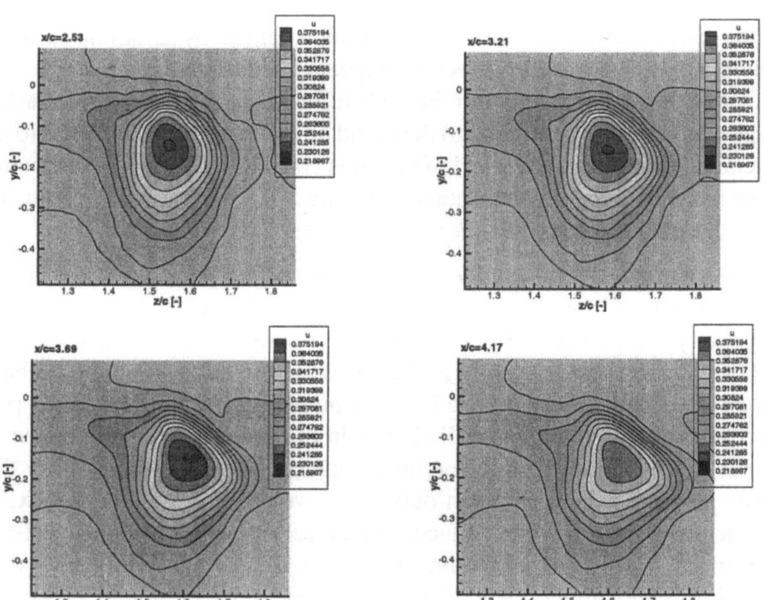

Figure 26 Distribution of the velocity component in the main flow direction u in the wake of the outboard jet for $u_{engine}/u_{\infty} = 2.5$ at four locations downstream of the wing

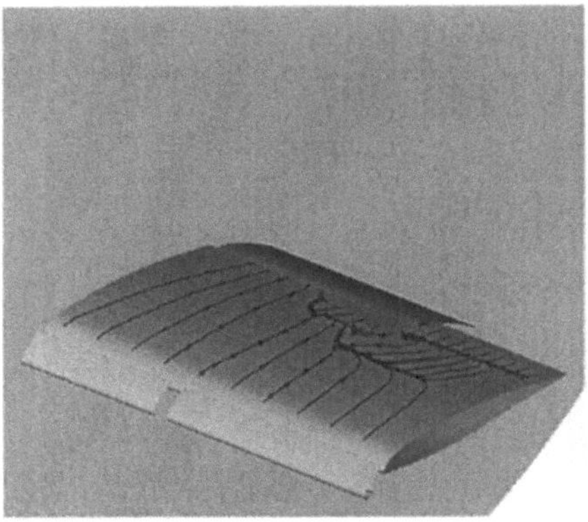

Figure 27 Near-wall streamlines on the suction side ($Ma_\infty = 0.198$ $Re_c = 8.0 \cdot 10^5$ $\alpha = 10°$)

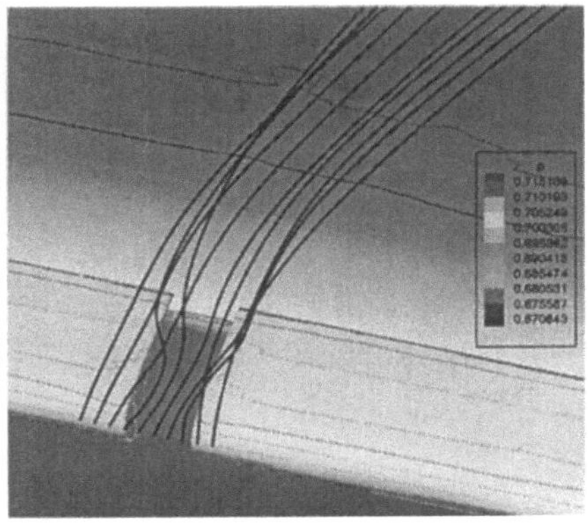

Figure 28 Streamlines near the slat interruption ($Ma_\infty = 0.198$ $Re_c = 8.0 \cdot 10^5$ $\alpha = 10°$)

the flap vortex is documented in Fig. 29. The streamlines in the plane at the first location ($x/c = 0.9$) confirm that the separation area affects the vortex formation.

Two simulations were initiated for the wake flow of the flapped wing segment with and without the simulation of the engine-jet flow. The engine jet was added explicitly at the inflow boundary of the wake integration domain by assuming an analytical distribution of the streamwise velocity component within the engine plume. The engine jet was prescribed parallel to the undisturbed flow at an angle of attack of 0° and with a velocity profile similar to the round turbulent jet flow [53, 54]. No bypass flow, temperature increase or swirl of the jet is considered here, which is consistent with the planned experimental setup. Pressure and density were left unchanged whereas the eddy viscosity was set to $\tilde{\nu} = 10.0\nu_l$.

The behaviour of the flap side-edge vortex and the engine jet is illustrated in Figs. 30 and 31. The figures show a top view of isosurfaces calculated for a constant streamwise vorticity component and a constant flow velocity u in the main flow direction. The vortex line oscillates slightly in the spanwise direction which could be interpreted as the vortex meander [4, 55, 56]. Comparison of the flap vortex paths in the two simulations reveals a small change in the spanwise position of the vortex especially in the region where the engine-jet approaches and is entrained by the vortex. The jet plume experiences an outward movement towards the vortex, which is a consequence of the vortex induced velocities. The vortex influence also leads to a deformation of the jet as documented in Fig. 32. It is also noticeable that this deformation mainly affects the border of the jet and not the engine-jet core. Within the limited domain of integration downstream of the wing, the region of decay or complete entrainment was not reached. As already mentioned for the wing/engine configutration this would require an extent of 5 to 10 spans behind the wing. Such a

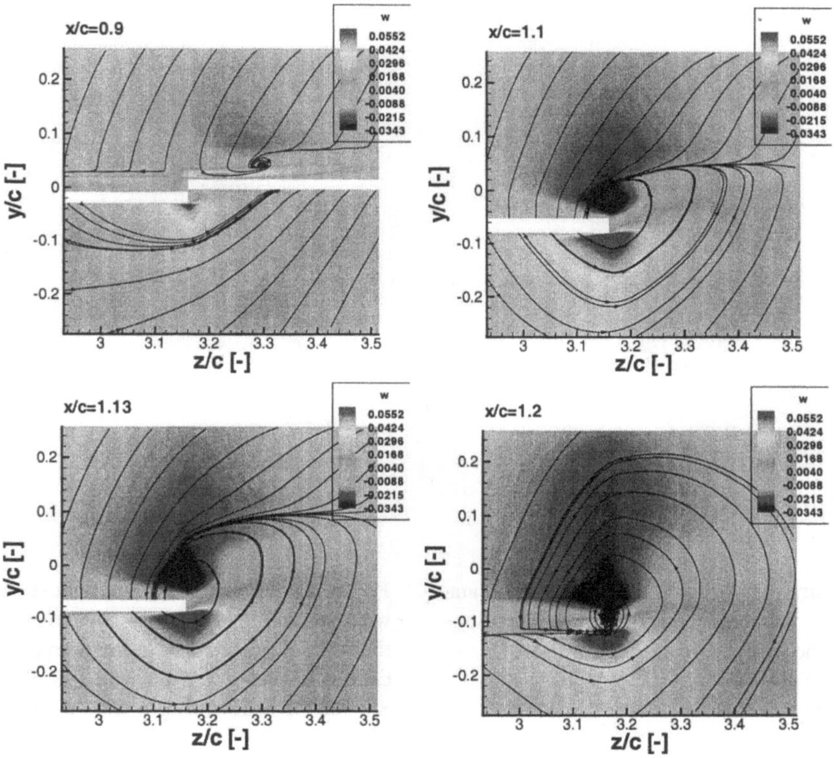

Figure 29 Development of the spanwise velocity component and projected streamlines at four locations downstream of the flap ($Ma_\infty = 0.198$ $Re_c = 8.0 \cdot 10^5$ $\alpha = 10°$)

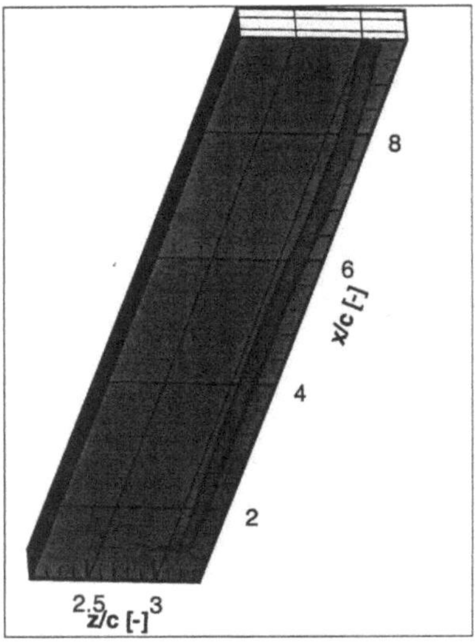

Figure 30 Isosurface of constant streamwise vorticity component in the wake of the flapped wing segment ($Ma_\infty = 0.198$ $Re_c = 8.0 \cdot 10^5$ $\alpha = 10°$)

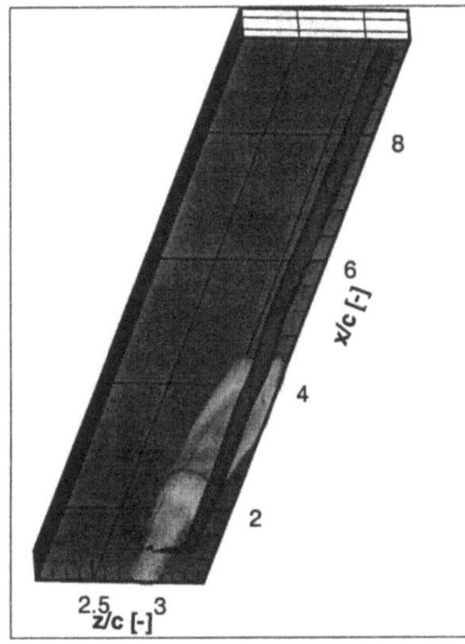

Figure 31 Isosurfaces of constant streamwise vorticity and velocity components in the wake of the flapped wing segment including engine-jet ($Ma_\infty = 0.198$ $Re_c = 8.0 \cdot 10^5$ $\alpha = 10°$ $u_{eng} = 3.0 \cdot u_\infty$)

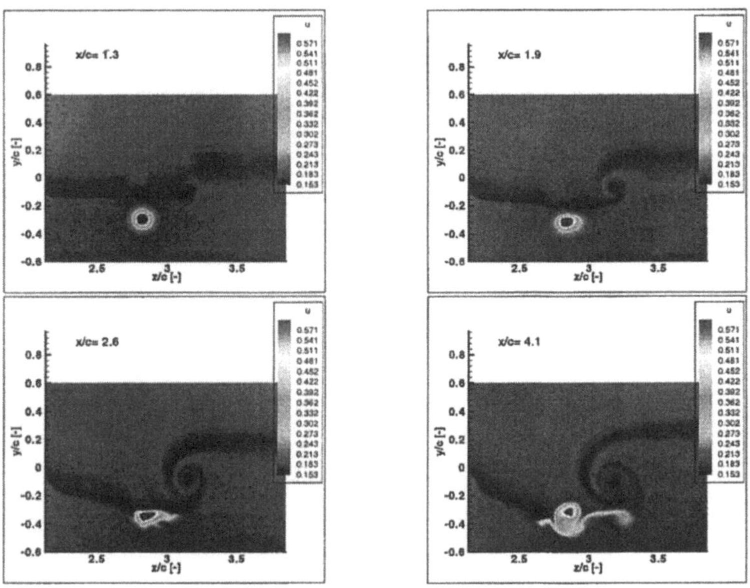

Figure 32 Development of the streamwise velocity component on four locations downstream in the wake of the flapped wing segment including engine-jet simulation ($Ma_\infty = 0.198$ $Re_c = 8.0 \cdot 10^5$ $\alpha = 10°$ $u_{eng} = 3.0 \cdot u_\infty$)

simulation will only be possible using local adaptive methods that are investigated presently.

6 Conclusion

Various aspects of the simulation of the near wake of a wing/engine configuration have been presented. The methods used were developed and enhanced based on several validation test cases. The introduced techniques include grid generation, turbulence modeling and the choice of the boundary conditions. Two main issues are shown to be crucial in the simulation of vortex dominated wakes generated by lift devices: the resolution of the vortex cores and the accurate prediction of turbulence. To account for these issues, a subdivision of the integration domain into a near wall and a wake region is suggested as well as a modified Spalart-Allmaras model. The behaviour of the flow in the near field could be simulated. The findings showed the influence of the flap vortex and the wingtip vortex on the path and the form of the engine plume. No relevant influence of the jet flow on the wingtip vortex formation or path was found within the simulated region for the three cases of the wing/engine configuration studied. A small impact of the jet flow on the flap vortex path was found for the wing segment case within the simulated region.

7 Acknowledgments

The authors thank Erol Özger from the Institut für Luft- und Raumfahrt (ILR) in Aachen for providing the experimental results.

References

[1] E. Fares, M. Meinke and W. Schröder. Numerical Simulation of the Interaction of Wingtip Vortices and Engine jets in the Near Field *AIAA Paper*, 2000-2222,2000.

[2] E. Fares, M. Meinke and W. Schröder. Numerical Simulation of the Interaction of Flap Side-Edge Vortices and Engine Jets. *ICAS Paper presented at ICAS 2000 Congress in Harrogate*, ICA 0212, 2000.

[3] P. R. Spalart. Airplane Trailing Vortices. Ann. Rev. Fluid Mechanics, Vol.30, pp.107-138, 1998.

[4] J. S. Chow, G. G. Zilliac and P. Bradshaw. Mean and Turbulence Measurements in the Near Field of a Wingtip Vortex. *AIAA Journal*, Vol.35, No.10:1561-1567,1997.

[5] J. Dacles-Mariani, G. G. Zilliac, J. S. Chow and P. Bradshaw. Numerical/Experimental Study of a Wingtip Vortex in the Near Field. *AIAA Journal*, Vol.33, No.9:1561-1568,1995.

[6] T. R. Quackenbush, M. E. Teske and A. J. Bilanin. Dynamics of Exhaust Plume Entrainment in Aircraft Vortex Wakes. *AIAA Paper*, 96-0747,1996.

[7] T. Gerz and T. Ehret. Wake Dynamics and Exhaust Distribution behind Cruising Aircraft. *AGARD-CP-584*,35,1996.

[8] L. Jacquin and F. Garnier. On the Dynamics of Engine Jets behind a Transport Aircraft. *AGARD-CP-584*,35,1996.

[9] W. Althaus, C. Brücker and M. Weimer. Breakdown of Slender Vortices. In S. I. Green, editor, *Fluid Vortices*, pp 373–426. Kluwer Academic Publishing, 1995.

[10] M. Meinke and E. Krause. Simulation of incompressible and compressible Flows on Vector-Parallel Computers. *Parallel Computational Fluid Dynamics: Proceedings of the Parallel CFD'98 Conference in Taiwan*, 25-34, 1999.

[11] E. Fares and W. Schröder. Zonal approach for the Solution of the Navier-Stokes Equations. *ZAMM 2000*, Heft 80, Bd.1,2000.

[12] M. S. Liou and Ch. J. Steffen. A New Flux Splitting Scheme. *Journal of Computational Physics*, 107:23-39, 1993.

[13] M. S. Liou. A Sequel to $AUSM$:$AUSM^+$. *Journal of Computational Physics*, 129:164-382, 1996.

[14] B. van Leer. Toward the Ultimate Conservative Difference Scheme V. A Second-Order *Journal of Computational Physics*, 32:101-136,1979.

[15] G. van Albada, B. van Leer and J. Roberts. A Comparative Study of Computational Methods in Cosmic Gas Dynamics. *Astron. Astrophysics*, 108:76-84, 1982.

[16] A. Jameson. Solution of the Euler equations for two-dimensional transonic flow by a multigrid method. *Applied Math. and. Comp.*, 13:327-355,1983.

[17] A. Wambecq. Rational Runge-Kutta Methods for Solving Systems of Ordinary Differential Equations. *Computing*, 20:333-342,1978.

[18] N. Satofuka, K. Morinishi, T. Takami and A. Shimizu. Computation of Two-Dimensional Transonic Cascade Flow using a new Navier-Stokes Solver. *AIAA paper*, 86-1381,1986.

[19] M. Meinke and D. Hänel. Time Accurate Multigrid Solutions of the Navier-Stokes Equations. *Multigrid Methods III. W. Hackbusch, U. Trottenberg ed.*, Birkhäuser Verlag, 64-88,1991.

[20] A. Jameson and T. J. Baker. Solution of the Euler Equations for Complex Configurations. *AIAA paper*, 83-1929,1983.

[21] A. Jameson and E. Turkel. Implicit Scheme and LU Decompositions. *Mathematics of Computation*, 81:385-397,1981.

[22] J. Häuser, P. Eiseman, Y. Xia and Z. Cheng. Parallel Multiblock Structured Grids. *Handbook of Grid Generation*, CRC Press,1999.

[23] J. H. Ferziger and M. Perić Computational Methods for Fluid Dynamics *Springer-Verlag*, 1996.

[24] D. C. Lewellen, W. S. Lewellen. Large-Eddy Simulation of the Vortex-pair Breakup in Aircraft Wakes. *AIAA Journal*, Vol.34, No.11:2337-2345,1996.

[25] S. Shen, F. Ding, J. Han, Y. Lin, S. P. Arya, F. Proctor. Numerical Modeling Studies of Wake Vortices: Real Case Simulations. *AIAA paper*, 99-0755,1999.

[26] Editors: W. Haase, E. Chaput, E. Elsholz, M. A. Leschziner and U. R. Müller. Validation of CFD Codes and Assessment of Turbulence Models. *Notes on Numerical Fluid Mechanics*, Volume 58,1997.

[27] J. E. Bardina, P. G. Huang and T. J. Coakely. Turbulence Modeling Validation, Testing and Development. *NASA Technical Memorandum*, 110446,1997.

[28] B. S. Baldwin and H. Lomax. Thin Layer Approximation and Algebraic Model for Separated Turbulent Flows. *AIAA Paper*, 78-257,1978.

[29] B. S. Baldwin and T. J. Barth. A One-Equation Turbulence Transport Model for High Reynolds Number Wall-Bounded Flows. *AIAA Paper*, 91-0610,1991.

[30] P. R. Spalart and S. R. Allmaras. A One-Equation Turbulence Model for Aerodynamic Flows. *AIAA Paper*, 92-0439,1992.

[31] W. P. Jones and B. E. Launder. The Prediction of Laminarization with a Two-Equation Model of Turbulence. *International Journal of Heat and Mass Transfer*, 15:301-314,1972.

[32] D. C. Wilcox. Reassessment of the Scale-Determining Equation for Advanced Turbulence Models. *AIAA Journal*, Vol.26, No.11, 1299-1310, 1988.

[33] F. R. Menter. Zonal Two Equation $k - \omega$ Turbulence Models for Aerodynamic Flows. *AIAA Paper*, 93-2906,1993.

[34] S. E. Rogers, F. R. Menter, P. A. Durbin, N. M. Mansour. A Comparison of Turbulence Models in Computing Multi-Element Airfoil Flows. *AIAA Paper*, 94-0291,1994.

[35] C. L. Rumsey and V. N. Vasta. A Comparison of the Predictive capabilities of Several Turbulence Models using Upwind and Central-Difference Computer Codes. *AIAA Paper*, 93-0192,1993.

[36] V. A. Sai and F. M. Lutfy. Analysis of the Baldwin-Barth and Spalart-Allmaras One-Equation Turbulence Models. *AIAA Journal*, Vol.33, No.10:1971-1974,1995.

[37] P. Godin and D. W. Zingg. High-Lift Aerodynamic Computations with One- and Two-Equation Turbulence Models *AIAA Journal*, Vol.35, No.2:238-243,1997.

[38] Q. Wang, S. J. Massey and K. S. Abdol-Hamid. Solving Navier-Stokes Equations with Advanced Turbulence Models on Three-Dimensional Unstructured Grids. *AIAA Paper*, 99-0156,1999.

[39] M. R. Khorrami, B. A. Singer and R. H. Radeztsky, Jr. Reynolds Averaged Navier-Stokes Computations of a Flap Side-Edge Flow Field. *AIAA Paper*, 98-0768,1998.

[40] W. R. C. Phillips. The Turbulent Trailing Vortex during Roll-Up. *Journal of Fluid Mechanics*, 105:451-467,1981.

[41] W. J. Devenport, M. C. Rife, S. I. Liapis and G. J. Follin. The structure and development of a wing-tip vortex *Journal Fluid Mechanics*, 312:67-106,1996.

[42] E. R. Hoffmann and P. N. Joubert. Turbulent line vortices. *Journal of Fluid Mechanics*, 16:395-411, 1963.

[43] G. K. Batchelor. Axial Flow in Trailing Line Vortices. *Journal of Fluid Mechanics*, 20:645-658, 1964.

[44] P. G. Saffman. Vortex Dynamics. *Cambridge Monographs on Mech. and Appl. Math.*, 1992.

[45] D. L. Whitfield. Three Dimensional Unsteady Euler Equation Solutions using Flux Vector Splitting. *Notes for Computational Fluid Dynamics User's Workshop, Tullahoma, Tennessee*, UTSI Publication No. E02-4005-023-84,1984.

[46] T. J. Poinsot and S. K. Lele. Boundary Conditions for Direct Simulations of Compressible Viscous Flows. *Journal of Computational Physics*,101:104-129,1992.

[47] O. Thomer, E. Krause, W. Schröder and M. Meinke. Computational Study of Normal and Oblique Shock-Vortex Interactions *European Congress on Computational Methods in Applied Sciences and Engineering, ECCOMAS 2000, Barcelona 11-14. September*, 2000.

[48] E. Özger, I. Schell and D. Jacob. On the Structure and Attenuation of an Aircraft Wake. *to be published as AIAA Paper*, 2000-4127,2000.

[49] P. R. Spalart. *private communication*,2000.

[50] I. R. M. Moir Measurements on a Two-dimensional Aerofoil with High-Lift Devices *AGARD-AR-303*, 2:58-59, 1994.

[51] E. Stanewsky, J. Délery, J. Fulker and W. Geissler. (Editors) Euroshock - Drag Reduction by Passive Shock Control. *Notes On Numerical Fluid Mechanics*, Vieweg, vol. 56, 1997.

[52] I. Hörschler. Entwicklung von Gitteroptimierungsmethoden f'ur die ELAC-1 Konfiguration *DGLR Jahrbuch 1998*,1998.

[53] W. Szabklewski. Turbulente Scherströmungen: Freie Turbulenz. *Schriftenreihe des Zentralinstituts für Mathematik un Mechanik der Akademie der Wissenschaften der DDR*, Akademie Verlag, 1979.

[54] C. G. Schulz. Grobstruktursimulation turbulenter Freistrahlen *Diss. RWTH-Aachen*, Verlag Mainz, 1998, Aachen.

[55] R. V. Westphal and R. D. Mehta. Interaction of an Oscillating Vortex with a Turbulent Boundary Layer *Experiments in Fluids*, Vol.7, No.6:405-411,1989.

[56] K. W. McAlister and R. K. Takahashi NACA 0015 Wing Pressure and Trailing Vortex Measurements *NASA TP-315*,Nov. 1991.

Aerodynamic Design of Transonic Adaptive Airfoils with Natural Transition

A. Meijering, W. Limberg, W. Schröder
Aerodynamisches Institut der RWTH Aachen
Wüllnerstraße zw. 5 und 7, D-52062 Aachen, Germany

Summary

To increase the aerodynamic performance of airfoils at different transonic flight conditions computational and experimental investigations for an airfoil with an adaptive upper side geometry were initiated. The main objective is to improve the lift-to-drag ratio by adapting the upper airfoil surface using numerical optimization such that the extent of the laminar boundary layer is enlarged without the occurrence of a separation bubble. Unlike in most airfoil optimization procedures the transition location is not fixed but is determined numerically using the semi-empirical e^N-method based on the linear stability theory. To verify the numerical results it is necessary to perform experiments with natural transition on a laminar-type airfoil at transonic mean flow conditions. The transition location and the separation area are determined at transonic speeds in the trisonic wind tunnel with an adaptive wall test section using the hot-film method with a 20 sensor array as well as the liquid-crystal technique. The methods will also be applied to a future airfoil model with an adaptive upper side geometry to experimentally verify the result of the numerical optimization.

1 Introduction

The flight performance of aircraft is governed by aerodynamics, structure mechanics, and propulsion. To improve the performance of lift generating configurations from an aerodynamic point of view, one focuses on the the cruise performance, which is the dominating segment of flight for long range aircraft. At these flight conditions the viscous drag accounts for approximately 50% of the total drag [17]. The two main sources of friction drag are the wings and the fuselage, which account for over 80% of the total amount of friction drag. The viscous drag depends mainly on the state of the boundary layer (laminar or turbulent). The laminar flow concept plays a critical role since a reduction of the airfoil drag by reducing the skin friction through an increased extent of the laminar boundary layer is very attractive, offering a potential for further fuel savings to be approximately 20% to 30% [6]. To be able to incorporate the drag reduction techniques in new wing designs, it is important to apply airfoil shape optimization techniques in the early stages of the design process.

In this work an optimization algorithm focuses on influencing the state of the boundary layer by modifying the airfoil cross section which is part of the natural laminar flow concept (NLF). Therefore, it is necessary to evaluate the transition location during the numerical computation. This is in contrast to most numerical optimizations of airfoil configurations which are performed on the assumption of a fully turbulent boundary layer or a fixed transition location close to the leading-edge. To reduce the computational effort required to optimize aerodynamic shapes the present research focuses on two-dimensional investigations. Thus, the number of possible instability mechanisms leading to transition as attachment-line transition (ALT), cross-flow instability (CFI), and the Tollmien-Schlichting instability (TSI) reduces to the TSI. To predict the transition location initiated by this instability mechanism the e^N-criterion [20, 24] based on the linear stability theory (LST) is used. Experiments have to be performed to empirically correlate the value N, the so-called N-factor, with the experimentally obtained transition location. That is, it is a must that the transition location and the size of separation bubbles can be accurately detected. For this reason, we combine various local and global measurement techniques to identify transition and separation on an existing laminar-type airfoil.

In the transonic testing regime a shock occurs on the suction side of the airfoil. Due to the pressure increase across the shock the boundary layer is very susceptible to separation. Depending on the strength of the shock and the trailing edge pressure different shock induced separation scenarios occur [14, 22]. In most cases the turbulent boundary layer separates directly downstream of the shock whereas the influence of the shock on the laminar boundary layer causes a separation of the boundary layer upstream of the shock leading to a λ-shock configuration. Due to the shock-boundary-layer interaction mechanism the advantages of the increased length of the laminar boundary layer in the reduction of wall friction could be annihilated on the one hand, due to the loss of lift and on the other hand, due to the unsteady buffet phenomenon.

The different physical phenomena yield a multi-modal, highly non-linear behavior of the design space which characterizes the optimization problem. Therefore, the evolutionary algorithms have been used since they perform well in solving complex optimization problems in aerodynamics due to their robustness and ease of implementation [12, 13, 15] thereby ensuring a maximum flexibility over the definition of the optimization problem as will be elaborated in the following section.

2 Numerical Methods

2.1 Evolution strategy

The evolution strategies (ES) are stochastic search algorithms based on the idea of natural selection during the evolution of a population as first proposed by Rechenberg [16] and Schwefel [18]. A key feature of them is the concept of a population based search, not from one single point at a time. The ES evolve only by using the value of the objective function. Moreover, no additional assumption on the continuity or differentiability of the objective function have to be made, and no derivatives of the objective function with respect to the design variables have to be calculated.

Due to the population based search the ES is implicitly suitable for parallel processing since no message passing between processors is necessary during the evaluation of the objective function, which implies the calculation of the entire flow field around the airfoil in the aerodynamic optimization. The aforementioned features make ES a robust algorithm suitable to tackle highly multi-modal or noisy objective functions like the separated transitional transonic airfoil flow investigated here.

In the present ES the population of search points, the individuals, are modified by the evolutionary operators mutation and recombination from the parents selected from the previous generation. An important feature of the ES is the capability to adapt strategy parameters, which drive the algorithm, during the optimization process itself. Mainly, the mutation operator σ_i, which can be interpreted as a mean step size, is adapted. This corresponds to large values of σ_i in the beginning of the search to be able to quickly scan the design space, and to small step sizes when the extremum is almost reached.

The ES scheme used in this study is a multimembered (μ, λ)-evolution strategy, where μ indicates the selected parents creating λ offspring based on normally distributed mutations and different recombination operators in such a way that $\lambda > \mu$. In this method the best μ individuals, according to the value of the objective function, are selected from the λ population without keeping the best solution from the previous generation. It seems odd to forget the best solution of the previous generation instead of keeping it as it is done in the elitist scheme $(\mu + \lambda)$ variant. Nevertheless, it has been recognized that the comma-scheme avoids getting stuck in a local extremum and effectively supports the self-adaptation mechanism of the evolution strategy [8] as it approaches the optimum.

The (μ, λ) scheme uses the deterministic selection where only the μ best individuals of the current generation will survive as parents for the next generation. This extinctive selection implies that individuals that perform worse will be rejected as parents for the following population. So the ratio of parents to offspring (μ/λ) can be interpreted as selection pressure. The smaller this ratio, the tougher the selection. The challenge remains to find the optimum selection pressure to reach 'convergence' to the global optimium as fast as possible without premature convergence to a local optimum. For multimodal objective function topologies that use the self-adaptivity of the scheme it has been recognized that the selection pressure should be approximately 1/7 [8, 19]. The moderate specific selection pressure also promotes the self-adaptation characteristics to determine the optimal mutation width of each variable during the optimization [8].

An individual is represented as a vector of continuous parameters $\vec{a} = (\vec{x}, \vec{\sigma})$ containing n object variables $\vec{x} = (x_1, ..., x_n)^T$ and n strategy parameters, here standard deviations, $\vec{\sigma} = (\sigma_1, ...\sigma_n)^T$. Mutation represents a main operator within the ES and is applied to every individual of the population. For the mutation from x_i to x_i' the strategy parameters σ_i are mutated and recombined. A mutation of an individual $\vec{a} = (\vec{x}, \vec{\sigma})$ to $\vec{a'} = (\vec{x'}, \vec{\sigma'})$ is implemented as follows [19]:

$$\begin{aligned} \sigma_i' &= \sigma_i \cdot e^{(\tau' \cdot N(0,1) + \tau \cdot N_i(0,1))} \\ \vec{x}' &= \vec{x} + N(0, \vec{\sigma}') \end{aligned} \tag{1}$$

where N denotes the Gaussian distribution whereas τ' and τ, the so-called learning rates, represent parameters of the method. This method in combination with the selection pressure enables the self-adaptive algorithm to learn the scaling of the mutations widths along the different dimension axes.

The recombination operator aims to combine the good properties of different parent individuals to obtain an even better offspring individual. So, recombination is only possible if the number of parents exceeds one. Its operator produces one new individual from two randomly selected parent individuals. The recombination is applied to the object variables as well as to the strategy parameters. In the present ES-algorithm two recombination mechanisms are implemented, the discrete recombination operator on the object variables and the intermediate recombination operator on the strategy parameters as proposed by Bäck et al. [3]:

$$x_i' = \begin{cases} x_{S,i} \quad or \quad x_{T,i} & \text{discrete recombination} \\ x_{S,i} + \chi \cdot (x_{T,i} - x_{S,i}) & \text{intermediate recombination} \end{cases} \tag{2}$$

The subscripts S and T denote two randomly chosen individuals ($\vec{x}$) from the parent pool. The intermediate factor χ is set to the fixed value $\chi = 0.5$.

To analyze the performance of the aforementioned optimization method, two test functions proposed by Rastrigin [25] and Ackley [2], f6 and f9, representing highly multi-modal objective function topologies were selected:

$$f_6(\vec{x}) = nA + \sum_{i=1}^{n}(x_i^2 - Acos(2\pi x_i)); \; with \; A = 10,$$
$$min(f_6(\vec{x})) = f_6(\vec{0}) = 0.0. \tag{3}$$

The feasible region is defined by $-5.12 \leq x_i \leq 5.11$. For f9 the generalized variant proposed by Bäck et al. [3] is used,

$$f_9(\vec{x}) = -20exp\left(-0.2\sqrt{\frac{1}{n}\sum_{i=1}^{n}x_i^2}\right) - exp\left(\frac{1}{n}\sum_{i=1}^{n}cos(2\pi x_i)\right) + 20 + e,$$

$$min(f_9(\vec{x})) = f_9(\vec{0}) = 0.0. \quad (4)$$

The feasible region for f9 is defined by $-30 \leq x_i \leq 30$. Both functions operate with n=30 variables. The ratio of parents to population size (μ/λ) has been set to $1/7$, which is suggested in the literature on ES and has been previously elaborated on this section. Two combinations have been considered, the (9,60)-ES denoting that nine parents will be chosen from a population of 60 individuals and the (30,200)-ES each of which has been run twenty times. The variables have been initialized by the values at the boundary of the feasible region. Both strategies always reached the minimum for both test functions defined by a best function value smaller than $f = 10^{-10}$. An example of the performance of the ES with respect to the number of function evaluations on both test functions for the (9,60)-ES is shown in figure 1. Although the (30,200)-ES did locate the minimum, more function evaluations were needed compared to the (9,60)-ES.

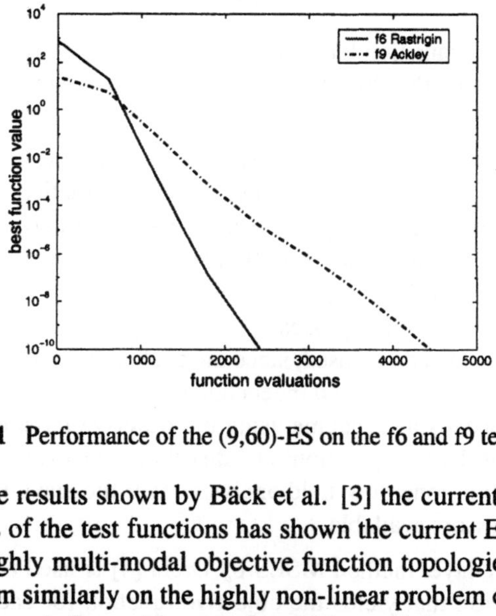

Figure 1 Performance of the (9,60)-ES on the f6 and f9 test cases.

Compared with the results shown by Bäck et al. [3] the current ES performs very well. The analysis of the test functions has shown the current ES to be suitable to find extrema in highly multi-modal objective function topologies. Therefore, it is expected to perform similarly on the highly non-linear problem of airfoil shape optimization in a transonic flow with natural transition and the presence of separation bubbles.

2.2 Airfoil shape optimization in transonic flow

At transonic free stream conditions the supersonic region on the suction side of the airfoil is mostly terminated by a shock. The pressure rise has an enormous impact on the state of the boundary layer. A turbulent boundary layer with a large amount of kinetic energy is less susceptible to this pressure rise and is able to remain attached to the airfoil contour over a longer distance than a laminar boundary layer. The laminar boundary layer separates upstream of the shock forming a separation bubble beneath the λ-shock pattern leading to the unsteady buffet phenomenon which has to be avoided. Therefore it is important that during the optimization the boundary layer is laminar over such a long distance that the friction drag is reduced and that the turbulent state is achieved just before the flow is likely to separate.

The mathematical implementation of the aforementioned physical considerations into the optimization procedure can be achieved by the use of constraints, e.g., constant lift or angle of attack, as well as the suitable objective function containing the important aerodynamic parameters describing the airfoil flow, e.g., transition location or lift-to-drag ratio. As mentioned before, the main objective of the optimization is to improve the lift-to-drag ratio by moving the transition location as far downstream as possible without the occurrence of a separation bubble. This yields the following objective function Q (quality function in ES) to be maximized:

$$max(Q(\vec{x})) = max\left\{\frac{C_L}{C_D}\cdot(x_{tr} - l_{sep})\right\}, \tag{5}$$

where x_{tr} denotes the transition location, l_{sep} represents the length of the separation bubble and C_D denotes the drag coefficient. The lift coefficient C_L is required constant.

One critical aspect of the optimization process using the evolution strategy is the high number of function evaluations needed, since one function evaluation for each individual of the population (one airfoil shape) has to be performed to obtain the objective function Q. In aerodynamic optimization, the evaluation of the objective function in Eq. 5 means the computation of the entire flow field over the airfoil, which is a very time consuming process. Fortunately, the implicit suitability of the ES to parallel processing can be used, since the evaluation of the objective function of each individual can be performed independently of each other. The current algorithm uses a master/slave scheme using the Message Passing Interface (MPI). The master performs the ES operations and distributes the evaluations of the objective function (i.e. one flow field analysis) over the slave-nodes representing the number of processors available.

The Euler-boundary-layer method MSES by Drela [4] is used to evaluate the flow field around the airfoil that yields the objective function for each individual. The numerical procedure incorporates a two-equation integral, boundary-layer method for laminar and/or turbulent flows. As pointed out by Drela [4] the analysis method accurately represents spatially limited separation regions that are sufficiently thin ($\mathcal{O}(\delta)$). Therefore, a suitable method to calculate shock-induced separation at transonic speeds is available. However, during the aerodynamic optimization, a number of airfoil shapes will inevitably be proposed by the optimizer that show massive separation, e.g., due to trailing-edge separation. Hence, the Euler-boundary-layer method will fail to converge. Obviously, the airfoil shape under consideration represents an infeasible result. To be able to return a value for the objective function of this individual to the optimizer, the methods for the constrained optimization based on the penalty function have been applied [11, 23]. The individual, which failed to converge, will be penalized. It has been recognized that good penalty functions can be constructed if the distance from the feasible region is taken into account [11]. However, in the specific case where convergence failed and no particular value for the objective function can be attributed to the individual, it will be assigned a fixed value that is even worse than the smallest possible objective function value obtained for an airfoil shape based on a flow solution that converged. Due to the selection pressure of the evolution stratey, this penalty method causes the rejection of the heavily penalized individual for the next generation.

These convergence specifics due to massive separation will primarily occur at the beginning of the optimization process in which the objective function topology is scanned using large values for the mutation widths, which could yield deteriorated airfoil shapes. However, since the length of the separation is also part of the objective function value, the airfoil shapes that reduce the separation will be preferred as parents for the new generation. Additionally, the values of the mutation widths will be reduced by the adaptive mutation width algorithm as the objective function value approaches the optimum. Therefore, the airfoil shapes assigned to the offspring of the new generation will be close to their parents geometries that showed no separation.

The transition location is very sensitive to the smoothness of the airfoil geometry. Therefore, the method of parameterization of the upper side of the airfoil geometry has to be chosen carefully. Since one function evaluation implies the computation of the entire flow field around the airfoil the number of design variables describing the airfoil geometry has to be as small as possible to reduce the overall optimization effort. However, a sufficient large range of possible airfoil geometries has to be covered by the parameterization. In this investigation the widely used method of a linear combination of k geometric modes is implemented to represent the airfoil shape. Preliminary investigations using approximated piecewise polynomials proved to yield too wavy airfoil shapes. Therefore, the parametrization using geometric modes has been prefered. The airfoil shape is defined by

$$y = y_o + \sum_{i=1}^{k} \eta_i f_i(s), \tag{6}$$

where y_o is the ordinate of the original airfoil, f_i denotes the geometrical mode function and η_i represents its amplitude. In the ES one individual defines one airfoil shape using k amplitude values $\vec{x} = (\eta_1, ..., \eta_k)^T$. In this study the number of k is set to $k = 8$ to define the upper surface of the airfoil distributed between 15% and 70% chord (Fig. 2).

The e^N-criterion is applied to predict the transition location during the calculation. The method assumes transition when the amplification factor of an unstable wave

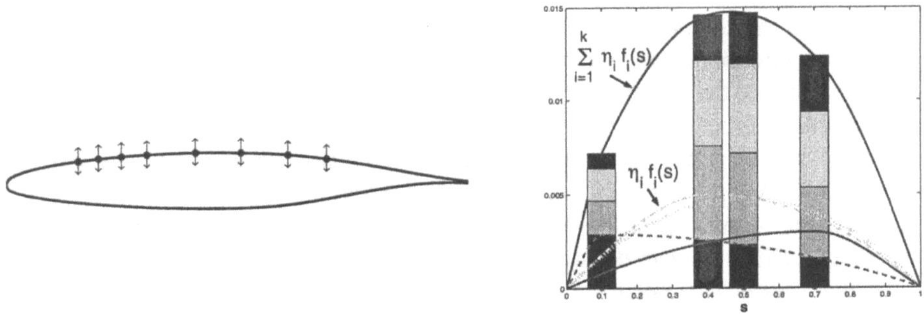

Figure 2 Location of each geometric mode on the airfoil upper surface (left). Example of the superposition of the geometric modes at four stations along the airfoil chord (right).

has reached a certain limiting value N_{crit}. The semi-empirical e^N-method depends on the choice of the value N, which, amongst others, depends on the turbulence level of the wind tunnel and the freestream conditions. Therefore, it is necessary to verify the performance of the applied numerical methods experimentally. The next section offers an overview of the applied experimental methods that allow the determination of the transition location as well as the location and extent of the separation bubble over a laminar-type airfoil.

3 Experimental Methods

3.1 Transition detection methods

To unambiguously determine the state of the boundary layer it is inevitable to combine different measurement techniques to provide as much information about transition and separation as possible. In this work the hot-film technique in conjunction with the liquid-crystal coating method have been used.

The hot-film technique is a local thermal measurement technique to determine the unsteady wall shear stress fluctuations and is able to determine the transition and separated areas. However, when a turbulent wedge reaches the sensor the corresponding signal could be misinterpreted as an indication of a complete turbulent field instead of a local turbulent phenomenon. This can be overcome by simultaneously using the shear sensitive liquid-crystal coating. This global method detects areas of high and low skin friction by color-coded information making it possible to measure turbulent areas (also turbulent wedges) as well as the location of separation and reattachment. However, when the liquid-crystal coating is used alone it is impossible to determine the transition regions inside the separated area. Hence, the technique has to be supplemented by the hot-film method. Thus, a detailed analysis of the shock-boundary-layer interaction characterized by a transitional separation bubble is only possible using a combination of the two methods. Additional pressure measurements yield quantitative results to determine aerodynamic coefficients, which are used to validate the numerical computations.

3.2 Wind tunnel and airfoil model

The experiments are performed in the trisonic wind tunnel of the Aerodynamisches Institut Aachen (AIA). The suction type wind tunnel with an adaptive wall test section of $0.4 \times 0.4\ m^2$ allows measurement times of approximately 3 seconds at transonic speeds. The chord Reynolds number, which is typically $Re_\infty \approx 2.6$ Million, is determined by the Mach number and the model chord length of 200 mm, since the ambient conditions determine the stagnation conditions.

For measurements with natural transition the freestream turbulence level of the wind tunnel plays a major role. In the trisonic wind tunnel the turbulence level ranges between $0.15\% \leq Tu \leq 0.4\%$, which is sufficient for transition measurements. To predict the transition location using the semi empirical e^N-criterion the N-factor depends on the turbulence level. Mack [9] correlates the turbulence level with the N-factor, which yields an N-factor of 6-7 in the trisonic wind tunnel.

The experiments have been performed on a laminar-type airfoil LVA-1A, designed by Airbus Germany. The flow field over this airfoil shape is characterized by the long extent of the laminar boundary layer up to 60% and 55% chord on the upper and the lower side of the airfoil, respectively. Therefore, the airfoil shape is very suitable to investigate transition, separation bubbles and shock-boundary-layer interaction.

To determine the aerodynamic coefficients the airfoil model is equipped with 60 pressure taps on the surface. The drag coefficient is determined by a wake rake consisting of 60 total pressure probes and four pressure probes determining the static pressure.

Figure 3 shows the positions of the hot-film sensors and the area covered by the liquid-crystal coating on the airfoil surface.

Hot-film arrays

The hot-film array has been designed and manufactured at AIA [7] . 20 sensors are operated in the constant temperature mode (CTA) and cover an area between x/c=0.49 to x/c=0.77 on the upper side of the model (Fig. 3).

The transition point is located where the RMS values reach a maximum and the skewness, a moment of higher order, which represents the deviation of the signal from the normal Gaussian distribution, changes sign with a negative gradient.

Using the unsteady sensor data the extent and location of separation bubbles can

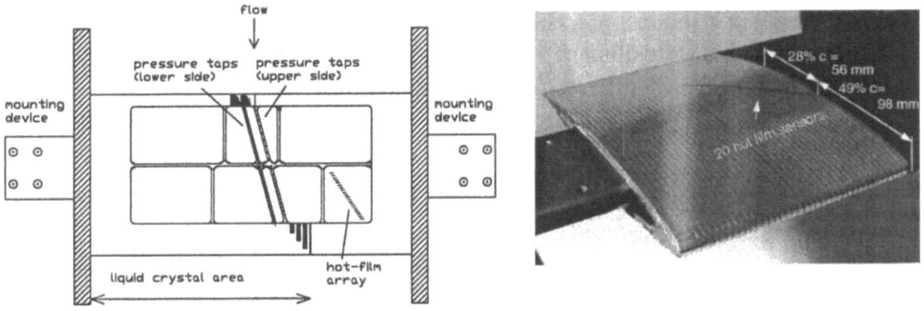

Figure 3 Arrangement of the instrumentation on the airfoil model (left). Surface mounted hot-film array with 20 sensors on the DA LVA-1A airfoil model (right).

be determined. As a criterion to detect separation the linear correlation coefficient of the time traces between neighboring sensors is used as proposed in [1, 5]. This method is more universal than the phase reversal criterion in low-frequency dynamic shear stress signals across the separation bubble which is only reliable at laminar separation [21].

Liquid-crystal coating

The local data on the location of transition and separation obtained from the hot-film arrays were verified using the liquid-crystal technique. The shear sensitive liquid-crystal coating is applied on the remaining part of the upper side of the airfoil next to the hot-film array in the spanwise direction. This allows the simultaneous measurement of the transition and the separation with the hot-film sensors and the liquid crystal coating.

A black coating has been applied on the airfoil surface to increase the contrast of the colors. To obtain an indirect white diffusive illumination of the liquid crystals on the model surface the matt white upper wall of the test section has been illuminated by a halogen light source located behind the trailing edge.

4 Results

4.1 Experimental results and validation of the numerical method

The objective of the optimization is to improve the aerodynamic performance by reducing the drag at a prescribed lift coefficient through airfoil shape design. The improvements will become especially evident in the high speed regimes. Therefore, the experiments have been performed at Mach-numbers ranging from subsonic to transonic flow speeds of $0.65 \leq Ma_\infty \leq 0.78$ at moderate angles of attack in the range of $0^o \leq \alpha \leq 4^o$ on a laminar-type airfoil LVA-1A.

Due to the pronounced accelerated region over the airfoil surface the boundary layer remains laminar up to the recompression point. In Fig. 5 the pressure distribution obtained from the experimental investigation is compared with the numerical computation using the Euler-boundary-layer method at freestream conditions $Ma_\infty = 0.72$, $Re_\infty = 2.6 \times 10^6$, and $\alpha = 1.0^o$. Neither in the experiment nor in the computation the transition has been fixed. From the pressure distribution it is not apparent wheter or not a separation bubble is present. Fig. 7 presents the corresponding liquid-crystal image showing no color change indicating a separation bubble. Additionally, the long extent of the laminar boundary layer up to 69% chord is apparent.

A more detailed analysis in Fig. 9 based on the hot-film measurements determines the transition location to be at 65% chord. By juxtaposing the liquid-crystal and the hot-film findings it can be stated that the color of the liquid-crystal changes at the point where the hot-film sensors indicate the beginning of the fully turbulent boundary-layer.

The flow characteristics become more complex by increasing the Mach number to $Ma_\infty = 0.74$ at the same angle of attack $\alpha = 1.0^o$. From the pressure distribution displayed in Fig. 6 it is obvious that the shock has an upstream influence on the boundary layer causing it to separate upstream of the shock and to form a recirculation underneath the λ-shock. The liquid-crystal image indicates that the boundary layer remains laminar up to the separation point at 55% chord. Through the combination of the liquid-crystal method (Fig. 8) with the hot-film-technique using the linear correlation coefficient the extent of the separation bubble is determined between 55% and 70% chord.

From the liquid-crystal image in Fig. 8 it is apparent that it is impossible to determine the transition location in a separation bubble flow using the just liquid-crystal technique. However, in [10] the possibility to identify the transition location using the hot-film technique even in the presence of a separation bubble was demonstrated. This is also shown in Fig. 10 using the same transition indication characteristics as in the flow without a separation bubble in Fig 9.

The good agreement of the experimentally and numerically obtained pressure distributions has been shown in Figs. 5 and 6. In the previous paragraph the possibility to measure the transition location with a combination of the liquid-crystal coating and the hot-film technique has been demonstrated. This allows the correlation of the numerically computed N-factors with the experimentally determined transition location. Fig. 4 shows the comparison of the experimentally and numerically determined transition location from the subsonic to the transonic Mach number range at an $\alpha = 1.0^o$ angle of attack.

Overall, the experimental and numerical results are in good agreement with respect

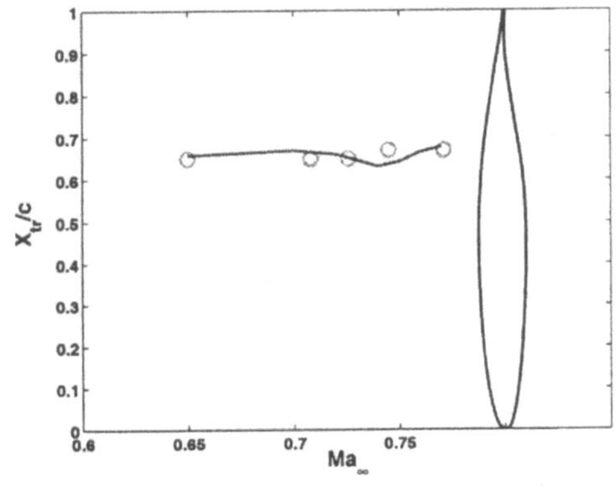

Figure 4 Comparison of the experimentally determined transition location using hot-film sensors (symbols) and the numerically computed transition location using the e^N-method with N=7 (solid line). ($\alpha = 1.0^o$, $Re_\infty = 2.4 \cdot 10^6$)

to the pressure distribution and the transition location. Therefore, it is justified to apply the numerical algorithm based on the Euler-boundary-layer in combination with the e^N-criterion to the airfoil shape design using natural transition. The experimental techniques will be applied in the near future to verify the optimized airfoil shapes using an airfoil model with an adjustable upper surface in wind tunnel experiments.

4.2 Numerical Results

The numerical optimization was performed using the tools just validated. For the investigation of the one-point optimization the freestream conditions at the design point have been set to $Ma_\infty = 0.76$ and $Re_\infty = 2.4 \cdot 10^6$ with natural transition. The lift coefficient is to be constant $C_L = 0.486$. The corresponding flow over the initial airfoil shows all aforementioned physical aspects such as an extended laminar boundary layer accompanied by a shock and a separation bubble. At this design point the potential of the ES in combination with the geometry parameterization to increase the lift-to-drag ratio by suspending the occurrence of a separation bubble can be investigated.

For the aerodynamic airfoil design different numbers of parents and offspring in the evolution strategy have been tested (see table 1). A comparison of the pressure distribution of the initial airfoil with the optimized airfoil shape is shown in figure 11. The flow around the initial airfoil is characterized by high wave drag, which is a common drawback of laminar-type airfoils. The favorable pressure gradient yields an extended laminar boundary layer up to 67% chord (see table 1). Unfortunately, the late transition leads to a separation bubble ranging from 60% to 69% chord on the suction side of the airfoil. According to the objective function Eqn. 5, the

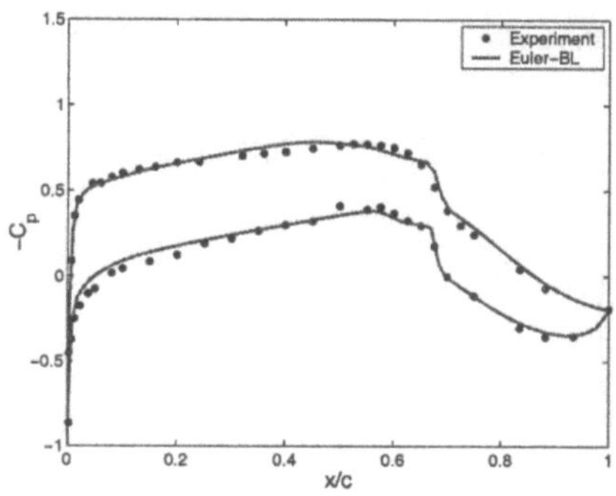

Figure 5 Comparison of the experimentally and numerically obtained pressure distributions (natural transition with N=7). ($Ma_\infty = 0.72; \alpha = 1.0^o$)

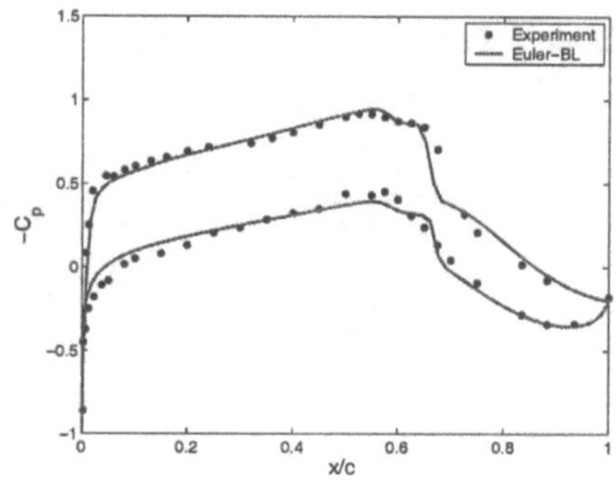

Figure 6 Comparison of the experimentally and numerically obtained pressure distributions (natural transition with N=7). ($Ma_\infty = 0.74; \alpha = 1.0^o$)

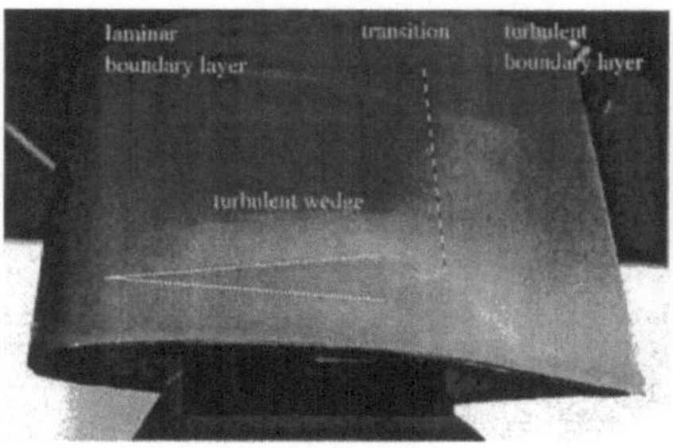

Figure 7 Liquid crystal image showing natural transition at 69% chord without previous separation. ($Ma_\infty = 0.72; \alpha = 1.0^o$)

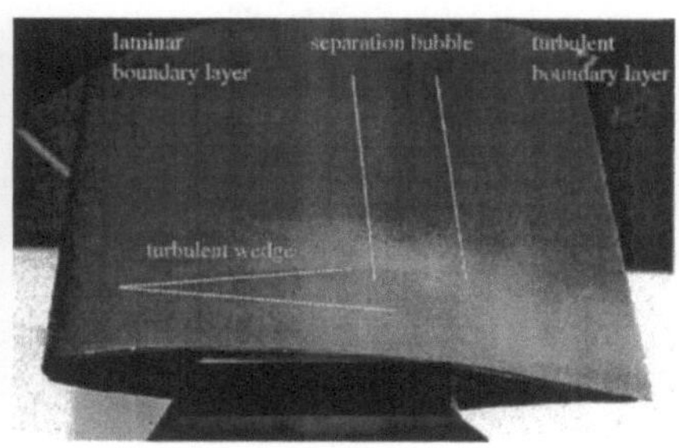

Figure 8 Liquid crystal image showing a separation bubble between 53% and 72% chord. ($Ma_\infty = 0.74; \alpha = 1.0^o$)

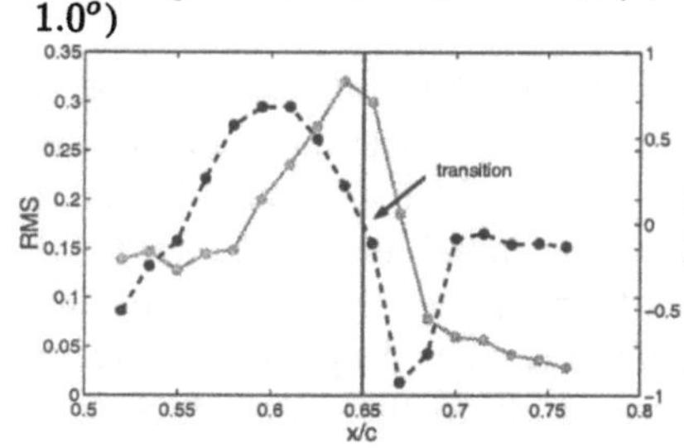

Figure 9 Determination of the transition location at 65% chord using the RMS (solid line) and the skewness (dashed line) of the hot-film time traces. ($Ma_\infty = 0.72; \alpha = 1.0^o$)

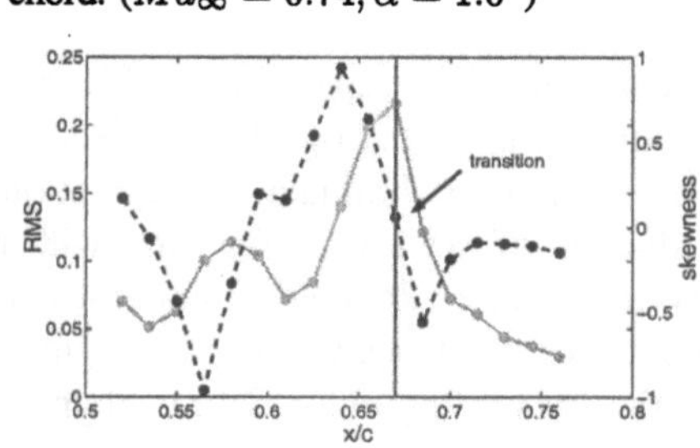

Figure 10 Determination of the transition location at 67% chord using the RMS (solid line) and the skewness (dashed line) of the hot-film time traces. ($Ma_\infty = 0.74; \alpha = 1.0^o$)

Table 1 Performance of different ES-schemes for the airfoil shape optimization at freestream conditions $Ma_\infty = 0.76,\ Re_\infty = 2.4 \cdot 10^6$ and $C_L = 0.486$.

scheme	best value	L/D	x_{tr}/c in %	l_{sep}/c in %
initial airfoil	26	46	67	9
(3,10)	53	89	60	-
(3,20)	56	92	61	-
(6,40)	54	91	60	-

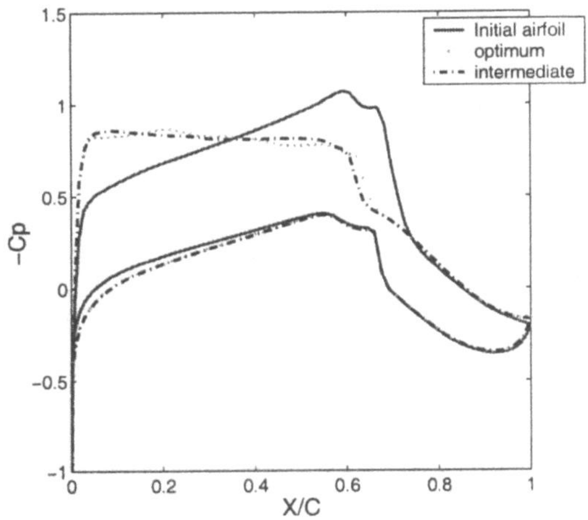

Figure 11 Comparison of the pressure distributions of the initial airfoil with two optimized airfoil shapes at freestream conditions $Ma_\infty = 0.76,\ Re_\infty = 2.4 \cdot 10^6$ and $C_L = 0.486$.

occurrence of the separation bubble implies that the initial airfoil shape is located in the infeasible region of the objective function topology. The main goal of the optimizer is to eliminate the separation bubble (l_{sep}), which even could yield a reduction of the lift-to-drag ratio with respect to the initial airfoil. However, at the investigated Reynolds number it is apparent from figure 11 that the algorithm yields airfoil shapes with a significant reduction of the wave drag and an increase of the lift-to-drag ratio. To maintain the prescribed lift coefficient the flow near the leading-edge undergoes stronger acceleration. Therefore, the favorable pressure gradient is no longer present, leading to an upstream movement of the transition location. However, the reduction of the extent of the laminar boundary layer yields an attached flow over the entire airfoil surface. The integral value of these positive effects is incorporated in the objective function value. As shown in table 1 the best objective function value is doubled for all evolution strategy schemes applied. A remarkable result from table 1 is that even the (3,10)-ES denoting that three parents will be chosen from a population of ten individuals is able to achieve a significant enhancement of the lift-to-drag ratio.

The airfoil shape possessing better flow characteristics than the initial airfoil is the result of a 'one-point' design in which a significant increase of the lift-to drag ratio could be achieved at $C_L = 0.486 = \text{const.}$ (see the dotted line in figure 12). This does not imply that the improved airfoil performs well under other flight conditions. To investigate the performance of the new airfoil shape in the off-design region, the drag polars of the initial and the improved airfoil shape are considered at the Mach number used for the optimization $Ma_\infty = 0.76$.

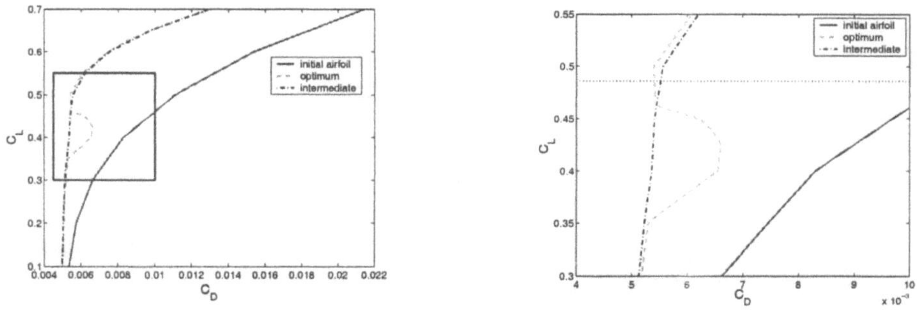

Figure 12 Drag polars of the initial airfoil shape with two adapted airfoil shapes at freestream conditions $Ma_\infty = 0.76$ and $Re_\infty = 2.4 \cdot 10^6$ (left). Enlargement of the area of the drag polar (right).

It is obvious from figure 12 that the improved airfoil shape reduces the total drag at a specified lift coefficient over a large portion of the drag polar. Within the range $0.35 \leq C_L \leq 0.46$ the transition location moves in the upstream direction leading to an increased drag. This shows the deficiency of the 'one-point' design. Figures 11 and 12 additionally show an intermediate result, which performs not as well as the aforementioned airfoil shape at the design point. Although the pressure distribution is very similar to that of the airfoil shape with the slightly better fitness, the performance at different flight conditions shown in the drag polar of figure 12 demonstrates a low drag regime over a larger portion of the polar compared to the 'one-point' best design.

Not only the airfoil performance at high flight Mach numbers is of interest. The off-design performance over a Mach number range of $0.65 \leq Ma_\infty \leq 0.77$ is depicted in figure 13 at freestream conditions $\alpha = 1.0^o$ and $Re_\infty = 2.4 \cdot 10^6$.

As discussed before, the considerable amount of wave drag produced at transonic speeds by the initial laminar-type airfoil is reduced by both adapted airfoils. In this flight regime the intermediate design induces lower drag than the 'one-point' optimum airfoil geometry. The drag reduction is obtained by the optimized airfoils over the whole transonic speeds regime, but a certain price is paid in the subsonic Mach number regime, in which the aerodynamic performance of the initial airfoil is clearly better than the airfoil geometries optimized for a design Mach number $Ma_\infty = 0.76$.

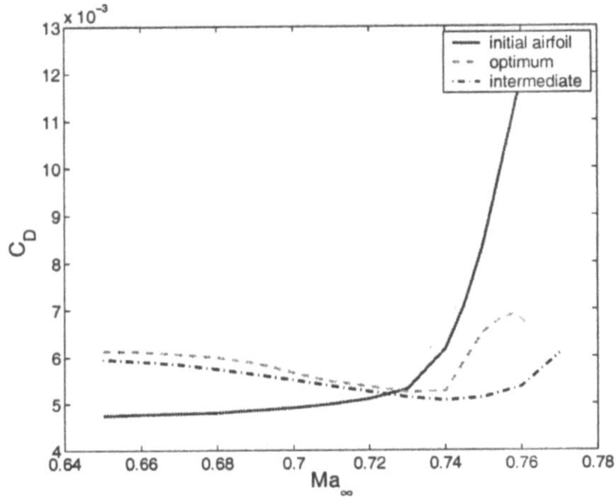

Figure 13 Total drag as a function of Ma_∞ of the initial airfoil shape and two adapted airfoil shapes. ($\alpha = 1.0^o$ and $Re_\infty = 2.4 \cdot 10^6$)

Instead of searching for one airfoil shape which performs reasonably well at all relevant flight conditions of a drag polar using a multi-point optimization, we focus on the optimization based on an airfoil with an adaptive upper surface. An adaptive surface shape represents the boundary of best possible aerodynamic performance using all geometric possibilities available. It consists of the envelope of drag polars of a family of optimum airfoil shapes at different lift coefficients as demonstrated in figure 14. This means that it will not be possible to reach a point on the left side of the adaptive airfoil polar by modifying the airfoil geometry or by varying the angle of attack. An example is presented using the polar of the airfoil shape which has been optimized for a lift coefficient $C_L = 0.1$.

The aerodynamic performance at the design point $C_L = 0.1$ is better than that of the initial airfoil. However, at higher angles of attack the performance deteriorates extremely, yielding a lift-to-drag ratio that falls behind the performance of the initial airfoil. For comparison, the drag polar of the intermediate airfoil shape is also displayed. The polar is characterized by an almost optimum performance over a large area of the drag polar, but it never surpasses the optimum drag polar of the adaptive airfoil. At high lift coefficients, which correspond to higher angle of attack, the drag increases more rapidly than for the adaptive airfoil. From figure 14 it is obvious that the overall performance of the intermediate airfoil shape at the higher flight Mach number and at the specified Reynolds number is very good concerning the off-design characteristics as well as the maximum lift-to-drag coefficient. For the adaptive airfoil shape this point is reached at $C_L = 0.55$, which is only slightly better than the maximum lift-to-drag coefficient of the intermediate airfoil shape.

Summarizing the above the potential of the optimization algorithm in the application to aerodynamic optimization has been illustrated. For aircraft flying at the design

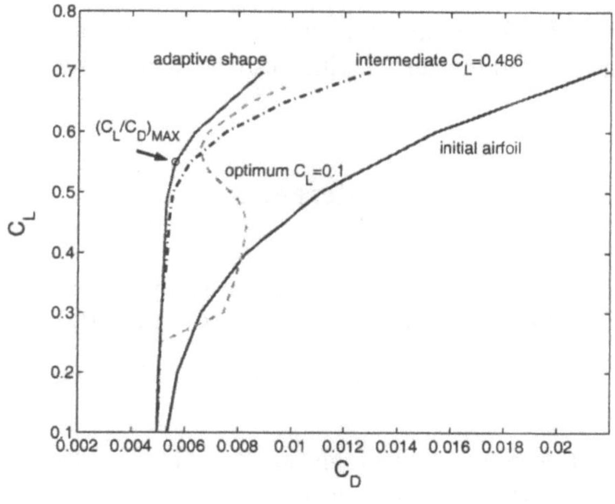

Figure 14 Comparison of the drag polar of the initial airfoil with the envelope of drag polars of an adaptive airfoil geometry at freestream conditions $Ma_\infty = 0.76$ and $Re_\infty = 2.4 \cdot 10^6$.

cruise speed an improvement of the aerdynamic performance can be obtained by a fixed airfoil shape. The use of an airfoil with an adaptive surface geometry only offers an additional gain of the performance at high lift coefficients. However, the adaptive airfoil is especially usefull to avoid the negative off-design performance at different Mach numbers.

4.3 Conclusions

A robust evolution strategy algorithm in combination with an Euler-boundary-layer method has been applied to a direct aerodynamic optimization of airfoil shapes. The investigation focused on airfoil geometries at transonic speeds with natural transition and possible separation bubbles. The measurement results served as data to validate the Euler-boundary-layer method in conjunction with the e^N-method. To detect the location of transition and separation bubbles the liquid-crystal coating technique as a global method in conjunction with the local surface mounted hot-film technique using 20 sensors have been used. The numerical computations are in good agreement with the experimental results concerning the pressure distribution as well as the transition location.

The validated optimization algorithm has been successfully applied to better through shape modifications the lift-to-drag ratio and eliminate separation bubbles of a laminar-type airfoil at transonic speeds. The numerical results show the significance to consider of natural transition during the optimization of airfoil shapes. The one-point optimization performed well at the design point. However, due to the upstream movement of the transition location the aerodynamic performance is reduced. To cope with a wide range of design points the potential of performance growth as well as the feasibility of adaptive airfoil technology has been demonstrated for a transonic laminar airfoil with an adaptive upper surface.

References

[1] R. Abstiens, S. Fühling, and W. Schröder. Boundary-layer measurements on the ELAC configuration at $Re = 20 \times 10^6$. In *ICTAM 2000 Chicago*, 2000.

[2] D. Ackley. *A Connectionist Machine for Genetic Hillclimbing*. Kluwer Academic Publishers, 1987.

[3] T. Bäck and H.-P. Schwefel. An overview of evolutionary algorithms for parameter optimization. *Evolutionary Computation*, 1(1):1–23, 1993.

[4] M. Drela and M. B. Giles. Viscous-inviscid analysis of transonic and low Reynolds number airfoils. *AIAA J.*, 25(10):1347–1355, Oct. 1987.

[5] S. Fühling. *Untersuchung transitioneller, ablösender und gestörter Grenzschichten mit der Multisensor-Heißfilmtechnik.* PhD thesis, Aerodyn. Inst. RWTH Aachen, 2000.

[6] E. Greff. Aerodynamic design for a new regional aircraft. In *17th ICAS Congress, Stockholm, Sweden*, ICAS-90-3.5.3, Sept. 1990.

[7] M. Kornberger and F. Feyzi. Transitionsbestimmung mit Multisensor-Heißfilmmeßtechnik in Windkanal und Freiflug. *DGLR-Report*, 90-06:76–80, 1990.

[8] F. Kursawe and H.-P. Schwefel. Künstliche Evolution als Modell für natürliche Intelligenz. In A. v. Gleich, editor, *Bionik – Ökologische Technik nach dem Vorbild der Natur?*, pages 65–89. Teubner, Stuttgart, 1998.

[9] L. M. Mack. Transition prediction and linear stability theory. AGARD Conf. Proc. 224, AGARD, 1977.

[10] A. Meijering and W. Schröder. Experimental analysis of separated, transitional transonic airfoil flow. In *31st AIAA Fluid Dynamics Conference, Anaheim*, AIAA 2001-2987, June 2001.

[11] Z. Michalewicz, D. Dasgupta, R. Le Riche, and M. Schoenauer. Evolutionary algorithms for constrained engineering problems. *Computers & Industrial Engineering Journal*, 30(2):851–870, September 1996.

[12] B. Naujoks, L. Willmes, W. Haase, T. Bäck, and M. Schütz. Multi point airfoil optimization using evolution strategies. In *ECCOMAS 2000*, Barcelona, 11.-14. Sept. 2000. Wiley & Sons Ltd, Chichester.

[13] A. Oyama, S. Obayashi, and K. Nakahashi. Wing design using real-coded adaptive range genetic algorithm. In *Proc. of IEEE Int. Conf. on System, Man, and Cybernetics*, Tokyo, 12.-15. Oct. 1999.

[14] H. H. Pearcey, J. Osborne, and A. B. Haines. The interaction between local effects at the shock and rear separation - a source of significant scale effects in wind-tunnel tests on aerofoils and wings. *AGARD Conf. Proc.*, 35:11–1 - 11–23, 1968.

[15] J. Periaux, B. Mantel, M. Sefrioui, B. Stoufflet, J. Desideri, S. Lanteri, and N. Marco. Evolutionary computational methods for complex design in aerodynamics. *AIAA Paper 98-0222*, 1998.

[16] I. Rechenberg. *Evolutionsstrategie: Optimierung technischer Systeme nach Prinzipien der biologischen Evolution.* Frommann-Holzboog, Stuttgart, 1973.

[17] J. P. Robert. Drag Reduction: An Industrial Challenge. In *AGARD Special Coarse on Skin Friction Drag Reduction*, volume 786, pages 2–1 - 2–15, 1992.

[18] H.-P. Schwefel. *Evolutionsstrategie und numerische Optimierung*. PhD thesis, TU Berlin, 1975.

[19] H.-P. Schwefel and T. Bäck. Künstliche Evolution – eine intelligente Problemlösungsstrategie? *KI – Künstliche Intelligenz*, 6(2):20–27, 1992.

[20] A. M. O. Smith and N. Gamberoni. Transition, pressure gradient and stability theory. TR ES 26388, Douglas Aircraft Co., 1956.

[21] J. P. Stack, S. M. Mangalam, and S. A. Berry. A unique measurement technique to study laminar-separation bubble characteristics on an airfoil. *AIAA Paper 87-1271*, 1987.

[22] E. Stanewsky. *Wechselwirkung zwischen Außenströmung und Grenzschicht an transsonischen Profilen.* PhD thesis, TU Berlin, 1981.

[23] G. N. Vanderplaats. *Numerical Optimization techniques for engineering design.* McGraw-Hill, New York, 1984.

[24] J. L. van Ingen. A suggested semi-empirical method for the calculation of the boundary layer transition region. Rep. VTH-74, TU Delft, 1956.

[25] D. Whitley, K. Mathias, S. Rana, and J. Dzubera. Evaluating evolutionary algorithms. *Artificial Intelligence*, 85:245 – 276, 1996.

Experiments in a Transonic Shock Tube at Subsonic Mach and High Reynolds Numbers

Th. Reichel, M. Zechner, H. Olivier
Stoßwellenlabor, RWTH Aachen
Schurzelterstraße 35, D-52074 Aachen, Germany

Summary

Modern civil transport aircraft cruise in the high transonic velocity region near the speed of sound. Mach numbers of up to 0.86 and Reynolds numbers based on the chord length of the wing of up to $Re_c = 50 \times 10^6$ and higher are operational for jet planes like the Airbus A340-600 and the Boeing 747-400. Even bigger aircraft are in the conceptual or design phase with higher passenger capacities e.g. the A380, resulting in larger wing chords to lift the increasing aircraft weights in order to accomodate more passengers or freight per aircraft. Another market trend tends to higher cruising Mach numbers to shorten flight time. So flight Reynolds and Mach numbers will increase in the future.

At the Shock Wave Laboratory (SWL) of the RWTH Aachen University a new wind tunnel was put into operation to perform airfoil testing, duplicating true flight Mach and Reynolds numbers simultaneously. After the calibration phase of the tunnel, extensive airfoil testing has been started. A reproducible test flow for stationary airfoil testing is achieved. Pressure measurements of high temporal resolution and flow visualization are conducted. The tunnel flow in the test section reaches Reynolds numbers of up to $Re_c = 40 \times 10^6$ at Mach numbers between $Ma_\infty = 0.6$ and 0.9.

1 Introduction

Wind tunnels are still a necessary tool during the conceptual and design phases of an aircraft project. The aerodynamic optimization of an aircraft design is one key point for the later economic success of a civil transport aircraft. Therefore experiments with high Reynolds number flow are necessary. This reduces uncertainties in the transferability of wind tunnel results to the free flight of future aircraft. In the project TP A7 of the Collaborative Research Center SFB 401 transonic flow around a two dimensional airfoil is investigated for the high Reynolds number region. The test flow is produced by a simple shock tube. Hereto a detonation tube for the investigation of dust detonations has been converted by the addition of a test section and other major parts into the transonic shock tube.

2 Working Principle and Facility Description

In contrast to continuously running wind tunnels of the Göttingen type the transonic shock tube is of an intermittent working principle. The test flow is produced by a moving planar shock wave running downstream in a shock tube, Figure 1. The

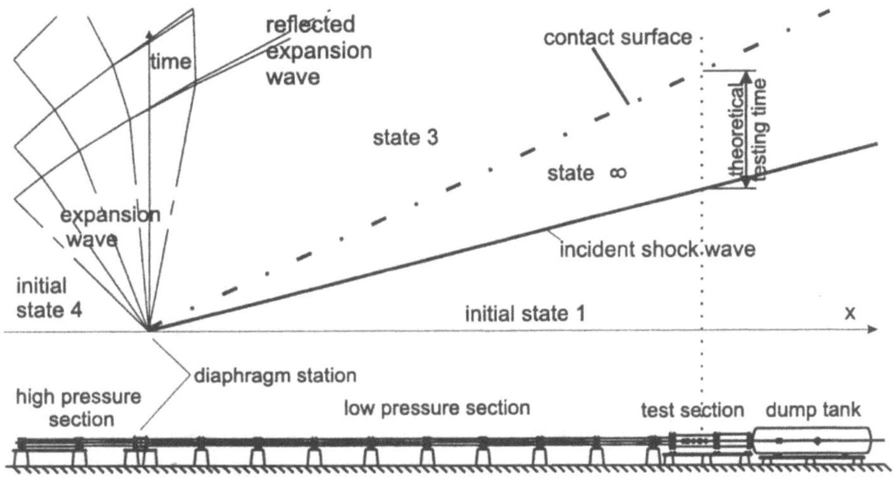

Figure 1 Principal sketch of the transonic shock tube

shock tube is divided into a high pressure and a low pressure section. In Figure 1 the working principle of the transonic shock tube is shown schematically. Before start of each test run, the high pressure section is filled with high pressure gas, in this case technical nitrogen or compressed air up to the initial pressure p_4 at the temperature T_4. The high pressure section is divided from the low pressure section by two metall diaphragms and inbetween them the diaphragm chamber. The low pressure section is filled with the test gas, typically also nitrogen or air at pressure p_1 and temperature T_1. To produce a shock wave the initial pressure p_4 has to be greater than p_1. In our case the temperatures T_1 and T_4 match ambient temperature. At $t = 0$ the diaphragms dividing the high and the low pressure section burst consecutively in a controlled manner. Due to the pressure difference a planar shock wave develops which runs downstream into the low pressure section with constant velocity u_s. Behind this shock wave the transonic flow condition establishes and serves as the model flow. This state is labelled with the index ∞. By gasdynamic shock relations, the initial pressure of the low pressure section p_1 and the shock velocity u_s determine the state p_∞, T_∞, ϱ_∞, u_∞, Ma_∞ and Re_∞ completely. The test gas is shock accelerated to the velocity u_∞ and compressed to the density ϱ_∞ ($T_\infty > T_1$, $p_\infty > p_1$). Compared to the more often used nozzle expansion to accelerate test flows here the great benefit of this technique becomes obvious. It allows to accomplish high Reynolds number flows without the cryogenic cooling often used in continuously running tunnels. An expansion wave runs upstream into the high pressure section. From the initial state 4 the high pressure gas changes its condition isentropically to the state 3 ($p_4 > p_3$). By the expansion the temperature of the high pressure gas decreases from T_4 to T_3, at the same time the density decreases ($\varrho_3 < \varrho_4$). The static pressures and the velocities of the states 3 and ∞ are equal ($p_\infty = p_3$, $u_\infty = u_3$). The dividing surface in between the gases of the states 3 and ∞ is called contact surface and forms a discontinuity in density and temperature ($\varrho_3 > \varrho_\infty$, $T_\infty > T_3$). The test section for the experiments is set up at the downstream end of the low pressure section. It consists of modular rectangular plates and forms a rectangular flow cross section. It is possible to install straight closed test section walls, longitudinally slotted or perforated wall plates. This allows to investigate the test section wall influence in detail.

Testing time starts with the arrival of the incident shock wave at the model station and ends with the arrival of either the contact surface or the reflected expansion wave travelling downstream. Test time duration depends on the initial pressure ratio p_4/p_1 and both the length of the high and low pressure sections. The shock tube flow is calculated with an inhouse developed Riemann solver called KASIMIR. The calculated test time is about 28 to 32 ms for the transonic cases of interest. With a semi-empirical approach of Mirels [1] the time reduction due to the influence of the shock tube boundary layer is estimated to result in an effective testing time between 17 and 19 ms. A survey of the experimental test data shows that in all tests the measured duration of stationary flow is about 12 ms. The actual test time ends due to the arrival of an expansion wave travelling upstream, which due to an area enlargement originates at the downstream junction between test section and connecting tube to the dump tank. The flange inbetween the test section and this connecting tube represents a sharp increase in cross section plus a change of initial filling gas pressure since it is used as a diaphragm station at the same time.

After the useful test time, the high pressure section gas of the state 3 reaches the test section. This gas was expanded from the state 4 to the state 3, thus it is of a lower temperature than the compressed test gas. Even with the same velocity $u_3 = u_\infty$, the flow Mach number is higher due to the lower speed of sound. With the lower temperature T_3 and equal pressure ($p_\infty = p_3$) the density, the pitot pressure and the Reynolds number increase. Due to this reason the aerodynamic models have to be designed for the mechanical loading of this flow condition. Figure 2 shows the theoretical Mach-Reynolds number envelope of the facility. The cross symbols represent flow conditions of some experiments performed so far. The test section has a maximum working pressure of 4 MPa and the shock tube of 25 MPa. Both limit the attainable Reynolds number based on the model chord length of 100 mm to $Re_c = 45\times10^6$, see Figure 2. In the transonic Mach number regime between $Ma_\infty = 0.8$ and 0.9 this affords initial filling pressures in the low pressure section

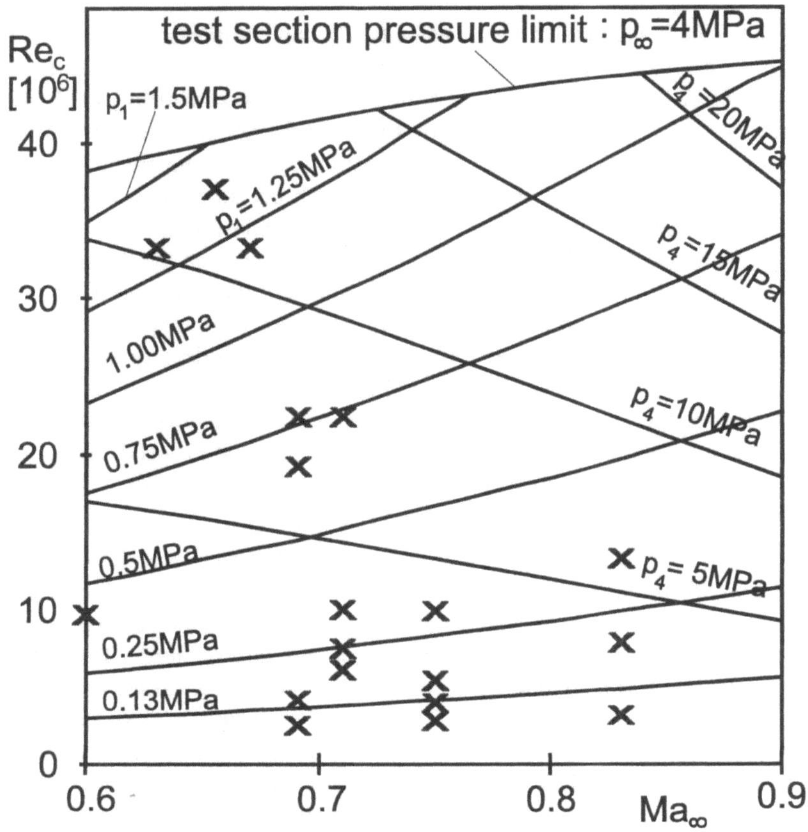

Figure 2 Mach-Reynolds number envelope

of the shock tube of p_1 = 1 to 1.2 MPa and in the high pressure section of up to p_4 = 25 MPa. This is the maximum pressure rating of the shock tube. In Figure 3

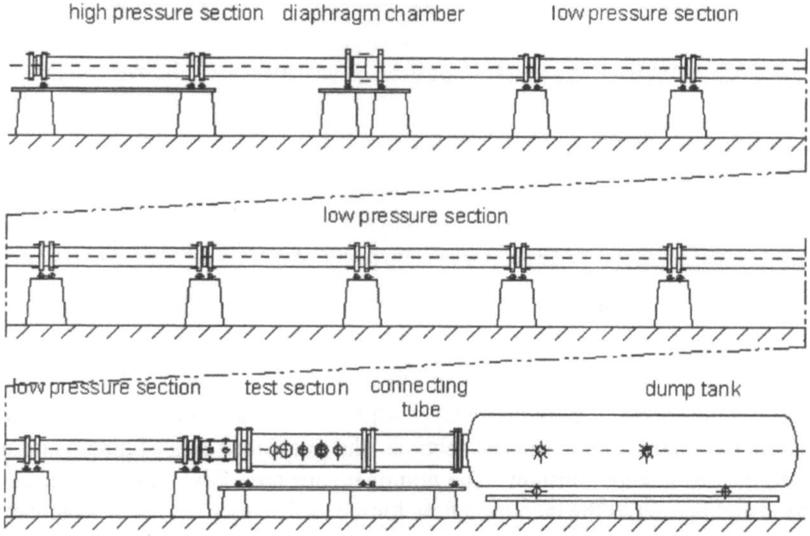

Figure 3 Sideview of the transonic shock tube

the transonic shock tube without the gas supply system is shown in a sideview. The overall length is 46 m. In the upper left part the high pressure section is shown, streamwise from left to right the double diaphragm chamber and the low pressure section follow. Downstream of the low pressure section, the test section and the connecting tube continue and the 8.7 m^3 dump tank completes the facility.

First preliminary work on the transonic shock tube was done at the SWL in 1995. Based on this work, the test section shown in Figure 4 in a longitudinal section view was designed for a working pressure of 4 MPa. The test section is designed as a double hull construction. The outer hull acts as the pressure vessel of the shock tube and enables complete sealing against the environment. Therefore it is possible to perform experiments with all sorts of gases. Experiments using heavy gases for example SF_6 or CF_4 can be performed easily. The inner test section walls are mechanically supported by the outer hull.

The pressurized cross section is rectangular with the dimensions 400 mm by 500 mm. The complete test section design is based on screwed joints. The inner cross section which contains the model station has a fixed width of 0.2 m and a stepwise adaptable heigth of 0.2 - 0.28 m. Entry and exit of the 2.2 m long test section are formed as boundary layer cookie cutters, Figure 4. The test flow which is generated in the circular shock tube of 300 mm inner diameter enters into the rectangular test section. There the sharp edges cut out the homogeneous core flow

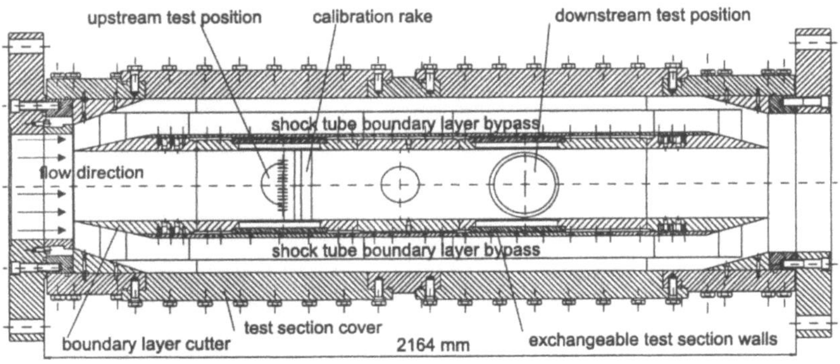

Figure 4 Longitudinal section of the test section

and divert the shock tube boundary layer which increases in thickness with time into separate by-passes around the test section. The total cross area of inner test section and by-passes is kept constant. The homogeneous core flow enters the measuring section and is used as the incoming flow for the airfoil tests. At the downstream end of the test section the core test section flow and the by-pass flows converge before they run through the connecting tube into the dump tank. At the upstream intake of the test section a fresh boundary layer develops with a running length of only 556 mm for the upstream test position or 1300 mm for the downstream position. At the upstream test position (see Fig. 4), measurements of the airfoil pressure distribution and the airfoil wake are performed. At the downstream position (see Fig. 4) optical windows allow flow visualization with a time resolved shadowgraph set up. The diameter of the visible access amounts to 169 mm. During the run the test gas expands through the connecting tube into the dump tank, see Figure 3. This leads to an expansion wave travelling upstream into the test section. It limits the test time to 12 ms. Current experiments make use of a convergent-divergent two dimensional nozzle at the downstream end of the test section. In the divergent nozzle part supersonic flow is generated to stop the expansion wave and disturbances to travel upstream into the measuring section.

3 Calibration and preparational work

The relatively small dimensions of the airfoil models used for the transonic shock tube, maximum chord length 100 mm with a maximum thickness of 11 mm, prevent the direct installation of a large number of pressure transducers into the models. For the high Reynolds number experiments the air loads are in the order of the strength of the model. Model or mounting failure should not lead to the loss of all transducers. Therefore, the transducers are mounted into the inner side walls of the test section and connected to the surface pressure taps by thin pressure tubings. To investigate the dynamic behaviour of this pressure measurement system preliminary to the main testing, a small shock tube was set up as a pressure impulse generator. The transfer behaviour of the complete pressure measurement system was investigated thoroughly, including pressure tap, tubing, pressure transducer, voltage amplifier and transient recorder. Length and inner diameter of the tubings were varied. Figure 5 shows exemplarily the response of five different pressure tubings to the stepwise pressure jump of the shock tube flow. The shorter the tubing, the higher the overshoot of the pressure signal. With reduced inner diameter of the tubes the damping increases so that exaggeration is limited. But at the same time the rise time of the signal is enlonged. In all cases shown, the signals reach about the same constant value after approximately 5 ms. Clearly visible are also the different delay times which mainly depend on the time needed for the first pressure wave travelling through the pressure tubing to reach the transducer at the end of this. These experiments proof that, choosing suitable dimensions of the pressure tubing the developed pressure measurement set up is capable of recording the true stationary values of the pressure distribution of the model and the wall pressure in the given testing time of the shock tube with the necessary accuracy [2]. After these preparatory tests, with

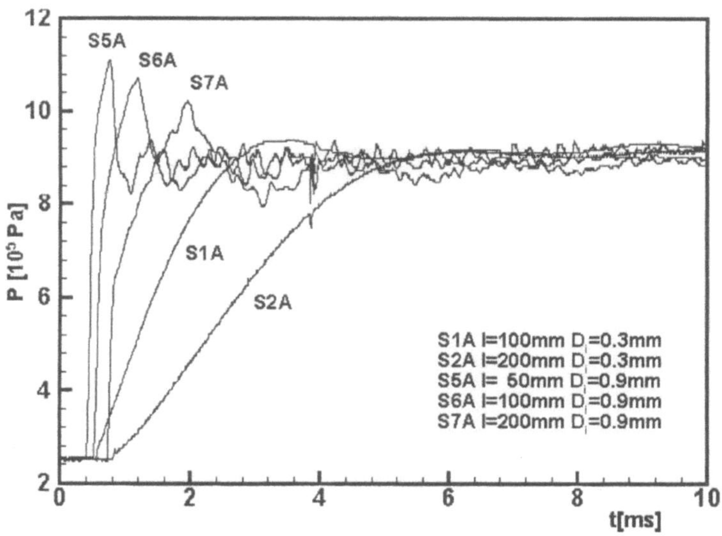

Figure 5 Response on a stepwise pressure jump measured with different pressure tubings

a pitot rake the test section flow was investigated and calibrated. In the first experiments this rake was installed in the upstream testing position vertically connected to 15 high frequency piezoresistive pressure transducers. It is shown in Figure 6. Reproducibility of pitot pressure readings in these experiments was good. In all experiments the pressure signals show a slightly higher level of noise starting at 10 ms. This could be the result of the shock tube boundary layer which after some time par-

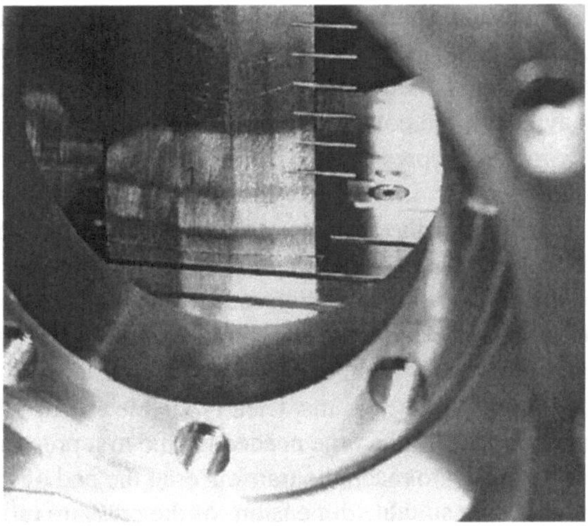

Figure 6 Pitot rake and longitudinally slotted test section wall

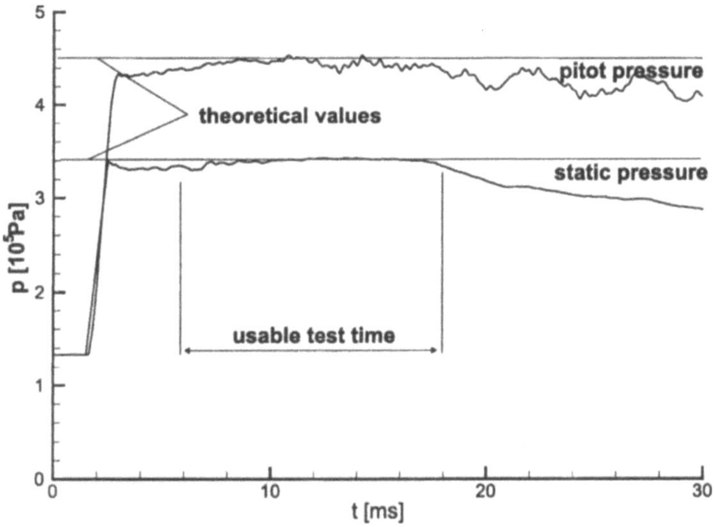

Figure 7 Experimental pitot and static pressure readings compared to their theoretical values, $Ma_{\infty} = 0.63$, $Re_c = 2.7 \times 10^6$

tially enters into the test section. This is visible in the pitot pressure reading shown in Figure 7. It shows the signal of a pressure gauge located close to the bottom test section wall. The signal starts to decrease approximately at 17 ms due to the arrival of the upstream travelling expansion wave, causing the static pressure to decrease at the same time. In addition to the experimental pressure readings the theoretical pitot and static pressure values are added in Figure 7. They show good agreement with the experimental data. The slight rise in the pressure signals with time especially visible for the Pitot pressure at the beginning of the run, results from weakening of shock strength with travelling distance along the shock tube. The test flow produced by the initially stronger shock wave enters the test section some time later than the test flow generated by the weaker shock near the test section. The calibration results

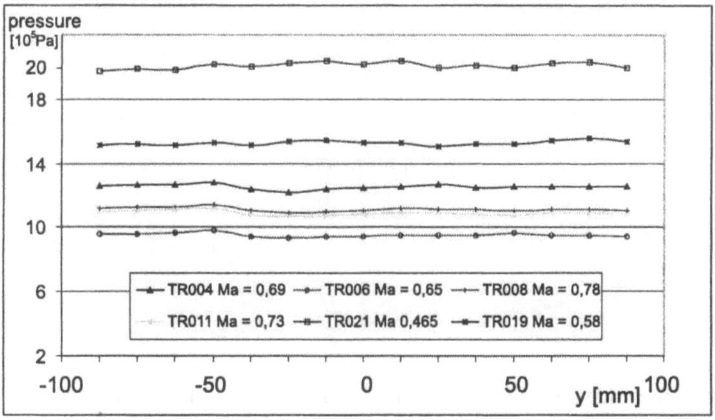

Figure 8 Horizontal pitot pressure variation in the test section

show a homogeneous flow in the test section. In Figure 8 the pitot pressure variation in the horizontal direction from the left to the right test section wall is shown for six experiments with different Mach numbers. The maximum deviation amounts to 3%.

4 Measurements and Flow Visualizations with Airfoil Models

For flow visualization a time resolved shadow system is set up which allows 24 single frames per shot. In the following only one representative frame for each experiment is shown which has been taken during the measuring time. For flow visualization the airfoil model is fixed by two thin side plates with a thickness of 4.5 mm. These plates are fixed at the two tunnel side walls. Therefore, either the upper or lower part of the airfoil and the corresponding flow field is visible. Due to the high mechanical loading caused by the high Reynolds number flow, the airfoil model of the BAC 3-11 shape [3] cannot be fixed by clamping between the tunnel windows. Exemplarily, in Fig. 9 the influence of the free stream Mach number is shown for an angle of attack of 2°.

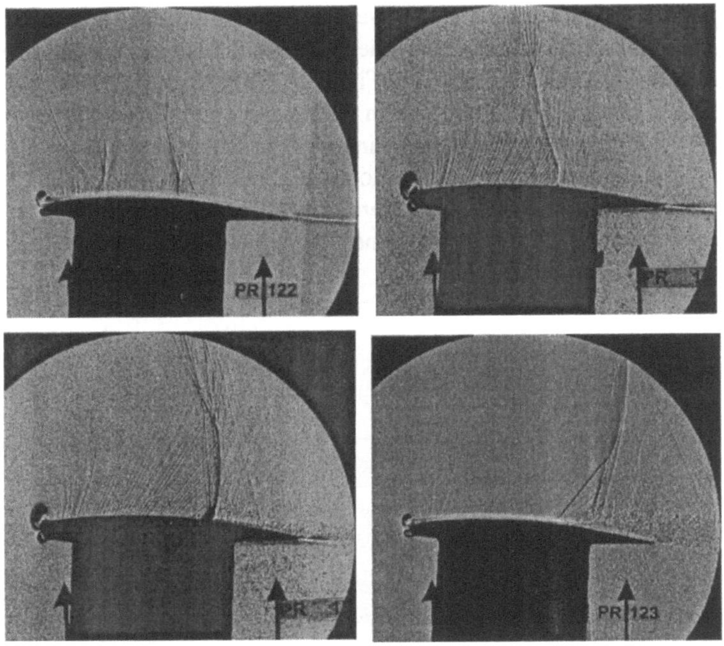

Figure 9 Upper side airfoil BAC3-11, $\alpha = 2°$, $Re_c = 3.2 \cdot 10^6$, top left: $M_\infty = 0.67$, top right: $M_\infty = 0.71$, bottom left: $M_\infty = 0.73$, bottom right: $M_\infty = 0.77$

For the low Mach number case (Fig. 9,top left) a weak two shock system develops. Due to the off axis arrangement of the 24 sparks and receiving optics a small vertical shift of the picture occurs which in Figs. 9,top left and bottom right is especially visible along the upper surface of the airfoil. With increasing Mach number the two shock system vanishes and a single shock develops (Fig. 9,top right). The Mach lines in the supersonic region between nose and shock are clearly visible as well as the expansion region around the nose. The shock is still too weak to cause boundary layer separation. For the slightly increased Mach number of 0.73 the shock becomes a little stronger and, as expected, its location is a little further downstream. For this Mach number a weak boundary layer separation behind the shock takes place. For the highest Mach number considered (Fig. 9,bottom right) a strong separation is visible behind the shock. Furtheron, a lambda shock configuration has been developed and the whole shock system has moved further downstream. In this figure the Mach lines in the supersonic region are not as clearly visible as in the case before because of a reduced sensitivity of the optical set up.

In addition to flow visualization, measurements of the pressure distribution around the airfoil have been performed. For this, a model of 200 mm span and 100 mm chord length has been built, where 24 pressure taps are located on the upper side and 19 on the lower side of the model [4]. As explained in chapt. 3 the relatively short running times of the facility do not allow very small inner diameters of the pressure taps and too long connecting tubes to the pressure transducers. Based on the experimental results discussed in chapt. 3 the inner diameter of the pressure taps was chosen to 0.6 mm, where the typical length of the pressure tubing of each pressure tap amounts to 152 mm. The pressure taps are located in the mid span area of the model. Commercial piezoresistive pressure transducers have been used. In

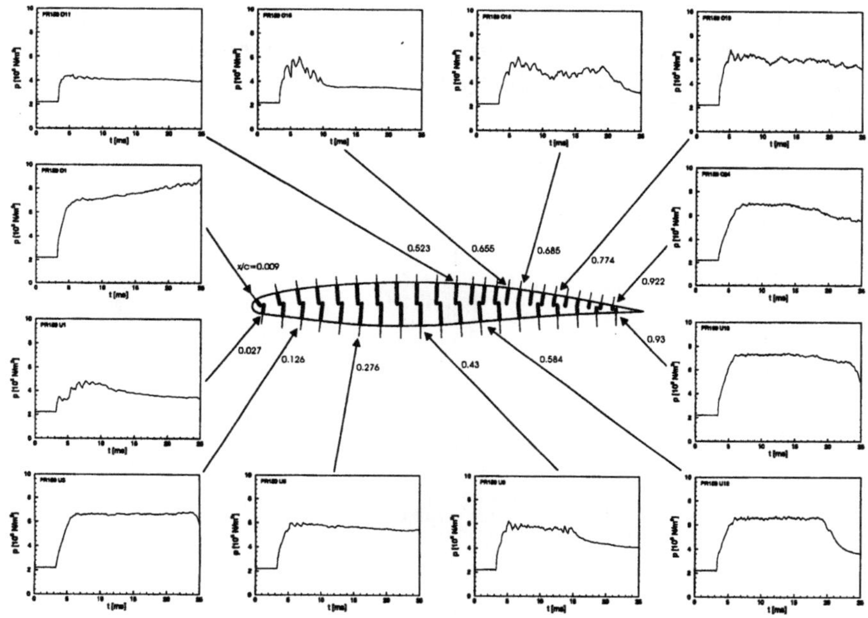

Figure 10 Pressure histories $p[10^5 N/m^2]$ for airfoil BAC3-11, $M_\infty = 0.7$, $Re_c = 7.2 \cdot 10^6$, $\alpha = 0°$

Fig. 10 some of the measured pressure histories are given for an experiment with a Reynolds number of 7.2 Mio. based on a chord length of 100 mm and a free stream Mach number of 0.74. In this case especially on the upper side a strong shock develops which is visible in the shadowgraphs as well as in the pressure distribution and time resolved pressure histories. Along the upper side of the airfoil from the nose up to the position of $x/c = 52.3$ %, after the flow establishment the pressure histories show a nice steady behaviour as it is exemplarily depicted in Fig. 10 for the position $x/c = 52.3$ %. Downstream of this position during the flow establishment a shock develops which then moves downstream to stabilize its position at about $x/c = 67$ %. This development and the downstream movement is clearly visible for the pressure history at $x/c = 65.5$ %. After the shock has passed this position, the pressure becomes steady at a lower level. At its final position the shock oscillates more or less significantly which causes the pressure fluctuations at $x/c = 68.5\ \%$. More downstream of the shock position the pressure histories become more and more steady. A much weaker shock is generated at the lower side which causes slight pressure fluctuations at the position $x/c = 43\ \%$.

For this flow condition in two experiments shadowgraphs were taken separately for the upper and lower side of the airfoil. They were put together to visualize the whole flow field (see Fig. 11). At the lower side downstream of the nose a first shock is generated at about x/c = 5 to 8 %. This shock of small lateral extend is visible at the end of the dark region at the nose (Fig. 11). Behind this shock the flow is subsonic and consequently no Mach lines are visible. But the flow accelerates again to supersonic speed. In a copy of high quality the supersonic region is clearly visible by the Mach lines. The end of this region is determined by a second shock at the lower surface.

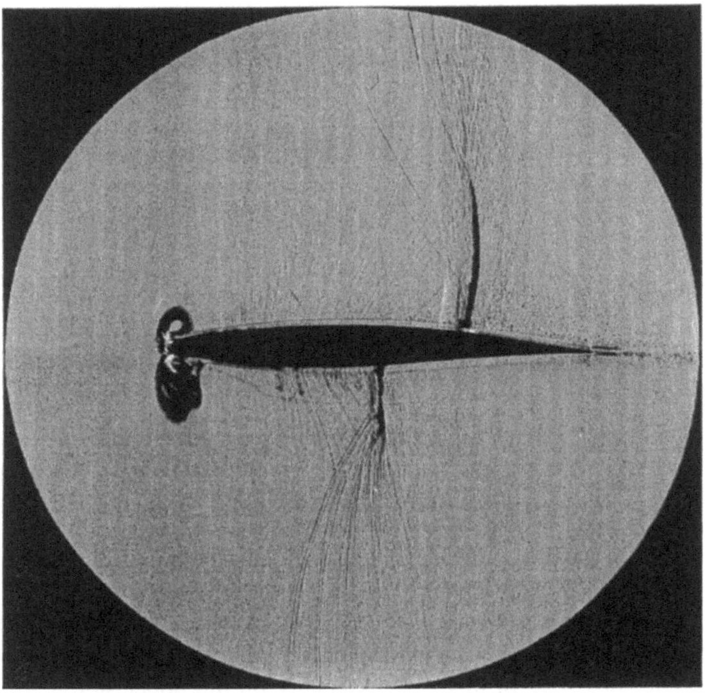

Figure 11 Shadowgraph from two experiments, lower and upper side of airfoil BAC 3-11, $M_\infty = 0.74$, $Re_c = 7.2 \cdot 10^6$

The most interesting feature of the facility is to achieve a high Reynolds number flow.Measurements have been performed typically for Reynolds numbers based on the model chord length of 20.5 Mio. for a Mach number of 0.7 [4]. For this flow condition Fig. 12 represents pressure histories obtained for the upper and lower side of the airfoil at an angle of attack of 0°.

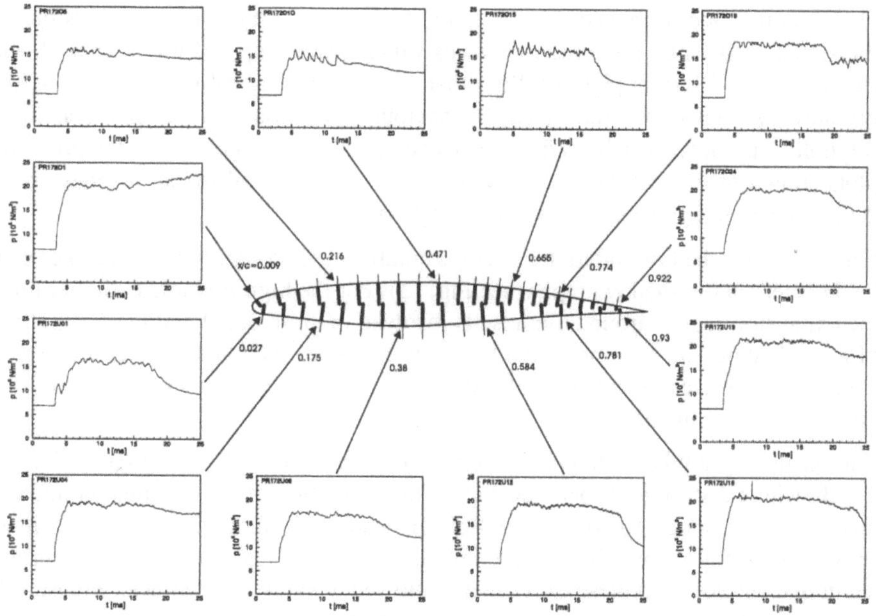

Figure 12 Pressure histories $p[10^5 N/m^2]$ for airfoil BAC3-11, $M_\infty = 0.7$, $Re_c = 20.5 \cdot 10^6$, $\alpha = 0°$

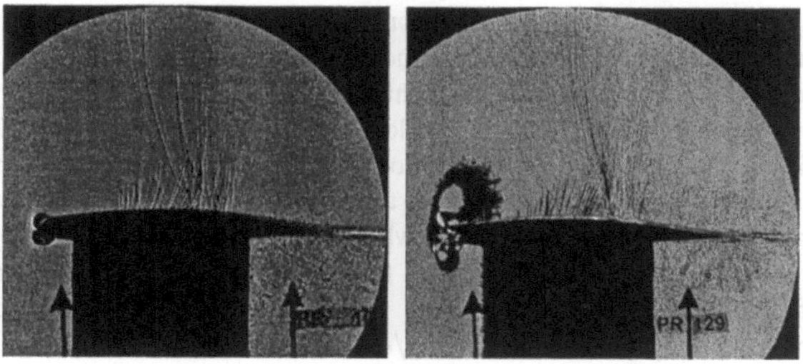

Figure 13 Influence of Reynolds number on airfoil BAC 3-11, upper side, M_∞ = 0.75, $\alpha = 0°$; left: $Re_c = 3.1 \cdot 10^6$, right: $Re_c = 20.7 \cdot 10^6$,

On the upper side from $x/c = 47$ % up to 65 % pressure fluctuations are visible in the signals caused by compression waves which steepen up to form a weak shock (see also Fig. 13, right picture). For other positions during the measuring time the pressure histories show a quite steady behaviour. In this transonic shock tunnel high Reynolds numbers are achieved by high static pressures. Therefore, in the pressure plots shown in Fig. 12 the maximum pressure on the airfoil model reaches up to 2.1 MPa which represents a remarkable high value.

For $\alpha = 0°$ for the low and high Reynolds number case no significant difference can be detected on the shadowgraphs (see Fig. 13). For the high Reynolds number flow the expansion region around the nose is resolved in more detail and the generation of a weak shock on the upper side becomes apparent. This weak shock or strong compression waves are responsible for the pressure fluctuations observed on the upper side of the airfoil in Fig. 12.

The Reynolds number influence becomes more pronounced for higher angles of attack, where a stronger shock develops. The interaction of the turbulent boundary layer with the shock for this special airfoil leads to a shock oscillation, as it is shown in the following. The high temporal resolution capability of the measuring technique employed in these experiments allowed to study these effects. Figure 14 gives the distributions of the pressure coefficient for 0° and 2° angle of attack for a low (square symbols) and a high Reynolds number flow (circles) for three different instants of time during the measuring time.

The experiments for $\alpha = 0°$ reveal an interesting phenomenon. If the flow approaches or slightly exceeds the critical state, i.e. if the flow becomes sonic, it tends to show an unsteady behaviour. This is visible for the pressure distributions on the upper side between $x/c = 35$ % and 60 %. The value c_p^* in Fig. 14a gives the pressure coefficient for which sonic speed is reached. The development of a local unsteady flow field is more pronounced for the high Reynolds number flow than for the lower Reynolds number. Upstream and downstream of this region the pressure coefficient shows a much more steady behaviour [5]. The situation changes if there is a strong shock imbedded in the flow field, as it is the case in Fig. 14b for 2° angle of attack. For the low Reynolds number flow during the time period between 7.5 ms and 9.5 ms the shock moves very slowly a little downstream. The velocity of this shock movement only amounts to 2 m/s which is a relatively small value compared to the free stream velocity of 290 m/s, i.e. there is enough time to establish a quasi-steady flow during the shock movement. For the high Reynolds number case the shock moves first upstream from about $x/c = 60$ % to 56 %. Then it moves downstream and during the remaining measuring time it becomes stationary at a position of about 60 %. In this case the shock position is located more upstream than for the low Reynolds number flow. Also the high Reynolds number leads to a lower suction pressure on the upper side but to a higher pressure on the lower side. Integration of the pressure distributions show that despite the different distributions for this Mach number the overall lift is practically not affected by the Reynolds number.

A first quantitative comparison of experimental and numerical results have been per-

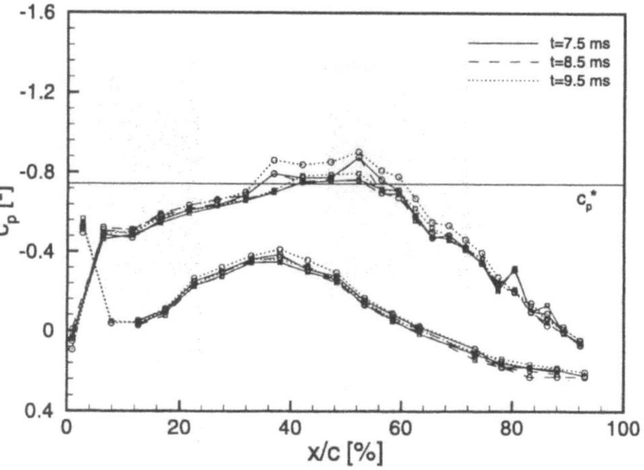

a)

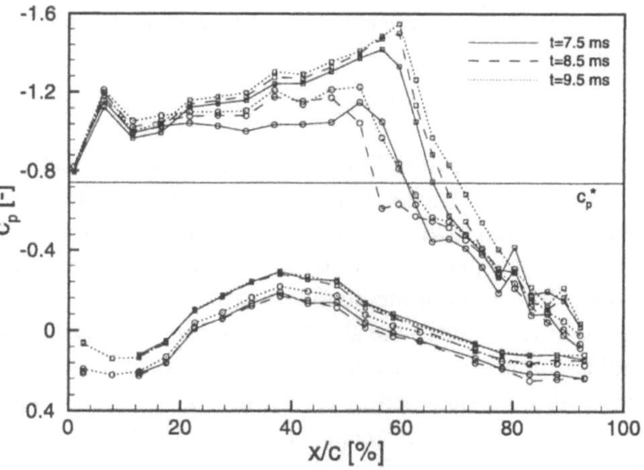

b)

Figure 14 Distribution of pressure coefficient, $M_\infty = 0.7$;
a) $\alpha = 0°$, $Re_c = 2.7 \cdot 10^6$ (square symbols), $Re_c = 20.5 \cdot 10^6$ (circles);
b) $\alpha = 2°$, $Re_c = 7.5 \cdot 10^6$ (square symbols), $Re_c = 20.3 \cdot 10^6$ (circles)

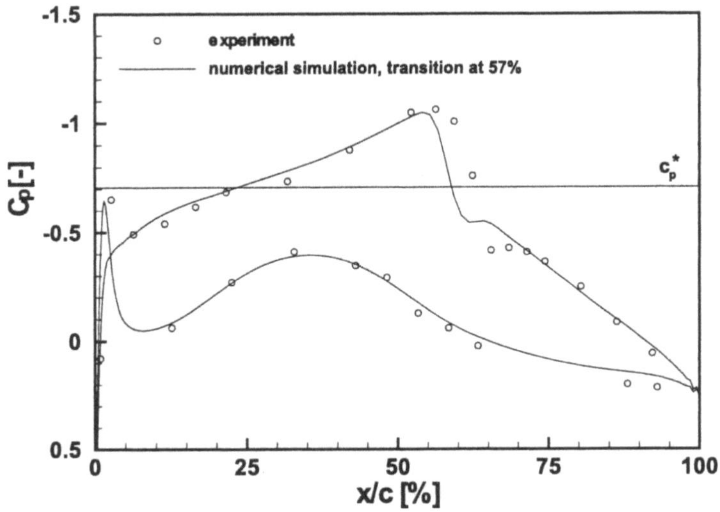

Figure 15 Comparison of experimental and numerical pressure distribution, $M_\infty = 0.72$, $Re_c = 2.5 \cdot 10^6$ $\alpha = 0°$

formed for a flow Mach number of 0.72 and a low Reynolds number of $2.5 \cdot 10^6$. For the numerical simulation a Navier-Stokes calculation was performed [6], taking into account the upper and lower test section walls which were closed as for all experiments described in this paper. For turbulence modelling the Baldwin-Lomax model has been used. Transition of the boundary layer was assumed to take place at $x/c = 57$ %. This corresponds to the shock position in the experiment. For the free stream Reynolds number considered, this assumption is realistic. In this case the pressure distributions are obtained as shown in Fig. 15. Good agreement is achieved for the overall distribution, except for the shock region where the experiment gives a more downstream shock position. Nevertheless, this first quantitative comparison gives evidence that beside of the short running time reliable results can be achieved with this type of facility. Figure 16 depicts experimental results showing two pressure coefficient distributions around two BAC 3-11 airfoil models of different size in two different wind tunnels for the same angle of attack. Both wind tunnels have different test section dimensions (AIA 400 mm x 400 mm, SWL 200 mm x 280 mm), and different test section walls (SWL straight closed walls, AIA adapted curved walls). The difference in the order of one magnitude of the Reynolds number: $36 \cdot 10^6$ (full diamond symbols) compared to $2.1 \cdot 10^6$ and fixed transition at 5% (open triangles) [7] should not influence the results significantly, since at 0° angle of attack and modest Mach number no strong shocks at the upper surface occur. This is confirmed by the critical pressure coefficient that is reached and only slightly exceeded for the AIA data [7] at the leading edge lower surface suction peak and for one pressure tab at the upper surface where a very small super

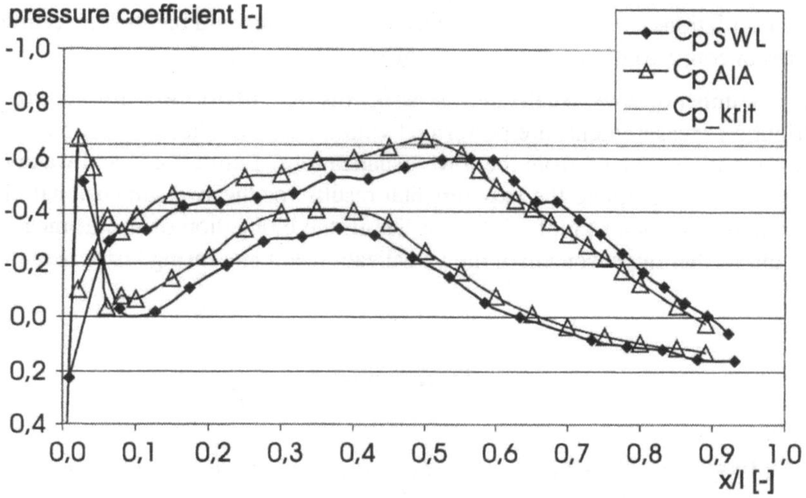

Figure 16 Comparison of two experimental pressure coefficient distributions, $M_\infty = 0.74$, $Re_c(\text{SWL}) = 36 \cdot 10^6$, $Re_c(\text{AIA}) = 2.1 \cdot 10^6$ (fixed transition at 5%), $\alpha = 0°$, c_L (both) = 0.247

sonic area exists. This critical pressure coefficient is never even reached by any of the SWL values. For the leading edge this could be because of not enough measuring positions in this area. Nevertheless the differences are limited and are most probably due to the different blockage ratios and different wall conditions (adapted and straight) of the tests mentioned above. To make the data more comparable, a wall correction method has to be applied for the straight wall case. But to provide validation data for numerical codes, experiments with straight closed walls are more suitable. Beside of other results shown in this paper, these tests for very high Reynolds numbers proved the concept of a transonic shock tube for high Reynolds number airfoil testing. Of course, in future, even more work has to be done to characterize the flow in the facility and to validate it with known test cases.

5 Conclusion

A transonic shock tube has been set up at the Shock Wave Laboratory of the RWTH Aachen University. Since end of 1998 experiments have been carried out in this facility. The test section flow was calibrated thoroughly and pitot pressure and static wall pressure measurements in the test section and in the low pressure section of the shock tube were performed.

A high frequency pressure measurement system and a high speed flow visualization system has been set up which allow to study highly transient phenomena, like shock oscillations etc., which are not detectable with standard measuring techniques applied in continuously running wind tunnels.

A large number of flow visualization experiments were performed. Surface pressure distribution measurements are performed with a 100 mm chord model and add to the information attained from the flow visualizations. Future force and wake measurements will complete the experimental results for the two dimensional airfoil experiments. Another point of interest lies in the elongation of useful measuring time and further improvement of the visualization and measuring instrumentation.

References

[1] MIRELS, H.: *Shock Tube Test Time Limitation Due to Turbulent-Wall Boundary Layer* AIAA Journal 1964 Vol. 2 No. 1 pp 84-99. (1964)

[2] REICHEL, T.; ZECHNER, M.; OLIVIER, H.: *Der Stoßrohr-Transschallkanal der RWTH Aachen, ein 2-D Windkanal zur gleich- zeitigen Einhaltung von Mach- und Reynoldszahlähnlichkeit* DGLR-Jahrestagung 2000 DGLR-JT2000-017 (2000)

[3] MOIR, I. R. M.: *Measurements on a two-dimensional aerofoil with high-lift devices* AGARD-AR-303 (1994)

[4] ZECHNER, M.: *Transsonische Profilumströmungen im Stoßrohr-Transschallkanal* Doctoral Thesis, RWTH Aachen University, Shock Wave Laboratory (2002)

[5] OLIVIER, H.; REICHEL, T.; ZECHNER, M.: *Flow Visualization and Pressure Measurements on an Airfoil in High Reynolds Number Transonic Flow* AIAA 2002-0301 (2002)

[6] BRAUN, C.: *private communication* Lehr- und Forschungsgebiet für Mechanik, RWTH Aachen University, Germany (2000)

[7] HILLENHERMS, C.: *private communication* Aerodynamisches Institut Aachen, RWTH Aachen University, Germany (2002)

3. Numerical Tools and Approaches for Flow Simulation with Local Adaptivity

H-Adaptive Multiscale Schemes for the Compressible Navier-Stokes Equations – Polyhedral Discretization, Data Compression and Mesh Generation

F. Bramkamp [1], B. Gottschlich-Müller [2], M. Hesse[1], Ph. Lamby [2], S. Müller [2], J. Ballmann[1], K.H. Brakhage[2], W. Dahmen [2]
[1] Lehr- und Forschungsgebiet für Mechanik, RWTH Aachen
Templergraben 64, D-52062 Aachen, Germany
[2] Institut für Geometrie und Praktische Mathematik, RWTH Aachen
Templergraben 55, D-52062 Aachen, Germany

Summary

In this paper we present the main conceptual ingredients and the current state of development of the new solver QUADFLOW for large scale simulations of compressible fluid flow and fluid–structure interaction. In order to keep the size of the discrete problems at every stage as small as possible for any given target accuracy, we employ a multiresolution adaptation strategy that will be described in the first part of the paper. In the second part we outline a new mesh generation concept that is to support the adaptive concepts as well as possible. A key idea is to understand meshes as parametric mappings determined by possibly few control points as opposed to store each mesh cell separately. Finally, we present a finite volume discretization along with suitable data structures which again is to support the adaptation concepts. We conclude with numerical examples of realistic applications demonstrating different features of the solver.

1 Introduction

The numerical simulation of non–stationary fluid–structure interactions poses enormous challenges to computing resources, data management strategies but also to the underlying mathematical concepts that contribute to keeping the computational complexity tractable. However, any increase in computing power is bound to be surpassed by the demands put forth by scientist and engineers through more and more realistic and consequently more complex models. Advancing the current frontiers in numerical simulation calls for combined efforts beyond traditional borders of scientific disciplines. Studying the interaction of aerodynamics and structural mechanics in complex geometries is a typical example. The fact that several severe obstructions such as time–dependency of the involved processes, varying complex geometries and the coupling of physical regimes with different characteristic features combined with the huge amount of data needed to model the process in an adequate way, indicates the principal limitations of conventional approaches. In order to resolve a typically singular behavior of the solution in complex geometries meshes with several millions of cells are required. Improved hardware or purely data oriented strategies such as parallel computing are not sufficient to overcome the arising difficulties. As important and necessary these aspects may be they have to be complemented in the long run by mathematical concepts that aim at minimizing in the first place the complexity and size of arising discrete problems.

This paper summarizes some recent attempts in this direction. It describes the present stage of an integrated development of dynamic adaptation strategies, mesh generation and discretization which is presented here for a still intermediate state of two–dimensional gas flow. The central objective is to realize adaptively generated discretizations that are able to resolve the physically relevant phenomena at the expense of possibly few degrees of freedom and correspondingly reduced storage demands. This requires a careful coordination of the core ingredients namely the discretization of the underlying system of partial differential equations, the generation and management of suitable meshes and the adaptation mechanisms, all three parts being based on suitable data structures. The following remarks are to indicate the main orientation. Accepting the Navier Stokes equations as the model of choice we give preference to quadrilateral and hexahedral meshes that still facilitate best boundary fitted anisotropic meshes. To retain sufficient geometric flexibility this is combined with block structuring. Complex geometry as well as the anticipated fine scale behavior of the solutions require extremely high resolution at least in parts of the computational domain.

To avoid storage demands which, in the light of the envisaged applications are prohibitive, we wish to employ local mesh refinement. A pivotal role is therefore played by reliable and efficient refinement strategies that are presented in Section 3.1. The main distinction from previous work in this regard lies in the fact that we employ here recent *multiresolution techniques*. The starting point is a proposal by Harten to transform the arrays of cell averages associated with any given finite volume discretization into a different format that reveals insight into the local behavior of the solution. The cell averages on a given highest level of resolution are represented as cell averages on some coarse level where the fine scale information is encoded in arrays of *detail coefficients* of ascending resolution. This requires a *hierarchy of meshes*. In contrast to Harten's idea the multiscale representation is not only used to avoid expensive flux evaluations in regions where the solution is smooth but to create *locally refined meshes*. As long as one works on a uniform mesh (even when in a major part inexpensive finite differences are employed) the computational complexity stays proportional to the number of cells which in 3d calculations with the above objectives is prohibitive. Thus a principal objective is to extract the inherent complexity of the problem by placing as few degrees of freedom so as to still capture the features of the searched for solution within a given tolerance. A central mathematical problem is then to show that the essential information to be propagated in time is still kept with sufficient accuracy when working on locally coarser meshes.

The adaptation strategy gives rise to locally refined meshes of quadtree respectively octree type. The second important ingredient is the generation of such meshes along with the information needed by the flow solver at any stage of a dynamical calculation. A key idea is to represent such meshes with as few parameters as possible while further successive refinements can be efficiently computed based on the knowledge of these parameters. This seems to be of vital importance with regard to (geometrically) non–stationary processes. Roughly speaking the mesh in each block results from evaluating a parametric mapping from the computational domain into the physical domain. Such mappings can be based on B-spline representations in combination with well established concepts from CAGD (computer aided geometric design). The quantities to be updated in time are the relatively few control parameters in those parametric representations, while mesh points on any level of resolution can be efficiently computed due to the locality of the B-spline representation. The fact that one needs indeed only relatively few control points in order to generate meshes of good quality is partly due to the variation diminishing property of B-splines. For this statement we refer to the discussion later in Section 4 of this paper.

Finally, Section 5 is devoted to the outline of a discretization scheme that meets the requirements of the adaptation concept and fits well with the mesh generation. To avoid complicated mesh management within each block and to keep the discretizations of the individual blocks as independent as possible and, in particular, to avoid global geometrical constraints, we insist on meshes with hanging nodes. Since the discretization is based on the finite volume concept this appeared to us as the best compromise. This requires the development of a finite volume scheme for fairly general cell partitions that can cope, in particular, with hanging nodes and possible unstructured parts in complicated regions of the flow domain. Section 5 offers a self contained account of the discretization scheme including data structures, realization of spatial second order, the treatment of convective and viscous fluxes, choice of limiters, the treatment of boundary conditions and the validation of the basic scheme by a proper selection of test cases.

We conclude with several applications to well–known numerical and fluid dynamical test cases that highlight the features of the whole flow solver.

2 Governing Equations

In the present study, laminar viscous fluid flow is described by the time dependent Navier–Stokes equations for a compressible gas. Neglecting body forces and volume supply of energy, the conservation laws for any control volume Ω with boundary $\partial\Omega$ and outward unit normal vector $\boldsymbol{n}$ on the surface element $dS \subset \partial\Omega$ can be written in integral form as:

$$\int_{\Omega} \frac{\partial \boldsymbol{u}}{\partial t} d\boldsymbol{x} + \oint_{\partial\Omega} \left(\boldsymbol{F}^c(\boldsymbol{u}) + \boldsymbol{F}^d(\boldsymbol{u}) \right) \boldsymbol{n}\, dS = \boldsymbol{0}\,. \tag{1}$$

To complete the posed problem initial values $\boldsymbol{u}\,(\boldsymbol{x}, t_0) = \boldsymbol{u}_0\,(\boldsymbol{x})$, $\boldsymbol{x} \in \Omega$ and boundary conditions $\boldsymbol{u}\,(\boldsymbol{x}, t)|_{\partial\Omega} = \mathcal{B}\,(\boldsymbol{x}, t)$, $\boldsymbol{x} \in \partial\Omega$ are to be prescribed.
$\boldsymbol{u} = (\varrho,\, \varrho\boldsymbol{v},\, \varrho e_{tot})^T$ denotes the vector of the unknown conserved quantities. $\boldsymbol{F}^c$ and $\boldsymbol{F}^d$ represent the physically conservative flux and the diffusive flux function, respectively. The physically conservative flux $\boldsymbol{F}^c$ is

$$\boldsymbol{F}^c = \begin{pmatrix} \varrho\boldsymbol{v} \\ \varrho\boldsymbol{v} \circ \boldsymbol{v} + p\,\boldsymbol{I} \\ \varrho h_{tot}\boldsymbol{v} \end{pmatrix} \tag{2}$$

where ϱ denotes the density, p the static pressure, $\boldsymbol{v}$ the velocity vector, e_{tot} the total energy and h_{tot} the total enthalpy. The symbol $\circ$ means the dyadic product.

The diffusive flux $\boldsymbol{F}^d$ contains the viscous stresses and heat conduction

$$\boldsymbol{F}^d = \begin{pmatrix} 0 \\ -\boldsymbol{T}^v \\ -\boldsymbol{v}\boldsymbol{T}^v + \boldsymbol{q} \end{pmatrix} \tag{3}$$

where the viscous stress tensor $\boldsymbol{T}^v$ for an isentropic Newtonian fluid is defined as

$$\boldsymbol{T}^v = \mu \left(grad\, \boldsymbol{v} + (grad\, \boldsymbol{v})^T \right) - \frac{2}{3}\mu\, (div\, \boldsymbol{v})\, \boldsymbol{I}\,. \tag{4}$$

Heat conduction is modeled by Fourier's law

$$\boldsymbol{q} = -\kappa\, grad\, T \tag{5}$$

where the thermal conductivity is assumed as

$$\kappa = c_p \frac{\mu}{Pr} \tag{6}$$

with Prandtl number $Pr = 0.72$. The variation of the molecular viscosity μ as a function of temperature is determined by the Sutherland formula

$$\mu = \mu_\infty \left(\frac{T}{T_\infty} \right)^{3/2} \frac{T_\infty + S}{T + S}\,, \tag{7}$$

where $S = 110°K$ denotes the Sutherland constant. The static pressure is related to the specific internal energy according to the equation of state for a perfect gas

$$p = \varrho(\gamma - 1)\left(e_{tot} - 1/2\,|\boldsymbol{v}|^2\right), \tag{8}$$

where γ is the ratio of specific heats, which is taken as 1.4 for air.

3 Adaptation Based on Multiscale Analysis

3.1 Problem Formulation and Preliminaries

The numerical schemes under consideration are to be applied to gas flow. Although the full Navier Stokes equations serve as the main model the behavior of the numerical scheme is largely determined by corresponding hyperbolic conservation laws describing the balance of quantities like mass, momentum and energy ignoring heat conduction and viscous effects. We will therefore confine the discussion first to this setting in order to bring out the basics of the concept. The same techniques carry over to viscous fluxes and source terms, see [14]. Omitting the upper index c in the conservative flux and denoting by $\Omega \subset \mathbb{R}^d$ the *computational domain* and by $\mathcal{D} \subset \mathbb{R}^m$ the *space of admissible states* hosting the *conserved quantities* $\boldsymbol{u} : [0, \bar{t}] \times \Omega \to \mathcal{D}$, the basic balance laws have the form

$$\int_V \frac{\partial\,\boldsymbol{u}(t,\boldsymbol{x})}{\partial\,t}\,d\boldsymbol{x} = -\int_{\partial V} \boldsymbol{F}(\boldsymbol{u}(t,\boldsymbol{x}))\,\boldsymbol{n}\,dS, \tag{9}$$

where V is an arbitrary but temporally fixed control volume (with sufficiently smooth boundary such that the Green–Gauss theorem works). Under the assumption that $\boldsymbol{u}$ is smooth one derives from (9) the usual first order system of conservation laws of the form

$$\frac{\partial\,\boldsymbol{u}(t,\boldsymbol{x})}{\partial\,t} + \sum_{i=1}^{d} \frac{\partial\,\boldsymbol{F}_i(\boldsymbol{u}(t,\boldsymbol{x}))}{\partial\,x_i} = \boldsymbol{0}, \tag{10}$$

which describes the temporal evolution of $\boldsymbol{u}$. Throughout this work the fluxes are assumed to be smooth in the state space, i.e., $\boldsymbol{F}_i \in C^2(\mathcal{D}, \mathbb{R}^m)$ and the Jacobian of the normal flux $\boldsymbol{F}(\boldsymbol{u})\,\boldsymbol{n}$ has m real eigenvalues $\lambda_i(\boldsymbol{u},\boldsymbol{n})$ as well as a complete system of m linearly independent right eigenvectors $\boldsymbol{r}_i(\boldsymbol{u},\boldsymbol{n})$ and left eigenvectors $\boldsymbol{l}_i(\boldsymbol{u},\boldsymbol{n})$, respectively, $i = 1, \ldots, m$, for all $\boldsymbol{u} \in \mathcal{D}$ and $\boldsymbol{n} \in \mathbb{R}^d$ with $\|\boldsymbol{n}\|_2 = 1$. The system (10) is then called *hyperbolic*.

As mentioned in Section 2, the solution of the conservation laws is subject to initial conditions and also to appropriate boundary conditions (see Section 5.1.6 for details). Since the choice of the latter conditions does not affect the discussion of adaptive concepts we will suppress this issue (which by itself is certainly a delicate one) and confine the discussion for the moment to pure initial value problems.

The system (10) is to be viewed as a model which has to be properly interpreted in order to cover all physically relevant features such as shocks and contact discontinuities. The common basis is provided by the notion of weak solutions of (10), which solve the variational problem obtained when multiplying (10) by smooth test functions and applying integration by parts. The quantities $\boldsymbol{u}$ are then no longer subjected to derivatives so that certain discontinuities are admissible. By properly varying the test functions one ultimately also recovers (9) which is therefore a natural starting point for discretizations.

Before explaining this one should note though that when passing to weak solutions uniqueness is generally lost even when proper side constraints are imposed. One has to resort to additional selection criteria usually referred to as *entropy conditions* (motivated by the thermodynamical entropy) that restore uniqueness see [22, 21]. This is well understood for scalar conservation laws and summed up by Kruzhkov's entropy concept. The situation for systems of conservation laws is by far less clear and several entropy concepts compete which are generally not equivalent, see for instance [9].

The presence of discontinuities does not only cause problems concerning existence and uniqueness but also crucially effects the design of numerical schemes. For instance, standard finite difference discretizations of (10) will produce oscillations near discontinuities. These schemes have to be stabilized by artificial viscosity which causes a significant loss in accuracy. In particular, discontinuities are smeared. Moreover, stability in the sense that the entropy solution is recovered by a numerical scheme severely limits its order of exactness. Again the design of such schemes is not the issue of the following discussion. We will rather assume at this point that we have a stable scheme at hand to explain then how to develop adaptive refinements based on such a given discretization. Concrete realizations that fit our particular application context best will be developed in Section 5. Therefore we content ourselves with describing the following necessary ingredients.

Recalling (9), a natural choice for the discretization of conservation laws are finite volume schemes. These schemes arise when observing that (9) describes the evolution of *cell averages*. In fact, integration over a time interval yields

$$\hat{\boldsymbol{u}}(t+\tau) = \hat{\boldsymbol{u}}(t) + \frac{\tau}{|V|}\,\boldsymbol{B}(t) \tag{11}$$

where the cell average $\hat{\boldsymbol{u}}(t)$ and the flux balance $\boldsymbol{B}(t)$ defined by

$$\hat{\boldsymbol{u}}(t) := \frac{1}{|V|}\int_V \boldsymbol{u}(t,\boldsymbol{x})\,dV, \quad \boldsymbol{B}(t) := \frac{1}{\tau}\int_t^{t+\tau}\int_{\partial V} \boldsymbol{F}(\boldsymbol{u})\,\boldsymbol{n}\,dS\,dt \tag{12}$$

refer to an arbitrary but fixed cell $V \subset \Omega$ with volume $|V| := \int_V 1\,dV$. Obviously, the cell average changes in time due to the flux balance, i.e., the information that is crossing the cell interface ∂V in the time interval $[t, t+\tau]$. Approximating the flux balance $\boldsymbol{B}(t)$ then leads to a finite volume scheme.

3.2 The Objectives

In the past, finite volume schemes have been applied not only to academic problems but also to real world problems arising, for instance, in engineering. In this context robustness has been a major issue. However, in order to guarantee the reliability of the computational results the discretization error has to be controlled as well. So far, error estimates are available only for restricted classes of numerical schemes applied to a single conservation law, see [5, 19]. The discretization error can then be shown to be bounded by a constant times h^α, $\alpha \leq 0.5$, where h denotes the largest cell diameter of the underlying mesh. Due to the lack of global smoothness, we cannot expect a convergence rate better than $\alpha = 1$ even for schemes that are higher–order accurate in regions where the solution is locally smooth. This is a severe drawback whenever a prescribed error tolerance is to be met. In fact, accuracy considerations may become increasingly important for nonstationary calculations or when the interaction of fluid flow and structure is to be simulated. In such cases one encounters relevant physical effects over a large range of length scales and different propagation speeds in the coupled regimes. Moreover, the presence of discontinuities, boundary layers, and other singular phenomena requires small mesh sizes at least in parts of the computational domain. In order to resolve these effects adequately by discretizations based on *quasi-uniform meshes* would require a prohibitively large number of cells. In particular, in 3D calculations the resulting necessary memory resources and computation time would by far exceed the capacity of modern computing facilities, e.g. parallel architectures or clusters. In order to make these numerical simulations feasible new algorithmic concepts have to be developed that reduce the computational complexity, i.e., computation time and memory requirements, to a degree that reflects the problem inherent degrees of freedom relative to a desired accuracy. This calls for concepts that automatically adapt the local resolution of a discretization to the local behavior of the solution. This means only in regions where small scale physical effects are present the discretization has to exhibit high resolution while elsewhere the mesh is to be kept as coarse as possible for still meeting given accuracy requirements. This is exactly the objective of adaptive schemes that aim at distributing the degrees of freedom as economically as possible based on current information obtained during the calculation. There are two major issues to be addressed in this context. First, the decisions made during the calculations have to be reliable, i.e., one has to be sure that no essential features are missed when keeping the mesh size moderate in certain areas, in particular, with regard to non–stationary processes. This task concerns mathematical analysis. Second, handling dynamic mesh adaptation requires the development of much more sophisticated data structures.

In the literature, several adaptive strategies have been discussed or are under current investigation. A standard strategy is to base local mesh refinements on *local indicators* which are typically related to strong gradients in a current approximation or local residuals in the context of a posteriori refinement indicators [27, 28, 18]. These are efficient. However, no reliable error control is offered by this concept.

For this purpose, reliable *a posteriori estimates* have been derived which aim at equilibrating local errors. So far, this type of error estimators are only available for scalar problems, see [20]. Here we will pursue a different strategy which originated from earlier proposals by Harten [17]. The key idea is to transform the arrays of cell averages produced by a given finite volume scheme into a different multiscale format. In this format only very few coefficients represent cell averages for a possibly coarse partition while the other coefficients encode the difference information corresponding to the intermediate length scales that is lost through the coarsening. Quite in the spirit of a *multiscale analysis* the size of these *detail coefficients* will be used as a basis for local mesh refinements. This gives rise to an adaptive mesh with respect to which the evolution step is performed. Of course, the crux in this context is to arrange this procedure in a way that at no stage of the computation there is ever made use of fully refined uniform mesh.

In the remainder of this section, we explain the main ingredients of such a multiscale analysis of an array of cell averages from which an adaptive mesh is derived. Then we outline the construction of the adaptive finite volume scheme and comment on the discretization error. Finally, some numerical results for model problems are presented, that give a first idea of the quantitative performance of the adaptation concept.

3.3 Multiscale Setting

Finite volume schemes are naturally related to cell averages of the solution. In order to detect singularities of the solution, we transform the array of cell averages into a new data format which allows for data compression.

3.3.1 Grid Hierarchy

The starting point for the multiscale setting is a hierarchy of nested grids $\mathcal{G}_j := \{V_{j,k}\}_{k \in I_j}$, $j = 0, \ldots, L$, such that each grid $\mathcal{G}_j$ is a *partition* of the computational domain, i.e., $\Omega = \bigcup_{k \in I_j} V_{j,k}$. Here the coarsest discretization is indicated by 0 and the finest discretization by L. Moreover, the grids are *nested* in the sense that the cells on the coarse grid can be represented as the union of cells on the next finer grid, i.e.,

$$V_{j,k} = \bigcup_{r \in \mathcal{M}^0_{j,k}} V_{j+1,r}, \quad k \in I_j. \tag{13}$$

The index set $\mathcal{M}^0_{j,k} \subset I_{j+1}$ corresponds to the cells on level $j+1$ resulting from the refinement of the cell $V_{j,k}$. A simple example is shown in Figure 1. We emphasize that the framework presented here can also be applied to unstructured grids and irregular mesh refinements. To keep the presentation as simple as possible we confine, however, the discussion to refinements of cells into a constant number of

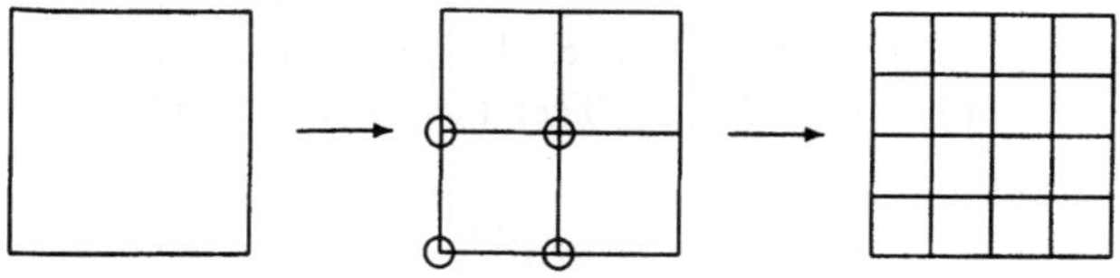

Figure 1 Sequence of nested grids

children, i.e.,

$$\# \mathcal{M}^0_{j,k} = M_r = const.$$

In the above example, $\mathcal{M}^0_{j,k}$ can be identified with the circled nodes in the middle part of Figure 1. With each cell in $\mathcal{G}_j$ we associate the *box function*

$$\tilde{\varphi}_{j,k}(\boldsymbol{x}) := \frac{1}{|V_{j,k}|} \chi_{V_{j,k}}(\boldsymbol{x}), \tag{14}$$

where the characteristic function χ_V corresponding to any set V takes the value one for $\boldsymbol{x} \in V$ and zero elsewhere. Then the cell average of a scalar, integrable function u can be expressed as

$$\hat{u}_{j,k} = \langle u, \tilde{\varphi}_{j,k} \rangle_\Omega, \tag{15}$$

where the scalar product is defined by $\langle u, v \rangle_\Omega := \int_\Omega u\, v\, d\boldsymbol{x}$. Note, that our framework is not restricted to scalar functions but can directly be applied to the components of vector–valued functions. Since the grids are nested, we infer the two–scale relation

$$\tilde{\varphi}_{j,k} = \sum_{r \in \mathcal{M}^0_{j,k}} m^{j,0}_{r,k} \tilde{\varphi}_{j+1,r}, \qquad m^{j,0}_{r,k} := \frac{|V_{j+1,r}|}{|V_{j,k}|}, \tag{16}$$

i.e., the coarse grid box function can be represented as a linear combination of the corresponding fine grid box functions. Consequently, the averages of two discretization levels are related to each other by

$$\hat{u}_{j,k} = \sum_{r \in \mathcal{M}^0_{j,k}} m^{j,0}_{r,k} \hat{u}_{j+1,r}. \tag{17}$$

Now the goal is to transform these data into a different format of cell averages corresponding to a sequence of resolution levels. This will be motivated by a simple univariate example.

3.3.2 A Univariate Example

We now consider the unit interval $\Omega = [0, 1]$ where the grid hierarchy is determined by a uniform dyadic partition of $[0, 1]$, i.e., $V_{j,k} = 2^{-j}[k, k+1]$,

$k \in I_j := \{0, \ldots, 2^j - 1\}$. Hence, the refinement sets are just given by $\mathcal{M}^0_{j,k} = \{2k, 2k+1\} \subset I_{j+1}$, $k \in I_j$. Then the box function has the form

$$\tilde{\varphi}_{j,k} = 2^j \, \chi_{[0,1]}(2^j \cdot -k),$$

and the two–scale relation (16) reads here

$$\tilde{\varphi}_{j,k} = \frac{1}{2}\,(\tilde{\varphi}_{j+1,2k} + \tilde{\varphi}_{j+1,2k+1}). \tag{18}$$

We explain now how to decompose the cell averages into averages of a coarser partition and details. To this end, we introduce the box wavelet (which in this case is better known as Haar wavelet)

$$\tilde{\psi}_{j,k} := \frac{1}{2}\,(\tilde{\varphi}_{j+1,2k} - \tilde{\varphi}_{j+1,2k+1}). \tag{19}$$

Then we can write any fine scale box function by means of the coarse box function $\tilde{\varphi}_{j,k}$ and the box wavelet $\tilde{\psi}_{j,k}$.

$$\tilde{\varphi}_{j+1,2k} = \tilde{\varphi}_{j,k} + \tilde{\psi}_{j,k}, \qquad \tilde{\varphi}_{j+1,2k+1} = \tilde{\varphi}_{j,k} - \tilde{\psi}_{j,k}. \tag{20}$$

These relations are motivated by the illustrations in Figure 2. In analogy to (17) we

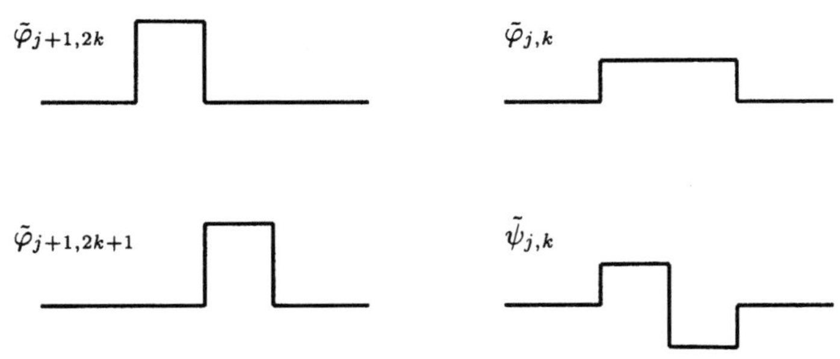

Figure 2 Box function and box wavelet

deduce from (18) the two–scale relations for the cell averages

$$\hat{u}_{j,k} = \frac{1}{2}\,(\hat{u}_{j+1,2k} + \hat{u}_{j+1,2k+1}).$$

On the other hand, (20) implies

$$\hat{u}_{j+1,2k} = \hat{u}_{j,k} + d_{j,k}, \qquad \hat{u}_{j+1,2k+1} = \hat{u}_{j,k} - d_{j,k}, \tag{21}$$

where the details are defined by

$$d_{j,k} := \langle u, \tilde{\psi}_{j,k} \rangle_{[0,1]} = \frac{1}{2} (\hat{u}_{j+1,2k} - \hat{u}_{j+1,2k+1}). \tag{22}$$

These relations show how to express fine scale averages in terms of coarse scale ones plus details and vice versa.

The above manipulations concern cell averages of some searched underlying function u. One might view these cell averages just as piecewise constant *approximations* to this function. Such approximations are, however, confined as piecewise constants to be at best of first order accurate. So this view would not carry very far. Instead one should view the cell averages just as *linear functionals* of u from which more accurate reconstructions of u could be derived. So there is a natural *duality* between the space hosting the cell averages and its dual where u itself lives. In order to make this distinction clear and to pave the way for later modifications, we have to introduce another system of functions

$$\varphi_{j,k} := 2^{-j}\, \tilde{\varphi}_{j,k} = \chi_{[0,1]}(2^j \cdot -k), \qquad \psi_{j,k} := 2^{-j}\, \tilde{\psi}_{j,k},$$

in which one should think of u being represented. At this point it differs from the earlier version only by normalizing factors but it will later be seen to be conceptually important to distinguish these systems. In fact, the $\varphi_{j,k}$ are the L_∞–normalized counterparts of the L_1-normalized box function and box wavelet, respectively. We will therefore refer to this system of functions as *primal system* (although we have actually started with the dual one). In particular, the dual system $\{\tilde{\varphi}_{j,k}\}_{k\in I_j} \cup \{\tilde{\psi}_{j,k}\}_{k\in I_j}$ is *biorthogonal* to the primal system $\{\varphi_{j,k}\}_{k\in I_j} \cup \{\psi_{j,k}\}_{k\in I_j}$, i.e.,

$$\begin{aligned} \langle \varphi_{j,k}, \tilde{\varphi}_{j,l} \rangle_{[0,1]} &= \langle \psi_{j,k}, \tilde{\psi}_{j,l} \rangle_{[0,1]} = \delta_{k,l}, \\ \langle \varphi_{j,k}, \tilde{\psi}_{j,l} \rangle_{[0,1]} &= \langle \psi_{j,k}, \tilde{\varphi}_{j,l} \rangle_{[0,1]} = 0. \end{aligned} \tag{23}$$

Of course, these functions satisfy two-scale relations analogous to (19) and (20) with slightly different normalizing factors.

It should already be noted now that one ultimately works only on the dual side while the primal system plays a conceptual role which is, however, important for the analysis and helpful for understanding later developments. To this end, note that given u, the function

$$u_j := \sum_{k\in I_j} \langle u, \tilde{\varphi}_{j,k} \rangle_{[0,1]}\, \varphi_{j,k} \tag{24}$$

is a projection of u onto the span of the level j functions $\varphi_{j,k}$ (which at this point happen to be still piecewise constants as well), where the expansion coefficients are, in view of (15), the cell averages of u with respect to a mesh of level j. It represents an approximate solution of the conservation law on a fixed time level corresponding to the discretization level j.

Note, that if u were (nearly) constant all coefficients in (24) would take (nearly) the same value and hence would be all equally significant. This is certainly not a very economical representation of a simple object such as a constant. We wish to transform the coefficients, namely the cell averages $\langle u, \tilde{\varphi}_{j,k}\rangle_{[0,1]}$ therefore into a format where such (near) redundancy becomes readily apparent. In fact, invoking the two-scale relation (19), (20) in combination with biorthogonality (23), u_j can be rewritten as

$$u_j = \sum_{k\in I_{j-1}} \langle u, \tilde{\varphi}_{j-1,k}\rangle_{[0,1]}\, \varphi_{j-1,k} + \sum_{k\in I_{j-1}} \langle u, \tilde{\psi}_{j-1,k}\rangle_{[0,1]}\, \psi_{j-1,k}. \tag{25}$$

Continuing in the same way with the first sum on the right hand side, eventually leads to a representation of u_j in terms of a coarse level cell averages complemented by a sum over all levels $l < j$ containing *detail coefficients* of the form $d_{l,k} := \langle u, \tilde{\psi}_{l,k}\rangle_{[0,1]}$. Clearly, raising the resolution level j would amount to adding further detail coefficients which explains their update character.

Next note that, in view of (22), the details $d_{j,k}$ vanish whenever the function u is a constant, i.e.,

$$\langle 1, \tilde{\psi}_{j,k}\rangle_{[0,1]} = 0. \tag{26}$$

Moreover, writing u on the support of $\tilde{\psi}_{j,k}$ as a constant plus a fluctuation (first order Taylor expansion) we see that $|d_{j,k}|$ is still small if that fluctuation is small. In fact, one easily verifies that

$$\begin{aligned} |d_{j,k}| &= |\langle u, \tilde{\psi}_{j,k}\rangle_{[0,1]}| = \inf_c |\langle u - c, \tilde{\psi}_{j,k}\rangle_{[0,1]}| \\ &\le \inf_c \|u - c\|_{L_\infty(\operatorname{supp}\tilde{\psi}_{j,k})} \|\tilde{\psi}_{j,k}\|_{L_1(\operatorname{supp}\tilde{\psi}_{j,k})} \le 2^{-j}\|u'\|_{L_\infty(\operatorname{supp}\tilde{\psi}_{j,k})}. \end{aligned}$$

Hence, the details may become small in smooth regions of u whereas they constitute significant contributions in (25) where u varies more strongly.

This suggests to neglect all sufficiently small details in order to keep only essential information on the function u thereby creating a *compressed (approximate) representation* of u. To pursue this line, however, at least two essential issues arise.

(i) For a compression to be really efficient one needs a more refined test than just annihilating constants as in (26) which only exploits first order smoothness. In fact, if one had instead of (26) $\langle p, \tilde{\psi}_{j,k}\rangle_{[0,1]} = 0$ for all polynomials p up to degree $m-1$ (order m) the same argument (now using higher order Taylor polynomials in place of the constant c above) would show that, when u has locally bounded mth order derivatives, the coefficients $d_{j,k}$ would be of the order 2^{-jm} thereby exhibiting a much stronger decay for higher smoothness. (This can be achieved by means of higher order biorthogonal systems instead of piecewise constants for the *primal system.* In case of the unit interval we refer to [7, 12].) Note that this is the same as saying that the functions $\varphi_{j,k}$ in the projection (24) reproduces all polynomials of order m exactly, i.e., is of higher order. But it is important to note that we actually never have to realize this higher order primal system as long as we manage to come up with dual wavelets having a higher order of vanishing moments. It will be explained below how to realize such higher order vanishing moments in an even more realistic setting of boundary fitted meshes.

(ii) Even if one realizes higher order vanishing moments, discarding small coefficients $d_{j,k}$ resulting from the corresponding multiscale transformation raises the question how this perturbation of the coefficients affects the underlying function in the relevant norm. This is the issue of *stability* of the wavelet expansions, which ultimately does involve knowledge about the primal system. Thus the primal system is *not* needed for the algorithmic realization but is important for the analysis of the scheme.

3.3.3 Higher Order Biorthogonal Systems

We will address now the above issues (i) and (ii) for a more realistic setting of mesh hierarchies. The first step is to transmit the above univariate case constructing box wavelets for an arbitrary nested mesh hierarchy. The resulting wavelets will serve as a starting point for the construction of modified box wavelets with improved cancellation properties with the aid of a concept introduced in [3].

For the construction of the generalized box wavelets we now proceed analogously to the univariate case. As a counterpart to (19) we introduce the box wavelet as a linear combination of the fine scale box functions $\tilde{\varphi}_{j+1,r}$, $r \in \mathcal{M}^0_{j,k}$, related to the refinement of the cell $V_{j,k}$,

$$\check{\psi}_{j,k,e} := \sum_{r \in \mathcal{M}^0_{j,k}} \check{m}^{j,e}_{r,k} \tilde{\varphi}_{j+1,r}, \quad e \in E^* := E \backslash \{0\}, \tag{27}$$

with $E := \{0, \ldots, M_r - 1\}$. The index e plays the same role as the distinction between even and odd indices earlier in (20). The (mask or filter) parameters $\check{m}^{j,e}_{r,k}$ are determined in such a way that the resulting system of box functions and box wavelets and their L_∞–normalized counterparts defined by

$$\varphi_{j,k} := |V_{j,k}| \, \tilde{\varphi}_{j,k}, \quad \psi_{j,k,e} := |V_{j,k}| \, \check{\psi}_{j,k,e}, \tag{28}$$

are biorthogonal in the sense of (23) and the box wavelets have one vanishing moment, i.e.,

$$\langle 1, \check{\psi}_{j,k,e} \rangle_\Omega = 0.$$

Moreover, one can then find in analogy to (20) parameters $\check{g}^{j,e}_{r,k}$, $r \in \mathcal{M}^0_{j,k}$, such that

$$\tilde{\varphi}_{j+1,k} = \sum_{e \in E} \sum_{r \in \mathcal{M}^0_{j,k}} \check{g}^{j,e}_{r,k} \check{\psi}_{j,r,e} \tag{29}$$

holds, where it is convenient to define $\check{\psi}_{j,k,0} := \tilde{\varphi}_{j,k}$. As in the univariate case (see (19), (20)), the two–scale relations (16), (27) and (29) allow one to realize a local change of basis between the systems formed by coarse scale functions $\varphi_{j,k}$ complemented by the oscillatory functions $\{\check{\psi}_{j,k,e}\}_{e\in E}$ and the fine scale functions $\{\varphi_{j+1,r}\}_{r\in\mathcal{M}^0_{j,k}}$. The corresponding details can be found in [24].

Introducing as in the univariate case the *details*

$$\check{d}_{j,k,e} := \langle u, \check{\psi}_{j,k,e}\rangle_\Omega, \qquad e \in E^*, \tag{30}$$

one then infers from (27) the two–scale relation

$$\check{d}_{j,k,e} = \sum_{r\in\mathcal{M}^0_{j,k}} \check{m}^{j,e}_{r,k}\, \hat{u}_{j+1,r}, \quad e \in E^*. \tag{31}$$

Note, that a similar relation holds for the cell averages, see equation (17).

Hence, in the same way as in the univariate case the resulting projection u_j of an integrable function u for the refinement level j

$$u_j := \sum_{k\in I_j} \langle u, \tilde{\varphi}_{j,k}\rangle_\Omega\, \varphi_{j,k} \tag{32}$$

can be split in analogy to (25) into a coarse scale part and a complementary detail part as

$$u_j = \sum_{k\in I_{j-1}} \langle u, \tilde{\varphi}_{j-1,k}\rangle_\Omega\, \varphi_{j-1,k} + \sum_{e\in E^*}\sum_{k\in I_{j-1}} \langle u, \check{\psi}_{j-1,k,e}\rangle_\Omega\, \psi_{j-1,k,e}. \tag{33}$$

Again the dual system consists still of piecewise constants. So by the very same arguments provided in the preceding section the decay of the detail coefficients obtained from a successive repetition of the above splitting into ever coarser parts and corresponding details would only be of order 2^{-j} on level j. As explained before, in order to get a better compression by exploiting higher order smoothness we have to raise the order of vanishing polynomial moments. The basic idea is to modify the box wavelet $\check{\psi}_{j,k,e}$ by some coarse grid box functions, leading to the ansatz

$$\tilde{\psi}_{j,k,e} := \check{\psi}_{j,k,e} + \sum_{l\in\mathcal{L}^e_{j,k}} l^{j,e}_{l,k}\, \tilde{\varphi}_{j,l}, \quad e \in E^*, \tag{34}$$

with parameters $l^{j,e}_{l,k}$ that are yet to be determined. Here the stencil $\mathcal{L}^e_{j,k} \subset I_j$ denotes the cells $V_{j,l}$ in the neighborhood of the cell $V_{j,k}$. Then the parameters $l^{j,e}_{l,k}$ are chosen such that

$$\langle p, \tilde{\psi}_{j,k,e}\rangle_\Omega = 0 \tag{35}$$

holds for all polynomials p of degree less than an arbitrary but fixed number M. The details of the construction can be found in [13, 24].

Note, that the modified wavelets $\tilde{\psi}_{j,k,e}$ still satisfy two–scale relations of the form (27) and (29). This can be seen by inserting (16) and (27) into (34), providing

$$\tilde{\psi}_{j,k,e} = \sum_{r\in\mathcal{M}^e_{j,k}} m^{j,e}_{r,k}\, \tilde{\varphi}_{j+1,r} \tag{36}$$

for some index set $\mathcal{M}^e_{j,k} \subset I_{j+1}$. Moreover, one can explicitly determine filter coefficients for reversing the two–scale relations (16) and (36) and express the fine scale functions through coarse scale functions and the wavelets $\tilde{\psi}_{j,k,e}$ by substituting (34) in (29), i.e.,

$$\tilde{\varphi}_{j+1,k} = \sum_{r\in\mathcal{G}^0_{j,k}} g^{j,0}_{r,k}\,\tilde{\varphi}_{j,r} + \sum_{e\in E^*}\sum_{r\in\mathcal{G}^e_{j,k}} g^{j,e}_{r,k}\,\tilde{\psi}_{j,r,e} \tag{37}$$

for some index set $\mathcal{G}^e_{j,k} \subset I_j$. Hence, the two–scale relations (16), (36) and (37) realize a change of basis between the systems $\{\tilde{\varphi}_{j,k}\}_{k\in I_j} \cup \{\tilde{\psi}_{j,k,e}\}_{k\in I_j, e\in E^*}$ and $\{\tilde{\varphi}_{j+1,k}\}_{k\in I_{j+1}}$.

The above manipulations are based upon the basic fact that if one forms a matrix $\check{M} = (\check{m}^{j,e}_{r,k})_{(k,e)\in E\times I_j, r\in I_{j+1}}$ the transpose of its inverse $\check{G}$ provides the filter coefficients for the biorthogonal primal system (here denoted by $\check{g}^{j,e}_{r,k}$). Therefore the above modifications of the wavelet filters $\check{m}^{j,e}_{r,k}$, $e \in E^*$ will change the $\check{G}$ to a new matrix G, where, however, now also the filter coefficients of the original scaled box function $\varphi_{j,k}$ have been modified. It should be emphasized that it is not clear a priori that such modifications will actually always produce a G such that now new modified functions $\varphi_{j,k}$ *exist* that satisfy the modified two scale relation. This will certainly depend on the choice of the coefficients $l^{j,e}_{l,k}$ in (34) and perhaps on the geometry of the cells. Of course, it is hoped that a modified system exists which will then actually have a corresponding higher accuracy providing better approximations to u on the primal side. That this is indeed the case can be shown under more restrictive assumptions on the mesh. For a general discussion of this issue and its relation to subdivision schemes see e.g. [10]. This is also closely related to the second issue (ii) in Section 3.3.2. As pointed out before the algorithmic realization relies only on the higher vanishing moments of the dual system and has so far never shown any indications that the primal system may fail to exist in any of the tests for realistic geometries.

3.3.4 Multiscale Transformation

As pointed out before, the reason for dealing with multiscale bases of the above type is merely to transform as in (28) an array of cell averages into a new data format that is more suitable to compression. Moreover, the degree of this compression as well as the analysis of the stability of the transformation in the sense of (ii) is still best analyzed through such bases. Now the transformation of data readily results from the two–scale relations (16), (36) and (37). In the same fashion as the relations (21), (22) have been derived in the univariate case, one substitutes the two–scale relations for the basis functions in the definition of the cell averages $\hat{u}_{j,k} := \langle u, \tilde{\varphi}_{j,k}\rangle_\Omega$ and the details $d_{j,k,e} := \langle u, \tilde{\psi}_{j,k,e}\rangle_\Omega$ to obtain in view of (17),

$$\hat{u}_{j,k} = \sum_{r\in\mathcal{M}^0_{j,k}} m^{j,0}_{r,k}\,\hat{u}_{j+1,r}, \quad d_{j,k,e} = \sum_{r\in\mathcal{M}^e_{j,k}} m^{j,e}_{r,k}\,\hat{u}_{j+1,r} \tag{38}$$

and

$$\hat{u}_{j+1,k} = \sum_{r\in\mathcal{G}^0_{j,k}} g^{j,0}_{r,k}\,\hat{u}_{j,r} + \sum_{e\in E^*}\sum_{r\in\mathcal{G}^e_{j,k}} g^{j,e}_{r,k}\,d_{j,r,e}. \tag{39}$$

Denoting by $\hat{\boldsymbol{u}}_j := (\hat{u}_{j,k})_{k\in I_j}$ and $\boldsymbol{d}_j := (d_{j,k,e})_{k\in I_j, e\in E^*}$ the arrays of cell averages, respectively details on level j, the array $\hat{\boldsymbol{u}}_L$ of cell averages on the finest level L can be decomposed into a sequence of coarse scale averages $\hat{\boldsymbol{u}}_0$ and details $\boldsymbol{d}_j$, $j = 0, \ldots, L-1$. This is done by applying the two–scale relations (38) successively from fine to coarse levels, see Figure 3. This decomposition is called the *multiscale*

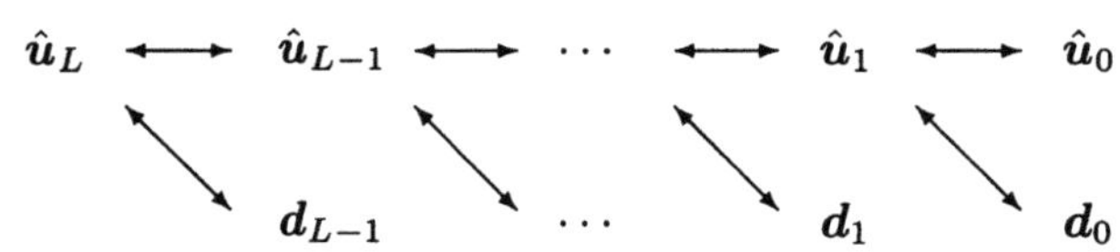

Figure 3 Pyramid scheme of multiscale transformation

transformation. It can be reversed by the *inverse multiscale transformation* based on the relations (39) ascending from coarser to finer levels. Note, that the multiscale transformation and its inverse require $\mathcal{O}(\#\, I_L)$ operations provided that the modified box wavelets are locally finite, i.e., the number of functions of level j that do not vanish in $\boldsymbol{x} \in \Omega$ is uniformly bounded. This is fulfilled whenever the stencils $\mathcal{L}^e_{j,k}$ are uniformly finite.

Recall that the multiscale transformation has been set up so as to realize a change of basis from a representation of the form (32) for ($j = L$ say) in single scale format into a representation with respect to the multiscale basis $\{\varphi_{0,k}\}_{k\in I_0} \cup \bigcup_{j=0}^{L-1} \{\psi_{j,k,e}\}_{k\in I_j, e\in E^*}$, i.e.,

$$u_L := \sum_{k\in I_L} \hat{u}_{L,k}\,\varphi_{L,k} = \sum_{k\in I_0} \hat{u}_{0,k}\,\varphi_{0,k} + \sum_{j=0}^{L-1}\sum_{k\in I_j}\sum_{e\in E^*} d_{j,k,e}\,\psi_{j,k,e}. \tag{40}$$

Whether small perturbations of the coefficients $d_{j,k,e}$ cause only small perturbations in u with respect to the L_∞-norm or of $\hat{\boldsymbol{u}}_L$ in a discrete L_1-norm relies on the *stability* of the multiscale basis which in turn depends on the existence of a regular primal basis, recall (ii) in Section 3.3.2. This basis is never used in any computation and hence is not needed explicitly. However, it is relevant for a stability analysis. In fact, if the multiscale basis can be shown to be a Riesz basis in L_2 this would imply that small perturbations of the detail coefficients over all levels measured in the Euclidean norm L_2 cause only small variations of u in L_2, or of $\hat{\boldsymbol{u}}_L$ in l_2, and vice versa, recall Section 3.3.2. Analogous statements for the L_∞- or L_1-norm cannot be expected to hold in uniform. One has to be content with scalewise stability estimations. A detailed discussion of these issues goes beyond the scope of this paper and the reader is referred to [24].

3.3.5 Local Grid Refinement

In the previous section we outlined the transformation of an array of cell averages corresponding to a finest uniform discretization level into a new data format. The new format is composed of cell averages on a coarsest discretization level and arrays of details describing the difference information between the cell averages on two successive discretization levels. At this point we assume that the underlying bases are stable. We now take advantage of the new representation exploiting the fact that details are guaranteed to decay at a certain rate depending on the local smoothness of the underlying function. The idea is simply to discard all coefficients $d_{j,k,e}$ in (40) that fall in absolute value below a certain threshold. For this purpose, we introduce the index set

$$\mathcal{D}_{L,\varepsilon} := \{(j,k,e)\ ;\ |d_{j,k,e}| > \varepsilon_j,\ e \in E^*,\ k \in I_j,\ j \in \{0,\ldots,L-1\}\}$$

corresponding to what will be referred to as *significant details*. Here $\varepsilon_j = 2^{j-L}\varepsilon$ is a level–dependent threshold value which is smaller on coarser levels.

It is important to distinguish this index set form an associated index set $\mathcal{G}_{L,\varepsilon}$ which determines an *adaptive grid* and is generated from $\mathcal{D}_{L,\varepsilon}$ as follows. First of all it selects cells $V_{j,k}$ from different levels in the grid hierarchy such that

$$\Omega = \bigcup_{(j,k)\in\mathcal{G}_{L,\varepsilon}} V_{j,k}.$$

The index set $\mathcal{G}_{L,\varepsilon}$ can be formed by traversing through the levels from coarse to fine. If there is a significant detail $d_{j,k,e} \in \mathcal{D}_{L,\varepsilon}$ for some $e \in E^*$ then the cell $V_{j,k}$ is locally refined according to (13), i.e., the index (j,k) is removed from $\mathcal{G}_{L,\varepsilon}$ and the refinement set $\mathcal{M}^0_{j,k}$ is added to $\mathcal{G}_{L,\varepsilon}$. Note, that the index set $\mathcal{G}_{L,\varepsilon}$ is initialized by all indices of the coarsest discretization, i.e., $(0,k) \in \mathcal{G}_{L,\varepsilon}$, $k \in I_0$. It is clear that the adaptive grid reflects local refinements and therefore involves *hanging nodes*, i.e., cell vertices of neighboring cells do not necessarily coincide, see Figure 4.

It is therefore important that the finite volume discretization employed in combination with the multiscale transforms copes well with hanging nodes.

We emphasize that this procedure only works provided that the index set of significant details correspond to a *graded tree*, the levels of neighboring cells differ at most by one. Since the set $\mathcal{D}_{L,\varepsilon}$ is in general not graded, we have to apply in addition a grading procedure to $\mathcal{D}_{L,\varepsilon}$. This will slightly inflate the index set but has so far

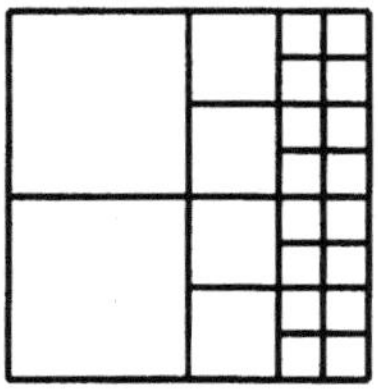

Figure 4 Locally refined grid with hanging nodes

been observed not to spoil the complexity reduction of floating point operations in any significant way. In fact, from the nature of singularities occurring in flow computations one expects the distribution of significant wavelet coefficients to exhibit at least nearly tree structure. If a high level wavelet overlaps a discontinuity this will be seen also by its coarser ancestors in the same region. For details on the grading procedure we refer to [24].

The actual time evolution will refer again to cell averages corresponding to the adaptive grid. Therefore it is important to interrelate the local cell averages and the significant details

$$\hat{\boldsymbol{u}}_{L,\varepsilon} := \{\hat{u}_{j,k}\}_{(j,k)\in\mathcal{G}_{L,\varepsilon}} \quad \boldsymbol{d}_{L,\varepsilon} := \{d_{j,k,e}\}_{(j,k,e)\in\mathcal{D}_{L,\varepsilon}}$$

corresponding to the projection of u_L on the *locally refined spaces* determined by the basis functions

$$\Phi_{L,\varepsilon} := \{\varphi_{j,k}\}_{(j,k)\in\mathcal{G}_{L,\varepsilon}}, \quad \Psi_{L,\varepsilon} := \{\varphi_{j,k}\}_{(j,k)\in I_0} \cup \{\psi_{j,k,e}\}_{(j,k,e)\in\mathcal{D}_{L,\varepsilon}},$$

respectively. Since the bases $\Phi_{L,\varepsilon}$ and $\Psi_{L,\varepsilon}$ span the same space, the projection u_L can equivalently be written as

$$u_L = \sum_{(j,k)\in\mathcal{G}_{L,\varepsilon}} \hat{u}_{j,k}\,\varphi_{j,k} = \sum_{k\in I_0} \hat{u}_{0,k}\,\varphi_{0,k} + \sum_{(j,k,e)\in\mathcal{D}_{L,\varepsilon}} d_{j,k,e}\,\psi_{j,k,e}. \tag{41}$$

The change of basis between the (locally) single–scale and the wavelet representation of u_L is realized by a *local* multiscale transformation analogously to (38) and (39), respectively. This transformation is to be realized with an optimal complexity, i.e., the number of operations should stay proportional to $\#\mathcal{D}_{L,\varepsilon}$. The algorithms of the local transformations are outlined in [24]. Here again, it turns out that feasibility of the algorithms is related to the gradedness of $\mathcal{D}_{L,\varepsilon}$. In particular, it is verified how the grading depends on the number of vanishing moments of the modified box wavelets.

3.4 Adaptive Finite Volume Schemes

We now outline the concept of adaptive finite volume schemes based on the above transformation concept. Of course, it will be crucial that at no stage of the algorithm the full uniform grid of level L is visited. Therefore, when evolving an approximate solution on an adaptive grid at time n to time level $n+1$ one has to guess a possibly economic new mesh which of course may vary when dealing with non–stationary processes. Here it turns out that the construction of the numerical fluxes on a locally refined grid and the *prediction* of the significant details on the new time level are the core ingredients which crucially influence the performance as well as the reliability of the adaptive scheme.

First work on this subject has been reported by Harten [16, 17]. His primary objective, however, has been to accelerate a *given* finite volume scheme on a grid of essentially *uniform* resolution by a *hybrid* flux computation. As a core ingredient he also used a *multiscale decomposition* (although derived differently) similar to that in Section 3.3. It can be utilized to distinguish smooth regions of the flow field from regions with local strong variations in the solution. In particular, the hybrid flux evaluation can be controlled by the decomposition, i.e., expensive upwind discretizations are only applied near discontinuities of the solution. Elsewhere cheaper linear combinations of already computed numerical fluxes on coarser scales are used instead. These correspond to finite difference approximations. In the meantime Harten's originally one–dimensional concept has been extended to multidimensional problems on Cartesian grids [2, 4], curvilinear patches [11] and triangulations [26, 1, 6].

The bottleneck of Harten's strategy is the fact that the *computational complexity*, i.e., the number of floating point operations as well as the memory requirements can at best be reduced by a fixed factor independent of the size of the mesh. Thus computational work and storage requirements still grow at least proportionally to a globally finest uniform grid. For multidimensional applications this is a severe obstruction which is even enhanced by additional requirements concerning varying domains and coupling with structure calculations. Recently, a genuine *adaptive* approach has been presented in [15] and has been investigated in [8]. Here the computational complexity can be kept proportional to the problem–inherent degrees of freedom. A selfcontained account of the adaptive concept for conservation laws can be found in [24].

3.4.1 Construction

Essentially any standard explicit or implicit finite volume scheme for a system of conservation laws (capable of handling hanging nodes) may serve as starting point for the adaptive scheme. Suppose that for a given (highest) discretization level L it can be written in the form

$$\boldsymbol{v}_{L,k}^{n+1} + \theta\,\lambda_{L,k}\,\boldsymbol{B}_{L,k}^{n+1} = \boldsymbol{v}_{L,k}^{n} - (1-\theta)\,\lambda_{L,k}\,\boldsymbol{B}_{L,k}^{n} \tag{42}$$

with the ratio $\lambda_{L,k} := \tau/|V_{L,k}|$ and a parameter $\theta \in [0,1]$. Here a fixed time interval $[0,T]$ is decomposed by a temporal mesh $0 = t_0 < \ldots < t_N = T$. To keep the exposition simple and since we focus here on spatial adaptation the temporal partition is assumed to be uniform, i.e., $t_{n+1} = t_n + \tau$ for all $n = 0,\ldots,N-1$.

This is no constraint imposed by the adaptive strategy. The numerical *flux balance* is defined by

$$\boldsymbol{B}_{L,k}^{n} := \sum_{l} |\Gamma_{k,l}^{L}|\, \boldsymbol{F}_{k,l}^{L,n}$$

approximating the exact flux balance (12) corresponding to the control volume $[t_n, t_{n+1}] \times V_{L,k}$. Here $\Gamma_{k,l}^{L}$ denotes the interface of the cell $V_{L,k}$ to the neighbor cell $V_{L,l}$ and $\boldsymbol{F}_{k,l}^{L,n}$ the corresponding numerical flux, see Figure 5. Note, that the numerical fluxes depend on some cell averages of the *finest* discretization level for reasons to be discussed later. Furthermore the numerical fluxes are assumed to

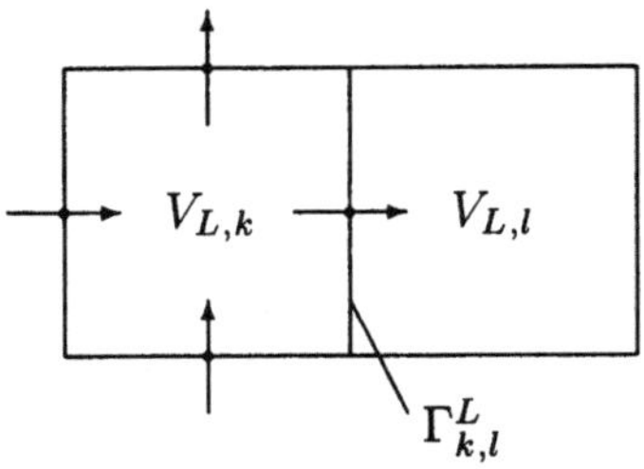

Figure 5 Finite volume scheme

be conservative, i.e.,

$$\boldsymbol{F}_{k,l}^{L,n} = -\boldsymbol{F}_{l,k}^{L,n}. \tag{43}$$

Then, for a given time level n, we introduce the cell averages $\boldsymbol{v}_{j,k}^{n}$ and details $\boldsymbol{d}_{j,k,e}^{n}$ on the coarser scales $j = L-1, \ldots, 0$. In fact, relations

$$\boldsymbol{v}_{j,k}^{n} := \sum_{r \in \mathcal{M}_{j,k}^{0}} m_{r,k}^{j,0}\, \boldsymbol{v}_{j+1,r}^{n}, \quad k \in I_j, \tag{44}$$

$$\boldsymbol{d}_{j,k,e}^{n} := \sum_{r \in \mathcal{M}_{j,k}^{e}} m_{r,k}^{j,e}\, \boldsymbol{v}_{j+1,r}^{n}, \quad k \in I_j,\ e \in E^*, \tag{45}$$

are derived from the two–scale relations for the box function and the modified box wavelet. Applying the multiscale transformation (38) to (42), we obtain discrete evolution equations for the cell averages and the details

$$\boldsymbol{v}_{j,k}^{n+1} + \theta\, \lambda_{j,k}\, \boldsymbol{B}_{j,k}^{n+1} = \boldsymbol{v}_{j,k}^{n} - (1-\theta)\, \lambda_{j,k}\, \boldsymbol{B}_{j,k}^{n}, \tag{46}$$

$$\boldsymbol{d}_{j,k}^{n+1} + \theta\, \lambda_{j,k}\, \boldsymbol{D}_{j,k}^{n+1} = \boldsymbol{d}_{j,k}^{n} - (1-\theta)\, \lambda_{j,k}\, \boldsymbol{D}_{j,k}^{n}, \tag{47}$$

where the flux balances $\boldsymbol{B}_{j,k}^{n}$ and their details $\boldsymbol{D}_{j,k}^{n}$ are determined by

$$\boldsymbol{B}_{j,k}^{n} := \sum_{r \in \mathcal{M}_{j,k}^{0}} \boldsymbol{B}_{j+1,r}^{n} = \sum_{V_{L,r} \subset V_{j,k}} \boldsymbol{B}_{L,r}^{n}, \tag{48}$$

$$\boldsymbol{D}_{j,k}^{n} := \sum_{r \in \mathcal{M}_{j,k}^{e}} \frac{m_{r,k}^{j,e}}{m_{r,k}^{j,e}} \boldsymbol{B}_{j+1,r}^{n}.$$

Here we employed the definition of the filter coefficients $m_{r,k}^{j,0}$ and $\lambda_{j,k} := \tau/|V_{j,k}|$.

So far, we only derived evolution equations for the averages and details corresponding to the *full* grids. The adaptive finite volume scheme is now determined by a selection of evolution equations corresponding to the adaptive grid $\mathcal{G}_{L,\varepsilon}^{n+1}$, i.e.,

$$v_{j,k}^{n+1} + \theta\,\lambda_{j,k}\,\boldsymbol{B}_{j,k}^{n+1} = v_{j,k}^{n} - (1-\theta)\,\lambda_{j,k}\,\boldsymbol{B}_{j,k}^{n}, \quad (j,k) \in \mathcal{G}_{L,\varepsilon}^{n+1}. \tag{49}$$

It remains to outline (i) the efficient computation of the local flux balances $\boldsymbol{B}_{j,k}^{n}$ and (ii) the construction of $\mathcal{D}_{L,\varepsilon}^{n+1}$ by means of $\mathcal{D}_{L,\varepsilon}^{n}$.

3.4.2 Local Flux Evaluation

According to the definition of the flux balances (48) the computation of $\boldsymbol{B}_{j,k}^{n}$ requires *all* flux balances $\boldsymbol{B}_{L,r}^{n}$ corresponding to the cells $V_{L,r} \subset V_{j,k}$ of the *finest* level. However, exploiting the conservation property of the fluxes (43) the right hand side of (48) can be rewritten as

$$\boldsymbol{B}_{j,k}^{n} = \sum_{l} |\Gamma_{k,l}^{j}|\,\boldsymbol{F}_{k,l}^{j,n} \tag{50}$$

with the local numerical fluxes

$$\boldsymbol{F}_{k,l}^{j,n} = \sum_{\Gamma_{k',l'}^{j+1} \subset \Gamma_{k,l}^{j}} \boldsymbol{F}_{k',l'}^{j+1,n} = \sum_{\Gamma_{k',l'}^{L} \subset \Gamma_{k,l}^{j}} \boldsymbol{F}_{k',l'}^{L,n}. \tag{51}$$

This is sketched in Figure 6 for a dyadic grid refinement. According to (48) we

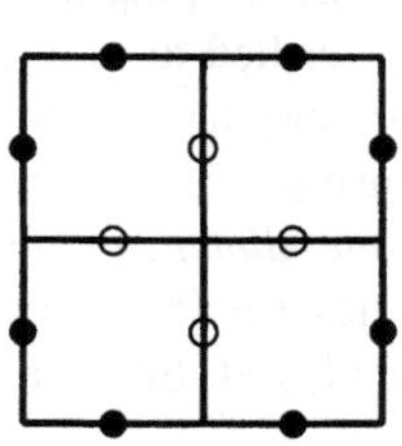

Figure 6 Computation of flux balance

have to compute all fluxes marked by • and ∘. However, the internal fluxes corresponding to ∘ cancel each other out due to the conservation property (43). Thus, an efficient computation of flux balances should be based on (50) rather than (48). Furthermore, we observe that the computation of the numerical fluxes $\boldsymbol{F}_{k',l'}^{L,n}$ requires the computation of cell averages on the *finest* discretization level which have to be provided by means of the inverse two–scale transformation (39). In general, this inflates the complexity by a logarithmic factor depending on the spatial dimension d. In order to treat multidimensional problems a less expensive approximation is desirable. In [24] two alternatives are discussed. The first one is to employ simply cell averages with respect to the nonuniform mesh. However, this requires flux computations for an unstructured grid. Therefore we prefer the following alternative. A local subdivision at places where cells of different levels meet creates *locally* uniform discretizations which eases flux computations and the dicretization of higher order terms. Although these alternatives are less expensive we have to be aware that an error is introduced. As investigations in [8] show, this error might become significant in case of a first order finite volume scheme. However, if the reference scheme is a higher–order accurate scheme employing a high–order reconstruction, then no significant loss in accuracy has been observed.

3.4.3 Prediction

Before the evolution step (49) can be performed, we have to determine the adaptive grid on the *new* time level. Since the corresponding averages, respectively details are not yet available, we have to *predict* all details that may become significant due to the evolution by means of the details on the *old* time level. In order to guarantee the adaptive scheme to be *reliable* in the sense that no significant future feature of the solution is missed, the prediction set $\tilde{\mathcal{D}}_\varepsilon^{n+1}$ has to satisfy

$$\mathcal{D}_{L,\varepsilon}^n \cup \mathcal{D}_{L,\varepsilon}^{n+1} \subset \tilde{\mathcal{D}}_{L,\varepsilon}^{n+1}, \tag{52}$$

where, of course $\mathcal{D}_{L,\varepsilon}^{n+1}$ is not known yet. In [17] Harten suggested a heuristic approach taking into account that (i) details in a local neighborhood of a significant detail may also become significant within one time step due to the finite speed of propagation and (ii) gradients may become steeper causing significant details on a higher refinement level due to the developing of discontinuities. So far Harten's approach could not be rigorously verified to satisfy (52). However, a slight modification of Harten's prediction strategy has recently been shown to lead to a reliable prediction strategy in the sense of (52), at least for a certain class of *explicit* finite volume schemes applied to *one–dimensional scalar* conservation laws on *uniform dyadic grids* as base hierarchies, see [8]. Here the explicit evolution equations ($\theta = 0$) for the details (47) are employed. So far an analogue for implicit schemes is not available yet.

3.5 Error Analysis

The above adaptation strategy based on compressing the arrays of detail coefficients is essentially a perturbation technique. Therefore the objective of the proposed adaptive scheme is to reduce for a given finite volume scheme computational complexity while preserving up to a fixed constant factor the accuracy provided by the reference scheme corresponding to the uniform finest discretization level. In order to make this precise we introduce the averages $\hat{\boldsymbol{u}}_L^n$ of the exact solution, the averages $\boldsymbol{v}_L^n$ determined by the finite volume scheme and the averages $\overline{\boldsymbol{v}}_L^n$ of the adaptive scheme. We stress, that the sequence $\overline{\boldsymbol{v}}_L^n$ is *not* generated by the adaptive scheme on the full uniform highest level L. In order to facilitate the intended accuracy comparisons these data are conceptually generated from the adaptive array $(v^n_{j,k})_{(j,k)\in\mathcal{G}^n_{L,\varepsilon}}$ by means of the inverse multiscale transformation (39) after assigning the value zero to the non–significant details.

An ideal strategy would be to prescribe an error tolerance *tol* and determine during the computation a possibly small number of refinement levels L such that the accuracy requirement is met, i.e.,

$$\|\hat{u}^n_L - \overline{v}^n_L\| \leq tol$$

for possibly small L. Here the error is measured in the weighted l_1–metric

$$\|u_L\| := \sum_{k\in I_L} |V_{L,k}|\,|u_{L,k}|,$$

i.e., the L_1–norm of a piecewise constant scalar function, which is usually applied in the error analysis of finite volume schemes. In order to estimate the discretization error we split the error into two parts corresponding to the *discretization error* $\tau^n_L := \hat{u}^n_L - v^n_L$ of the reference finite volume scheme and the *perturbation error* $e^n_L := v^n_L - \overline{v}^n_L$, i.e., it would suffice to make sure that the sum of the unperturbed discretization error and the perturbation stays below the tolerance

$$\|\hat{u}^n_L - \overline{v}^n_L\| \leq \|\tau^n_L\| + \|e^n_L\| \leq tol. \tag{53}$$

Note, that e^n_L describes indeed a perturbation of the reference scheme because the adaptive scheme coincides with the reference scheme when the threshold value ε is zero. At the current stage we are not able to extract a reliable value of L from the computation yet. Therefore we now assume that there is an a priori error estimate of the discretization error such that

$$\|\tau^n_L\| \leq C\,2^{-\alpha L}. \tag{54}$$

Here the cell diameters are assumed to be proportional to 2^{-L} and α denotes the convergence order of the reference scheme. Then the number of necessary refinement levels is of course determined by $2^{-\alpha L} \sim tol$. In order to preserve the accuracy of the reference scheme we now may admit a perturbation error which is proportional to the discretization error.

So far, this ideal concept can only be rigorously founded for scalar conservation laws because rigorous error estimates are only available for this class of problems, see [5, 19]. In this case it remains to estimate the perturbation error. To this end, we first note that in each time step a threshold error is introduced that accumulates in time. Therefore the best we can hope for is an estimate of the form

$$\|e^n_L\| \leq C\,n\,\varepsilon \tag{55}$$

where the constant C is independent of L, n, τ and ε. Since the threshold error may in addition be amplified in each evolution step, the existence of such a constant is not obvious. However, if the multiscale transformation is stable, the explicit scheme is l_1–*contractive* and if the prediction is reliable then one can show that the constant C is indeed independent of the number of refinement levels L, the threshold value ε and the number of time steps n, [8]. From (54) and (55) we then deduce that the perturbation error stays proportional to the discretization error provided the threshold value is chosen as

$$\varepsilon \sim 2^{-(1+\alpha)\,L}. \tag{56}$$

One should be aware that the above perturbation analysis, based on the assumption that errors accumulate, may be to pessimistic. Already the studies in [8] concerning the scalar case where the true error can be assessed indicate that the actual error is overestimated. However, to play save this is not incorporated in the numerical realizations.

3.6 Numerical Results

In the previous sections we have outlined a new concept for the construction of adaptive finite volume schemes employing multiscale techniques. For scalar conservation laws this could be verified to be reliable in the sense of (53). In [8] parameter studies have been presented for scalar one–dimensional problems which confirm the analytical results. In the sequel, we verify numerically that the adaptive concept works also reliably for *systems of conservation laws*, i.e., (i) all kinds of singularities, e.g., discontinuities and kinks, corresponding to physically relevant effects are detected by means of the multiscale decomposition and adequately resolved; (ii) the computational complexity of the adaptive scheme with respect to computational costs and memory requirements is proportional to the number of significant details and (iii) the perturbation error is stable in the sense of (55). For this purpose, we present several computations for the two–dimensional Euler equations for perfect air, see also [24]. These have been carried out on PC's with a 600 MHz processor (Pentium III). Further numerical results for real life problems treated in the collaborative research center SFB 401 will be presented and discussed later in Section 6. They have been produced by the solver QUADFLOW incorporating the above adaptive concepts.

Below we consider a two–dimensional Riemann problem where the initial configuration is determined by four states corresponding to the four quadrants of the coordinate system, see Figure 1. Away from the origin of the coordinate system, the solution exhibits a one–dimensional wave pattern consisting of a rarefaction wave, a contact surface and a shock wave. Close to the origin the different one–dimensional waves interact forming a genuinely two–dimensional wave pattern. Note, that the solution is non–stationary.

The reference finite volume scheme for the problem solved here is an essentially non–oscillatory (ENO) scheme characterized by a one–dimensional second order accurate reconstruction technique, and the numerical fluxes are determined by Roe's approximate Riemann solver. For more details on the scheme we refer to [23].

Table 1 2D Riemann problem: Initial configuration

II	*I*
I	*II*

	State I	State II
ϱ	0.125	1.000
u_x	1.000	0.000
u_y	1.000	0.000
p	0.100	1.000

The computational domain $\Omega = [-0.4, 0.4]^2$ is discretized by a uniform Cartesian grid with spatial step size $h_L = 0.04 \times 2^{-L}$ where we perform $n_L = 150 \times 2^{L-2}$ time steps on the time interval $[0, 0.15]$ with a temporal step size $\tau_L = 0.001 \times 2^{2-L}$. Note that the discretization is chosen such that the CFL condition holds on the *finest* discretization level.

The multiscale analysis is based on the modified box wavelets, see Section 3.3.3, where the degree of vanishing moments is chosen as $M = 3$. For details on the choice of the stencils $\mathcal{L}^e_{j,k}$ in case of Cartesian grids see [24].

In particular, the thresholding procedure is slightly modified for vector–valued functions. Here we take into account that the conservative quantities may vary by several orders of magnitude. In principle, we could choose a specific threshold value for each quantity. In practice, a different strategy turned out to be preferable. To this end, we introduce for a vector valued detail $\boldsymbol{d}_{j,k,e}$ its counterpart $\boldsymbol{d}^*_{j,k,e}$ by

$$(\boldsymbol{d}^*_{j,k,e})_i := \begin{cases} (\boldsymbol{d}_{j,k,e})_i & , \quad u_i^\infty < tol \\ \frac{(\boldsymbol{d}^*_{j,k,e})_i}{u_i^\infty} & , \quad \text{elsewhere} \end{cases}$$

where each component is scaled by the maximum

$$u_i^\infty := \max_{(j,k)\in\mathcal{G}_{L,\varepsilon}} |(\hat{u}_{j,k})_i|, \quad i = 1, \ldots, m.$$

Here *tol* is a machine depending tolerance to avoid division by 0. Alternatively, we could have rewritten the equations (10) in dimensionless form. Then we call a detail significant, if there is an $e \in E^*$ such that $|(\boldsymbol{d}^*_{j,k,e})_i| > \varepsilon_j$ for at least one component $i = 1, \ldots, m$.

Finally, we have to specify the local flux evaluation. Here we do not apply the exact computation according to (51) but the cheaper alternative based on the evaluation of the numerical fluxes with respect to a locally structured grid because the reference numerical flux is only defined on a *structured* grid.

For the test configuration under consideration we have carried out a parameter study with respect to an increasing number of refinement levels L where the threshold value is kept fixed by $\varepsilon = 0.001$ for all computations. This is not quite in agreement with the ideal strategy outlined in Section 3.5 but we are lacking reliable a-priori estimates in the present cases and focus therefore on the *stability* of the scheme, a point that will be taken up again later. In Figure 7 the density is presented for $t = 0.15$ [s]. The corresponding adaptive grid is shown in Figure 8. These figures reflect the computational results for the highest number of refinement levels listed in the Table 2. We observe that the higher resolution levels are only accessed near discontinuities. This verifies that the adaptation criterion based on significant details is able to detect the relevant physical effects in the flow field and to resolve them adequately.

Table 2 2D Riemann problem: Parameter study

L	N_L	$N_{\mathcal{G}}$ %	C_{MS} [min]	Mem [MB]	S_{MS}	$\lVert e_L \rVert_*$
3	25600	58	4.53E+0	16	0.98	1.1E−3
4	102400	34	2.33E+1	39	1.62	1.2E−3
5	409600	19	1.12E+2	90	2.67	1.2E−3
6	1638400	10	4.99E+2	186	4.79	
7	6553600	5	1.98E+3	337	9.67	

Furthermore we are interested in the computational complexity of our scheme with regard to computation time and memory that is documented in Table 2. First of all, we consider the number $N_{\mathcal{G}} := \#\,\mathcal{G}_{L,\varepsilon}/N_L$ representing the number of cells corresponding to the largest adaptive grid determined during the computation relative to the number $N_L = 400 \times 4^L$ of all cells corresponding to the full grid on level L. We note that the relative number of cells reduces exponentially almost by a factor 2. In fact, the total number of cells increases by a factor 2 for the adaptive grid instead of 4 for the uniform grid. This is also reflected by the memory size Mem which corresponds to the peak of memory that has been allocated in one time step. Comparing the numbers Mem for the different computations corresponding to an increasing number of refinement levels, we conclude that memory size is increasing accordingly by a factor 2. Analogously, this is reflected in the computation times C_{MS} used by the adaptive scheme and C_{FVS} of the reference scheme on the full grid of level L. The computation time C_{MS} increases with each additional refinement level because the number of time steps is doubled according to the CFL condition for the finest level. Since the number of cells in the adaptive grid increases in addition by a factor 2, the computation time C_{MS} increases by a factor 4 instead of 8 for the reference scheme. This results in an exponential behavior of the speedup rates $S_{MS} := C_{FVS}/C_{MS}$, i.e., the computation time becomes significantly smaller for the adaptive scheme in comparison to the computation of the reference scheme on the full grid. These observations confirm that the number of floating point operations as well as the memory size of the resulting scheme stays proportional to the number of cells in the adaptive grid.

The realization of such optimal computational complexity behavior depends heavily on appropriate data structures. In our implementation we use hash tables. For details on the design and the implementation of the data structures see [25, 24].

Of course, the objective of the adaptive strategy is to reduce the computational complexity while maintaining (up to a constant) the accuracy in comparison with the reference scheme on a uniform grid. Since for systems of conservation laws bounds on the global discretization error are not available we focus in the following just on the stability of the perturbation error in the sense of (55). That is, instead of balancing discretization and perturbation error we fix the threshold value ε and explore the dependence of the perturbation error on the level L, the number of time steps n and the temporal step size τ. For this purpose, the perturbation error is also listed in Table 2 as far as the computation for the reference scheme on the uniform discretization level could be performed on the available computers. Here the perturbation error $\boldsymbol{e}_L$ is measured in the weighted l_1–metric where each conservative quantity is scaled by its global maximum in the computational domain, i.e.,

$$\|\boldsymbol{e}_L\|_* := \max_{i=1,\dots,m} \|\boldsymbol{e}_{L,i}\| / \|\overline{\boldsymbol{v}}_{L,i}\|_\infty$$

where the vectors $\boldsymbol{e}_{L,i} := ((\boldsymbol{e}_{L,k})_i)_{k\in I_L} = ((\overline{\boldsymbol{v}}_{L,k})_i) - (\boldsymbol{v}_{L,k})_i)_{k\in I_L}$ and $\overline{\boldsymbol{v}}_{L,i} := ((\overline{\boldsymbol{v}}_{L,k})_i)_{k\in I_L}$ correspond to the perturbation error and the results of the adaptive scheme for a single conservative quantity $i = 1,\dots,m$. Recall that the thresholding is also applied with respect to the scaled details. For instance, see Figure 9 where the contours of the perturbation error are shown for the density. We observe that the error is proportional to the threshold value ε, in particular, it does not increase with increasing number of refinement levels L.

We conclude the discussion on the parameter studies with a remark on the "ideal" computation as discussed in Section 3.5. Whenever the perturbation error is smaller than the discretization error, we waste performance in the sense of computation time and memory size, because the discretization error is still dominating. Conversely, we loose accuracy if the perturbation error is larger than the discretization error. Therefore the ideal computation of the parameter study corresponds to the refinement level where both the perturbation error and the discretization error are balanced. Since in our parameter studies the threshold value ε is fixed, we cannot directly conclude on the discretization error. If we *assume* that the discretization error for a uniform grid of level L behaves like $2^{-\alpha L}$ we see from Table 2 at which level the recorded perturbation error is of comparable size. This gives an impression of the interrelation between the overall accuracy and the amount of computational work. However, the computations on the full uniform grid of level L can only be realized for a small number of refinement levels L due to the huge amount of memory needed. This constitutes a significant advantage of adaptive schemes over non–adaptive ones on uniform meshes, i.e., under relevant accuracy requirements simulations become computationally tractable that would not be possible to perform with the aid of quasi–uniform meshes.

3.7 Summary of Requirements

We conclude the discussion of adaptation strategies with briefly summarizing the main requirements on the finite volume discretizations and the corresponding mesh generators that are put forward by the above concepts.

(i) The multiscale setting as outlined in Section 3.3.1 is based on a *grid hierarchy* of *nested grids*. Although we never access to the *full* grids of this hierarchy the grid generator has to provide an efficient representation of this hierarchy depending only on a small number of parameters in comparison to the full grids. Moreover, this representation has to allow for a *fast computation of higher order moments* with regard to the construction of higher order biorthogonal systems, see Section 3.3.3.

(ii) The efficiency of the flux computation crucially depends on the assumption that the numerical fluxes of the underlying finite volume discretization are *conservative*, see Section 3.4.2. In addition, the computation of the numerical flux balances can be further improved when computing the incorporated numerical fluxes by means of the averages corresponding to the adaptive grid instead of the averages on highest resolution level. This introduces an error that can be compensated when using a *higher order reconstruction* in the finite volume scheme.

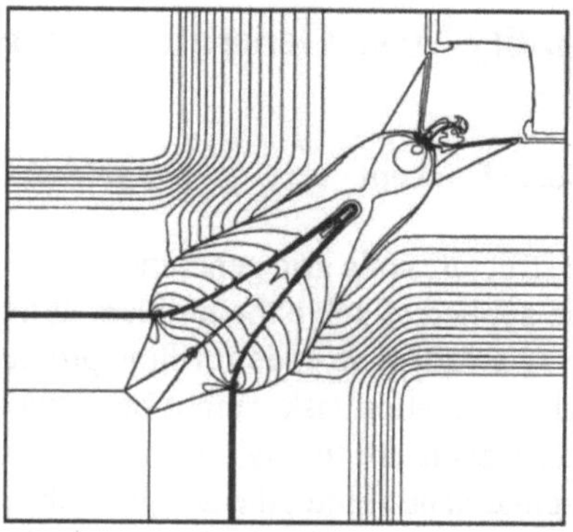

Figure 7 2D Riemann problem: Density contours ($L = 7$)

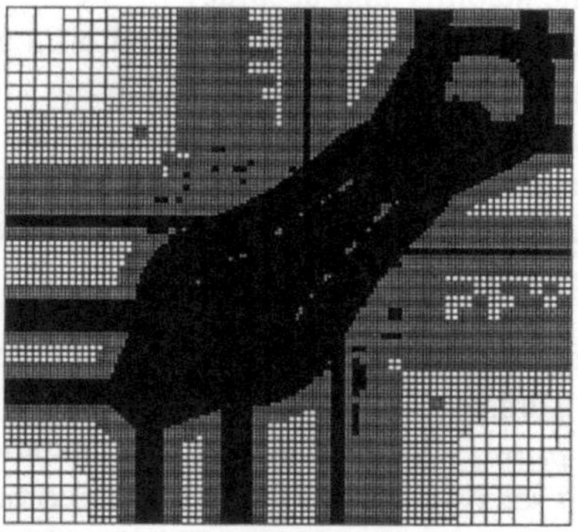

Figure 8 2D Riemann problem: Adaptive grid ($L = 7$)

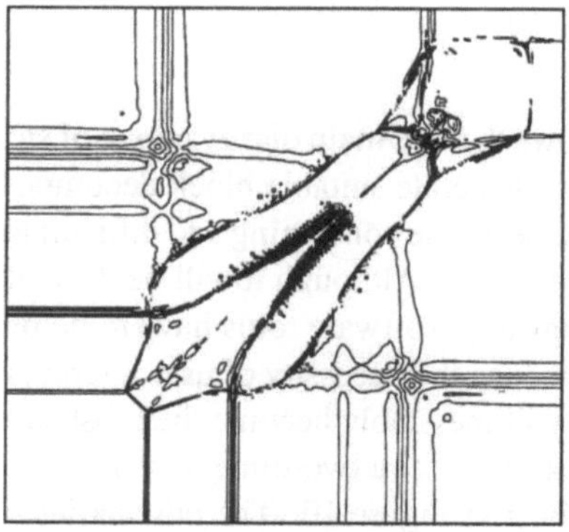

Figure 9 2D Riemann problem: Contours of perturbation error in density ($L = 5$)

4 High Quality Mesh Generation Using B-Splines

4.1 Objectives and Basic Concept

The algorithms and data structures of a grid generation code have to reflect the needs of the numerical scheme applied to solve the problem at hand. The QUADFLOW project aims at developing an adaptive code for flow problems in time varying geometries. Hence the grid generation task cannot be viewed as preprocessing, because the grid generation aspects are tightly connected to the solution process due to transformations and refinements. Here, the design of the grid generation concept used within QUADFLOW is motivated by the following objectives.

(i) With regard to the stability and accuracy of the flow calculations the usual grid quality measures such as smoothness and orthogonality have to be met. It is widely accepted that hexahedral based meshes are preferable for the discretization of viscous flows, because they facilitate best the generation of boundary fitted anisotropic meshes. In particular this is important for the resolution of boundary layers. Thus, *block structured*, *boundary conforming* grids have been employed.

(ii) The grid generator has to fulfill the requirements posed by the multiscale technique, which are summarized in Section 3.7. Due to the dynamic adaptation it is not known a priori, where a fine scale discretization is needed in the flow field. Therefore it seems useful to separate the geometric aspect of grid generation from the discretization aspect. This is achieved by representing the logically Cartesian grids by their analytic counterparts, namely, by parametric mappings from the unit square or unit cube to the physical domain, see Figure 11.

Common experience shows that the main disadvantage of structured grid generation consists in the problem to generate suitable block decompositions around complex geometries. This tends to be a time consuming and difficult task and usually requires human expertise and interaction. Although it will be difficult or even impossible to automate this work completely, software tools have to be provided, that are simple and intuitive to use and reduce the necessity of user interaction as much as possible. Otherwise this problem will inevitably become the most severe obstacle to realistic applications. A typical example for a two-dimensional block structured grid as used in the current work is given in Figure 10. The boundaries of the blocks have been defined interactively.

In the grid generation community a lot of work has been devoted to exploiting the excellent abilities of B-Splines and existing CAD programs, see for example [29]. Many user interfaces of grid generation programs provide B-Splines or NURBS as tools for modeling boundary curves or surfaces. In the present approach we go

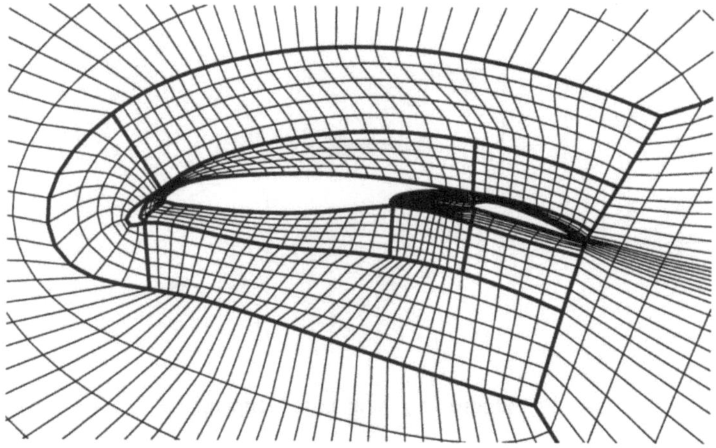

Figure 10 Block decomposition around high lift configuration

even one step further and aim at representing the blocks as independent curvilinear coordinate systems using B-Spline tensor products.

In the following we describe and analyze the above concept in more detail and present some tools that have been developed to realize this concept.

4.2 Representation of Grids as Parametric Mappings

The discussion in the previous section indicates that the grid generation has to meet the following requirements.

(R1) *Boundary conforming* grids which are appropriate for finite volume calculations of flows based on the Euler and Navier-Stokes equations, have to be generated.

(R2) A hierarchy of nested grids for the multiscale algorithm has to be provided.

(R3) The local adaptation of the grid during the solution process has to be efficient.

(R4) The transformation of the grids due to moving boundaries has to be cheap with respect to computational time.

In the following, it will be shown that these requirements suggest to represent grids as parametric mappings and to realize these mappings using B-Splines.

Requirement (R2) states that each cell of a coarse grid must be the union of its subcells. This is indeed a severe constraint, because it implies that the geometric approximation of the physical domain is determined completely by the shape of the coarse grid. This conflicts with the demand, that the shape of the body must be

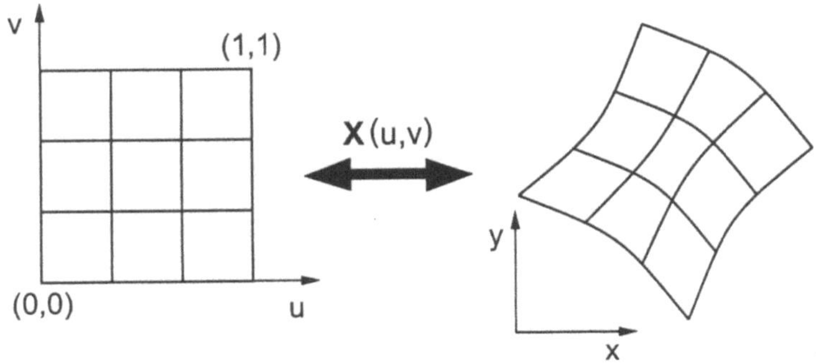

Figure 11 Parametric Mappings

represented precisely. Indeed, one cannot define the coarse cells simply as quadrilaterals in 2D respectively hexahedra in 3D connecting its corner points with straight lines, because this would result in a poor approximation of the boundary curve or surface. Another method would be to calculate the fine grid cells first and to define the coarse grid cells as unions of its subcells. But this would lead to complicated non-smooth shapes of the coarse grid cells, what makes the proper computation of fluxes across the cell interfaces more expensive, and requires the calculation of a refined grid in a preprocessing step. This would in turn contradict the inherent philosophy of adaptive codes, where calculations are initialized on the coarse grid and the grid is refined only where necessary. However, representing grids as curvilinear coordinates allows to define the cells as images of the corresponding cells in computational space. Then grid refinement is easily performed by evaluating the grid function. This method has the following three advantages:

(i) The approximation of the contour does not depend on the refinement level, but only on the quality of the geometric model.

(ii) The grid generation process becomes completely independent of the discretization parameters, for example the number of grid points in the coordinate directions, as the flow solver receives its data simply by evaluating the grid function, see Figure 11.

(iii) From the same representation, grids suitable for both Euler and Navier-Stokes flows can be computed, because the latter only requires the application of an appropriate blending function to resolve the boundary layer.

Note, that the grid cells now have curved edges. This complicates the computation of some geometric quantities, for instance cell volumes, see Section 4.4.5.

Since in realistic problems the geometry will not be given in terms of elementary analytic mappings, such as Cartesian or polar coordinates, one has to approximate given free form geometries. B-Splines seem to be a very appropriate tool for this task, because

(i) they posses excellent approximation properties,

(ii) modeling with B-Splines is intuitive and fast,

(iii) evaluation of B-Splines is fast and numerically stable.

In particular, the first property is important, because it guarantees that one has to compute and store only few parameters to represent a grid with sufficient accuracy. This is of interest with respect to requirement (R4), because one has only to modify these parameters, when the mesh has to be frequently updated in time.

Of course, it is also important that the well established grid generation methods can be integrated into our concept. Indeed, due to the development of fast interpolation and approximation algorithms, the B-Spline representation can be used as post-processing to existing grid generating tools. We feel though, that such a point of view is too restrictive. In fact, the present approach makes use of the genuine advantages of the parametric B-Spline representation within a large variety of algorithms.

A possible alternative are NURBS ("Non-Uniform Rational B-Splines"), because they are even more flexible and are capable of representing typical geometric objects as for example circles and ellipses exactly. But the evaluation of the derivatives of a rational function is rather expensive. This can become a time critical aspect in algorithms that depend on geometrical parameters as for example the curvature of a curve or surface. Therefore we restrict ourselves to integral B-Splines.

4.3 Block Decomposition

The crucial point in structured grid generation is the definition of the block structure. The shape of the blocks is often decisive for the quality of the grid and the calculations performed on it. For the QUADFLOW grid generator the following aspects are of special interest:

(B1) In the vicinity of body contours high quality boundary conforming grids are required in order to resolve the boundary layers.

(B2) In the far-field as many blocks as possible should be Cartesian, because this will enhance the performance of the multiscale algorithm significantly. In fact, the mask coefficients in the multiscale transformation can then be derived by dilatation and translation of corresponding quantities for a single reference volume.

(B3) There should be as few topological constraints on the decomposition as possible in order to support the automation of the block decomposition.

With regard to the last point, we permit geometrically non-conforming block partitions, i.e. faces of adjacent blocks need not match, and do not impose any continuity conditions across the block borders. Here of course, we exploit the ability of the flow solver to handle hanging nodes.

For the description of the grid topology we have adopted the connectivity hierarchy from blocks to faces to edges to vertices proposed in [48] to the case of parametric mappings. In this concept the geometric description of all topological elements, namely, the blocks, faces, edges and vertices is stored separately. Blocks are viewed as parametric mappings of the unit cube, faces as mappings of the unit square and edges as mappings of the unit interval. In addition we have to store the following *connectivity relations*:

(i) for each block by which six faces it is bordered,

(ii) for each face by which four edges it is bordered, and

(iii) for each edge by which two vertices it is spanned.

For two-dimensional grids we just omit the first kind of relations. Nonconforming block interfaces are realized by allowing faces and edges to be divided into subfaces respectively subedges. The information about the relations between faces and their subfaces respectively edges and their subedges has also to be stored. From this four types of connectivity relations all other topological information can be derived.

To fulfill requirements (B1), we have implemented methods that automatically generate smooth offset curves around the body – see below in Section 4.6. These offsets are curvature dependent and can roughly be interpreted in the context of hyperbolic grid generation. They are generated in a first step, before the rest of the block decomposition is defined. This strategy is illustrated in Figure 12.

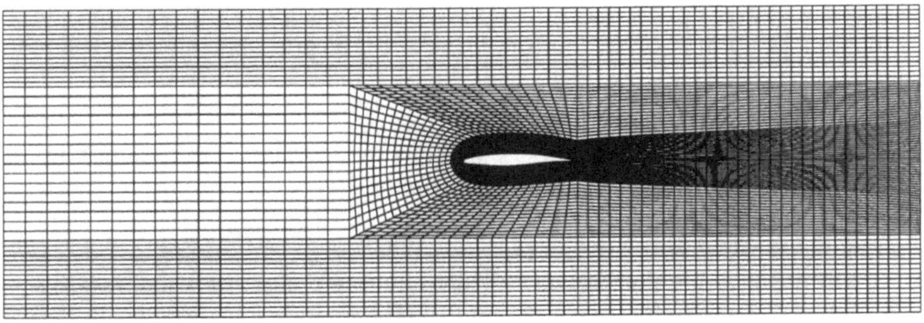

Figure 12 Grid in 12 blocks around reference configuration in a channel

So far only a traditional, user interactive concept of block decomposition has been realized. By means of an existing CAD program [32, 33], which provides the tools based on the B-Spline techniques introduced in Section 4.4, the curves representing the body contour, the offset curves and the block interfaces are defined interactively. These curves serve as input for the grid generation code, that applies the algebraic and/or elliptic methods described in Section 4.5 to the individual blocks. The integration of block-decomposition and grid generation in a common user interface and the automation of these tasks is work in progress.

4.4 B-Spline Basics

In this section we summarize the basic properties of B-Splines needed in this paper. We start with B-Spline functions. Then we show how to build curves and planar grids, surfaces and volume grids. After that we turn to the essential interpolation and approximation algorithms. Details and proofs can be found in many textbooks about numerical analysis and computer graphics, e.g. [34], [38], [43]. In CAD (Computer Aided Design) usually curves and surfaces are treated. It is easy but technical to extend these techniques to volume grids as needed in the 3D-case. For this purpose we have developed fast algorithms for interpolation and approximation. They make extensive use of the tensor-product structure of the B-Splines, leading to an enormous reduction of computation time.

4.4.1 B-Spline Functions

Let $T = (t_0, t_1, \ldots, t_m)$ be a nondecreasing and non-stationary sequence of real numbers, i.e. $t_i \le t_{i+1}, i = 0, 1, \ldots, m-1$. and $t_i < t_{i+p}, i = 0, 1, \ldots, m-p$. T is called "knot vector". Then for $i = 0, \ldots, m-p$ the B-Spline functions $N_{i,p,T}(t) = N_{i,p}(t)$ of order p are recursively defined as

$$N_{i,1}(t) := \chi_{[t_i,t_{i+1})}(t) = \begin{cases} 1 & \text{if } t_i \le t < t_{i+1} \\ 0 & \text{otherwise} \end{cases}, \tag{57}$$

$$N_{i,p}(t) := \frac{t - t_i}{t_{i+p-1} - t_i} N_{i,p-1}(t) + \frac{t_{i+p} - t}{t_{i+p} - t_{i+1}} N_{i+1,p-1}(t). \tag{58}$$

They are piecewise polynomials of degree $p-1$, have compact support $[t_i, t_{i+p}]$, are nonnegative and together form a partition of unity.

In the case that T contains multiple knots equation (58) can yield the quotient $0/0$; we define this quotient to be zero. B-Spline functions of order p are C^{p-l-1}-continuous at a knot of multiplicity l.

A p-th order B-Spline curve is now defined by

$$\boldsymbol{x}(t) = \sum_{i=0}^{m-p} \boldsymbol{p}_i N_{i,p}(t). \tag{59}$$

The $\boldsymbol{p}_i$ are called *control points* or *de Boor points*. They form in ascending order the control polygon. Evaluation of the curve can be performed very efficiently applying the *de Boor algorithm* which is a recursion formula derived from equation (58).

Geometric modeling with B-Splines is very intuitive, because the shape of the control polygon roughly represents the shape of the B-Spline curve. For planar curves the point $\boldsymbol{x}(t)$, with $t_i < t < t_{i+1}$ always lies in the convex hull of the control points $\boldsymbol{p}_{i-p+1}, \ldots, \boldsymbol{p}_i$ and its position is only influenced by these p control points. Accordingly, moving the control point $\boldsymbol{p}_i$ changes $\boldsymbol{x}(t)$ only in the interval $[t_i, t_{i+p}]$. This is especially convenient for modeling curves at a graphics terminal, because changing the position of one control point has only a local effect on the curve.

In addition B-Splines possess the following *variation diminishing property:* A planar B-Spline curve crosses any straight line at most as many times as its control polygon does. From this it follows for example, that a convex control polygon will necessarily define a convex spline curve.

It is often useful to insert additional control points to an existing B-Spline curve without modifying the shape of the curve itself. This is easily accomplished by *knot insertion*. Successive knot insertion causes the control polygon to converge towards the curve.

In practice we concentrate on cubic B-Spline curves (order $p = 4$) and therefore write in the following $N_i(t)$ instead of $N_{i,4}(t)$. In addition, we use only knot vectors that start and end with p equal knots. In this case the spline function is defined on the interval $[t_{p-1}, t_{m-p+1}] =: [a, b]$. This has the advantage that the first and the last point of the curve coincide with the first and last polygon vertices. Furthermore, the slope of the B-Spline curve at the first and last polygon vertices is equal to the slope of the first and last polygon spans.

Planar grids and surfaces are represented as B-Spline tensor products, taking a net $\boldsymbol{p}_{i,j}$ of control points, two knot vectors U, V and building products of the univariate B-Spline functions:

$$\boldsymbol{x}(u, v) = \sum_{i=0}^{m-p} \sum_{j=0}^{n-q} \boldsymbol{p}_{i,j}\, N_{i,p,U}(u)\, N_{j,q,V}(v), \tag{60}$$

see Figure 13. The subscripts U and V are usually omitted. Bearing this simplification in mind, one defines volume grids in a similar way:

$$\boldsymbol{x}(u, v, w) = \sum_{i,j,k} \boldsymbol{p}_{i,j,k}\, N_{i,p}(u)\, N_{j,q}(v)\, N_{k,r}(w). \tag{61}$$

The derivative $\boldsymbol{x}'(t)$ of a p-th order B-Spline curve is again a B-Spline curve but of order $p - 1$. For surfaces and volumes partial differentiation reduces the order for corresponding *directions*.

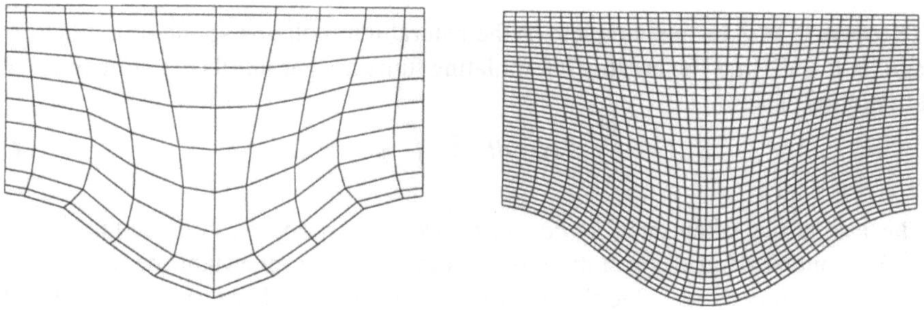

Figure 13 Control Points and Evaluation of Grid Function

4.4.2 Interpolation of Curves

Often the geometric configuration is not given as a spline representation but only as sequences of points. In this case the first task is to transform this point-description into a spline representation via interpolation or approximation.

In the case of cubic B-Spline interpolation we search a curve

$$\boldsymbol{x}(t) = \sum_{i=0}^{m-4} \boldsymbol{p}_i N_i(t) \tag{62}$$

that interpolates the given curve at its knots

$$\boldsymbol{x}(t_{i+3}) = \boldsymbol{x}_i,\ i = 0, 1, \ldots, n := m - 2p + 2 = m - 6. \tag{63}$$

Here we assume, that an appropriate data adapted knot vector with p-fold knots at its ends and no multiple interior knots has been priorly determined. For suitable strategies see for example [38]. So there are $n + 1$ conditions for the $n + 3$ control points $\boldsymbol{p}_i$. Thus we have to pose two more conditions. Usually one prescribes the derivatives at the beginning and the end of the curve ($\boldsymbol{x}'(a) = \boldsymbol{v}_0$, $\boldsymbol{x}'(b) = \boldsymbol{v}_n$) ("complete spline interpolation") or if such data is not available demands vanishing curvature at the beginning and the end of the curve ($\boldsymbol{x}''(a) = \boldsymbol{x}''(b) = 0$). ("natural spline interpolation"). In any case it follows from the theorem of Schoenberg and Whitney that the interpolation problem has a unique solution. Because of the locality of the B-Spline basis the interpolation problem leads to a tridiagonal matrix, that is totally positive due to a result of Karlin and therefore can be solved effectively by Gauss elimination without pivoting.

The following theorems, formulated for scalar functions, give additional reasons for the popularity of the complete spline interpolation [42].

Theorem: Let s be the complete interpolation spline of a function $f \in C^4[a, b]$ with respect to the knots t_i with $h := max_i|t_{i+1} - t_i|$. Then

$$||f - s||_\infty \le \frac{5}{384} h^4 ||f^{(4)}||_\infty \tag{64}$$

Theorem: Let s be the complete or the natural interpolation spline of a function f and let $y \in C^2[a,b]$ be another interpolating function that fulfills the same boundary conditions. Then

$$\int_a^b s''(t)^2 \, dt \leq \int_a^b y''(t)^2 \, dt. \tag{65}$$

The last result, often cited as *best approximation property*, is of certain interest in the context of grid generation as smoothness is an important measure of grid quality. Of course, its value should not be overestimated, because it does not make any statement about the geometric shape of the interpolation curve.

4.4.3 Approximation of curves

Indeed, despite these results, it turns out that interpolation generally is not a sufficient tool for generating suitable representations of boundaries

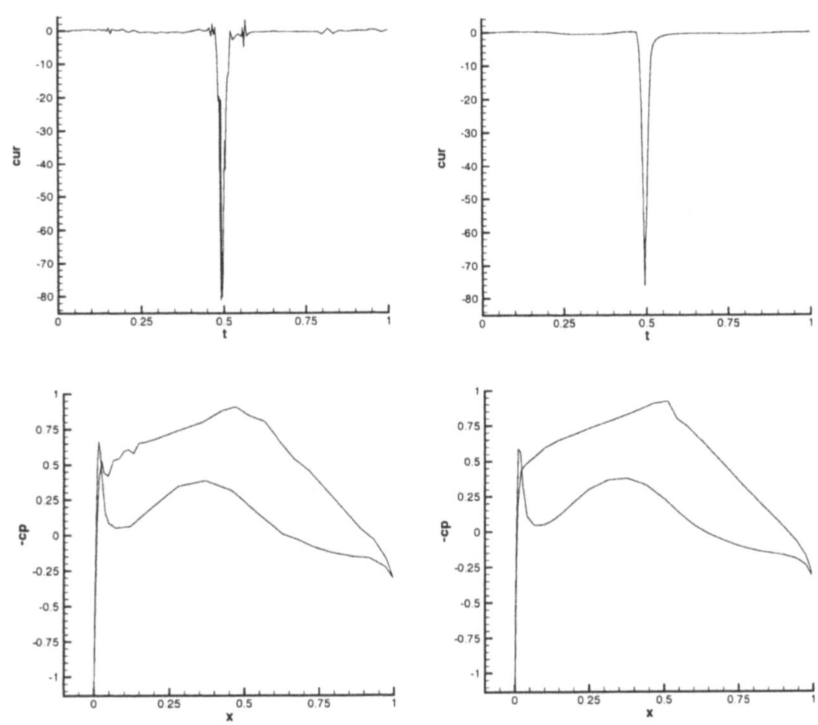

Figure 14 Influence of boundary-approximations

The top graphs of Figure 14 show curvature plots of two B-Spline representations of the same air-foil, the SFB 401 cruise configuration. The first one was interpolated from a set of 193 discrete points. Although there were no deviations of these points from the original description of the configuration the interpolated spline shows oscillations in the curvature. This effect is typical and may be explained by the fact, that it is not easy to find an appropriate knot vector, and with it a good parameterization of the spline corresponding to the distribution of the given data. This effect can be reduced by approximating more points $\boldsymbol{x}_i$ than control points $\boldsymbol{p}_j$. The second representation was produced by approximation of an enlarged data set of 2000 points with 200 control points. To estimate the influence on the flow solution in the bottom row the pressure distribution on the air-foil calculated by QUADFLOW is plotted.

This example shows, that an adaptive code will – in contrast to a method on a fixed grid – react sensibly to even small scale inaccuracies in the description of the geometry. In order to avoid the necessity to work on large numbers of control-points and large data-sets, we implemented fairing methods based on the algorithms given in [40],[37] and [36]. Indeed using these methods, the above mentioned profile can be precisely and smoothly approximated with less than 30 control points.

In the basic approximation algorithm we have to choose a knot vector for the spline and to assign suitable parameter values τ_j to the data points $\boldsymbol{x}_j = \boldsymbol{x}(\tau_j)$. This gives rise to the "least squares problem"

$$||A\boldsymbol{P} - \boldsymbol{X}||_2 \longrightarrow \min \tag{66}$$

where $A = (N_i(\tau_j))_{i,j}$. Since we have more equations than control points we have conditions in between the knots. This enlarges the bandwidth of the matrix A from three to four and it is no longer quadratic. Usually the least squares problem is solved via solving the normal equations, see for example [34]. In view of the extension to the multidimensional case and the worse condition of the normal equations we prefer the solution via QR-decomposition.

4.4.4 Tensor Product Interpolation and Approximation

In the context of grid generation typically the problem arises to find a B-Spline tensor product of the form (60) that approximates a given net $\boldsymbol{x}_{r,s}$ of grid points at predefined parameters (u_r, v_s), $r = 0, \dots, N$, $s = 0, \dots, M$. In the case of interpolation, i.e. $M = m - p$ and $N = n - q$ we obtain the following matrix equation:

$$\left(\sum_{i,j} A_{r,i} B_{s,j} \boldsymbol{p}_{i,j}\right)_{\begin{pmatrix} r=0,1,\dots,M \\ s=0,1,\dots,N \end{pmatrix}} = A\boldsymbol{P}B^T = \boldsymbol{X}. \tag{67}$$

Here we have assembled the control points $\boldsymbol{p}_{i,j}$ to be computed in the matrix $\boldsymbol{P}$ and the interpolation points in the matrix $\boldsymbol{X}$. Note that the collocation matrices $A = (A_{r,i}) = (N_r(u_i))$ and $B = (B_{s,j}) = (N_s(v_j))$ have the same structure as the corresponding 1D collocation matrices. Therefore equation (67) can be solved easily by LR-decomposition of these matrices.

If M and N are large it is often preferable not to interpolate but to approximate the given points for sake of data reduction . In the following we assume $M > m - p$ and $N > n - q$. Although it is possible to set up and solve the standard least squares problem

$$\sum_{r,s}(\boldsymbol{x}(u_r, v_s) - \boldsymbol{x}_{r,s})^2 \longrightarrow \text{min}, \tag{68}$$

this task is computationally rather expensive, because it requires the solution of fairly large linear systems. Since it is sometimes necessary to solve the approximation problem repeatedly, for example if one wants to approximate the given data within a given accuracy but with minimal number of control points, we want to avoid this computation. Instead we measure the approximation error in another norm in order to exploit the tensor product structure of the problem. To do this, we note that the least squares problem (68) can be formulated equivalently as the problem to minimize the *Frobenius norm* of the matrix $R := A\boldsymbol{P}B^T - \boldsymbol{X}$, where A, B, $\boldsymbol{P}$, and $\boldsymbol{X}$ are the same matrices as in equation (67).

$$||R||_F = \left(\sum_{i,j} R_{i,j}^2\right)^{\frac{1}{2}} \longrightarrow \text{min}. \tag{69}$$

The matrix R contains the approximation errors in every data point in their natural order. But instead of the Frobenius norm we propose to minimize of the 2-norm, i.e. the *spectral norm* of R.

This can be justified by the following elementary norm equivalence, that holds for every matrix $M \in \mathbb{R}^{m\times n}$ with $m > n$:

$$||M||_2 \leq ||M||_F \leq \sqrt{n}||M||_2. \tag{70}$$

This lemma states that the solution of the modified minimization problem delivers reasonable approximations of the solution of the least squares problem, as the two norms differ at most by a factor that equals the square root of the number of data points in one coordinate direction.

The advantage of the spectral norm is that it is invariant under multiplication with orthogonal matrices. So we can solve the problem analogously to the interpolation case but using QR instead of LR-decomposition. In matrix notation this reads as

$$\begin{aligned} ||A\boldsymbol{P}B^T - \boldsymbol{X}||_2 &= ||Q_A R_A \boldsymbol{P}(Q_B R_B)^T - \boldsymbol{X}||_2 \\ &= ||R_A\, \boldsymbol{P}\, R_B^T \ - \ Q_A^T \boldsymbol{X}\, Q_B\,||_2 \longrightarrow \text{min}. \end{aligned} \tag{71}$$

Note that again A and B are 1D collocation matrices, which are generally banded matrices with low bandwidth, for bicubic approximation, for example, the bandwidth is ≤ 4. So the computation of the QR-decompositions needed is cheap. In every case this results in computational costs proportional to the number of given points.

In the case of 3D interpolation we are given a 3-dimensional array $\boldsymbol{X} = (\boldsymbol{x}_{i,j,k})$ of grid points and have to determine an interpolating B-Spline tensor product in form (61). We can easily generalize the notation used above to

$$\left(\sum_{i,j,k} A_{r,i} B_{s,j} C_{t,k} P_{i,j,k}\right)_{\begin{pmatrix} r=0,1,\ldots,M \\ s=0,1,\ldots,N \\ t=0,1,\ldots,L \end{pmatrix}} = B(\boldsymbol{A}\boldsymbol{P}\boldsymbol{C}^T) = \boldsymbol{X}. \tag{72}$$

Again solving this system is reduced to the task to compute LR-decompositions of the 1D-collocation matrices A, B, and C. Since $\boldsymbol{P} \in (\mathbb{R}^3)^{L\times M\times N}$ the different matrix products work on the first, second and third index of $\boldsymbol{P}$, that is running on the other ones. Analogously we treat the 3D approximation case, just by employing the QR-decompositions instead of LR decompositions. In this case we cannot give a motivation based on equivalence of matrix norms similar to the 2D case, but the algorithm is fast and yields satisfactory results.

4.4.5 Computation of Volumes and Moments

For the multiscale analysis (here in 2D) the computation of the cell moments $M_{n,m} := \int_V x^n y^m \, dV$ is required, e.g. for $n = m = 0$ one gets the volume of V. As we have defined a grid-cell to be the image of a cell in computational space, the calculation of this integral is expensive, because the cells have curved edges. To simplify the evaluation of the integrals we use the Gaussian integral theorem in the following way. Setting

$$\boldsymbol{f}(x) := \left(\frac{1}{n+1} x^{n+1} y^m, 0\right) \tag{73}$$

we can reduce the cell integral to an integral over its boundary:

$$M := \int_V x^n y^m dV = \int_V div \boldsymbol{f}(x) dx = \int_{\partial V} \boldsymbol{f} \cdot \boldsymbol{n} ds. \tag{74}$$

Here $\boldsymbol{n}$ denotes the outer unit normal on the boundary of V. Assuming V to be the image of the cell $[u_1, u_2] \times [v_1, v_2]$ and using the parameterization of the cell boundaries implicitly given by the grid function, we obtain from the substitution rule

$$
\begin{aligned}
M = & \int_{v_1}^{v_2} \boldsymbol{f}(\boldsymbol{x}(u_1,\tau))\boldsymbol{n}(u_1,\tau)||\boldsymbol{x}_v(u_1,\tau)||_2 \, d\tau \\
& + \int_{v_1}^{v_2} \boldsymbol{f}(\boldsymbol{x}(u_2,\tau))\boldsymbol{n}(u_2,\tau)||\boldsymbol{x}_v(u_2,\tau)||_2 \, d\tau \\
& + \int_{u_1}^{u_2} \boldsymbol{f}(\boldsymbol{x}(\tau,v_1))\boldsymbol{n}(\tau,v_1)||\boldsymbol{x}_u(\tau,v_1)||_2 \, d\tau \\
& + \int_{u_1}^{u_2} \boldsymbol{f}(\boldsymbol{x}(\tau,v_2))\boldsymbol{n}(\tau,v_2)||\boldsymbol{x}_u(\tau,v_2)||_2 \, d\tau.
\end{aligned} \tag{75}
$$

Noticing that the product of the outer unit normal and the norm of the derivative is except of the sign just the normal to the tangent itself, the above equation simplifies to (with $(v_1, v_2)^\perp = (-v_2, v_1)$)

$$
\begin{aligned}
M = & \int_{v_1}^{v_2} \boldsymbol{f}(\boldsymbol{x}(u_1,\tau))\boldsymbol{x}_v^\perp(u_1,\tau) \, d\tau \\
& - \int_{v_1}^{v_2} \boldsymbol{f}(\boldsymbol{x}(u_2,\tau))\boldsymbol{x}_v^\perp(u_2,\tau) \, d\tau \\
& - \int_{u_1}^{u_2} \boldsymbol{f}(\boldsymbol{x}(\tau,v_1))\boldsymbol{x}_u^\perp(\tau,v_1) \, d\tau \\
& + \int_{u_1}^{u_2} \boldsymbol{f}(\boldsymbol{x}(\tau,v_2))\boldsymbol{x}_u^\perp(\tau,v_2) \, d\tau.
\end{aligned} \tag{76}
$$

We now see that the integrands are products of piecewise univariate polynomials and that the integrals therefore can be evaluated exactly. But due to the high degree of these polynomials this calculation turns out to be rather expensive. Therfore it should be performed only where it is really necessary, i.e. near the body contours. In the far field the use of simpler grid models is preferable.

4.5 Generation of Boundary Conforming Grids

4.5.1 Basic Procedure

Consider a simply connected bounded domain Ω in two dimensional space. Suppose that Ω is bounded by four edges and that these edges are given as B-Splines $\boldsymbol{x}_{u0}(u)$, $\boldsymbol{x}_{u1}(u)$, $\boldsymbol{x}_{0v}(v)$ and $\boldsymbol{x}_{1v}(v)$. The task is to generate a grid function $\boldsymbol{x}(u,v)$, that interpolates the given curves along its boundaries:

$$\boldsymbol{x}(u,0) = \boldsymbol{x}_{u0}(u), \qquad \boldsymbol{x}(0,v) = \boldsymbol{x}_{0v}(v), \tag{77}$$

$$\boldsymbol{x}(u,1) = \boldsymbol{x}_{u1}(u), \qquad \boldsymbol{x}(1,v) = \boldsymbol{x}_{1v}(v). \tag{78}$$

As we want to represent the grid by a B-Spline tensor product we have to assure first, that the opposite edges possess the same knot vector. This can easily be achieved by knot insertion. Since we only use knot vectors with p-fold knots at their ends the control polygons of the boundary B-Splines are always equal to the corresponding border lines of the control net for any tensor product grid function conforming to the above boundary conditions.

Most standard methods cannot manipulate the B-Spline control points directly, for example the finite difference discretization of an elliptic grid generator uses point representations of the grid. In this case we evaluate the B-Spline at its knots and after successful grid generation interpolate the points again. Since the solution of the interpolation is unique it is guaranteed that the original boundary curve is reproduced.

4.5.2 Algebraic Grid Generation

We employ several variants of transfinite interpolation, which is based on Gordon's method [41].

We only mention the standard linear blend. Denoting the four corner points of the given domain with $\boldsymbol{x}_{00} = \boldsymbol{x}(0,0)$, $\boldsymbol{x}_{10}$, $\boldsymbol{x}_{01}$ and $\boldsymbol{x}_{11}$ it can be written in the form

$$\begin{aligned} \boldsymbol{x}(u,v) = \quad & (1-u \quad u)\begin{pmatrix} \boldsymbol{x}_{0v}(v) \\ \boldsymbol{x}_{1v}(v) \end{pmatrix} + (\boldsymbol{x}_{u0}(u) \quad \boldsymbol{x}_{u1}(u))\begin{pmatrix} 1-v \\ v \end{pmatrix} \\ - \quad & (1-u \quad u)\begin{pmatrix} \boldsymbol{x}_{00} & \boldsymbol{x}_{01} \\ \boldsymbol{x}_{10} & \boldsymbol{x}_{11} \end{pmatrix}\begin{pmatrix} 1-v \\ v \end{pmatrix}. \end{aligned} \tag{79}$$

This method is especially easy to apply, because, as proven in [39], a linear blend obtained from four cubic B-Spline boundary curves is indeed itself a bicubic tensor product B-Spline and its control net can be obtained directly from the boundary control polygons, just by interpreting all the control points $\boldsymbol{p}_{ij}$, $i = 0, \ldots, N$, $j = 0, \ldots, M$ as values $\boldsymbol{p}(i/N, j/M)$ of a virtual grid function $\boldsymbol{p}$, and applying the formula defining the linear blend to the inner control grid points.

The other basic method is to prescribe in addition the normal derivatives to the given curves and to interpolate these data with a bicubic Coons-Patch.

4.5.3 Elliptic Grid Generation

It is common experience that in practical application the use of an elliptic grid generator is inevitable, because elliptic methods are superior to all alternatives with respect to robustness of the algorithm and smoothness of the generated grids. The disadvantage of elliptic methods is that they require the very time consuming solution of a nonlinear PDE and are therefore not well suited for time varying geometries or interactive use.

We implemented the grid generator by Spekreijse [46], [47], because his concept of coupling elliptic with algebraic transformations is theoretically well founded and easily integrated into our approach, that already provides a variety of algebraic methods.

To speed up the solution of the PDE we begin with generating very coarse discrete grids with the aid of the elliptic generator. Then this grid will be converted into a B-Spline representation. If the quality of the grid achieved is satisfactory we stop here, if not we evaluate the function at a higher number of points and use this grid as initial guess for a new iteration with the elliptic solver. Since this starting grid is then usually much better than for example an initial guess based on transfinite interpolation, the number of Newton or fix-point iterations used for the solution of the nonlinear PDE can be reduced. But it turns out that usually it is sufficient to compute and interpolate very coarse elliptic grids to achieve a sufficient quality for flow calculations, as long as enough grid-points are chosen to prevent discrete grid folding. For planar blocks we use typically 30×30 points. This makes interactive elliptic grid generation possible, at least in many practical cases.

Without laying claim to a rigorous mathematical foundation of this method, one may justify this concept by the remark, that basically elliptic grid generation is founded on the minimization of the functional

$$\int_\Omega u_{xx}^2 + v_{yy}^2 \longrightarrow \min \tag{80}$$

and, as pointed out above, B-Splines fulfill a similar variation diminishing property, at least in the functional 1D-case.

The following example is taken from [44]. Using the algorithms proposed by Spekreijse we first generated an elliptic grid in this domain demanding orthogonality and control of the first layer of cells along the curved borders.We have discretized the domain with 100×100 grid-points (only every second line is plotted for the sake of better visualization) and then compared it with the grid obtained when we discretize the elliptic equation with 30×30 points, and evaluate the interpolating spline again at 50×50 points. The smoothness of both grids is nearly the same, only the high resolution elliptic grid produced a somewhat thinner cell layer at the right boundary of the domain.

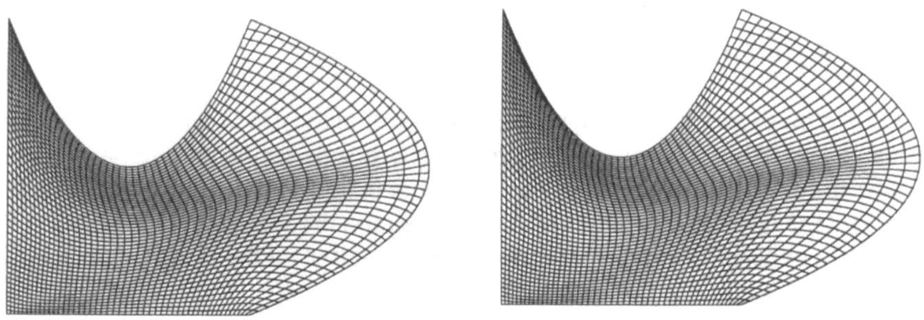

Figure 15 Fine and coarse elliptic grid

4.5.4 Reproduction of Grids

The approximation tools enable us to reproduce grids from external sources automatically, for example pointwise given grids used for calculation with other solvers than QUADFLOW. Within this project for example we are often asked to compare the results of QUADFLOW with former results produced by the FLOWer code. In order to avoid the work of generating block-decompositions ourselves, we have automated the task of converting the pointwise given grids into B-Spline representations. Thereby only one constraint must be respected. In practical grid generation often grids are given that contain non-differentiable corners in one boundary. The simplest example is a ramp that is connected at a certain angle with a plain surface.

This is a technical problem, because cubic B-Splines without multiple interior knots are C^2-continuous. It is possible to model such corners with $p - 1$ coalescing control points [35]. But then the derivative in this point would vanish what may result in grid folding near this point. Another method is to use knot vectors with multiple knots, but in this case further manipulation of this spline is more difficult, because evaluating the spline at its knots and reproducing it again by interpolation as proposed at the top of this section is not possible if the knot vector contains multiple interior knots. So the simplest method is to split the block in question into two and to apply the methods described above in both blocks separately. Since it is not easy to decide from the discrete set of points automatically, where such cusps occur, a certain extent of user interaction remains.

4.5.5 Extensions

It is clear that the detour described above, namely first to evaluate the spline at certain points and then to interpolate these points again, can only serve as provisional strategy. It has been chosen in order to make use of already existing tools. Future developments will aim at directly manipulating the B-Spline control points which opens particularly promising perspectives with regard to temporally moving meshes.

The extension of this concept to 3D is straight forward. Here we assume that the simply connected domain Ω is bounded by six faces, given as B-Spline surfaces. The volume grid is represented as 3-dimensional tensor product. So the whole geometry is processed within a unified framework.

4.6 Offset-Curves

In this section we describe how we compute the offset curves required for the boundary fitted blocks, see also [30].

Let a regular planar curve C be explicitly/parametrically given by

$$\boldsymbol{x}(u) = \begin{pmatrix} x(u) \\ y(u) \end{pmatrix} : [u_0, u_1] \longrightarrow \mathbb{R}^2 \tag{81}$$

The curvature of the curve C is given by (for simplicity we omit the arguments)

$$\kappa = \frac{\dot{x}\ddot{y} - \dot{y}\ddot{x}}{(\dot{x}^2 + \dot{y}^2)^{\frac{3}{2}}} = \frac{\begin{vmatrix} \dot{x} & \dot{y} \\ \ddot{x} & \ddot{y} \end{vmatrix}}{(\dot{x}^2 + \dot{y}^2)^{\frac{3}{2}}}. \tag{82}$$

Now we can define curvature dependent offset curves. Consider the curve as a front $\boldsymbol{x}(u,t)$ propagating along its normal direction with curvature dependent speed $F(\kappa)$.

$$\boldsymbol{x}(u,0) = \boldsymbol{x}(u), \tag{83}$$

$$\frac{\partial}{\partial t}\boldsymbol{x}(u,t) = F(\kappa(u))\boldsymbol{n}(u). \tag{84}$$

This idea was introduced to grid generation by Sethian [45]. To compute the evolution of the curve, Sethian embeds the interface as the zero level set of a higher dimensional function and solves a partial differential equation to follow its propagation. The advantage of the level set approach is primarily that topological changes such as breaking and merging of several fronts are well defined and performed without additional effort. But since Sethian's method leads to complex and time consuming algorithms, and topological changes cannot occur in the present context, we use instead the parametric B-Spline representations for the offset surfaces. So we achieve a stable, easy to implement and fast algorithm that is more suited to the needs of grid generation.

We assume, that a B-Spline representation $\hat{\boldsymbol{x}}_{t_i}(u)$ of the propagating front at time level t_i is given. We want to calculate a B-Spline representation $\hat{\boldsymbol{x}}_{t_{i+1}}$ at the next time-level $t_{i+1} = t_i + \Delta t$. Therefore we just choose an appropriate, i.e. curvature dependent set of parameter values u_j – they may differ in every time-step – and compute sampling points for the next offset curve by

$$\hat{\boldsymbol{x}}_{t_{i+1}}(u_j) = \hat{\boldsymbol{x}}_{t_i}(u_j) + \Delta t\, F(\kappa(u_j))\boldsymbol{n}(u_j). \tag{85}$$

Remember that the relevant geometric quantities, curvature and normals, can be exactly computed from the B-Spline representation. These sampling points are approximated afterwards by a new B-Spline using the approximation tools and fairing methods from Section 4.4.3, in order to guarantee that all level sets become optimally smooth.

We have tested the same functions F as used in [45], namely

(i) $F(\kappa) = 1 - \varepsilon\kappa$

(ii) $F(\kappa) = e^{-\varepsilon\kappa}$

(iii) $F(\kappa) = \max\{1 - \varepsilon\kappa, F_{threshold}\}$.

Here $F_{threshold} > 0$ is chosen to prevent the front from moving inwards. In [45] ε and $F_{threshold}$ are computed only once. With our method their computation is very cheap so we decided to adapt them in every step to use an optimal step size Δt.

Two examples are given in Figure 16. The left plot shows several offset curves around the SFB 401 reference profile, but not the final grid. The lines in cross direction only symbolize curvature dependent sampling points. To achieve good grid quality after the computation of the offset curves one can add smooth iso-lines orthogonal to the surfaces. For this purpose a hyperbolic method based on the solution of ordinary differential equations has been developed in [31]. The right plot demonstrates the robustness of the method with the aid of an academic but more difficult example. It is clear, that because of the concave parts of the boundary the usual techniques from CAD to generate parallel curves would fail here. Additionally the flexibility of splines with regard to boundary conditions was used to enforce $GC2$-periodicity across the cut, yielding a smooth closed curve.

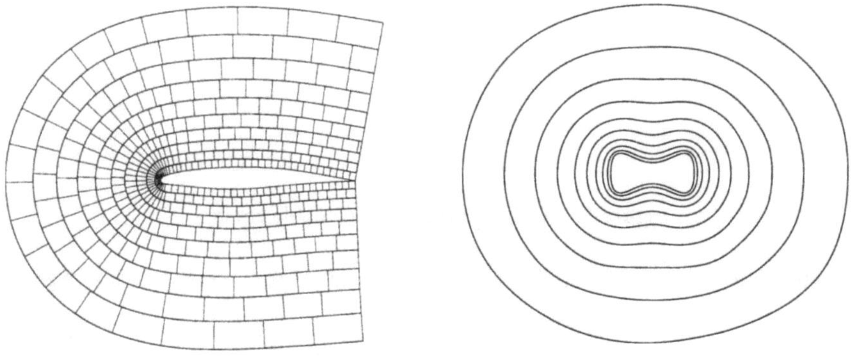

Figure 16 Offset-Curves around cruise configuration (left) and artifical example (right)

The last example given by Figure 17 demonstrates that the B-Spline method is robust enough to handle even non-differentiable corner points. The initial configuration consists of two circles connected by a straigth line. This example might seem to be somewhat artificial, but really is of practical relevance, because similar situations occur in practice, when one wants to generate offset curves around multiple connected domains. In a first step a smooth curve approximating the intitial configuration is defined applying the knot insertion algorithm near the corner points.

Approximatively 100 control points are needed to represent this curve. After that the above algorithm is applied. Here it is particulary important, that one can choose an adopted, optimally large stepsize independently in every time step, because in the beginning only very small timesteps can be performed.

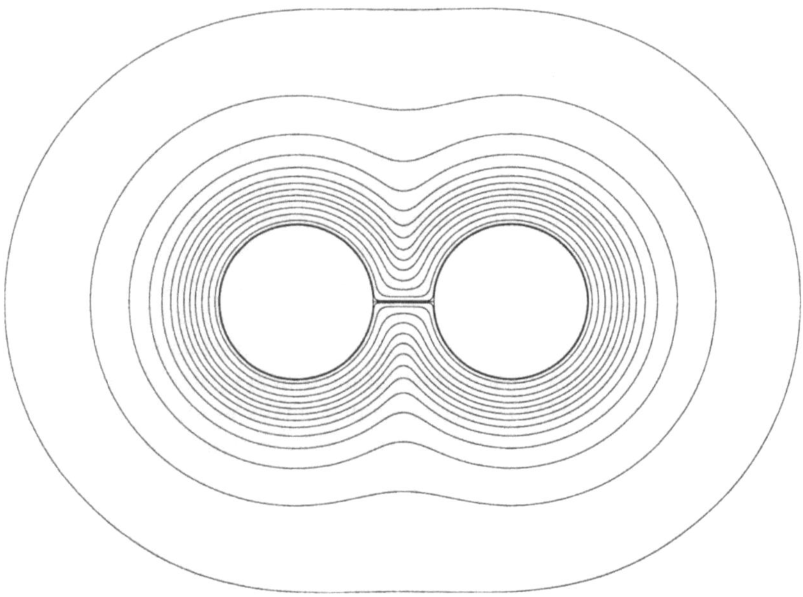

Figure 17 Offset-Curves around double circle

The extension of this method to 3D is straight forward. From a parametrically given surface

$$\boldsymbol{x}(u,v) : [u_0, u_1] \times [v_0, v_1] \longrightarrow \mathbb{R}^3 \tag{86}$$

one can first compute the relevant geometric quantities, then the sampling points for the next offset curve, and after that approximate these by a smooth B-Spline surface. In this case we have more freedom, as one can choose between several notions of curvature, for example mean curvature, Gaussian curvature or principal curvatures. Of course, the primary selection criterion for a suitable curvature notion is to prevent grid folding. First results with this method are promising and will be subject of further investigations.

5 Discretization of the Navier-Stokes Equations

5.1 Finite Volume Method

The occurance of hanging nodes due to local mesh adaptation poses particular difficulties concerning the discretization of the governing equations. In the following section we present a finite volume method, the special merits of which are its flexibility to operate on meshes of any arbitrary topology. This approach offers a unified way to incorporate hanging nodes. Its main ingredients will be discussed in detail, including data structures, realization of spatial second order accuracy, treatment of convective and viscous fluxes, choice of limiters, treatment of boundary conditions and implicit time integration. The section will be concluded by the validation of the basic scheme applying it to a proper selection of test cases for inviscid and viscous flows.

In the present study, we will concentrate on the discretization of the governing equations in two space dimensions. Its extension to three space dimensions is subject to current work.

5.1.1 Data Structure

The discretization of the governing equations (1) is based on a cell centered finite volume scheme. The grid is treated as a fully unstructured mesh, composed of simply connected elements with otherwise arbitrary topology. In particular, the flow solver is not restricted to operate on locally refined quadrilateral cells, but may be combined with any grid topology, e.g. mixed element type grids. This flexibility is crucial for supporting the adaptive concept.

Different element types are processed in a unified manner, rather than being distinguished. This property is often related to as *grid transparency.* Hanging nodes, which occur due to local adaptation, do not require any special treatment. The use of transition elements at refined interfaces is not necessary.

The data structure of the flow solver is primarily based on the faces of the grid. A face based data structure has the advantage that there are no limitations on the number of faces, which can be connected to a cell, see Figure (18) for illustration. The evaluation of fluxes and their contribution to cells can be efficiently implemented by sweeps over the faces.

The mesh is composed of the basic grid objects: cells, faces and nodes, see Figure (19). Grid objects are related to each other via connectivity lists. Two type of pointers are required: A link between faces and cells, that share the face in question, and a connection between faces and nodes, that belong to the face:

	Pointer type	Data structure
•	Face to cell	`face_2_cell(iface,iC)`
•	Face to node	`face_2_node(iface,iN)`

where `iC` $\in [1,2]$ denotes the left and right neighbour of the face and `iN` $\in [1,2]$ represents the adjoining nodes, respectively. See also Figure (20) for illustration.

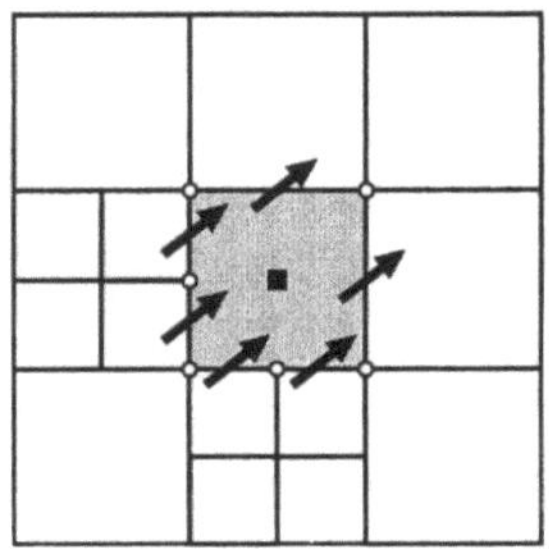

Figure 18 Collection of fluxes for polygonally bounded control volume in 2D

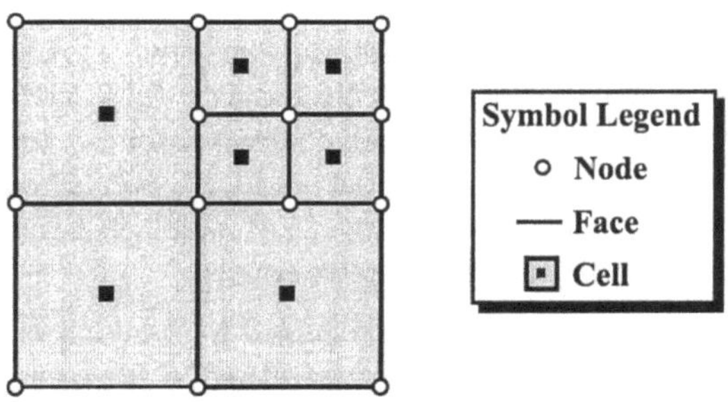

Figure 19 Basic grid objects of unstructured mesh

a) Connectivity between faces and cells **b)** Connectivity between faces and nodes

Figure 20 Logical relation between different grid objects

5.1.2 Discretization of Conservative Fluxes

The conservative fluxes are determined by solving quasi-one-dimensional Riemann problems at cell interfaces. The rotational invariance of the Euler equations

$$\boldsymbol{F}^c(\boldsymbol{u})\,\boldsymbol{n} = \boldsymbol{R}^T(\boldsymbol{n})\,\boldsymbol{F}^c(\tilde{\boldsymbol{u}}) \tag{87}$$

is utilized to express the projection of the conservative flux into normal direction, where $\tilde{\boldsymbol{u}} = \boldsymbol{R}(\boldsymbol{n})\,\boldsymbol{u}$ and $\boldsymbol{R}$ represents the rotational matrix

$$\boldsymbol{R}(\boldsymbol{n}) = \begin{pmatrix} 1 & 0 & 0 & 0 \\ 0 & n_x & n_y & 0 \\ 0 & -n_y & n_x & 0 \\ 0 & 0 & 0 & 1 \end{pmatrix}. \tag{88}$$

Within the present paper, the conservative flux function is based on the HLLC scheme proposed by Batten and Leschziner [49]. The method is capable of exactly preserving isolated shocks, contact, and shear waves, which are desired attributes to predict viscous flows accurately. Further, the entropy condition is inherently enforced by the scheme, i.e. neither an entropy correction nor a shock fix is required.

To define the HLLC scheme, we first introduce the following quantities:

$$S_L = min\left[\lambda_1(\boldsymbol{u}_l),\, \lambda_1(\boldsymbol{u}^{Roe})\right],\; S_R = max\left[\lambda_m(\boldsymbol{u}^{Roe}),\, \lambda_m(\boldsymbol{u}_r)\right], \tag{89}$$

$$S_M = \frac{\varrho_r v_{n,r}(S_R - v_{n,r}) - \varrho_l v_{n,l}(S_L - v_{n,l}) + p_l - p_r}{\varrho_r(S_R - v_{n,r}) - \varrho_l(S_L - v_{n,l})}, \tag{90}$$

$$\Omega_l = (S_L - S_M)^{-1}, \tag{91}$$

$$p^* = \varrho_l(v_{n,l} - S_L)(v_{n,l} - S_M) + p_l\,. \tag{92}$$

The subscripts l and r denote the left and right state of the Riemann problem. $\lambda_1(\boldsymbol{u}^{Roe})$ and $\lambda_m(\boldsymbol{u}^{Roe})$ represent the smallest and largest eigenvalues of the Roe matrix [50] and v_n, v_t are the velocity components normal and tangential to a cell interface, respectively.

Introducing the intermediate vector of state $\boldsymbol{u}_l^*$:

$$\boldsymbol{u}_l^* = \begin{pmatrix} \varrho_l^* \\ (\varrho v_n)_l^* \\ (\varrho v_t)_l^* \\ (\varrho e_{tot})_l^* \end{pmatrix} = \Omega_l \begin{pmatrix} \varrho_l(S_L - v_{n,l}) \\ (S_L - v_{n,l})(\varrho v_n)_l + p^* - p_l \\ (S_L - v_{n,l})(\varrho v_t)_l \\ (S_L - v_{n,l})(\varrho e_{tot})_l - p_l v_{n,l} + p^* S_M \end{pmatrix}, \tag{93}$$

the numerical flux function is expressed as follows:

$$
\boldsymbol{F}_{HLLC} = \begin{cases} \boldsymbol{F}_l & \text{if } S_L > 0 \\ \boldsymbol{F}\left(\boldsymbol{u}_l^*\right) & \text{if } S_L \leq 0 < S_M \\ \boldsymbol{F}\left(\boldsymbol{u}_r^*\right) & \text{if } S_M \leq 0 \leq S_R \\ \boldsymbol{F}_r & \text{if } S_R < 0 \end{cases} \tag{94}
$$

with

$$
\boldsymbol{F}\left(\boldsymbol{u}_l^*\right) = \begin{pmatrix} \varrho_l^* S_M \\ (\varrho v_n)_l^* S_M + p^* \\ (\varrho v_t)_l^* S_M \\ \left((\varrho e_{tot})_l^* + p^*\right) S_M \end{pmatrix} . \tag{95}
$$

$\boldsymbol{F}_l$ and $\boldsymbol{F}_r$ represent the conservative flux $\boldsymbol{F}^c(\boldsymbol{u})$, evaluated at the corresponding state. The relevant expressions for $\boldsymbol{F}\left(\boldsymbol{u}_r^*\right)$ are obtained by interchanging the subscripts $l \leftrightarrow r$ and $L \leftrightarrow R$.

5.1.3 Higher Order Method in Space

In order to obtain second-order accuracy in space, for each control volume Ω_i a linear reconstruction of the primitive flow variables $w \in \{\varrho, u, v, p\}$ is determined:

$$
w\left(\boldsymbol{x}\right)|_{\Omega_i} = w_i + \varphi_i \left(\boldsymbol{x} - \boldsymbol{x}_i\right)^T \cdot \nabla w_i \, , \tag{96}
$$

where w_i represents its mean value at the centroid $\boldsymbol{x}_i$ of Ω_i and φ_i denotes a limiter function. The approximate gradient ∇w_i is calculated by a multi-dimensional, second order truncated Taylor series expansion around the centroid of Ω_i

$$
w_j = w_i + \left(\boldsymbol{x}_j - \boldsymbol{x}_i\right)^T \cdot \nabla w_i, \, j \in N_i \, . \tag{97}
$$

N_i is the set of neighbouring cells of Ω_i, which support the reconstruction of w. The fixed stencil N_i consists of cells, that share a face with Ω_i (face neighbours).

In practice, equation (97) represents an overdetermined system of equations for the gradient of w_i. It is solved in a least squares sense, using normal equations. To improve the condition of the problem, the difference vector $(\boldsymbol{x}_j - \boldsymbol{x}_i)$ is locally rescaled independently in x and y direction

$$
x_j - x_i \rightarrow \left(x_j - x_i\right)/\tilde{x}_i \, , \quad y_j - y_i \rightarrow \left(y_j - y_i\right)/\tilde{y}_i \, .
$$

$\tilde{x}_i$ and $\tilde{y}_i$ denote the width of the cell in the respective coordinate direction, which is measured as the maximum distance between the cell centroid and the corresponding midpoint of the faces, that form the control volume in question.

Numerical experiments have demonstrated, that solving the least squares problem by means of a numerically more robust Householder transformation does not yield higher accuracy for the considered applications, but is significantly more expensive.

Alternatively to the least squares method, the gradient ∇w_i may be computed by using the Green-Gauss theorem

$$\nabla w_i = \frac{1}{\Gamma_i} \oint_{\partial \Gamma_i} w\, \boldsymbol{n}\, dS_i \tag{98}$$

where Γ_i denotes an auxiliary control volume, which is defined by the centroids of the neighbouring cells of Ω_i, which support the reconstruction.

In the current study, the Green-Gauss technique is only used in case of the so-called Ringleb flow, discussed in Section 5.2.1. Everywhere else in the paper, reconstruction is based on the least squares method.

5.1.4 Monotonicity Enforcement

At local extrema and discontinuities, the recovery polynomial may generate new extrema and therefore yield oscillations in the numerical solution. In order to circumvent this problem, limiter functions with TVD (total variation diminishing) property are used. The slope limiter by Venkatakrishnan [51] is employed. which is defined as follows:

$$\varphi_{i,g} = \begin{cases} \dfrac{w_{i,g}^{+2} + 2w_{i,g}^{+} w_{i,g}^{-} + \varepsilon}{w_{i,g}^{+2} + w_{i,g}^{+} w_{i,g}^{-} + 2w_{i,g}^{-2} + \varepsilon}, & \text{if } w_{i,g}^{-} \neq 0; \\ 1 & , \text{if } w_{i,g}^{-} = 0; \end{cases} \tag{99}$$

with

$$\begin{aligned} w_{i,g}^{-} &= w_g - w_i, \\ w_{i,g}^{+} &= \begin{cases} w_i^{max} - w_i, & \text{if} \quad w_{i,g}^{-} > 0 \\ w_i^{min} - w_i, & \text{if} \quad w_{i,g}^{-} < 0 \end{cases} \\ w_i^{max} &= max_{j=1,\ldots,N_i} \{w_i, w_j\}, \\ w_i^{min} &= min_{j=1,\ldots,N_i} \{w_i, w_j\}. \end{aligned}$$

w_g denotes the unlimited reconstructed value of w at the Gauss quadrature point g. The coefficient ε is typically set to $\varepsilon = 10^{-4}$ in our applications. Since condition (99) has to be fulfilled in every quadrature point of the faces, the final limiter associated with the control volume Ω_i is taken as the corresponding minimum value $\varphi_i = min\,(\varphi_{i,g})$.

5.1.5 Discretization of Viscous Fluxes

At cell interfaces, the gradients of the velocity and temperature are determined by utilizing the divergence theorem. As proposed by Coirer [52], a secondary volume is introduced, the edges of which are formed by connecting the two vertices of an edge with the centroids of the cells, that share the interface in question, see Figure 21(a). It is commonly referred to as "diamond path". At refined interfaces, the standard diamond path due to Coirer leads to a non-symmetric control volume, which may result in a loss of accuracy. Therefore, in case of locally adapted meshes we use a modified diamond path proposed by Delanaye [53], which maintains a regular symmetric shaped control volume even in the presence of hanging nodes. To apply the divergence theorem, an interpolation is required to evaluate the flow quantities at the vertices. The interpolation is supported by the set of cells N_{v_j} that share the vertex v_j in question, see Figure 21(b). A linearity-preserving procedure based on the pseudo-Laplacian formula according to Holmes and Connell [54] is used to determine the averaged quantity $\widetilde{w}$ at the vertex v_j:

$$\widetilde{w}_{v_j} = \frac{\sum_{i=1}^{N_{v_j}} \xi_i w_i}{\sum_{i=1}^{N_{v_j}} \xi_i}, \tag{100}$$

where w_i represents the corresponding flow quantity at the cell centroids $i \in N_{v_j}$. The dimensionless weights ξ_i are given by

$$\xi_i = 1 + \lambda_x(x_i - x_{v_j}) + \lambda_y(y_i - y_{v_j}) \tag{101}$$

with

$$\lambda_x = \frac{I_{xy}R_y - I_{yy}R_x}{I_{xx}I_{yy} - I_{xy}^2}, \tag{102}$$

$$\lambda_y = \frac{I_{xy}R_x - I_{xx}R_y}{I_{xx}I_{yy} - I_{xy}^2} \tag{103}$$

where the moments I_{xx}, I_{yy}, I_{xy}, R_x and R_y are defined as:

$$R_x = \sum_{i=1}^{N_{v_j}} (x_i - x_{v_j}), \tag{104}$$

$$R_y = \sum_{i=1}^{N_{v_j}} (y_i - y_{v_j}), \tag{105}$$

$$I_{xx} = \sum_{i=1}^{N_{v_j}} (x_i - x_{v_j})^2, \tag{106}$$

$$I_{yy} = \sum_{i=1}^{N_{v_j}} (y_i - y_{v_j})^2, \tag{107}$$

$$I_{xy} = \sum_{i=1}^{N_{v_j}} (x_i - x_{v_j})(y_i - y_{v_j}). \tag{108}$$

At the midpoint of each face, the velocity and temperature are averaged from the corresponding quantities at the vertices.

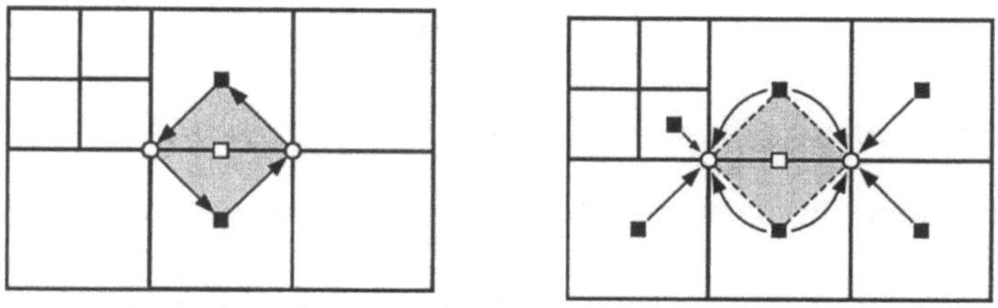

(a) Diamond path due to Coirer (b) Averaging procedure at grid nodes

Symbol legend: ■ cell center, ○ node, □ face midpoint

Figure 21 Calculation of gradients at cell interfaces, using divergence theorem

5.1.6 Boundary Conditions

Euler Walls

At solid walls at rest, the contact condition $\boldsymbol{v} \cdot \boldsymbol{n} = 0$ is imposed via the modification of the conservative fluxes through the boundary face. The contact condition yields the normal flux formula

$$\boldsymbol{F}^c \cdot \boldsymbol{n}\big|_{wall} = (0,\, p \cdot \boldsymbol{n},\, 0)^T\,, \tag{109}$$

where the static pressure p is extrapolated from the interior domain, using the multi-dimensional reconstruction.

Navier–Stokes Walls

The treatment of Navier–Stokes walls is based on an approach proposed by Anderson and Bonhaus [55], which has also successfully been used e.g. by Geuzaine [56] in the context of turbulent flows. At wall boundary cells, the coordinate location that is associated with the vector of unknowns, is shifted from the cell centroid to the midpoint of the corresponding wall edge. The governing equations directly express the temporal evolution of the flow quantities on the wall itself. This approach offers several advantages: First, at every time step the boundary conditions are strongly enforced for the wall boundary cells, rather than just being considered through the modification of the fluxes. Second, the boundary conditions are implemented in such a way that no assumption on the pressure distribution is required. The latter is often utilized in other approaches, which are based on modification of the fluxes or some mirror principle.

No-slip condition: The no-slip condition $\boldsymbol{v} = \mathbf{0}$ on a wall at rest is imposed by modification of the momentum equations

$$\int_{\Omega} \left.\frac{\partial \varrho \boldsymbol{v}}{\partial t}\right|_{wall} dV = \mathbf{0} \tag{110}$$

with initial conditions $\varrho \boldsymbol{v}\,(t = 0,\, \boldsymbol{x}_{wall}) = \mathbf{0}$.

Isothermal wall: In the present study, only isothermal walls are considered. In order to impose a constant wall temperature T_{wall}, the energy equation is modified in such a way that during the temporal evolution of the flow quantities, a constant wall temperature is automatically maintained. This approach does not require any assumption on the pressure distribution, e.g. $\partial p/\partial n = 0$, which is actually a predicate of a component of the gradient of pressure appearing explicitly in the local balance of moment and therefore not truly appropriate as a boundary condition in a boundary value problem.

The temperature at the wall is related to the internal energy of the fluid via

$$T_{wall} = \frac{1}{c_v} e_{wall} = \frac{1}{c_v} \left.\frac{\varrho e_{tot}}{\varrho}\right|_{wall}, \tag{111}$$

because here the specific internal energy e is locally equivalent to the total energy e_{tot}, due to the no-slip condition. By means of eq. (111), the temporal evolution of the energy is expressed in terms of the temporal evolution of the density:

$$\int_{\Omega} \left.\frac{\partial \varrho e_{tot}}{\partial t}\right|_{wall} dV = c_v T_{wall} \int_{\Omega} \left.\frac{\partial \varrho}{\partial t}\right|_{wall} dV\,, \tag{112}$$

where the temporal evolution of the density is determined by the continuity equation.

Far-Field Boundaries

At artificial boundaries of the computational domain which replace the true far–field boundaries, the type of boundary condition of each individual face may either be explicitly prescribed, or fully automatically detected, depending on the local flow solution. The latter approach is particularly useful for C-type grids.

Two different approaches are used here to treat subsonic inflow and outflow conditions. The first one is based on standard characteristic boundary conditions [57]. Outgoing Riemann invariants are determined by quantities being extrapolated from the interior domain, while incoming Riemann invariants are set by free stream conditions.

The second approach, which is especially suited for stationary planar flow, is based on a point vortex correction according to Thomas and Salas [58]. In this case, the flow quantities at far–field are specified by a solution according to full potential flow theory. The corrected velocity components are prescribed by

$$u = |\boldsymbol{v}_\infty| \, cos\alpha + V_\Gamma sin\theta \, , \tag{113}$$

$$v = |\boldsymbol{v}_\infty| \, sin\alpha - V_\Gamma cos\theta \, . \tag{114}$$

The vortex induced velocity magnitude V_Γ is given by

$$V_\Gamma = \frac{\Gamma}{2\pi r} \frac{\sqrt{1 - M_\infty^2}}{1 - M_\infty^2 sin^2 (\theta - \alpha)} \tag{115}$$

where the circulation Γ of the vortex is determined by $\Gamma = 1/2 c_{ref} |\boldsymbol{v}_\infty| C_L$. c_{ref} denotes the chord length and C_L is the lift coefficient. r and θ are the radius and the polar angle, measured from the airfoil quarter–chord to the individual boundary face at the far–field, respectively. The remaining flow quantities are determined under the assumption of isentropic flow and constant total enthalpy.

5.1.7 Time Integration

For steady fluid flow time plays the role of an iteration parameter to achieve asymptotically stationary flow in the computation. The solution is advanced in time by an implicit Euler method. Local time steps are applied. For the flux balance of the cell Ω_i we define the so-called residual vector

$$\boldsymbol{R}_i := \oint_{\partial\Omega_i} \boldsymbol{F}^c (\boldsymbol{u}) \, \boldsymbol{n} \, dS \, . \tag{116}$$

Introducing the definition

$$\widehat{\boldsymbol{R}}^{n+1} \left(\boldsymbol{u}^{n+1}\right) := \boldsymbol{R}^{n+1} \left(\boldsymbol{u}^{n+1}\right) + \frac{\Omega}{\Delta t} \left(\boldsymbol{u}^{n+1} - \boldsymbol{u}^n\right) \tag{117}$$

the conservation laws (1) are expressed in semi-discrete form as

$$\widehat{\boldsymbol{R}}^{n+1} \left(\boldsymbol{u}^{n+1}\right) = \boldsymbol{0} \, . \tag{118}$$

Equation (118) is solved by a Newton scheme. For stationary flows we take one Newton iteration per physical time step. In this case the Newton scheme is:

$$\widehat{\boldsymbol{J}} \left(\boldsymbol{u}^n\right) \Delta \boldsymbol{u}^n = -\boldsymbol{R} \left(\boldsymbol{u}^n\right) \tag{119}$$

with

$$\widehat{\boldsymbol{J}} = \frac{\Omega}{\Delta t} \boldsymbol{I} + \left. \frac{\partial \boldsymbol{R}}{\partial \boldsymbol{u}} \right|^n , \quad \Delta \boldsymbol{u}^n := \boldsymbol{u}^{n+1} - \boldsymbol{u}^n \, . \tag{120}$$

The linear system of equations (119) is solved by an iterative Krylov subspace method. In this study, we employ the GMRES algorithm [59], preconditioned by an incomplete LU-factorization. Typically, a Krylov subspace method requires only the inner product of a Jacobian matrix $\boldsymbol{J}$ and a vector $\boldsymbol{\nu}$, and not $\boldsymbol{J}$ explicitely [60]. This product can also be approximated by a difference quotient of the form

$$\boldsymbol{J}(\boldsymbol{u})\,\boldsymbol{\nu} \approx \frac{\boldsymbol{R}(\boldsymbol{u}+\varepsilon\boldsymbol{\nu}) - \boldsymbol{R}(\boldsymbol{u})}{\varepsilon}, \quad \text{with} \quad \varepsilon \in \mathbb{R}. \tag{121}$$

The difference parameter ε is chosen as

$$\varepsilon = \hat{\varepsilon}\frac{\sqrt{1.0 + ||\boldsymbol{u}||_2}}{||\boldsymbol{\nu}||_2} \tag{122}$$

with $\hat{\varepsilon} = 10^{-8}$ in our application. The advantage of using a matrix free version of a Krylov subspace method is, that the matrix vector product may consistently be discretized according to the evaluation of the residual vector, that represents the right hand side of equation (119). Nevertheless, a Jacobian is still required for the application of the $ILU(n)$ pre-conditioner. The pre-conditioning matrix is based on a first order accurate scheme in space. The Jacobian is derived numerically [61], using one sided difference operators of the form:

$$\frac{\partial \boldsymbol{F}_i(\boldsymbol{u})}{\partial \boldsymbol{u}_j} = \lim_{\tilde{\varepsilon}\to 0} \frac{\boldsymbol{F}_i(\boldsymbol{u}+\tilde{\varepsilon}\boldsymbol{e}_j) - \boldsymbol{F}_i(\boldsymbol{u})}{\tilde{\varepsilon}}, \quad \text{with} \quad \tilde{\varepsilon} \in \mathbb{R}, \tag{123}$$

where $\boldsymbol{F}_i$ is the i-*th* component of the numerical flux vector and $\boldsymbol{e}_j$ is the j-*th* unit vector, associated with the dependent variable $\boldsymbol{u}_j$.

Presently, viscous terms are not included within the Jacobian. For the computation of inviscid flows presented in this paper, we employ the standard GMRES algorithm using a Jacobian derived by numerical means. For viscous flows, we utilize a matrix–free GMRES, which consistently accounts for the viscous terms within the matrix vector product. In this case, only the pre-conditioner suffers from neglecting the viscous terms.

The infrastructure related to the Newton-Krylov method in this paper is based on the PETSc [62] library of Argonne National Laboratory. For further details of the numerical method we refer to [63], [64].

5.2 Validation of the Basic Scheme

In the following, various test cases are considered using regular, non-adaptive structured grids in order to validate the basic finite volume scheme. Particular emphasis is put on to the spatial accuracy of the method. For inviscid flows, the Ringleb flow is investigated, for which an analytic solution is available. For viscous flows, the laminar flow over an isothermal flat plate is considered, for which the similarity solution due to Blasius serves as a reference. Further, the transonic flow about an airfoil according to the SFB 401 cruise configuration is used to verify the method in the presence of shocks. In this case, results obtained with the FLOWer code are used for comparison. Finally, convergence acceleration to steady state using implicit time integration is investigated.

5.2.1 Ringleb flow

The so-called Ringleb flow [65] has been used as a test case to assess the spatial accuracy of the basic scheme for inviscid flows. It is an exact solution of the Euler equations obtained by hodograph transformation. The flow depends on the inverse of the stream function k and the velocity magnitude q. In our application we choose the values $k = 0.4$ and $k = 0.8$ to define the rigid walls, and $q = 0.3$ to define the inlet- and outlet boundaries. The flow is fully subsonic within the domain. The inlet Mach number is 0.3027 and the maximum Mach number reaches a value of 0.8567 at the center of the bottom wall. The error E_{h_j} and the spatial order of the scheme *EOC* (*Experimental Order of Convergence*) are estimated in the L_1 norm of the Mach number:

$$E_{h_j} = \frac{1}{N_j} \sum_{i=1}^{N_j} \left| f_{h_j} - 1 \right|, \; EOC = \frac{log\left(\dfrac{E_{h_j}}{E_{h_{j+1}}}\right)}{log\left(\dfrac{h_j}{h_{j+1}}\right)}$$

where f_{h_j} represents the ratio of the numerical solution and the exact solution on a grid with the characteristic length scale $h_j = 1/\sqrt{N_j}$; N_j is the number of grid cells. The accuracy analysis is carried out on four different grids with a resolution of 16x4, 32x8, 64x16 and 128x32 cells, respectively. In addition, the grids are distorted randomly in order to assess the sensitivity of the solution with regard to mesh quality, see Figure 22. Figure 23 illustrates the Mach number distribution along the bottom wall for three different regular grids. The linear Green-Gauss reconstruction method according to eq. (98) has been employed. For the highest resolution, the maximum error between the numerical solution and the exact solution is about 0.02%. Figure 24 shows the error E_{h_j} in the Mach number, depending on the characteristic length scale h_j of the mesh. A constant and a linear reconstruction, based on the Green-Gauss method, have been considered. For the linear reconstruction, the sensitivity of the error with regard to grid distortion is weak. The order of accuracy can be estimated between 2.5-2.9 for the linear reconstruction, see Table 1. The constant reconstruction scheme is numerically less then first order accurate for the considered grids.

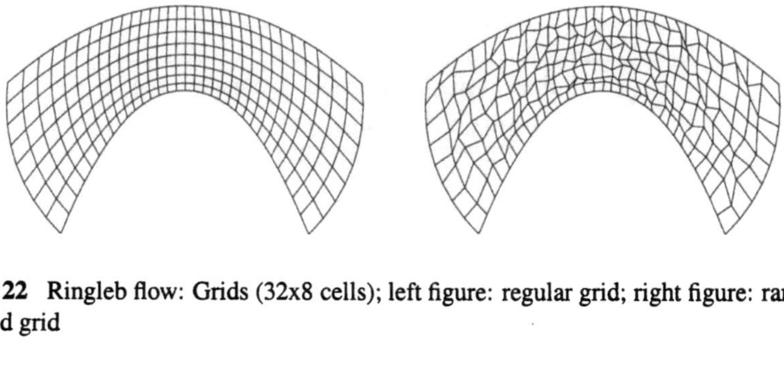

Figure 22 Ringleb flow: Grids (32x8 cells); left figure: regular grid; right figure: randomly distorted grid

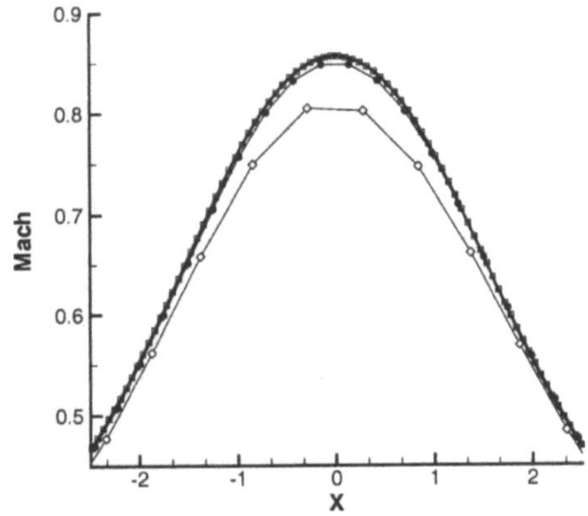

Figure 23 Ringleb flow: Mach number distribution at bottom wall (partial view); linear Green-Gauss reconstruction on regular grids; – exact solution, □128x32 cells, •32x8 cells, ◇16x4 cells

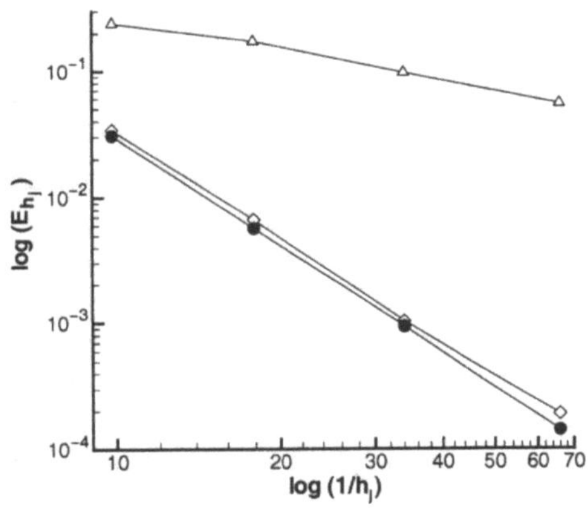

Figure 24 Ringleb flow: Error in the Mach number as function of the characteristic length scale h_j of the mesh. Linear Green-Gauss reconstruction: • regular grid, ◇ distorted grid ; △ constant reconstruction (regular grid)

Table 3 Ringleb flow: experimental order of convergence (EOC) for linear reconstruction and constant reconstruction on regular and distorted grids

	EOC		
grid	linear rec. regular	linear rec. distorted	constant rec. regular
32x8	2.79	2.70	0.543
64x16	2.83	2.93	0.909
128x32	2.84	2.53	0.862

5.2.2 Laminar Boundary Layer

The next test case considers the laminar flow over an isothermal flat plate. In order to reduce compressibility effects, a free stream Mach number of $M_\infty = 0.2$ is chosen. The Reynolds number is $Re_\infty = 10^4$, based on unit length. The wall is considered as isothermal, with $T_{wall} = T_\infty = 273.0K$. A linear dependence of the molecular viscosity on the temperature is assumed. The Prandtl number is $Pr = 1$. For purpose of validation, the similarity solution according to Blasius [66] for an incompressible laminar fluid flow serves as a reference.

A structured, non-adaptive grid with 176 × 56 cells (9856 cells all together) is considered, see Figure 25. The plate extends on the x-axis between $x = 0.0$ and $x = 2.0$, with 96 cells located on the plate itself. Upstream of the leading edge, the lower boundary of the domain is modeled as an inviscid impermeable wall. The first grid spacing normal to the wall is 10^{-3}, which equals $y^+ \approx 1$ at $Re = 10^4$. The grid is clustered about the leading edge, measuring a first grid spacing of 10^{-3} in streamwise direction. Far-field boundary conditions are applied as follows: At inflow, the temperature and velocity vector are prescribed, while the static pressure is extrapolated from the interior domain. At outflow, the static pressure is prescribed and the density as well as the velocity vector are extrapolated from the interior domain. The upper boundary of the domain is treated as an outflow.

Figure 26(a) and Figure 26(b) present a comparison between the computed u-velocity and v-velocity profiles with the theoretical solution at $Re_x = 9169$, respectively. Both velocity components are predicted accurately, compared with the theoretical solution. At the outer part of the boundary layer ($\eta > 4$) the v-velocity is slightly overpredicted by the numerical scheme.

Figure 27 illustrates the evolution of the skin friction coefficient along the plate (*please note*: the graph has been truncated for $Re < 1000$). The computed skin friction agrees well with the Blasius solution.

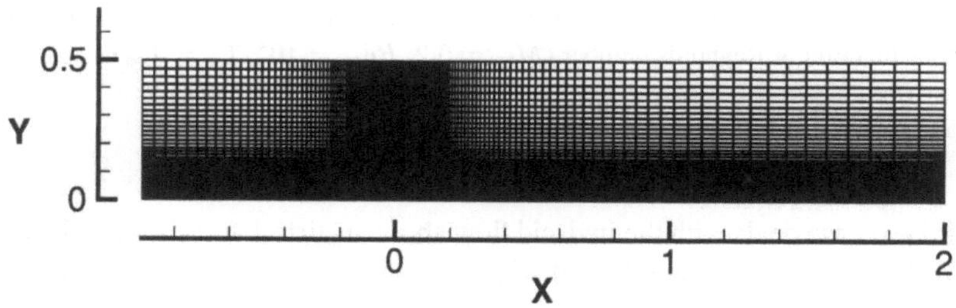

Figure 25 Structured, non-adapted grid for laminar flow over a flat plate. Plate extension: $x \in [0, 2]$. Grid resolution: 176x56 cells, 96 cells located on the plate.

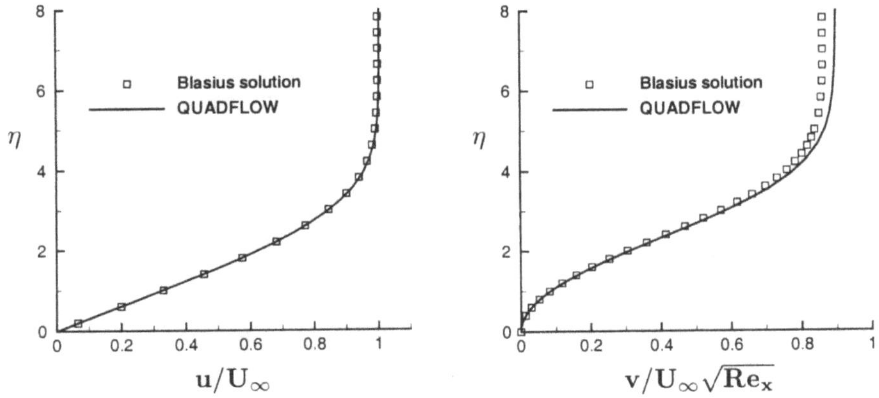

(a) u-velocity profile **(b)** v-velocity profile

Figure 26 Boundary layer profiles at $Re_x = 9169$ for the laminar flow over an isothermal flat plate ($M_\infty = 0.2$, $Re_\infty = 10^4$, $T_w = T_\infty$)

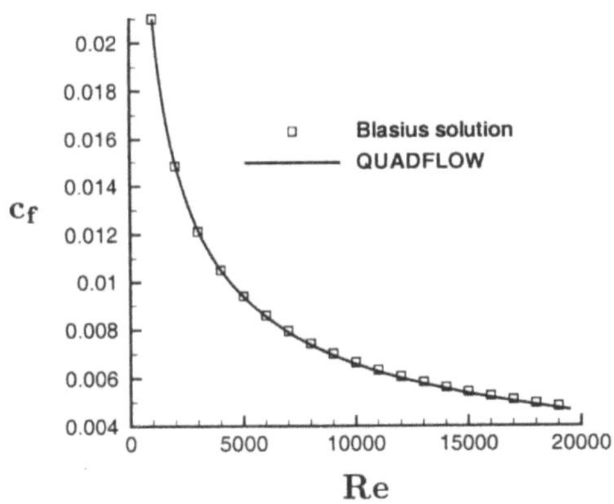

Figure 27 Distribution of the skin friction coefficient for the laminar flow over an isothermal flat plate as function of Reynolds number ($M_\infty = 0.2$, $Re_\infty = 10^4$, $T_w = T_\infty$)

5.2.3 SFB 401 Cruise Configuration

The third test case deals with the inviscid flow about an airfoil according to the transonic SFB 401 cruise configuration at $M_\infty = 0.85$, $\alpha = 0°$. A regular multiblock structured grid consisting of 256x32 cells is employed, with 192 cells located on the airfoil itself. The far-field boundary is placed about 20 chord lengths away from the airfoil. Far-field conditions are determined by using the point vortex correction according to eq. (113) and eq. (114). In Section 6.2, we will revisit this test case, using the fully adaptive algorithm.

Figure 28 presents the computational grid and the isobars for the present flow conditions. The flow exhibits a strong shock at the trailing edge of the upper surface. On the lower surface of the airfoil, two shocks can be identified. The shock located near the leading edge is very smeared, due to insufficient grid resolution, see Figure 28(b).

Figure 29 shows the pressure distribution on the airfoil surface. For the purpose of validation, numerical computations conducted with the FLOWer code [67] serve as a reference. In the latter case, the spatial discretization is based on central differences, stabilized by artificial dissipation of Jameson type. Vortex correction included. Both computational results are in good agreement. As expected, the upwind scheme (QUADFLOW) provides sharper shock resolution than the spatial discretization based on central differences (FLOWer). The aerodynamic lift coefficient is $C_L = 0.463527$ determined by the FLOWer code, and $C_L = 0.497806$ by QUADFLOW, which differ by about 7.4%.

(a)

(b)

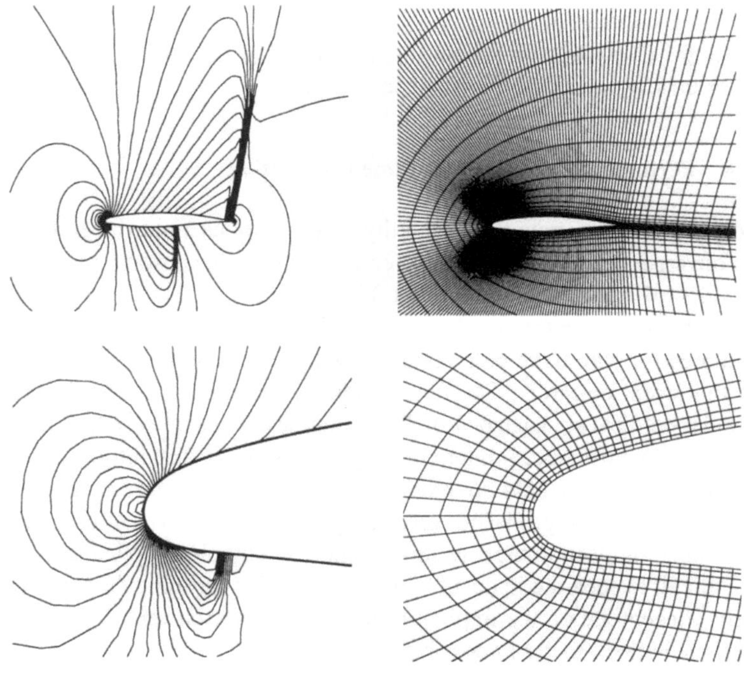

Figure 28 SFB 401 cruise configuration, $M_\infty = 0.85$, $\alpha = 0°$. *Left Figure*: contours of c_p distribution, $c_{p,min} = -1.35$, $c_{p,max} = 1.2$, $\Delta c_p = 0.075$. *Right Figure*: Non-adapted multiblock structured grid, 256x32 cells. **a)** Total view of airfoil. **b)** Partial view of leading edge.

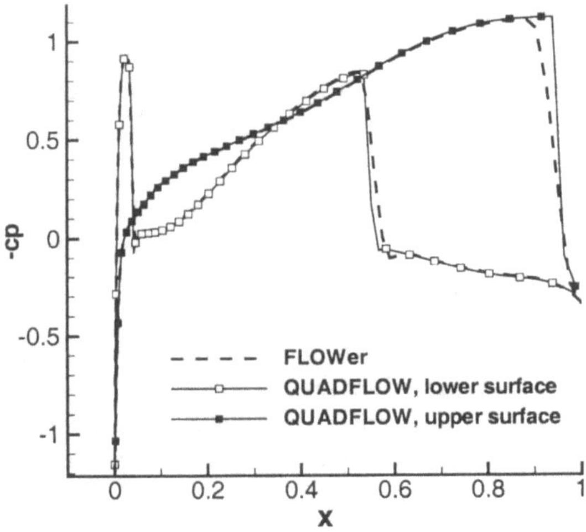

Figure 29 Distribution of the pressure coefficient for the SFB 401 cruise configuration, $M_\infty = 0.85$, $\alpha = 0°$. Grid resolution: 256x32 cells.

5.2.4 Performance of Implicit Time Integration

In order to demonstrate the feasibility of the implicit time integration to accelerate convergence to steady state, we consider the fully subsonic inviscid flow about the SFB 401 cruise configuration at $M_\infty = 0.5$, $\alpha = 0°$. The analysis of the implicit scheme concentrates only on inviscid flow, since viscous terms have not been accounted for within the preconditioner yet. Its extension for viscous flows is subject to current research. The flow conditions have been chosen in order exclude secondary effects associated with discontinuities, which may reduce the rate of convergence. In particular, limiter clipping in vicinity of discontinuities, reflection of entropy waves at far-field boundaries in the presence of shocks, and CFL restrictions due to the non-global convergence property of the Newton scheme may have a significant impact on the rate convergence. In complex applications, these effects are difficult to separate in detail. Therefore, a fully subsonic flow is investigated to solely concentrate on the implicit time integration itself and to exclude the above mentioned effects as far as possible.

Computations have been carried out on a structured, non-adaptive mesh consisting of 256x32 cells like in the previous test case. Cell ordering is arranged in a quasi-lexicographic manner, such that a small bandwidth is achieved. For the solution of the linear system, the residual is converged three orders of magnitude. The Krylov subspace has a dimension of 30 search directions. No restart was required for the considered test cases.

Figure 30 illustrates the convergence history of the non-linear residual for three different CFL numbers, namely $CFL = 10$, 10^2 and 10^3. The residual is monitored in the L_1 norm of the density. $ILU(2)$ serves as preconditioner. The convergence is greatly accelerated with increased CFL number. *Remark*: The scheme remained stable even for $CFL = 10^8$, but did not exhibit further acceleration.

Next, four different levels of fill-in for the $ILU(n)$ preconditioner are considered, namely $n \in [0, 1, 2, 4]$. $CFL = 1000$ is chosen, since it has demonstrated the best rate of convergence, according to Figure 30. Figure 31 illustrates that in this case $ILU(2)$ is the best choice, in terms of computation time.

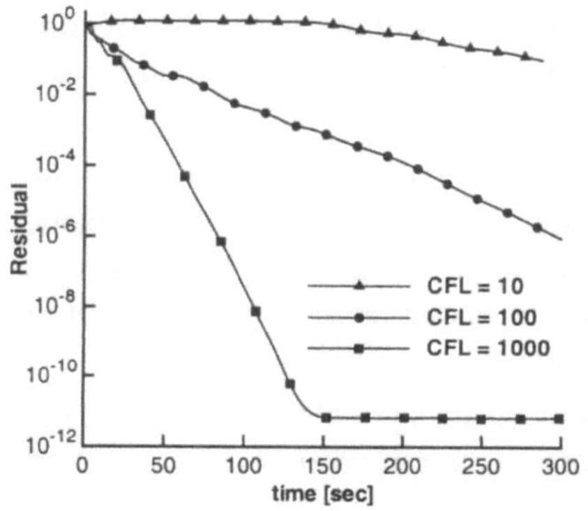

Figure 30 Convergence histories for SFB 401 cruise configuration, $M_\infty = 0.5$, $\alpha = 0°$. $ILU(2)$ preconditioner

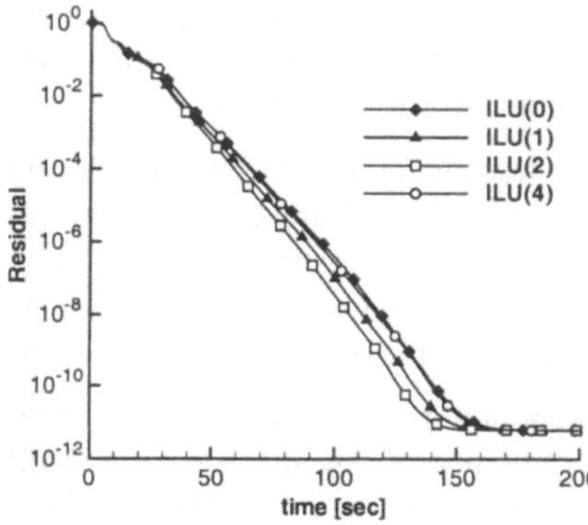

Figure 31 Convergence histories for different levels of $ILU(n)$, $CFL = 1000$. SFB 401 cruise configuration, $M_\infty = 0.5$, $\alpha = 0°$

6 Numerical Results of the Adaptive Scheme

After validation of the basic scheme, we now employ the fully adaptive algorithm. First, the subsonic flow about a NACA0012 airfoil at $M_\infty = 0.63$, $\alpha = 2°$ is investigated. We can demonstrate, that the adaptive algorithm is capable of providing highly accurate flow solutions. Next, we revisit the transonic flow about an airfoil according to the SFB401 cruise configuration, that has been studied in Section 5.2.3 using regular structured grids. Further, the transonic flow over a NACA0012 airfoil at $M_\infty = 0.95$, $\alpha = 0°$ is considered in order to demonstrate the advantage of adaptation to highly resolve complex shock configurations. Finally, the laminar flow over a flat plate is investigated, employing the fully adaptive scheme.

6.1 Subsonic NACA0012 airfoil

The first test case considers the subsonic flow about a NACA0012 airfoil at $M_\infty = 0.63$, $\alpha = 2.0°$. For this configuration, the aerodynamic lift coefficient may be predicted as $C_{L,pot} = 0.333$, by utilizing a full potential method.

Accurate prediction of the aerodynamic coefficients greatly depends on grid resolution and far–field conditions. Both aspects are investigated here. The computational C-type grid extends about 20 chords away from the airfoil. At the far–field, two types of boundary conditions have been applied: The first one utilizes characteristic based boundary conditions. Riemann invariants that enter the computational domain are determined by free–stream conditions, i.e. without application of vortex correction. In the second case, far–field boundary conditions are specified by using the point–vortex correction.

To assess the influence of grid resolution, two different threshold values $\varepsilon = 5 \cdot 10^{-2}$ and $\varepsilon = 3 \cdot 10^{-3}$ are employed for adaptation. The sensitivity of the adaptation primarily influences the grid resolution in the vicinity of the airfoil. In particular, adequate resolution of the stagnation area of the leading edge is of significant importance for the overall prediction of the aerodynamic coefficients.

Computations are initialized on a coarse mesh, consisting of 400 cells, with 10 cells on each side of the airfoil surface itself. A maximum number of $L_{max} = 9$ grid levels are permitted. 15 cycles of adaptation have been performed. The actual grid level that has locally been reached during the computation is $L = 7$ for $\varepsilon = 5 \cdot 10^{-2}$ and $L = 9$ for $\varepsilon = 3 \cdot 10^{-3}$. The highest refinement levels are located within the stagnation area at the leading edge. The residual decreased five orders of magnitude, measured in the L_1-norm of the density-residual. No limiter has been applied.

Figure 32 and Figure 33 show the isolines of the Mach number and the Mach distribution on the airfoil, respectively. Vortex correction has been applied, $\varepsilon = 3 \cdot 10^{-3}$. Table 4 summarizes the results for the considered parameter. The aerodynamic lift coefficient, its derivation from the potential flow solution and the maximum Mach number on the airfoil surface are presented. The variation of the threshold parameter has a large influence on the number of grid cells, which almost increases by a factor of three. Without vortex–correction, the aerodynamic lift is underestimated by about 3%, measured on the finest mesh. Application of the vortex–correction greatly enhances the solution. On the fine grid, the potential flow solution is reached very accurately. On the coarser grid, the aerodynamic lift is underpredicted by about 2%, even with applied vortex–correction. This deficit results from insufficient resolution about the stagnation area at the leading edge. The maximum Mach number only reaches a value of $M_{max} \approx 0.964$, compared with $M_{max} \approx 0.987$ for the fine grid.

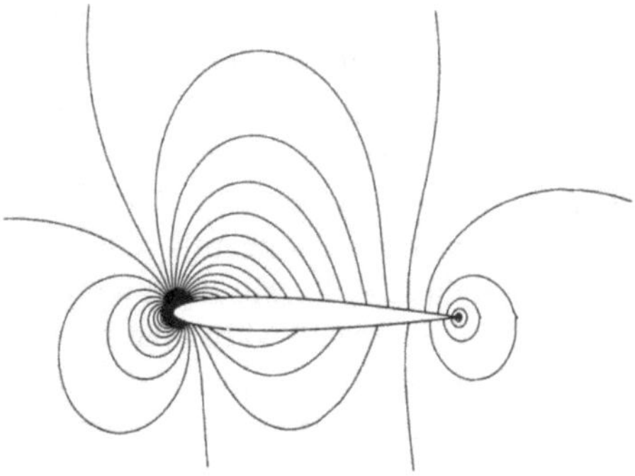

Figure 32 Mach number contours for NACA0012 airfoil, $M_\infty = 0.63$, $\alpha = 2.0°$. $M_{min} = 0.0$, $M_{max} = 0.975$, $\Delta M = 0.025$

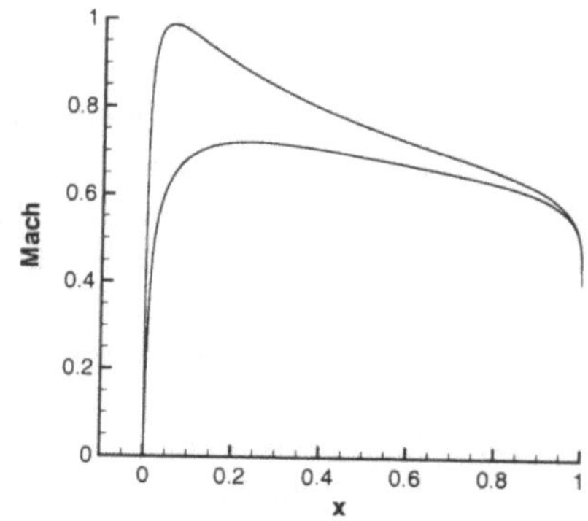

Figure 33 Mach distribution for NACA0012 airfoil, $M_\infty = 0.63$, $\alpha = 2.0°$

Table 4 Aerodynamic coefficients for NACA0012 airfoil, $M = 0.63$, $\alpha = 2.0°$

	$\varepsilon = 5.0 \cdot 10^{-2}$		$\varepsilon = 3.0 \cdot 10^{-3}$	
vortex correction	**no**	**yes**	**no**	**yes**
number of cells	5638	5962	17035	17083
C_L	0.3138246	0.3266950	0.3223484	0.3329005
1-$C_L/C_{L,pot}$ [%]	5.7583783	1.8933933	3.1986786	0.0298798
M_{max}	0.9553167	0.9641491	0.9787976	0.9872640

6.2 Transonic SFB 401 Cruise Configuration

We revisit the inviscid flow around the SFB 401 cruise configuration at $M_\infty = 0.85$, $\alpha = 0°$. In Section 5.2.3, this test case has been investigated using a regular, non-adaptive structured grid. In the following study, the fully adaptive algorithm is utilized. Computations are initialized on a grid consisting of four blocks, each with a resolution of 16x16 cells, see Figure 34. Adaptation is carried out each time the residual decreased to $2.5 \cdot 10^{-4}$, based on the initial residual, measured in the L_1 norm of the density. 14 adaptation cycles have been conducted, with a maximum refinement level of $L_{max} = 9$. Adaptation is applied to the set of primitive variables. The threshold value for the multiscale analysis is $\varepsilon = 8 \cdot 10^{-2}$.

Figure 35 and Figure 36 present computational grids and corresponding pressure distributions for three different stages of adaptation. Namely, for grid no. 1, 4 and 14. Figure 37 shows the corresponding pressure distribution on the airfoil surface for grid no. 1 and grid no. 14. On the coarsest mesh, the shock at the trailing edge of the upper surface can already be identified. On the lower surface, the shock at $x \approx 0.53$ is very smeared while the shock near the leading edge cannot be identified at all. After 3 cycles of adaptation, all three shocks are present (Figure 35, Figure 36). The mesh contains 7405 cells, with a maximum grid level $L = 4$. The locally finest resolution of the grid approximately corresponds to the resolution of a structured, non-adaptive mesh with 256x32 cells, as used in Section 5.2.3. With further adaptation, the shock resolution can still be improved significantly. On grid no. 14, all shocks are highly resolved. The mesh contains 28288 cells, with a maximum grid level $L = 9$. A large number of cells is concentrated within the shock regions. Although the number of grid cells of the finest mesh appears to be rather high, compared with the structured grid used in Section 5.2.3, it has to be emphasized that the shock resolution is greatly improved by the adaptive scheme. Further, to obtain the same solution quality, the utilization of a structured, non-adaptive grid would prospectively require a significantly higher number of cells than the locally adapted grid. In addition, the process of adjusting a structured grid to the shock location requires a detailed a-priori knowledge of the solution to provide adequate grid resolution.

The aerodynamic lift coefficient is $C_L = 0.514092$, measured on the finest mesh (grid no. 14). In Section 5.2.3 we determined a lift coefficient of $C_L = 0.463527$ (FLOWer) and $C_L = 0.497806$ (QUADFLOW), respectively. This reflects a difference of 10% (FLOWer) and 3% (QUADFLOW), compared with the locally adapted solution.

Figure 38 presents the evolution of the number of grid cells for different stages of adaptation. During the initial phase, the mesh size rapidly increases with each adaptation step. With further adaptation, the number of cells converges against a constant. It is important to note, that large regions of the shocks have not been resolved by the highest refinement level possible ($L_{max} = 9$), but have only reached a lower level of adaptation $L < L_{max}$. I.e., convergence of the mesh size has not been enforced by reaching the highest grid level. Even in the presence of shocks, the adaptation process converges within the range of the prescribed threshold value.

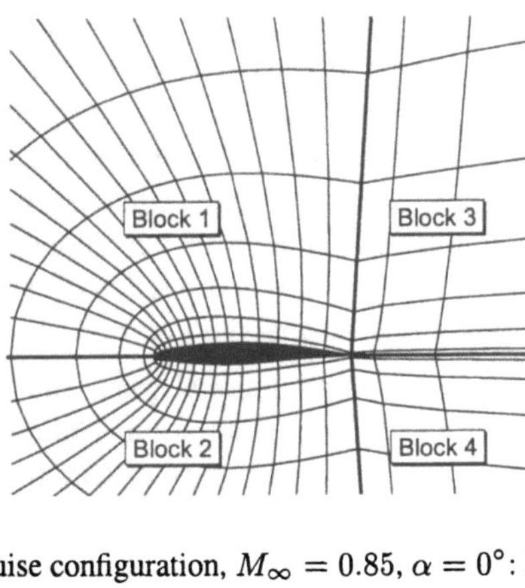

Figure 34 SFB 401 cruise configuration, $M_\infty = 0.85$, $\alpha = 0°$: Initial grid (partial view)

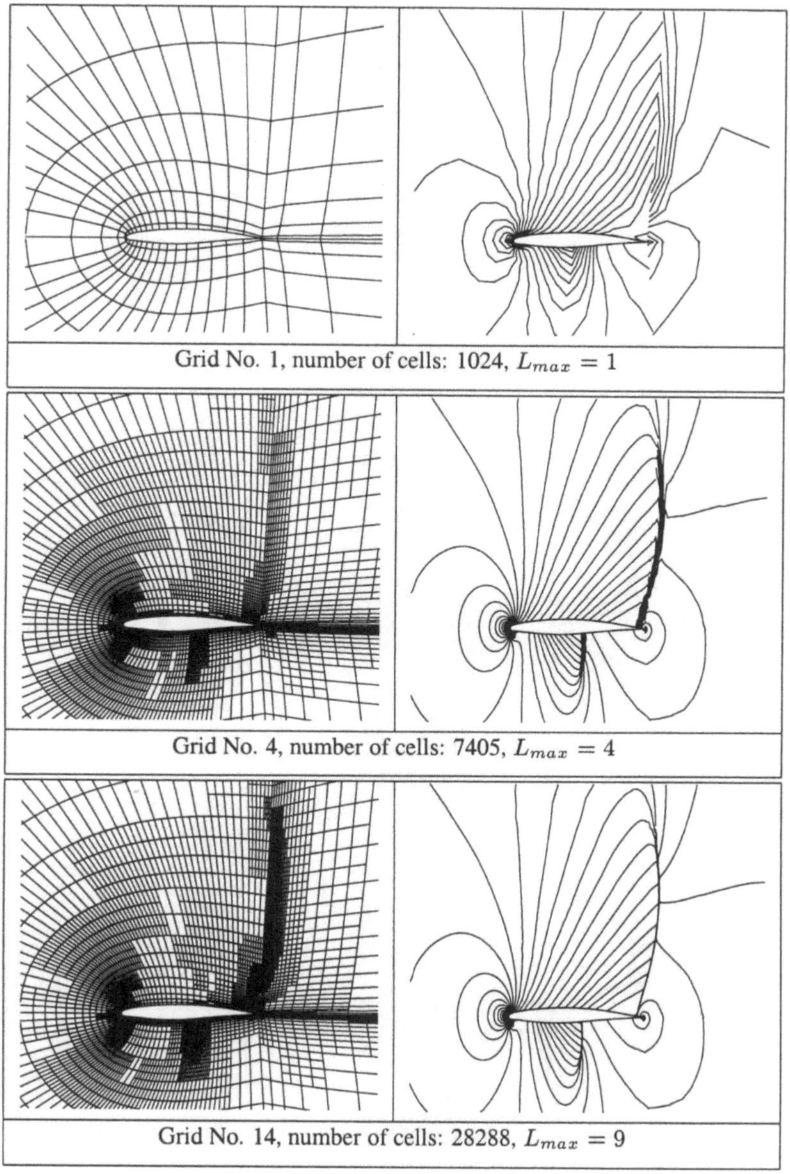

Figure 35 SFB 401 cruise configuration, $M_\infty = 0.85$, $\alpha = 0°$. *Left Figure*: Computational grid. *Right Figure*: c_p distribution, $c_{p,min} = -1.35$, $c_{p,max} = 1.2$, $\Delta c_p = 0.075$

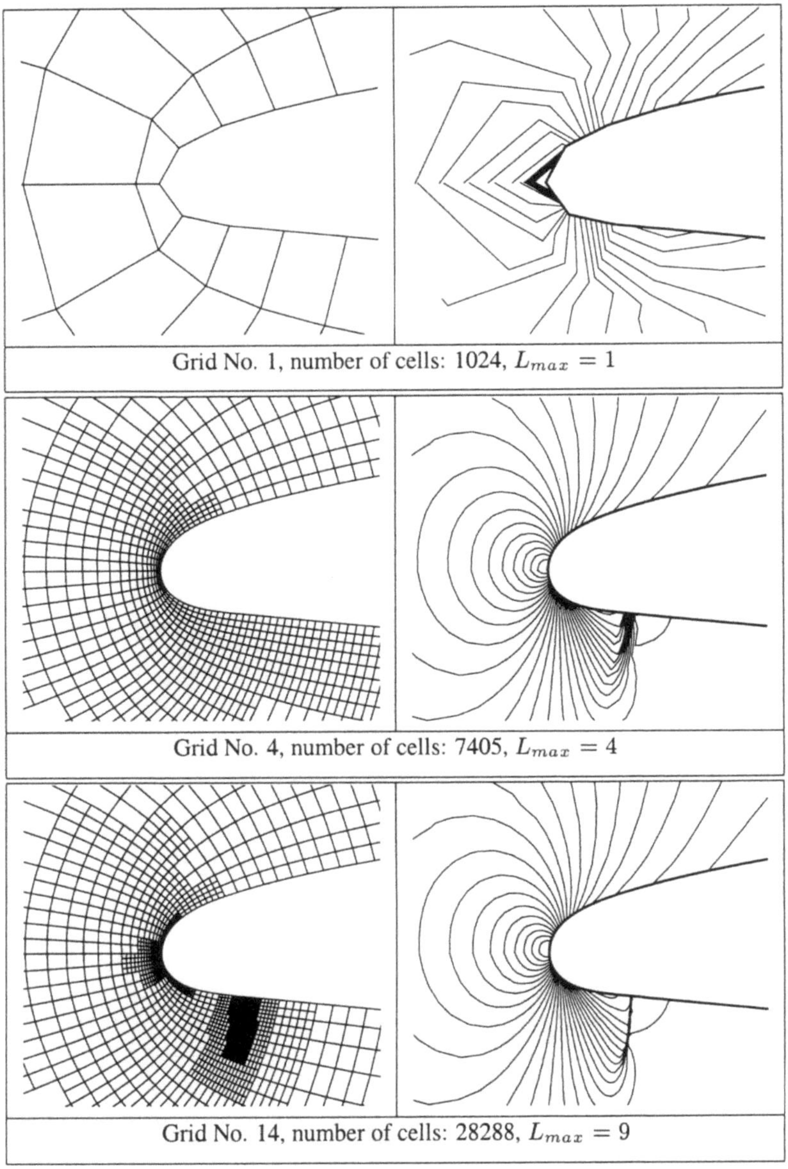

Figure 36 SFB 401 cruise configuration: Detailed view of stagnation area, $M_\infty = 0.85$, $\alpha = 0°$. *Left Figure*: Computational grid. *Right Figure*: c_p distribution, $c_{p,min} = -1.35$, $c_{p,max} = 1.2$, $\Delta c_p = 0.075$

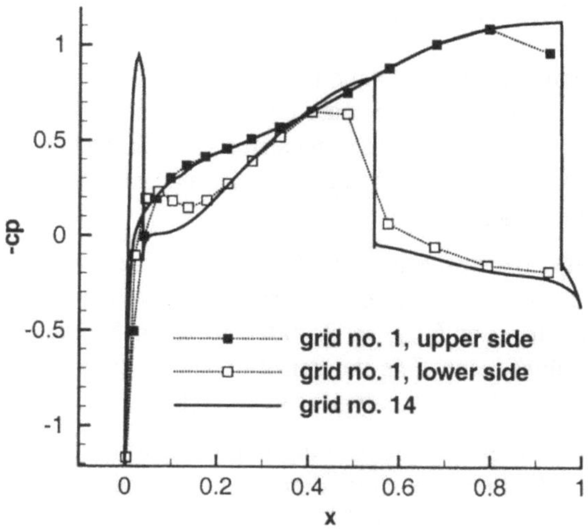

Figure 37 Pressure distribution for SFB 401 cruise configuration, $M_\infty = 0.85$, $\alpha = 0°$

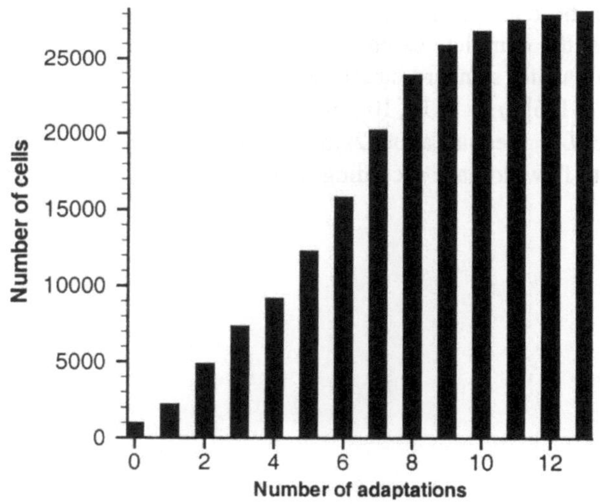

Figure 38 Number of grid cells during adaptation process of SFB 401 cruise configuration, $M_\infty = 0.85$, $\alpha = 0°$

6.3 Transonic NACA0012 airfoil

The next test case concerns the transonic flow over a NACA0012 airfoil at $M_\infty = 0.95$, $\alpha = 0°$ (AGARD reference test case 03 [68]). The far-field boundary is located about 20 chord lengths away from the airfoil. Standard characteristic boundary conditions are applied at the far-field, where the incoming Riemann invariant is set by free stream quantities.

Computations are initialized on a structured grid consisting of 4 blocks with a resolution of 20x20 cells each. 13 cycles of adaptation have been performed with a maximum refinement level of $L_{max} = 8$. Adaptation has been carried out on the set of primitive variables, each time the density residual decreased to 10^{-4}. The threshold value for the multiscale analysis is $\varepsilon = 4 \cdot 10^{-2}$.

The flow pattern downstream of the trailing edge is characterized by a complex shock configuration. Two oblique shocks are formed at the trailing edge. The remaining supersonic region behind the oblique shocks is closed by a further normal shock. This configuration is often related to as so-called *fish-tail*. Figure 39 presents the locally adapted grid and the corresponding Mach distribution in the vicinity of the airfoil, after 13 cycles of adaptation. The grid consists of 55084 cells. All three shocks are highly resolved by the adaptive grid. The position of the normal shock is located at $x \approx 2.1721$ chord length behind the trailing edge of the airfoil.

Figure 40 presents a total view of the shock configuration. The oblique shocks extend about 10 to 12 chord lengths into the flow domain. The adaptive grid provides high resolution over the complete extent of the shocks. Such a high shock resolution is not feasible using standard structured grids. Discretization of the shock region between $x \in [1, 5]$, $y \in [-10, 10]$, by a uniform structured mesh according to a refinement level $L = 8$ equals about $29.5 \cdot 10^6$ grid cells. A uniform discretization of the complete flow domain according to $L = 8$ would result in about 10^8 cells.

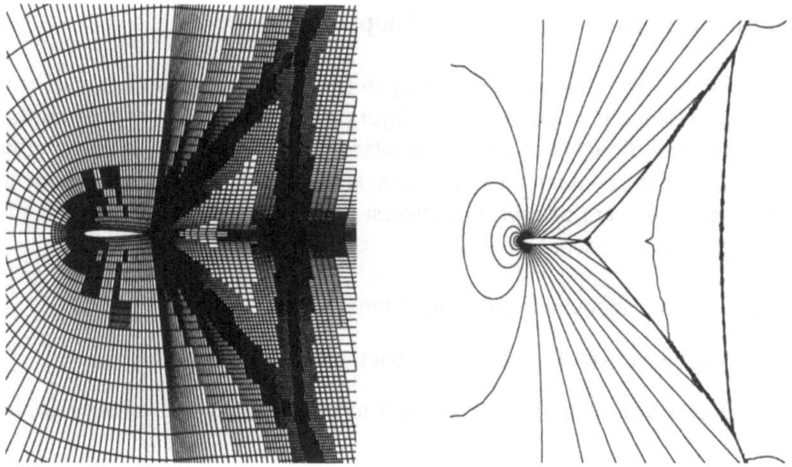

Figure 39 Partial view of NACA0012 airfoil, $M = 0.95$, $\alpha = 0.0°$. *Left Figure*: Computational grid. *Right Figure*: $Mach$ distribution, $M_{min} = 0.0$, $M_{max} = 1.45$, $\Delta M = 0.05$

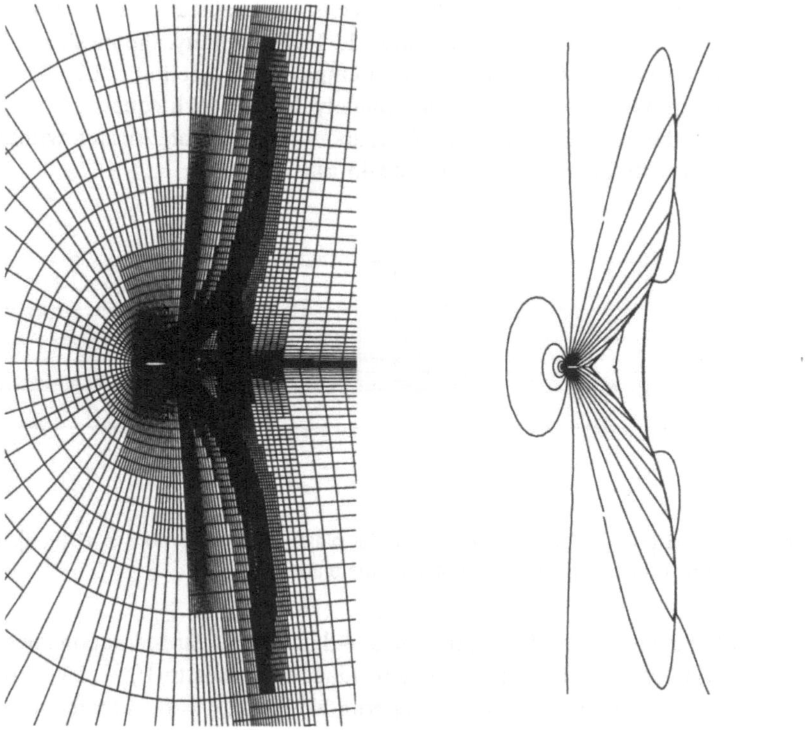

Figure 40 Total view of NACA0012 airfoil, $M = 0.95$, $\alpha = 0.0°$. *Left Figure*: Computational grid. *Right Figure*: $Mach$ distribution, $M_{min} = 0.0$, $M_{max} = 1.45$, $\Delta M = 0.05$

6.4 Laminar Boundary Layer using Adaptation

In this section, we will demonstrate the capability of the multiresolution scheme to automatically resolve laminar boundary layers. The application of the multiscale analysis to viscous flows is of particular interest, since the method has originally been developed in the context of hyperbolic systems of partial differential equations. The following issues are of essential interest for the successful application of the adaptive algorithm to viscous flows:

1. detection and adequate resolution of the boundary layer
2. appropriate choice of the first grid spacing normal to the wall
3. convergence of the multiscale analysis in presence of steep velocity gradients

All three issues will be addressed here.

We reconsider the laminar flow over a flat plate extending on the x-axis in the domain $x \geq 0$. Flow conditions are equivalent to that used in Section 5.2.2, where a regular structured, non-adaptive grid has been employed for validation of the basic scheme. The initial grid consists of 20×8 cells (160 cells all together), with 12 cells located on the plate itself, see Figure 41. For simplicity, the mesh has been generated by coarsening the structured grid employed in Section 5.2.2. The first grid spacing normal to the wall is approximately 10^{-2}, which equals $y^+ \approx 10$ at $Re = 10^4$. At the trailing edge of the plate, about four cells are located within the boundary layer with thickness δ, see Figure 43(a).

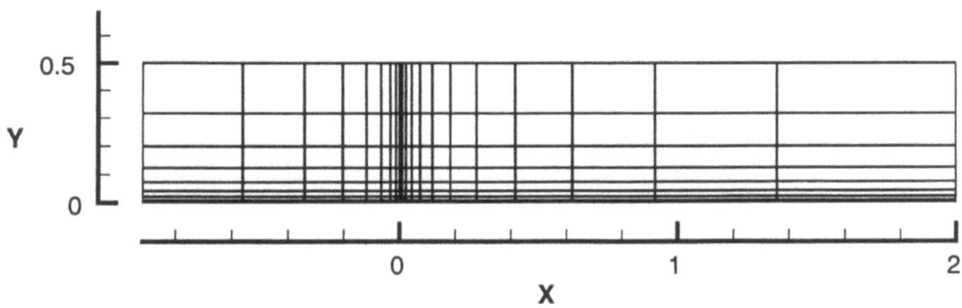

Figure 41 Initial grid for laminar flow over a flat plate. Plate extension: $x \in [0, 2]$. Grid resolution: 20×8 cells with 12 cells located on the plate

10 cycles of adaptation have been performed. Adaptation is carried out on the set of primitive variables, each time the density residual decreased to 10^{-6}, based on its initial value. The highest refinement level permitted is $L_{max} = 8$. A threshold value of $\varepsilon = 8 \cdot 10^{-3}$ is chosen for the multiscale analysis. Figure 42 shows the locally adapted grid after 10 cycles of adaptation. The grid consists of 3595 cells, which is about 36% of the number of cells being used for the structured grid employed in Section 5.2.2. The boundary layer is automatically detected by the present adapta-

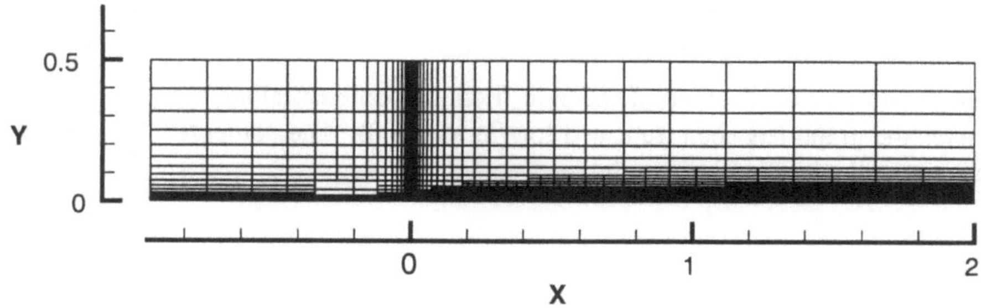

Figure 42 Locally adapted grid for laminar flow over a flat plate. Plate extension: $x \in [0, 2]$. Grid resolution: 3595 cells

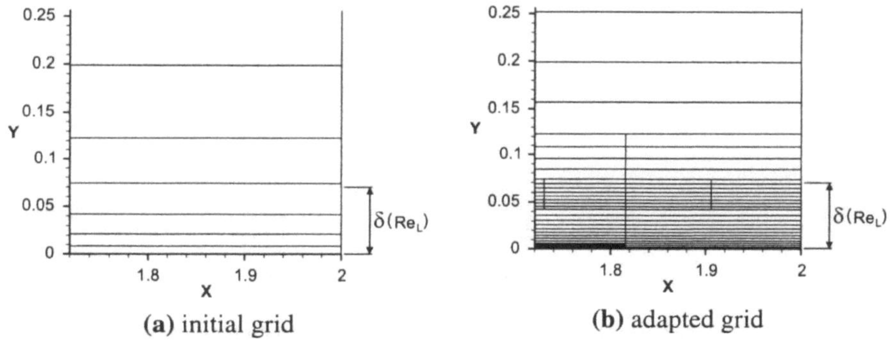

Figure 43 Partial view of trailing edge of flat plate. δ = boundary layer thickness according to Blasius solution

tion criteria. The detailed view of the trailing edge (Figure 43(b)) shows that after reaching the border of the boundary layer ($y > \delta$), the mesh resolution is immediately decreased. The boundary layer is resolved by about 25 cells, at the trailing edge. The first spacing normal to the wall is about 10^{-3}, which equals $y^+ \approx 1$. It is important to note, that the resolution of the first grid spacing corresponds to a refinement level of $L = 4$. I.e., the choice of the first grid spacing has not been enforced by reaching the highest refinement level possible ($L_{max} = 8$). The highest grid level has only been reached in vicinity of the stagnation point at the leading edge of the plate. This fact indicates that even in the presence of steep gradients within the boundary layer, the multiscale analysis converges within the range of the prescribed threshold parameter. The number of grid cells remains bounded during the adaptation process.

Figure 44(a) and Figure 44(b) present a comparison between the computed u-velocity and v-velocity profiles with the Blasius solution at $Re_x = 9188.75$, respectively. Both velocity components are in good agreement with the theoretical solution. In accordance with the results obtained in Section 5.2.2, the v-velocity is slightly overpredicted for $\eta > 4$ by the numerical scheme. We conclude that this discrepancy is independent of the grid resolution.

Figure 45 shows evolution of the skin friction coefficient along the plate (*please note*: the graph has been truncated for $Re < 1000$). The computed skin friction agrees well with the Blasius solution.

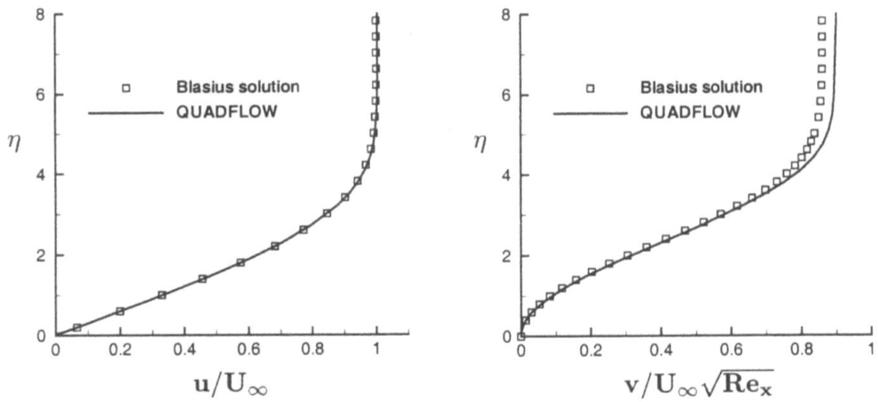

(a) u-velocity profile

(b) v-velocity profile

Figure 44 Boundary layer profiles at $Re_x = 9188.75$ for the laminar flow over an isothermal flat plate after 10 cycles of adaptation ($M_\infty = 0.2$, $Re_\infty = 10^4$, $T_w = T_\infty$)

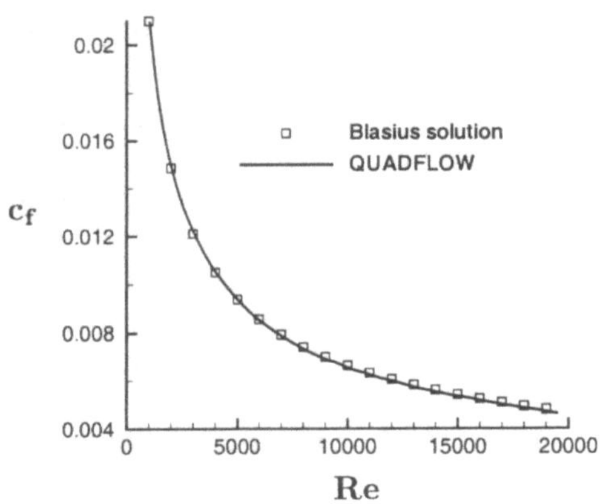

Figure 45 Distribution of the skin friction coefficient for the laminar flow over an isothermal flat plate as function of Reynolds number after 10 cycles of adaptation ($M_\infty = 0.2$, $Re_\infty = 10^4$, $T_w = T_\infty$)

References

[1] R. Abgrall, *Multiresolution analysis on unstructured meshes: Applications to CFD*, in Experimentation, modeling and computation in flow, turbulence and combustion, eds. Chetverushkin and al., John Wiley & Sons, 1997

[2] B. Bihari, A. Harten, *Multiresolution schemes for the numerical solution of 2–D conservation laws I*, SIAM J. Sci. Comput., 18, no. 2 (1997), 315–354

[3] J.M. Carnicer, W. Dahmen, J.M. Peña, *Local decomposition of refinable spaces and wavelets*, Appl. Comput. Harmon. Anal., 3 (1999), 127-153

[4] G. Chiavassa, R. Donat, *Point value multiresolution for 2D compressible flows*, SIAM J. Sci. Comput., 23, no. 3 (2001), 805-823

[5] B. Cockburn, F. Coquel, P. LeFloch, *An error estimate for finite volume methods for multidimensional conservation laws*, Math. Comp., 63 (1994), 77-103

[6] A. Cohen, N. Dyn, S.M. Kaber, M. Postel, *Multiresolution finite volume schemes on triangles*, J. Comp. Phys., 161 (2000), 264-286

[7] A. Cohen, I. Daubechies, J. Feauveau, *Bi–orthogonal bases of compactly supported wavelets*, Comm. Pure Appl. Math., 45 (1992), 485–560

[8] A. Cohen, S.M. Kaber, S. Müller, M. Postel, *Fully adaptive multiresolution finite volume schemes for conservation laws*, Math. Comp., posted on December 5, 2001, PII S 0025-5718(01)01391-6 (to appear in print)

[9] C.M. Dafermos, *Hyperbolic Conservation Laws in Continuum Physics*, Springer Verlag, Berlin, Grundlehren der Mathematischen Wissenschaften, 2000

[10] W. Dahmen, *Some remarks on multiscale transformations, stability and biorthogonality*, in Wavelets, Images and Surface Fitting, eds. P.J. Laurent, A. Le Méhauté, L.L. Schumaker, A.K. Peters, Wellesley, 157–188, 1994

[11] W. Dahmen, B. Gottschlich–Müller, S. Müller, *Multiresolution schemes for conservation laws*, Numer. Math., 88 (2001), 399-443

[12] W. Dahmen, A. Kunoth, K. Urban, *Biorthogonal spline-wavelets on the interval – stability and moment conditions*, Appl. Comput. Harmon. Anal., 6 (1999), 132-196

[13] B. Gottschlich–Müller, *Multiscale Schemes for Conservation Laws*, PhD thesis, RWTH Aachen, Shaker–Verlag, 1998

[14] Gottschlich–Müller, B., Müller, S., *Application of Multiscale Techniques to Hyperbolic Conservation Laws*, in: Computational Mathematics, Lecture Notes in Pure & Applied Mathematics, **202**, eds. Z. Chen, Y. Li, Ch. A. Micchelli, Y. Xu, Harry–Dekker–Verlag, New York, 1998, 113–138

[15] Gottschlich–Müller, B., Müller, S., *Adaptive Finite Volume Schemes for Conservation Laws based on Local Multiresolution Techniques*, in: Hyperbolic Problems: Theory, Numerics, Applications, eds. M. Fey and R. Jeltsch, Birkhäuser, 1999, 385–394

[16] Harten, A., *Adaptive multiresolution schemes for shock computations*, J. Comp. Phys., 115 (1994), 319–338

[17] Harten, A., *Multiresolution algorithms for the numerical solution of hyperbolic conservation laws*, Comm. Pure Appl. Math. 12, vol. 48 (1995), 1305–1342

[18] P. Houston, J.A. Mackenzie, E. Süli, G. Warnecke, *A posteriori error analysis for numerical approximations of Friedrichs systems*, Numer. Math., 82 (1999), 433-470

[19] D. Kröner, S. Noelle, M. Rokyta, *Convergence of higher-order upwind finite volume schemes on unstructured grids for scalar conservation laws in several space dimensions*, Numer. Math., 71 (1995), 527-560

[20] D. Kröner, M. Ohlberger, *A posteriori error estimates for upwind finite volume schemes for nonlinear conservation laws in multi dimensions*, Math. Comp., 69 (1999), 25-39

[21] S.N. Kruzhkov, *First order quasilinear equations with several space variables*, Math. USSR Sb., 10 (1970), 217–243 (In Russian)

[22] N.N. Kuznetsov, *The weak solution of the Cauchy problem for a multi–dimensional quasilinear equation*, Mat. Zametki, 2 (1967), 401–410 (In Russian)

[23] Müller, S., *Erweiterung von ENO–Verfahren auf zwei Raumdimensionen und Anwendung auf hypersonische Stauspunktprobleme*, PhD thesis, RWTH Aachen, Shaker–Verlag, 1993

[24] Müller, S., *Adaptive Multiscale Schemes for Conservation Laws*, Habilitation thesis, RWTH Aachen, 2001

[25] S. Müller, A. Voss, *A Manual for the Template Class Library* `igpm_t_lib`, IGPM–Report 197, RWTH Aachen, 2000

[26] F. Schröder–Pander, T. Sonar, O. Friedrich, *Generalized multiresolution analysis on unstructured grids*, Numer. Math., 86 (2000), 685-715

[27] T. Sonar, V. Hannemann, D. Hempel, *Dynamic adaptivity and residual control in unsteady compressible flow computation*, Math. and Comp. Modelling, 20 (1994), 201-213

[28] T. Sonar, E. Süli, *A dual graph–norm refinement indicator for finite volume approximations of the Euler equations*, Num. Math., 78 (1998), 619-658

[29] *Grid Generation, Finite Elements and Geometric Design*, eds. B. Hamann, R.F. Sarraga, special issue, CAGD 12 (1995)

[30] K.H. Brakhage, *High Quality Mesh Generation and Sparse Representation Using B-Splines*, in Proceedings of the 7th International Conference on Numerical Grid Generation in Computational Field Simulations, eds. B.K. Soni, J. Häuser, J.F. Thompson, P. Eiseman, 753-762, Chateau Whistler Resort, British Columbia, 2000

[31] K.H. Brakhage, S. Müller, *Algebraic-Hyperbolic Grid Generation with Precise Control of Intersection of Angles*, Int. J. Numer. Math. Fluids 33(2000), 89-123

[32] K.H. Brakhage, Ein menügesteuertes intelligentes System zur zwei- und dreidimensionalen Computergeometrie, VDI Verlag, series 20: Rechnergestütze Verfahren, no. 26, 1990

[33] K.H. Brakhage, CAG — Computer Aided Geometry, user manual (in German), Institut für Geometrie und Praktische Mathematik, 1986-1999

[34] C. de Boor, *A practical Guide to Splines*, Springer-Verlag, New York, 1978

[35] S.A. Coons, *Surface Patches and B-Spline Curves*, in Computer Aided Geometric Design, eds. R.E. Barnhill, R.F. Riesenfeld, Academic Press, 1974

[36] M. Eck, J. Hadenfeld, *Local Energy Fairing of B-Spline Curves*, Computing Supplementum 10, Geometric Modelling, 129-147, Springer, 1995

[37] M. Eck, J. Hadenfeld, *Knot Removal for B-Spline Curves*, CAGD 12(1995), 259-282

[38] G. Farin, *Curves and Surfaces for Computer Aided Geometric Design*, 2nd edition, Academic Press, 1990

[39] G. Farin, *Commutativity of Coons and Tensor Product Operations*, Rocky Mountain Journal of Mathematics 22(1992), 541-546

[40] G. Farin, G. Rein, N.S. Sapidis, A.J. Worsey, *Fairing cubic B-Spline curves*, CAGD 4(1987), 91-103

[41] W.J. Gordon, C.A. Hall, *Construction of Curvilinear Coordinate Systems and Applications to Grid Generation*, International Journal for Numerical Methods in Engineering, 7(1973), 461-477

[42] C.A. Hall, W.W. Meyer, *Optimal Error Bounds for Cubic Spline Interpolation*, Journal of Approximation Theory 16(1976), 105-122

[43] J. Hoschek, D. Lasser, *Grundlagen der geometrischen Datenverarbeitung*, 2nd edition, B.G. Teubner, Stuttgart, 1992

[44] P. Knupp, S. Steinberg, *Fundamentals of Grid Generation*, CRC Press (1994)

[45] J.A. Sethian, *Curvature Flow and Entropy Condition Applied to Grid Generation*, J. Comp. Phys. 115(1994), 440-454

[46] S.P. Spekreijse, *Elliptic Grid Generation Based on Laplace Equations and Algebraic Transformations*, J. Comp. Phys. 118(1995), 38-61

[47] S.P. Spekreijse, J.W. Boerstoel, *Multiblock Grid Generation. Part 1: Elliptic Grid Generation Methods for Structured Grids*, in Computational Fluid Dynamics, VKI Lecture Series 1996-06, ed. H. Deconinck, von Karman Institut for Fluid Dynamics, 1-48, 1996

[48] S.P. Spekreijse, J.W. Boerstoel. *Multiblock Grid Generation. Part 2: Multiblock Aspects*, in Computational Fluid Dynamics, VKI Lecture Series 1996-06, ed. H. Deconinck, von Karman Institut for Fluid Dynamics, 1-39, 1996

[49] P. Batten, M.A. Leschziner, U.C. Goldberg, *Average-State Jacobians and Implicit Methods for Compressible Viscous and Turbulent Flows*, J. Comp. Phys., 137, 1997, 38–78

[50] P.L. Roe, *Approximate Riemann Solvers, Parameter Vectors, and Difference Schemes*, J. Comp. Phys, 43, 1981, 357–372

[51] V. Venkatakrishnan, *Convergence to Steady State Solutions of the Euler Equations on Unstructured Grids with Limiters*, J. Comp. Phys., 118, 1995, 120–130

[52] W.J. Coirer, *An Adaptively Refined, Cartesian, Cell-Based Scheme for the Euler and Navier-Stokes Equations*, Ph.D. Thesis, University of Michigan, 1994

[53] M. Delanaye, M.J. Aftosmis, M.J. Berger, Y. Liu, T.H. Pulliam, *Automatic Hybrid–Cartesian Grid Generatin for High–Reynolds Number Flows around Complex Geometries*, AIAA Paper 99-0777, 1999

[54] D.G. Holmes, S.D. Connell, *Solution of the 2D Navier-Stokes Equations on Unstructured Adaptive Grids*, AIAA Paper 89-1932, 1989

[55] W.K. Anderson, D.L. Bonhaus, *An Implicit Upwind Algorithm for Computing Turbulent Flows on Unstructured Grids*, Computers & Fluids, vol 23, no 1, 1994, 1-21

[56] P. Geuzaine, *An Implicit Upwind Finite Volume Method for Compressible Turbulent Flows on Unstructured Meshes*, PhD Thesis, Universitè de Liège, 1999

[57] C.H. Hirsch, *Numerical Computation of Internal and External Flows: Computational Methods for Inviscid and Viscous Flows*, volume 2, Wiley, 1988

[58] J.L. Thomas, M.D. Salas, *Far–Field Boundary Conditions for Transonic Lifting Solutions to the Euler Equations*, AIAA J., 24, 1986, 1074–1080

[59] Y. Saad, *Iterative Methods for Sparse Linear Systems*, PWS, Boston, 1996

[60] I. Grotowsky, J. Ballmann, *Efficient Time Integration of Navier-Stokes Equations*, Computers & Fluids, vol 28, no 2, 1999, 243–263

[61] K.J. Vanden, P.D. Orkwis, *Comparison of Numerical and Analytical Jacobians*, AIAA Journal, vol 34, no 6, 1996

[62] S. Balay, W. Gropp, L.C. McInnes, B.F. Smith, *PETSc 2.0 Users Manual*, Tech. Rep. ANL-95/11 - Revision 2.0.28, Argonne National Laboratory, 2000, `http://www-fp.mcs.anl.gov/petsc/`

[63] F. Bramkamp, J. Ballmann, S. Müller, *Development of a Flow Solver Employing Local Adaptation Based on Multiscale Analysis on B-Spline Grids*, in: Proceedings of 8th Annual Conf. of the CFD Society of Canada, 2000, 113-118

[64] F. Bramkamp, J. Ballmann, *Solution of the Euler Equations on Locally Adaptive B-Spline Grids*, Notes on Numerical Fluid Mechanics, Vol. 77, 2002, pp. 281-288

[65] F. Ringleb, *Exakte Lösung der Differentialgleichung einer adiabatischen Gasströmung*, ZAMM, 20(4), 1940, 185–198

[66] L.G. Loitsianski, *Laminare Grenzschichten*, Akademie Verlag, Berlin, 1967

[67] K. Becker, N. Kroll, C.C. Rossow, F. Thiele, *Numerical Flow Calculations for Complete Aircraft - the Megaflow Project*, DGLR-JT98-225, DGLR Jahrbuch 1998, 355–364

[68] *Test Cases for Inviscid Flow Field Methods*, AGARD Advisory Report No. 211, 1985

Automated Gradient Calculation

Christian H. Bischof, H. Martin Bücker, Bruno Lang, Arno Rasch
Lehrstuhl für Hochleistungsrechnen, RWTH Aachen
Seffenter Weg 23, D-52074 Aachen, Germany

Summary

Automatic differentiation is a powerful technique for evaluating derivatives of functions given in the form of a high-level programming language such as Fortran, C, or C++. The program is treated as a sequence of elementary statements to which the chain rule of differential calculus is applied mechanically. A key feature of this technique is its ability to generate accurate derivatives rather than approximations obtained from numerical differentiation like divided differences. We survey automatic differentiation and report on preliminary results that have been obtained applying the automatic differentiation tool ADIFOR to a large-scale computational fluid dynamics solver, TFS, developed at Aerodynamisches Institut, Aachen University of Technology. This solver comprises approximately 24,000 lines of Fortran 77 and is able to resolve steady and unsteady laminar and turbulent flows in two and three dimensions.

1 Introduction

Given a numerical simulation of a large-scale engineering problem arising from, for instance, computational fluid dynamics (CFD), one is typically interested in a qualitative and quantified analysis of the dependences of certain program outputs on certain input parameters. Such information is invaluable for optimizing main output objectives with respect to suitable decision variables or for the identification of model parameters with respect to given experimental data. One particular topic of investigation within the collaborative research center (SFB) 401 at Aachen University of Technology is the optimization of take-off and landing frequencies of aircrafts [10, 11]. A detailed understanding of the near wake of a wing/engine configuration is needed. The simulation of the wake flow field and the pressure field is therefore investigated by aeronautical engineers.

When optimizing such output objectives of a large CFD simulation, one alternative is to use evolutionary- or genetic-type algorithms. This class of algorithms is often used because of their ability to escape local minima and because they do not re-

The original version of this chapter was revised. Author name in this chapter was incorrectly captured as H. Martin Bicker. The author name should read as H. Martin Bücker. This has been corrected. A correction to this chapter can be found at DOI 10.1007/978-3-540-44866-2_16.

quire any derivative information. The latter argument is often brought forward if the function is too complicated to calculate its derivatives by hand as in the case where the function is defined by the solution of a CFD problem. However, derivatives are a key ingredient to achieve rapid convergence in most sophisticated algorithms for local optimization [22]. We will demonstrate that the technique of automatic differentiation (AD) is capable of providing exact derivatives—as opposed to approximations of derivatives obtained from numerical differentiation based on finite differencing—for arbitrarily complicated functions in an efficient way. To this end, we apply automatic differentiation to a large-scale computational fluid dynamics solver, TFS, developed at Aerodynamisches Institut of Aachen University of Technology. More precisely, we use the automatic differentiation system ADIFOR [3, 4] to compute the derivatives of the density with respect to the angle of attack. Automatic differentiation in the context of computational fluid dynamics is considered in more detail in [21].

The goal of this new project is twofold. On the one hand, derivatives provide valuable sensitivity information that lead to a better understanding of the simulation results and can be used for improving the robustness of the design. On the other hand, they allow the combination of genetic-type optimization techniques and local optimization methods, thus yielding hybrid optimization algorithms.

The structure of this article is as follows. In Section 2, several traditional approaches for computing derivatives are reviewed. A less widely known alternative, automatic differentiation, is surveyed in Section 3. To illustrate the differences between the approaches, a deliberately simple function is considered in Section 4. The purpose of Section 5 is to introduce two different ways of implementing the AD technology: operator overloading and source transformation. In Section 6, available AD tools are briefly mentioned. Preliminary results of automatically differentiating TFS are reported in Section 7. Finally, we outline future research directions in Section 8.

2 Calculating Derivatives

Derivatives are not only useful in the optimization problems mentioned above. They are also ubiquitous in various areas of computational techniques including the solution of nonlinear systems, stiff ordinary differential equations, partial differential equations, differential-algebraic equations, and sensitivity analysis, to name a few. Such computational techniques require the evaluation of derivatives of some objective function. In large-scale scientific applications, the objective function

$$f : \mathbb{R}^n \to \mathbb{R}^m \tag{1}$$

is typically not available in analytic form but is given by a computer code written in a high-level programming language such as Fortran, C, or C++. Think of f as the function computed by, say, a (parallel) computational fluid dynamics code consisting of hundreds of thousands lines that simulates the flow in a complex three-dimensional geometry. Given such a representation for evaluating the objective function

$$f(x) = \left(f_1(x), f_2(x), \ldots, f_m(x)\right)^T$$

at some point of interest $x \in \mathbb{R}^n$, computational methods often demand the evaluation of the Jacobian matrix

$$f'(x) := \begin{pmatrix} \frac{\partial}{\partial x_1} f_1(x) & \cdots & \frac{\partial}{\partial x_n} f_1(x) \\ \vdots & \ddots & \vdots \\ \frac{\partial}{\partial x_1} f_m(x) & \cdots & \frac{\partial}{\partial x_n} f_m(x) \end{pmatrix} \in \mathbb{R}^{m \times n} \tag{2}$$

at the same point.

In the following three subsections, we describe different approaches available for evaluating derivatives. We start with numerical differentiation and also briefly point out symbolic and analytic differentiation. Another alternative is automatic differentiation, which is outlined in the next section.

2.1 Numerical Differentiation

A well-known and widely used approach for the approximation of the Jacobian matrix is divided differences (DD). For the sake of simplicity, we mention only first-order forward DD but stress that the following discussion applies to DD as a technique of numerical differentiation in general. Using first-order forward DD, one can approximate the ith column of the Jacobian matrix (2) by

$$\frac{f(x + h_i e_i) - f(x)}{h_i}, \tag{3}$$

where h_i is a suitably-chosen step size and $e_i \in \mathbb{R}^n$ is the ith Cartesian unit vector. An advantage of the DD approach is that only the function f needs to be evaluated at some suitably chosen points. Roughly speaking, f is used as a black box evaluated at some points. The main disadvantage of DD is that the accuracy of the approximation depends crucially on a suitable choice of these points, specifically, of the step sizes h_i. There is always the dilemma that the step size should be small in order to decrease the truncation error of (3) and that, on the other hand, the step size should be large to avoid cancellation errors using finite-precision arithmetic when evaluating the numerator in (3).

2.2 Symbolic Differentiation

Another option is symbolic differentiation by computer algebra packages such as Macsyma [23] or Mathematica [25]. Here, the term "symbolic differentiation" is used in the traditional sense, i.e., computations based on expressions using formulae. Note that some computer algebra packages also implement automatic differentiation. For instance, programs written in Maple [20] syntax can be differentiated automatically by Maple's `codegen` package.

In computer algebra packages, functions are represented by expressions so that symbolic differentiation manipulates these expressions to obtain expressions for the derivatives. Unfortunately, because of the rapid growth of the underlying explicit expressions for the derivatives, traditional symbolic differentiation is currently inefficient [15]. The example given in Section 4 illustrates this issue further.

2.3 Analytic Differentiation by Hand

Another alternative is to first deduce an analytic expression for the corresponding derivative and then implement a computer program for its actual computation by hand. Deriving an analytic expression for $f'(x)$ is often inadequate. Moreover, implementing such an analytic expression by hand is challenging, error-prone, and time-consuming. Hence, other approaches are typically preferred.

3 Automatic Differentiation

Another technique for computing derivatives of a function is automatic differentiation (AD) [16, 24]. Given a program in virtually any high-level programming language like Fortran, C, or C++ that evaluates a function f of type (1) at an arbitrary point $x \in \mathbb{R}^n$, automatic differentiation generates another program evaluating not only $f(x)$ but also the $m \times n$ Jacobian matrix $f'(x)$ at the same point. The key concept behind AD as we will explain below is the fact that every computation, no matter how complicated, is executed on a computer as a—potentially very long—sequence of a limited set of elementary arithmetic operations such as additions, multiplications, and intrinsic functions such as `sin()` and `cos()`. By applying the chain rule of differential calculus over and over again to the composition of those elementary operations, $f(x)$ and $f'(x)$ can be evaluated almost up to machine precision. Besides the advantage of accuracy, AD requires minimal human effort and has proven to be more efficient than divided differences under a wide range of circumstances [2, 8, 17].

3.1 Forward Mode

We will illustrate the so-called forward mode of AD in its simplest form using a very short piece of code. It should, however, be stressed that the AD technology scales up to programs with several hundreds of thousands lines. Our simple example computes the arithmetic mean a and the geometric mean g of n numbers $x_1, \ldots, x_n$ (thus, $f(x) = (a, g)^T$ and $m = 2$) and is given by

```
a ← 0
g ← 1
for i = 1 to n do
   a ← a + x_i
   g ← g * x_i
a ← a/n
g ← exp(ln(g)/n)
```

From a conceptual point of view, AD works as follows. In a first step, complex expressions in assignments are decomposed by introducing intermediate quantities until only elementary operations of the form

$v \leftarrow c$		(constant)
$v \leftarrow v_i + v_j$	or $v \leftarrow v_i - v_j$	(previously computed quantities)
$v \leftarrow v_i * v_j$	or $v \leftarrow v_i / v_j$	
$v \leftarrow \mathrm{func}(v_i)$		(standard function, e.g., exp, sin, ...)

remain. Proceeding this way, the last line of the above algorithm might be replaced with the sequence

$$v_1 \leftarrow \ln(g)$$
$$v_2 \leftarrow v_1/n$$
$$g \leftarrow \exp(v_2)$$

As the derivatives of each elementary operation are known, e.g., in the case of the multiplication $v \leftarrow v_i * v_j$

$$\mathbf{v}' = v_i \mathbf{v_j}' + v_j \mathbf{v_i}'$$

by virtue of the product rule, the "only" thing automatic differentiation has to do is to accumulate these derivatives of the elementary operations according to the chain rule in order to give the derivatives of the overall function f. To this end, each intermediate quantity v of the algorithm is associated with a *derivative object* $\mathbf{v}'$ that is updated whenever v changes.

Since in our example we are interested in the derivatives with respect to $x_1, \ldots, x_n$, all derivative objects are length-n row vectors. In the resulting code, which is given below, boldface is used for the derivative objects to indicate that loops or subroutine calls may be hidden within the respective statements.

```
initialize x_i' for i = 1,...,n
a ← 0 ; a' ← 0
g ← 1 ; g' ← 0
for i = 1 to n do
```

$$
\begin{aligned}
&\quad a \leftarrow a + x_i \ ; \ \mathbf{a}' \leftarrow \mathbf{a}' + \mathbf{x_i}' \\
&\quad \mathbf{g}' \leftarrow g\mathbf{x_i}' + x_i\mathbf{g}' \ ; \ g \leftarrow g * x_i \\
&a \leftarrow a/n \ ; \ \mathbf{a}' \leftarrow \mathbf{a}'/n \\
&v_1 \leftarrow \ln(g) \ ; \ \mathbf{v_1}' \leftarrow \mathbf{g}'/g \\
&v_2 \leftarrow v_1/n \ ; \ \mathbf{v_2}' \leftarrow \mathbf{v_1}'/n \\
&g \leftarrow \exp(v_2) \ ; \ \mathbf{g}' \leftarrow g\mathbf{v_2}'
\end{aligned}
$$

Execution of the code yields the desired derivatives

$$
f'(x) = \begin{pmatrix} \mathbf{a}' \\ \mathbf{g}' \end{pmatrix} .
$$

Note that in the sixth line the code for updating $\mathbf{g}'$ must precede the update of g because the old value of g is used in the product rule, whereas the rule for exponentiation (last line) makes use of the new value g, so that g must be updated before $\mathbf{g}'$. In the remaining lines both orders are valid.

The cost for evaluating the derivatives depends strongly on the initial values for the $\mathbf{x_i}'$, which we have not yet discussed.

- Consider the most obvious choice in which $\mathbf{x_i}'$ are initialized with the gradient ∇x_i of x_i with respect to x. In this case, the ith unit vector is taken, i.e., $\mathbf{x_i}' \leftarrow e_i^T = (0, \ldots, 0, 1, 0, \ldots, 0) \in \mathbb{R}^{1 \times n}$, and the above sequence of operations will evaluate the full *Jacobian* $f'(x) \in \mathbb{R}^{2 \times n}$ along with $f(x)$. Here the updates of the derivative objects involve length-n vectors, resulting in a roughly n-fold increase of the work and memory as compared to the evaluation of f alone.
- Setting $\mathbf{x_i}' \leftarrow u_i = \nabla x_i \cdot u \in \mathbb{R}$, where $u \in \mathbb{R}^n$ is a given vector, will yield $f'(x) \cdot u$, the *product of the Jacobian with the vector* u *without actually computing the full Jacobian*. In this case all derivative objects are scalars so that the cost for obtaining the derivatives is reduced to a modest multiple of the cost for evaluating f. This is a much more efficient way for computing the directional derivative of f in direction u than first determining $f'(x)$ and then multiplying with u.

We mention once more that AD is not only applicable to small pieces of code as given above but scales up to arbitrarily complicated codes consisting of several hundreds of thousands lines [7]. Software tools such as ADIFOR [3, 4], ADIC [5], or ADOL-C [18] implement the technology of automatic differentiation and can by applied in a black-box fashion. A survey of AD tools will soon be available at

`http://www.autodiff.org.`

Besides the forward mode sketched above, there are further alternatives for the way how derivatives are accumulated in automatic differentiation. They all lead to the same numerical results but they may differ dramatically in their cost in terms of number of operations and storage requirements. Other issues such as the structure of the problem or the sparsity of the Jacobian matrix may also significantly influence the performance of the AD-generated code.

In the context of automatic differentiation, a program input variable is called *independent* if derivatives with respect to that variable are to be computed. An output variable is called *dependent* if its derivatives with respect to the independent variables are desired. Finally, *active* variables are those that must have an associated derivative object because they depend on some of the independent variables and also contribute to some of the dependent variables. So, in our example $x_1, \ldots, x_n$ are the independent variables, a and g are the dependent variables, and all non-integer variables are active. The initialization of the derivative objects for the independent variables is called *seeding*. As we have seen, proper seeding can reduce the time and memory requirements in cases where the full Jacobian is not required. Some AD tools are able to insert the initialization code for standard situations (*automatic seeding*); in other cases the seeding must be done by hand.

3.2 Reverse Mode

The forward mode starts out with the known derivatives $\partial x_j / \partial x$ and then propagates the derivatives $\partial v / \partial x$ of the intermediate results v with respect to the independent variables x, until $\partial f / \partial x$ is reached when the evaluation of f is completed.

By contrast, the reverse mode propagates the partial derivatives $\partial f / \partial v$ of the final values with respect to the intermediate results v (the so-called *adjoints* $\overline{v}$), starting with the known values $\overline{f} = \partial f / \partial f$ and working *backwards* through all intermediate results occurring in the evaluation of f, until $\overline{x} = \partial f / \partial x$ is reached. For an elementary operation

$$v \leftarrow v_i \text{ op } v_j \quad ,$$

where op is one of $+$, $-$, $*$, or $/$, and v_i and v_j have been computed before v, this reverse propagation is done by incrementing the adjoints $\overline{v_i}$ and $\overline{v_j}$ of the operands by $\overline{v} \cdot \partial v / \partial v_i$ and $\overline{v} \cdot \partial v / \partial v_j$, respectively. Considering the multiplication $z \leftarrow u * w$ as an example, the adjoint updates read

$$\overline{u} \leftarrow \overline{u} + \overline{z} * w \quad ; \quad \overline{w} \leftarrow \overline{w} + \overline{z} * u \quad .$$

Standard functions are handled similarly.

In order to proceed this way,

- the evaluation of f must be completed before the backward derivative computation can start, and
- all intermediate results v must be stored (or recomputed later) because they are needed in the update of the adjoints. (If the code for evaluating f involves

branches then we must also keep track of the actual branch taken in each case.)

We will illustrate the reverse mode with the computation of the arithmetic mean from the above example. We stress that the geometric mean can be handled similarly and that it is only omitted for the sake of brevity.

To simplify matters for the arithmetic mean, overwriting of a is disabled by introducing an additional index:

$a_0 \leftarrow 0$
$\texttt{for}\ i = 1\ \texttt{to}\ n\ \texttt{do}$
$\quad a_i \leftarrow a_{i-1} + x_i$
$a \leftarrow a_n / n$

Once these operations have been carried out we can compute the adjoints as follows:

$\overline{a} \leftarrow 1\ ;\ \overline{a_i} \leftarrow 0\ \texttt{for}\ i = 0, \ldots, n\ ;\ \overline{x_i} \leftarrow 0\ \texttt{for}\ i = 1, \ldots, n$
$\overline{a_n} \leftarrow \overline{a_n} + \overline{a}/n$
$\texttt{for}\ i = n\ \texttt{to}\ 1\ \texttt{step}\ -1\ \texttt{do}$
$\quad \overline{a_{i-1}} \leftarrow \overline{a_{i-1}} + \overline{a_i}\ ;\ \overline{x_i} \leftarrow \overline{x_i} + \overline{a_i}$

If the function f produces m results then the adjoints are length-m vectors. Thus the cost for obtaining the derivatives increases roughly linearly with m. In the particular case of computing the gradient of a scalar-valued function, the adjoints are also scalars, and therefore *the cost for evaluating the gradient is only a moderate multiple of the cost for the function evaluation itself, independently of the gradient's length*. This low complexity cannot be achieved with divided differences, symbolic differentiation, or forward mode AD.

Again, proper seeding can be used to extract exactly the desired derivative information at minimum cost. In reverse mode, any linear row combinations $S^T \cdot f'(x)$ of the Jacobian $f'(x)$ can be computed, whereas in forward mode any linear column combinations $f'(x) \cdot S$ may be obtained. Here, the *seed matrix* S comprises the initial values for the $\mathbf{x_i}'$ and $\overline{f}$ in forward and reverse mode, respectively.

Implementing a reverse mode AD tool is quite a bit more tricky than the forward mode, but for the AD user this complexity is not visible.

3.3 Some Additional Remarks

The following issues should be considered when applying the AD technology to a particular code.

Availability of the complete source code for f is indispensable for applying AD. A call from within the code of f to some obscure library function whose source code is not available makes application of AD to that library function impossible. However, if it is known what the library function does you may write derivative code for the library function by hand or, if everything else fails, approximate those derivatives by divided differences. Combining this code with the code generated by applying AD to the remainder of the code for f would also work.

It depends on the values of n and m and on the structure of the problem which of the two basic modes of AD should be applied. In general, if $n \gg m$ the reverse mode is adequate, while the forward mode is more appropriate if $n \ll m$. This choice is, however, not straight forward, as other issues, such as sparsity of the Jacobian, also have to be considered. However, AD tools, such as described in Section 6, allow one to easily compare the performance of different approaches. Note that these choices only affect computational performance. Up to machine roundoff, the derivative values are identical.

Finally it should be pointed out that similar techniques exist for computing higher-order derivatives (e.g., Hessians or the coefficients in the Taylor expansions of scalar-valued functions). Such derivatives are needed whenever linearized models are not adequate.

4 Comparing the Differentiation Approaches: An Example

Automatic differentiation is often confused with traditional symbolic differentiation or numerical differentiation by divided differences. To illustrate the differences, assume that one is interested in derivatives of the function

$$f(x) = \prod_{i<j} \sin(x_i - x_j) \tag{4}$$

where, as before, $x \in \mathbb{R}^n$. Note that f is pretty simple and that an advantage of AD is its feasibility to handle arbitrarily complex codes involving loops, branches, and similar control structures as well as complicated data structures. Our implementation of f in Fortran 77 consists of 13 lines of code without comments and blank lines. A routine to compute f and its n-dimensional gradient generated by an AD tool is only moderately longer, namely 54 lines of code. We stress the fact that the code length is independent of n. Given specific values for the input, AD generates derivative *values* by accumulating intermediate values along a trace of a particular run of the program.

Derivatives computed by computer algebra systems are based on *expressions* rather than on values. The length of an expression for f in our example depends on n. For the sake of simplicity, let $n = 8$ in the sequel. Our symbolic implementation of f in

Maple expands to the expression

$$\sin(x_1 - x_2)\sin(x_1 - x_3)\sin(x_1 - x_4)\sin(x_1 - x_5)\sin(x_1 - x_6)\sin(x_1 - x_7)$$
$$\sin(x_1 - x_8)\sin(x_2 - x_3)\sin(x_2 - x_4)\sin(x_2 - x_5)\sin(x_2 - x_6)\sin(x_2 - x_7)$$
$$\sin(x_2 - x_8)\sin(x_3 - x_4)\sin(x_3 - x_5)\sin(x_3 - x_6)\sin(x_3 - x_7)\sin(x_3 - x_8)$$
$$\sin(x_4 - x_5)\sin(x_4 - x_6)\sin(x_4 - x_7)\sin(x_4 - x_8)\sin(x_5 - x_6)\sin(x_5 - x_7)$$
$$\sin(x_5 - x_8)\sin(x_6 - x_7)\sin(x_6 - x_8)\sin(x_7 - x_8).$$

The representation consists of 5 lines, and differentiating with respect to x_1 leads to an expression extending over 28 lines. So, the representation of the full gradient consists of about 220 lines. Note that the representation of the derivatives increases by a factor of 44 for symbolic differentiation and by a factor of only 4 for automatic differentiation. Furthermore, the expansion factor for symbolic differentiation grows further when n is increased whereas the expansion factor is fixed for automatic differentiation. This is one reason for symbolic differentiation being limited to rather simple functions.

Recall from (3) that the accuracy of divided differences (DD) depends crucially on a suitable step size. There is always the dilemma that the step size should be small in order to decrease the truncation error and that, on the other hand, the step size should be large to avoid cancellation errors using finite-precision arithmetic. In Table 1, the dependence of the DD approximation with respect to a step size h is shown for the function given in (4) and $n = 8$. More precisely, the table gives the DD approximation to $\partial f/\partial x_1$ evaluated at $x = (1, 2, 3, 4, 5, 6, 7, 8)^T$ when using various step sizes. Evaluating the corresponding symbolic expression by a computer algebra package using 10 digits yields $-1.199084436 \cdot 10^{-6}$. Starting with a large step size $h = 1$, the DD approximations are getting more accurate when decreasing the step size down to $h = 10^{-5}$. However, when the step size is reduced further cancellation error becomes dominant and the DD approximations start to get less accurate.

A fundamental difference between divided differences (DD) and either symbolic or automatic differentiation is that derivatives generated by DD are approximations due to the truncation error. *There is no truncation error when using AD.* In the above example, the value for $\partial f/\partial x_1$ at the same point x is given by $-1.1990844347159 \cdot 10^{-6}$ using an AD tool. Like divided differences, AD can also be used in a black-box fashion but it eliminates the need for experimenting with an appropriate step size.

5 Operator Overloading versus Source Transformation

Automatic differentiation can be viewed as a particular semantic transformation problem: Given a code C for evaluating a function, we would like to generate a

Table 1 Divided difference (DD) approximations to $\partial f/\partial x_1$ with f according to (4) evaluated at $x = (1,2,3,4,5,6,7,8)^T$ using various step sizes h. (The correct value is $-1.199084436 \cdot 10^{-6}$.)

Step size h	DD Approximation to $\partial f/\partial x_1$
1	$+0.140214655 \cdot 10^{-6}$
10^{-1}	$-1.123935235 \cdot 10^{-6}$
10^{-2}	$-1.199775180 \cdot 10^{-6}$
10^{-3}	$-1.199238600 \cdot 10^{-6}$
10^{-4}	$-1.199099000 \cdot 10^{-6}$
10^{-5}	$-1.199080000 \cdot 10^{-6}$
10^{-6}	$-1.199000000 \cdot 10^{-6}$
10^{-7}	$-1.197000000 \cdot 10^{-6}$
10^{-8}	$-1.190000000 \cdot 10^{-6}$
10^{-9}	$-1.000000000 \cdot 10^{-6}$
10^{-10}	$+1.000000000 \cdot 10^{-6}$
10^{-11}	$+10.00000000 \cdot 10^{-6}$
10^{-12}	$+100.0000000 \cdot 10^{-6}$
10^{-13}	$+1000.000000 \cdot 10^{-6}$
10^{-14}	$+10000.00000 \cdot 10^{-6}$

code C' that evaluates the function *and* its derivatives. To affect this semantic transformation, two approaches have been employed that are introduced in the next two subsections. A more detailed comparison of these two approaches is given in the survey by Bischof and Bücker [6].

5.1 Operator Overloading

Modern computer languages such as C++ or Fortran 90 support *operator overloading*, thus making it possible to redefine the meaning of elementary arithmetic operators and intrinsic functions with special routines that propagate derivatives in addition to performing the original operation. We can, for example, define a type for floating-point numbers that have gradient objects associated with them. This new type is called `adouble`, say. Then, we can redefine the meaning of an operator for variables of type `adouble` in such a way that, e.g., an assignment $z = x * y$ involving the overloaded operator "$*$" for x and y both of type `adouble` not only computes the product of x and y but also updates the gradient object $\mathbf{z}'$ associated to z according to $\mathbf{z}' \leftarrow x\mathbf{y}' + y\mathbf{x}'$. So, each occurrence of a multiplication of two `adoubles` in the code will also effect the update of the associated derivatives in a completely transparent fashion.

A user of an AD tool based on operator overloading typically has to perform the following small modifications to the original code C in order to prepare it for differentiation:

- All active variables have to be redeclared to the new type `adouble`.
- The beginning and the end of the code to be differentiated must be marked.
- Independent and dependent variables have to be specified.

Finally, the code must be recompiled and linked to an appropriate AD library.

An advantage of the operator overloading approach is that the potentially complicated class definition implementing the derivative calculations is completely hidden from the user of an AD tool. Moreover, the user's code is not affected by changes in the class definition. For instance, switching from first to second order derivatives is reflected in the class definition but not in the code being differentiated. However, there are also drawbacks of the operator overloading approach. Debugging may be difficult because the user's code does not change and the meaning of the operations is buried in the associated class definitions. The runtime overhead of the operator overloading technique can also be substantial due to the large number of function calls. Tools based on operator overloading are also likely to be sensitive to the choice of the compiler and its flags.

5.2 Source Transformation

Another way of changing the semantics of a code C is to rewrite it explicitly. *Source transformation* uses a preprocessor to generate an explicit code C' by augmenting the code C with statements for the computation of derivatives. For example, the assignment $z = x * y$ is rewritten into a piece of code that contains not only the computation of z but also an implementation of the vector linear combination $\mathbf{z}' = x\mathbf{y}' + y\mathbf{x}'$, either in a loop or via a subroutine call.

A user of an AD tool based on source transformation does not need to make any modifications to the original code C. The user only has to specify, typically in a control file, the independent and dependent variables as well as the routine whose invocations causes the evaluation of the function to be differentiated. Then the preprocessor is invoked generating a code C'. Finally, the code C' must be compiled and linked to an appropriate AD library.

An advantage of the source transformation approach is the fact that the generated code allows a human inspector to follow the control flow. In contrast to operator overloading working on the level of single operations (unless more sophisticated techniques like expression templates are employed), source transformation is capable of detecting and exploiting higher level constructs such as statements, loops or nested loops for which more efficient derivative code can be produced. In addition, code can be generated in a context-sensitive fashion; see [1].

Disadvantages of the source transformation approach are a potentially significant code expansion and its high implementation complexity. Of course, the latter issue

Table 2 Characteristics of AD tools.

Tool	Reference	Language	Mode	Order	Approach
ADIC	[5]	ANSI C	F	1, 2	ST
ADIFOR 2.0	[4]	F77	F	1	ST
ADIFOR 3.0	[9]	F77	F or R	1, 2	ST
ADOL-C	[18]	C/C++	F or R	$1,\ldots,k$	OO
TAF	[14]	F90	F or R	1	ST
TAMC	[13]	F77	F or R	1	ST

is only of interest to developers of AD tools.

6 Available AD Tools

The AD tools to be described in in this section are based on the principles just described in Section 3, but they work in a much more sophisticated way. Some of these optimizations may be found in the book by Griewank [16] and in proceedings of AD workshops [17, 2, 8]. We briefly survey existing software packages implementing the technology of automatic differentiation. Our selection of AD tools is, of course, personally-colored. The discussion contains only tools actively under development. A more comprehensive list of AD tools is currently being compiled and will soon be available at

`http://www.autodiff.org.`

In this note, the focus is on the six AD tools listed in Table 2, and their features will be discussed in turn. In the table, the letters F and R denote the forward and reverse mode, respectively. The symbol k is used to refer to derivatives of arbitrary order. **S**ource **T**ransformation and **O**perator **O**verloading are abbreviated by ST and OO, respectively.

6.1 ADIC

The ADIC system [5], implementing **A**utomatic **DI**fferentiation of **C** programs, is developed by the Mathematics and Computer Science Division at Argonne National Laboratory. It is currently the only AD tool for programs written in ANSI C that is based on source transformation. The system is available without fee for educational and non-profit research purposes from

`http://www.mcs.anl.gov/adic.`

ADIC essentially implements the forward mode for the computation of first-order derivatives. More precisely, it treats a program as a sequence of assignment statements applying the reverse mode at the statement level and the forward mode overall. For second-order derivatives, there is a so-called Hessian module capable of generating code using the forward mode.

ADIC accepts arbitrary ANSI C programs, e.g., it supports functions, dynamic memory allocation (`malloc` and `free`), pointers, and user-defined data types. It can also handle a limited set of C++ features.

Support for MPI and reverse mode are under development. Work is also in progress for developing an ADIC Application Server [19] bringing the power of AD to the World Wide Web. Users of this server will be able to submit a code and have it differentiated by the server and retrieve the differentiated code. Thus, users will be able to compile and run the differentiated code locally when linking to appropriate libraries.

6.2 ADIFOR

Developed in a collaborative project between the Mathematics and Computer Science Division at Argonne National Laboratory and the Center for Research on Parallel Computation at Rice University, the ADIFOR system [3, 4] implements **A**utomatic **DI**fferentiation of **FOR**tran 77 programs by source transformation. Version 2.0 of the software is accessible free of charge at

`http://www.mcs.anl.gov/adifor` or

`http://www.cs.rice.edu/~adifor`

for educational and non-profit research or for the purpose of commercial evaluation. For first-order derivatives, ADIFOR 2.0 uses the same hybrid forward/reverse mode approach as ADIC.

ADIFOR 2.0 provides full Fortran 77 support, e.g., it can handle `common` blocks, `complex` arithmetic, arbitrary `function` and `subroutine` calling sequences as well as common extensions such as `include` statements, `double complex`, and `implicit none`. Furthermore, ADIFOR 2.0 reports exceptions such as the differentiation of `sqrt(x)` when x is zero. It also allows the exploitation of sparsity in Jacobians by its SparsLinC (**Sparse Lin**ear **C**ombination) library.

The ADIFOR 3.0 system [9], which is currently under development, retains the full Fortran 77 support of its predecessor ADIFOR 2.0 while at the same time adding significant new functionality. In particular, ADIFOR 3.0 involves the forward *and* reverse mode for computing first-order derivatives. Together with ADIC, ADIFOR 3.0 shares a Hessian module implementing the forward mode for second-order derivatives. Version 3.0 also supports automatic seeding as well as a subset of the MPI (**M**essage **P**assing **I**nterface) specification for programming parallel computers. At the time of writing, ADIFOR 3.0 is not yet publicly available.

6.3 ADOL-C

At the Institute of Scientific Computing at Technical University of Dresden, the ADOL-C system [18] is being developed facilitating **A**utomatic **D**ifferentiation by **O**ver**L**oading in **C**++. The package implements automatic differentiation of programs written in C or C++ by operator overloading. ADOL-C is available free of charge from

`http://www.math.tu-dresden.de/wir/project/adolc.`

It is capable of evaluating derivatives of any order in the forward or reverse mode. For convenience, easy-to-use drivers for solving optimization problems and nonlinear equations are distributed with the package. Exploitation of sparsity of the Jacobian is also supported. To improve performance, users may use particular classes for active vectors and matrices called `adoublev` and `adoublem`, respectively.

6.4 TAMC and TAF

The TAMC system [13] was developed at the Max Planck Institute for Meteorology. The acronym stands for **T**angent Linear and **A**djoint **M**odel **C**ompiler. TAMC is intended to be used for programs written in Fortran 77 and is based on source transformation. For non-commercial use it is available free of charge from

`http://puddle.mit.edu/~ralf/tamc.`

The system implements both forward and reverse mode for evaluating first-order derivatives. TAMC supports almost the full Fortran 77 standard and certain Fortran 90 extensions such as array assignments, `where` and `use` statements, and modules. A commercial successor of TAMC called TAF (**T**ransformation of **A**lgorithms in **F**ortran) is currently being developed by FastOpt [12]. TAF supports the full Fortran 90 standard with the exception of operator overloading.

7 Preliminary Results with TFS

In this section, we report on preliminary results obtained from applying the ADIFOR [3, 4] system to the CFD package TFS developed at Aerodynamisches Institut of Aachen University of Technology. TFS is able to resolve steady and unsteady laminar and turbulent flows in two and three dimensions. The code consists of approximately 24,000 lines of mostly Fortran 77 spread over 220 subroutines. As a test problem, we were interested in obtaining derivatives of the density ϱ in a simple two-dimensional numerical simulation. More precisely, the airfoil whose profile is depicted in Figure 1 was investigated with TFS. Note that we abstract from any details of the numerical methods being used. From the AD point of view, the density ϱ

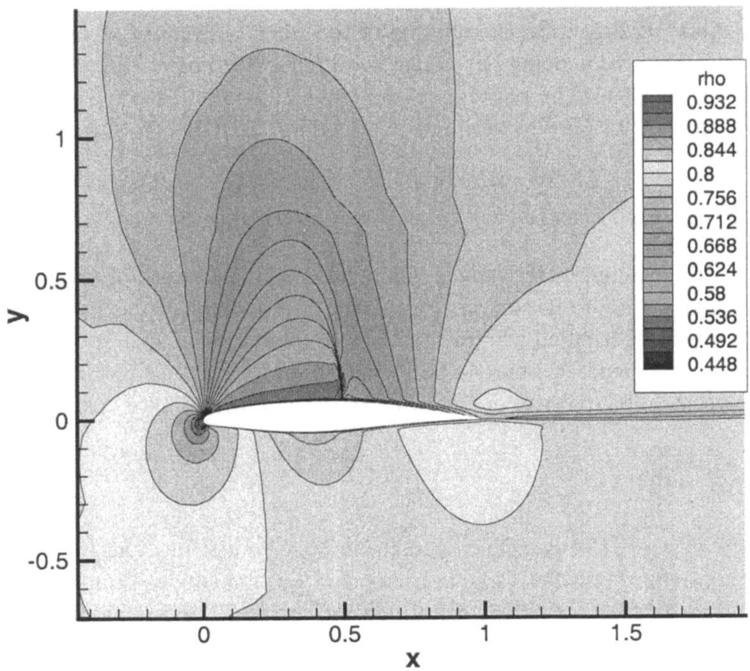

Figure 1 The density ϱ evaluated at $\alpha = 3°$.

computed by the actual simulation is a certain function of the angle of attack α; that is,

$$\varrho = f(\alpha).$$

This function is depicted in Figure 1 for $\alpha = 3°$.

Automatic differentiation is applicable to compute the derivatives

$$f' := \frac{\mathrm{d}\,\varrho}{\mathrm{d}\,\alpha}$$

in a completely mechanical fashion. Sophisticated AD tools are capable of analyzing the complete code and detecting data dependencies in a fully automated way. In this specific example, approximately 220 subroutines were fed into the ADIFOR system. ADIFOR recognized about 100 subroutines to be transformed. The code generated automatically by ADIFOR consists of approximately 20,000 lines of Fortran 77. The derivatives $\mathrm{d}\,\varrho/\mathrm{d}\,\alpha$ computed by the generated code are compatible with results obtained from appropriate finite differences approximations. However,

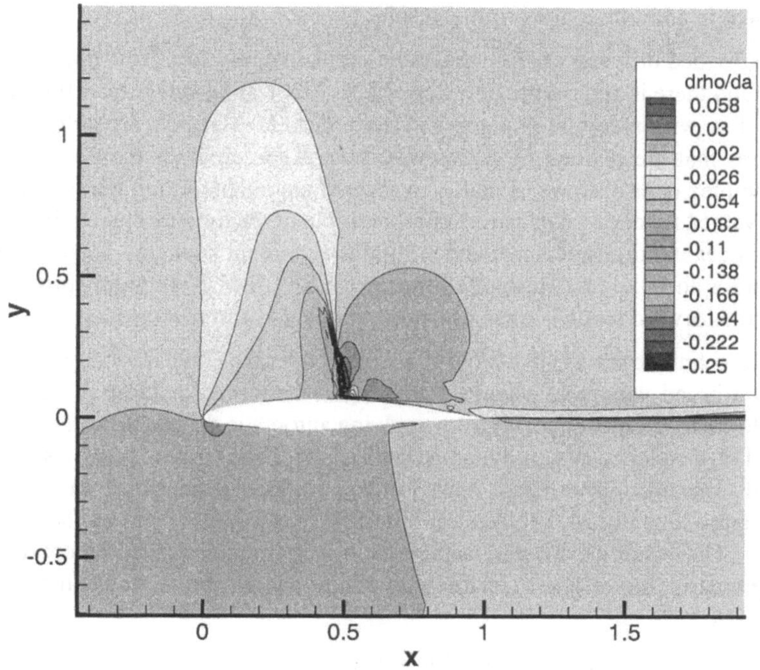

Figure 2 The derivatives $f' := \mathrm{d}\varrho/\mathrm{d}\alpha$ evaluated at $\alpha = 3°$.

automatic differentiation eliminates the need for experimenting with different step sizes. The AD-computed derivatives are given in Figure 2. Compared to the execution time of the original simulation, the execution time of the AD-generated version is increased by a factor of roughly 2.5 which is moderate keeping in mind that, in addition to the original function, its derivative is computed. The storage requirement increases by a factor of approximately 1.4.

8 Concluding Remarks and Future Research Directions

Automatic differentiation (AD) is an efficient technique to obtain accurate derivatives of functions given in the form of a computer program written in a high-level language such as Fortran, C, or C++. We reviewed the principles and basic properties of AD. This technology is used to extend the functionality of existing code such that derivatives of selected output variables w.r.t. selected input variables are computed in addition to the original results.

Unlike divided differences, AD-generated derivatives are free from approximation errors, and there is no concept of a step size in AD. Unlike symbolic differentiation by hand or with a computer algebra system, current AD tools are able to handle functions that are defined by codes with several hundreds of thousands of lines and complex control flows. Finally, in reverse mode computing the gradient of a scalar-valued function takes only a small multiple of the time that is required for the function evaluation itself, independently of the gradient's length. More generally, the cost for an $m \times n$ Jacobian of a function $f : \mathbb{R}^n \to \mathbb{R}^m$ is roughly proportional to n if the forward mode is used, and proportional to m in reverse mode.

The AD technique scales up to large simulation codes that are used today as a crucial part in a broad variety of scientific and engineering investigations. We reported on preliminary results obtained from applying automatic differentiation to the CFD solver TFS developed at Aerodynamisches Institut. TFS is able to resolve steady and unsteady laminar and turbulent flows in two and three dimensions. In the present study, approximately 24,000 lines of Fortran 77 were transformed by the ADIFOR system. The resulting differentiated version was produced in an automated way by augmenting the original version with additional statements for computing the derivatives.

For a very simple two-dimensional test configuration, we reported on derivatives of the density with respect to the angle of attack. We are planning to extend the AD process to obtain derivatives of the wake flow field, the pressure field, the lift coefficient, and the Mach number with respect to the angle of attack, and to compute derivatives of all these quantities with respect to further parameters. Work is also in progress to use more realistic test cases than the simple problem configuration considered in this note.

Acknowledgements

The authors would like to thank Jakob Risch for his notable contribution during the initial phase of this project.

References

[1] J. Abate, C. Bischof, A. Carle, and L. Roh. Algorithms and design for a second-order automatic differentiation module. In *Proc. Int. Symposium on Symbolic and Algebraic Computing (ISSAC '97)*, pages 149–155, New York, 1997. Association of Computing Machinery.

[2] M. Berz, C. Bischof, G. Corliss, and A. Griewank. *Computational Differentiation: Techniques, Applications, and Tools*. SIAM, Philadelphia, 1996.

[3] C. Bischof, A. Carle, G. Corliss, A. Griewank, and P. Hovland. ADIFOR: Generating derivative codes from Fortran programs. *Scientific Programming*, 1(1):11–29, 1992.

[4] C. Bischof, A. Carle, P. Khademi, and A. Mauer. ADIFOR 2.0: Automatic differentiation of Fortran 77 programs. *IEEE Computational Science & Engineering*, 3(3):18–32, 1996.

[5] C. Bischof, L. Roh, and A. Mauer. ADIC — An extensible automatic differentiation tool for ANSI-C. *Software–Practice and Experience*, 27(12):1427–1456, 1997.

[6] C. H. Bischof and H. M. Bücker. Computing Derivatives of Computer Programs. In J. Grotendorst, editor, *Modern Methods and Algorithms of Quantum Chemistry: Proceedings, Second Edition*, volume 3 of *NIC Series*, pages 315–327. NIC Directors, Jülich, 2000.

[7] C. H. Bischof, H. M. Bücker, B. Lang, A. Rasch, and J. W. Risch. On the Use of a Differentiated Finite Element Package for Sensitivity Analysis. In V. N. Alexandrov, J. J. Dongarra, B. A. Juliano, R. S. Renner, and C. J. K. Tan, editors, *Computational Science – ICCS 2001, Proceedings of the International Conference on Computational Science, San Francisco, USA, May 28–30, 2001. Part I*, volume 2073 of *Lecture Notes in Computer Science*, pages 795–801, Berlin, 2001. Springer.

[8] G. Corliss, C. Faure, A. Griewank, L. Hascoët, and U. Naumann. *Automatic Differentiation of Algorithms: From Simulation to Optimization*. Springer, New York, 2002.

[9] M. Fagan and A. Carle. Adifor 3.0 overview. Technical Report CAAM–TR00–03, Rice University, Department of Computational and Applied Mathematics, 2000.

[10] E. Fares, M. Meinke, and W. Schröder. Numerical simulation of the interaction of flap side-edge vortices and engine jets. In *Proceedings of the 22nd International Congress of Aeronautical Sciences, Harrogate, UK, August 27–September 1, 2000*, ICAS 0212, 2000.

[11] E. Fares, M. Meinke, and W. Schröder. Numerical simulation of the interaction of wingtip vortices and engine jets in the near field. In *Proceedings of the 38th Aerospace Sciences Meeting and Exhibit, Reno, NV, USA, January 10–13, 2000*, AIAA Paper 2000–2222, 2000.

[12] FastOpt: Fast Optimization and Sensitivities, Apr. 2001. Homepage at `http://www.fastopt.de`.

[13] R. Giering and T. Kaminski. Recipes for Adjoint Code Construction. *ACM Trans. Math. Softw.*, 24(4):437–474, 1998.

[14] R. Giering and T. Kaminski, 2001. Private communication.

[15] A. Griewank. On automatic differentiation. In *Mathematical Programming: Recent Developments and Applications*, pages 83–108, Amsterdam, 1989. Kluwer Academic Publishers.

[16] A. Griewank. *Evaluating Derivatives: Principles and Techniques of Algorithmic Differentiation*. SIAM, Philadelphia, 2000.

[17] A. Griewank and G. Corliss. *Automatic Differentiation of Algorithms*. SIAM, Philadelphia, 1991.

[18] A. Griewank, D. Juedes, and J. Utke. ADOL-C, a package for the automatic differentiation of algorithms written in C/C++. *ACM Trans. Math. Softw.*, 22(2):131–167, 1996.

[19] P. Hovland and B. Norris. A distributed application server for automatic differentiation. In *Proceedings of the 15th International Parallel and Distributed Processing Symposium*, pages 1849–1856. IEEE, 2001.

[20] M. Kofler. *Maple: An Introduction and Reference*. Addison-Wesley, 1997.

[21] B. Mohammadi and O. Pironneau. *Applied Shape Optimization for Fluids*. Oxford University Press, Oxford, 2001.

[22] J. Nocedal and S. J. Wright. *Numerical Optimization*. Springer, New York, 1999.

[23] R. J. Petti. *Introduction to MACSYMA*. Publishers, Inc., 1997.

[24] L. B. Rall. *Automatic Differentiation: Techniques and Applications*, volume 120 of *Lecture Notes in Computer Science*. Springer Verlag, Berlin, 1981.

[25] S. Wolfram. *The Mathematica Book*. Cambridge University Press, 4th edition, 1999.

4. Computational and Experimental Aeroelasticity

On Well-Posedness and Modelling for Nonlinear Aeroelasticity

Ralf Massjung[1], Jörg Hurka[2], Wolfgang Dahmen[1], Josef Ballmann[2]
[1]Institut für Geometrie und Praktische Mathematik, RWTH Aachen
Templergraben 55, D-52062 Aachen, Germany
[2]Lehr- und Forschungsgebiet für Mechanik, RWTH Aachen
Templergraben 64, D-52062 Aachen, Germany

Summary

We consider general theoretical issues of the aeroelastic fluid-structure-interaction. In particular we have chosen an elastic panel in transonic flow as our model problem for computational experiments. First we consider the question of well-posedness of the equation system, giving a uniqueness result. Numerical simulations address the choice of the structural model and the choice of the time-marching-scheme that couples the equations of fluid and structure.

1 Introduction

We consider the aeroelastic problem of a plate, clamped or hinged at its edges into a smooth surface, over which a compressible fluid flows. For our numerical experiments we reduce the spatial dimension of the problem by one, so that we look at a 2D-flow over a 1D-structure. In the transonic and supersonic flow regime, this system may exhibit aeroelastic instabilities such as divergence and flutter.

In the first part we give a precise description of the initial boundary value problem (IBVP) governing the continuous problem. Here we employ the compressible Navier-Stokes Equations and a modified version of the nonlinear von-Kármán plate model. A function space is given such that a solution lying in that space is unique. The proof of the theorem is based on the energy method, using techniques from [20], and we describe only the proof's starting point, referring to [16] for all the analytical details.

In the second part of this paper, the aeroelastic problem is solved by numerical integration of the corresponding dynamic equations in the time domain, where we use the Euler equations to model the fluid motion, while Reissner-Mindlin or von Kármán equations are employed to model the plate dynamics.

One objective is to investigate the influence of the mathematical model used to de-

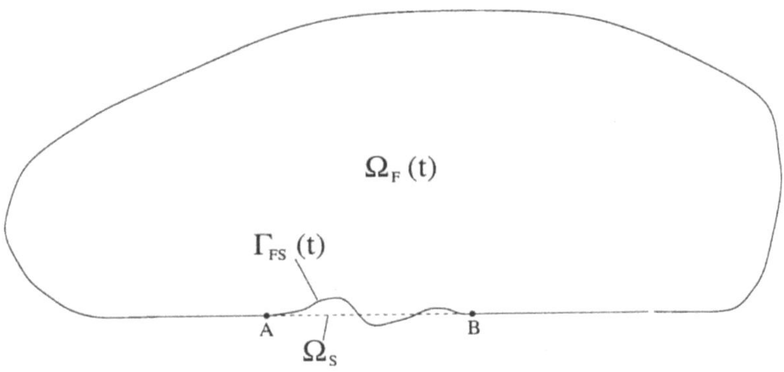

Figure 1 2D-projection of problem geometry.

scribe the dynamics of the structure and its influence on the solution of the aeroelastic problem. In particular, we wish to study the influence of shear deformation. Classical plate models, which do not account for shear deformation, yield parabolic partial differential equations (PDEs) which allow for infinite (and hence, unphysical) information propagation. Reissner-Mindlin (RM) type plate models, which incorporate shear deformation, on the other hand, result in hyperbolic PDEs which exhibit physically adequate finite information propagation.

Furthermore, we compare different schemes to couple fluid and structural dynamics methods in order to compute aeroelastic solutions. Three coupling schemes are considered in this work.

2 Uniqueness for fluid-structure-interaction

We look at an initial boundary value problem (IBVP), defined by a distribution of density ϱ, velocity $\boldsymbol{v}$ and temperature θ of a gas initially placed in the domain $\Omega_F(0) \subset \mathbb{R}^3$ and an initially undeflected thin plate occupying a domain $\Omega_S \subset \mathbb{R}^2$ in the xy-plane. The domains are placed such that Ω_S is part of the fluid's boundary $\partial\Omega_F(0)$. At the initial moment the interface of the two media is $\Gamma_{FS}(0) \equiv \{(x, y, 0) : (x, y) \in \Omega_S\} \subset \partial\Omega_F(0)$. Fig. 1 shows a projection of the problem in the xz-plane. The dotted line between the points A and B, at which the plate is clamped, represents the undeflected plate points.

Assigning an initial velocity distribution to the plate, the fluid-structure-interaction is triggered, causing the interface to move. Denoting the plate deflection by $u(t, x, y)$, the interface at time t is defined by $\Gamma_{FS}(t) = \{(x, y, u(t, x, y)) : (x, y) \in \Omega_S\}$ and the fluid domain changes it's shape, such that $\Gamma_{FS}(t) \subset \partial\Omega_F(t)$. The solid curve between A and B in Fig. 1 shows the plate in a deflected position. Assuring this connectivity of the two media, we have to add appropriate interface conditions to the different balance equations that have to be satisfied in the two domains. Further we have to assign boundary conditions at the remaining part of the fluid boundary and boundary conditions at $\partial\Omega_S$ to define how the plate is fixed.

In our analytical investigations, we fix a set of fluid material, i.e. look at the fluid in Lagrangian coordinates. This makes inflow and outflow of fluid material impossible, but fluid movement may be imposed by a corresponding boundary deflection $u^w(t, \boldsymbol{x})$ as boundary condition.

Initial, boundary and interface conditions may be posed to be physically meaningful and the requirement that the resulting set of equations is mathematically well-posed. For many well-established models a complete answer to the question of well-posedness is not yet available. Below the IBVP is given for which we later formulate a uniqueness result. Remarks to the meaning of the numbered equations will folllow.

The IBVP:

The Navier-Stokes Equations in $\Omega_F(t)$

$$\frac{D}{Dt}\varrho + \varrho\,(\nabla \boldsymbol{v}) \;=\; 0$$

$$\varrho\,\frac{D}{Dt}\boldsymbol{v} \;=\; \nabla S$$

$$\varrho\, c_v(\theta)\,\frac{D}{Dt}\theta \;=\; S : D + \sigma\,\nabla(\nabla\theta)$$

with $D \;=\; 1/2\cdot(\nabla \boldsymbol{v} + (\nabla \boldsymbol{v})^T)\,,\; S \;=\; -p(\varrho,\theta)I + 2\mu\, D + (\zeta - \frac{2}{3}\mu)\,(div\,\boldsymbol{v})\cdot I$

where p is a thermodynamic state relation expressing the pressure in terms of the density and temperature, and ζ and μ are constant viscosity coefficients, I is the identity matrix, $c_v(\theta)$ the specific heat at constant volume and σ the heat conduction coefficient

Boundary conditions on $\partial\Omega_F(t)\backslash\Gamma_{FS}(t)$, formulated for $\boldsymbol{x} \in \partial\Omega_F(0)\backslash\Gamma_{FS}(0)$

$$\boldsymbol{v}\big(t, \boldsymbol{x} + \boldsymbol{u}^w(t, \boldsymbol{x})\big) = \boldsymbol{u}_t^w(t, \boldsymbol{x}) \tag{1}$$

$$\theta\big(t, \boldsymbol{x} + \boldsymbol{u}^w(t, \boldsymbol{x})\big) = \theta^w(t, \boldsymbol{x})\,. \tag{2}$$

Initial conditions in $\Omega_F(0)$

$$\varrho(0, \boldsymbol{x}) \;=\; \varrho^0(\boldsymbol{x})\,, \qquad \boldsymbol{v}(0, \boldsymbol{x}) \;=\; \boldsymbol{v}^0(\boldsymbol{x})\,, \qquad \theta(0, \boldsymbol{x}) \;=\; \theta^0(\boldsymbol{x})\,.$$

Modified von-Kármán Equation in Ω_S

$$u_{tt} - \alpha \cdot \Delta u_{tt} + \beta \cdot \Delta^2 u - [u, w] = p_\infty - g \tag{3}$$

$$\gamma \cdot \Delta^2 w + [u, u] = 0$$

$$\text{with} \qquad [u, w] \equiv u_{xx} w_{yy} - 2 u_{xy} w_{xy} + u_{yy} w_{xx}$$

where α, β, γ are positive constants and p_∞ is the constant pressure below the plate.

Boundary conditions on $\partial\Omega_S$

$$u(t,x,y) = 0\,, \quad \frac{\partial u(t,x,y)}{\partial n_S} = 0 \qquad \text{and} \qquad w(t,x,y) = 0\,, \quad \frac{\partial w(t,x,y)}{\partial n_S} = 0. \tag{4}$$

Initial conditions in Ω_S

$$u(0,x,y) = 0\,, \quad u_t(0,x,y) = v_S^0(x,y)\,.$$

Interface conditions on $\Gamma_{FS}(t)$, formulated for $(x,y) \in \Omega_S(0)$, respectively $\boldsymbol{x} = (x,y,0) \in \Gamma_{FS}(0)$

$$-u_t(t,x,y) \cdot \boldsymbol{n}_F(0) = \boldsymbol{v}\big(t, \boldsymbol{x} - u(t,x,y) \cdot \boldsymbol{n}_F(0)\big) \tag{5}$$

$$g(t,x,y) = -\boldsymbol{n}_F^T(0) \cdot S\big(t, \boldsymbol{x} - u(t,x,y) \cdot \boldsymbol{n}_F(0)\big) \cdot \boldsymbol{n}_F(t)\,. \tag{6}$$

Remarks: In [16] some more structural models are treated, e.g. the linear thermoelastic body. Let's give some comments to the equations above.

(1) is a prescribed boundary movement, with the boundary's deflection given by u^w.

(2) prescribes a boundary temperature θ^w; it may be replaced by other common boundary conditions for θ, see [16].

(3) is a modification of the original von-Kármán equation [18], which is recovered by setting $\alpha = 0$, but for the original von-Kármán equation uniqueness cannot be proved [15]; w is the Airy stress function characterizing the plate's planar stress distribution

(4) is the condition of a clamped plate, here $n_S \in \mathbb{R}^2$ is the outward normal to Ω_S lying in the xy-plane.

(5) insures the connectivity of fluid and structure, it is a no-slip condition for the fluid; due to our choice to place Ω_S in the xy-plane, the outward normal of the fluid domain at the interface, $\boldsymbol{n}_F(0) = (0,0,-1)^T$, represents the negative direction of the plate points' movement.

(6) contains $S \cdot \boldsymbol{n}_F(t)$, the surface force on the fluid at the interface $\Gamma_{FS}(t)$, and $-\boldsymbol{n}_F^T(0)\, S\, \boldsymbol{n}_F(t)$ is its projection into the direction $-\boldsymbol{n}_F(0)$.

Lagrangian coordinates

The transformation to Lagrangian coordinates is defined via the particle paths $\boldsymbol{\eta}$:

$$\boldsymbol{\eta}_t(t,\boldsymbol{x}) = \boldsymbol{v}(t,\boldsymbol{\eta}(t,\boldsymbol{x}))\,, \quad \boldsymbol{\eta}(0,\boldsymbol{x}) = \boldsymbol{x} \qquad for \quad \boldsymbol{x} \in \Omega_F(0)\,.$$

Then, the fluid quantities in Lagrangian coordinates are given by

$$\overline{\varrho}(t,\boldsymbol{x}) := \varrho(t,\boldsymbol{\eta}(t,\boldsymbol{x}))\,, \quad \overline{\boldsymbol{v}}(t,\boldsymbol{x}) := \boldsymbol{v}(t,\boldsymbol{\eta}(t,\boldsymbol{x}))\,, \quad \overline{\theta}(t,\boldsymbol{x}) := \theta(t,\boldsymbol{\eta}(t,\boldsymbol{x}))\,.$$

In the sequel of this paragraph we remove the bar and whenever fluid quantities appear they are taken in Lagrangian coordinates. Further we introduce an abbreviation for the determinant of the particle path gradient,

$$\mathcal{D}(t,\boldsymbol{x}) := det(\nabla\boldsymbol{\eta})\,.$$

Now the IBVP may be transformed to Lagrangian coordinates, which is only a formal act, so we ommit it here. But in Lagrangian coordinates the fluid equations hold on $\Omega_F \equiv \Omega_F(0)$, a domain fixed in time, and the movement of the domain is represented by the particle paths $\boldsymbol{\eta}$. This system is easier to handle analytically, and we formulate our result for the fluid in Lagrangian coordinates.

Uniqueness result

Let $Q_T \equiv [0,T] \times \Omega_F$ and denote by $D\varrho$ any spacial derivative of order 1 of ϱ and by $D^2\varrho$ any spacial derivative of order 2 of ϱ and so on. The uniqueness result reads as follows:

Theorem: Let Ω_F and Ω_E be bounded domains with Lipschitz boundary, and let the constants $\mu,\ \zeta,\ \sigma > 0$ and the thermodynamic relations $p(\varrho,\theta) \in C^1$ and $c_v(\theta) \in C^1$, so that there is a constant $c_{v*} > 0$ with $c_v \geq c_{v*}$. In this case, the IBVP has at most one solution in the following class of functions.

$$\begin{aligned} u, w \in C(0,T;W^{2,2}(\Omega_S)) \quad &\text{and} \quad u_t \in C(0,T;W^{1,2}_o(\Omega_S)),\\ \varrho \in L^\infty(Q_T),\ D\varrho \in L^2(0,T;L^\infty(\Omega_F)),&\\ \boldsymbol{v} \in L^\infty(Q_T),\ D\boldsymbol{v} \in L^2(0,T;L^\infty(\Omega_F)) \quad &\text{and} \quad D^2\boldsymbol{v} \in L^2(0,T;L^\infty(\Omega_F)),\\ \theta \in L^\infty(Q_T),\ D\theta \in L^2(0,T;L^\infty(\Omega_F)) \quad &\text{and} \quad D^2\theta \in L^2(0,T;L^\infty(\Omega_F)),\\ \boldsymbol{\eta} \in L^\infty(Q_T),\ D\boldsymbol{\eta} \in L^\infty(Q_T) \quad &\text{and} \quad D^2\boldsymbol{\eta} \in L^\infty(Q_T) \end{aligned}$$

and there are positive constants $\mathcal{D}_*, \varrho_*$ such that $\mathcal{D} \geq \mathcal{D}_*$ and $\varrho \geq \varrho_*$ in Q_T

and $\boldsymbol{\eta}(t,\cdot\,) : \overline{\Omega}_F \to \mathbb{R}^n$ is injective for every $t \in [0,T]$.

The proof of the Theorem initiates from the assumption that two solutions u, w, η, ϱ, $\boldsymbol{v}$, θ and $\tilde{u}$, $\tilde{w}$, $\tilde{\eta}$, $\tilde{\varrho}$, $\tilde{\boldsymbol{v}}$, $\tilde{\theta}$ of the IBVP are given. Then energy estimates are derived, i.e. for the question of uniqueness we consider the differences

$$u' \equiv \tilde{u} - u\,, \quad w' \equiv \tilde{w} - w\,, \quad \eta' \equiv \tilde{\eta} - \eta\,, \quad \text{... and so on}$$

and estimate an expression $E(t)$ in these quantities that is related to the physical energy of the system: In [16] we chose

$$\begin{aligned} E(t) \;\; &= \;\; \frac{1}{2}\int_{\Omega_S} {u'_t}^2 + \alpha|\nabla u'_t|^2 + \beta(\Delta u')^2 \; dx\, dy \\ &+ \;\; \int_{\Omega_F} |\eta'|^2 + |\nabla\eta'|^2 + |\varrho'|^2 + \frac{1}{2}\tilde{\mathcal{D}}\tilde{\varrho}|\boldsymbol{v}'|^2 + \frac{1}{2}\tilde{\varrho}\tilde{c}_v\tilde{\mathcal{D}}|\theta'|^2 \; dx\,. \end{aligned}$$

Then plugging in the relations from the IBVP that hold for the two assumed solutions and using the assumption from the Theorem, in [16] a lengthy analysis leads to an estimate of the form (7) as assumed in Gromwall's Lemma below. Here we used the techniques of [21] and [20] and an estimate for the plate from [15]. From [4] we have

Gronwall's Lemma: Given functions $E \in L^\infty(0,T)$ and $c \in L^1(0,T)$ with $E(t) \geq 0$ and $c(t) \geq 0$ almost everywhere (a.e.) on $[0,T]$ then for an arbitrary constant E_o the inequality

$$E(t) \;\; \leq \;\; E_o + \int_0^t c(\tau)\, E(\tau)\; d\tau \quad \text{a.e. on } [0,T] \tag{7}$$

implies the inequality

$$E(t) \;\; \leq \;\; E_o \cdot exp\left(\int_0^t c(\tau)\; d\tau\right) \quad \text{a.e. on } [0,T]\,.$$

In our estimate we have $E_o = E(0)$, which vanishes due to the initial conditions. Thus from Gronwall's Lemma $E(t) = 0 \;\; \forall t \in [0,T]$ follows, which implies that the two assumed solutions are identical.

3 Numerical Simulations

3.1 Structural method

If rotary inertia and shear deformation are included into the analysis, the governing equations for the rotation of the plate's cross section, φ, and the lateral deflection,

u, are [7]

$$\frac{\varrho_s h^3}{12}\frac{\partial^2\varphi}{\partial t^2} - D\frac{\partial^2\varphi}{\partial x^2} - \kappa G h\left(\frac{\partial u}{\partial x} - \varphi\right) = 0, \tag{1}$$

$$\varrho_s h\frac{\partial^2 u}{\partial t^2} - \kappa G h\left(\frac{\partial^2 u}{\partial x^2} - \frac{\partial\varphi}{\partial x}\right) - N\frac{\partial^2 u}{\partial x^2} + \Delta p = 0, \tag{2}$$

where $N = N_0 + N_x$ and

$$N_x = \frac{Eh}{2c}\int_0^c (\partial u/\partial x)^2\,dx. \tag{3}$$

Here N_0 and N_x denote in-plane loads due to externally applied forces and lateral deflection, respectively, while $\Delta p = p - p_\infty$, D, E, G and ϱ_s denote the aerodynamic pressure loading on the panel, the plate's bending stiffness, Young's modulus, shear modulus and mass density, respectively. In this work, we will take $N_0 = 0$. Equations (1) and (2) represent a 1D nonlinear version of the Reissner-Mindlin plate equations [17].

If the effects of rotary inertia and shear deformation are neglected, Eqs. (1) and (2) maybe combined to yield the following single equation for the plate's lateral deflection u,

$$D\frac{\partial^4 u}{\partial x^4} - N\frac{\partial^2 u}{\partial x^2} + \varrho_s h\frac{\partial^2 u}{\partial t^2} + \Delta p = 0, \tag{4}$$

where N is as given above. Equation (4) represents a 1D version of von Kármán's plate equation.

We use the two-noded Timoshenko beam finite element of Friedman & Kosmatka [13] to discretize Eqs. (1) and (2). The Friedman-Kosmatka element makes use of cubic shape functions and is free of shear locking. Note that for a certain choice of a single parameter Φ the Friedman-Kosmatka element is switched to the classical Hermitian beam finite element, which is particularly suited to discretize Eq. (4). In addition to RM plate dynamics, our FE implementation is thus able to capture the dynamics of a classical plate, as given by Eq. (4). We use a slightly modified version of the conservative load projection algorithm proposed by Cebral & Löhner [3] to convert the pressure distribution on the fluid grid into a corresponding pressure distribution on the solid grid. Second order accurate implicit time integration is performed by application of Bossak's α-method [9] in combination with Newton-Raphson subcycling. The emerging systems of linear algebraic equations are solved by direct methods via Lapack library routines. Further datails of our implementation may be found in [11].

In addition to the FE solver, a modal Galerkin method has been implemented to solve Eq. (4). Here, we expand u in terms of the eigenmodes of the linear Kirchhoff plate

$$u(x,t) = \sum_n q_n(t)\sin(n\pi x/c) \tag{5}$$

whereby we restrict ourselves to the boundary conditions of hinged support. Inserting Eq. (5) into (4), multiplying by $\sin(r\pi x/c)$ and integrating the resulting equation between the limits 0 and c yields the following coupled set of nonlinear ordinary differential equations (ODEs) for the modal amplitudes q_r, as given, e.g., by Dowell [6],

$$\begin{aligned}\varrho_s h\ddot{q}_r =& - D\left(\frac{r\pi}{c}\right)^4 q_r \\ & - \left(\frac{r\pi}{c}\right)^2 \frac{Eh}{4}\left\{\sum_k \left(\frac{k\pi q_k}{c}\right)^2\right\} q_r - 2p_r,\end{aligned} \tag{6}$$

where

$$p_r = \frac{1}{c}\int_0^c \Delta p(x)\sin(r\pi x/c)\,dx. \tag{7}$$

We assume that the pressure on the panel surface varies linearly between the discrete pressure values computed by the flow solver and use Simpson's rule to evaluate the integral in Eq. (7). In order to compute the time evolution of the modal amplitudes q_r, Eq. (6) is rewritten as a system of first order ODEs. We implemented Jameson's second order accurate implicit dual time stepping method [12] and employed a standard 4-step Runge-Kutta (RK) method [19] to achieve convergence during each implicit time step.

3.2 Fluid method

We use the DLR FLOWer code to compute the flow field over the deforming panel. FLOWer is a block structured finite volume solver to the three-dimensional unsteady Euler and Reynolds averaged Navier Stokes equations, with the option of a dynamically deforming mesh. A central space discretization with artificial viscosity is employed. Explicit Runge-Kutta time stepping and several convergence accelerating methods such as multigrid may be used for the calculation of steady flow fields. Jameson's implicit dual time stepping method [12] is used for time accurate calculations of unsteady flows [8]. For a detailed discussion of the FLOWer code, see, e.g., Kroll et al. [14].

3.2.1 Grid deformation algorithm

We implemented an algebraic grid deformation algorithm, which is a modification of the method described in [2] and [22], to adjust the fluid grid to the deforming plate surface during each time step. The algorithm may be sketched as follows.

1. For each grid line leaving the panel surface, we perform a translation and rigid body rotation such that the new grid lines are normal to the deformed panel surface.

2. We calculate the minimal distance of points, where adjacent new grid lines intersect, to the panel surface, and call it a_0. Finally, we take $a_0 = \min(a_0, \varepsilon)$, where ε is some prescribed threshold value.

3. We add the displacement of the grid line foot point on the panel surface, weighted by some function g, to the points of the old grid line. The weight function is an exponential of the form $g = \exp\{-(\kappa a/a_0)^2\}$, where a is the arc length along the grid line and κ is some numerical factor depending on the grid dimensions. We choose κ such that g is close to unity for $a \ll a_0$, i.e., near the panel surface, and $g \approx 0$ for $a \geq a_0$.

4. Finally, we interpolate between the rotated grid lines, taken from step 2, and the modified old grid lines from step 3 by use of a weight function f, such that the final grid lines leave the deformed panel surface in a nearly perpendicular direction, while far away from the panel the old and new grid points are identical. The weight function f is of the same generic type as g with a slightly different numerical value for κ. Note that the parameter a_0 is recalculated in each time step to avoid intersection of grid lines.

For grid lines leaving the fixed wall in the neighborhood of the panel end points, in step 2 of our algorithm, we perform rotations appropriate to the panel end points weighted by some smoothly decaying function. Again, this is done to avoid grid line intersection.

3.3 Fluid structure coupling scheme

Three different schemes to couple fluid and structural solvers have been implemented and are briefly discussed in this section.

The first of these schemes, scheme A (Fig. 3), is a conventional loose coupling scheme which allows for a lag of one time step between fluid and structure motion. Given numerical solutions W_n and U_n for fluid and structure, respectively, the fluid pressure p_n is used to advance the structure to time level $n+1$ (while strictly, the pressure at time level $n+1$, p_{n+1}, would be required here). The structural displacements u_{n+1} thus gained are subsequently used to update the fluid boundary and mesh. Finally, the fluid solution is advanced to time level $n+1$.

In the second scheme, scheme B (Fig. 4), we start the coupling procedure by extrapolating the fluid solution W_n to time level $n+1$ according to $W_{n+1}^{ex} = W_n + \Delta W_n$, where ΔW_n is a first or second order backwards difference. From W_{n+1}^{ex} we obtain an estimate for the pressure on the fluid-structure interface at time level $n+1$, p_{n+1}^{ex}, and use it to compute the structural solution, U_{n+1}, at level $n+1$. The structural displacements u_{n+1} are finally used to update the fluid boundary and mesh and the flow solution is advanced from W_n to W_{n+1}.

In the third scheme, scheme C (Fig. 5), a tight coupling between fluid and structural solvers is achieved by essentially iterating the steps of the loose coupling scheme A until a convergence criterion has been met.

3.4 Results

We present results of numerical simulations of 2D Euler flows over flat aluminium panels with hinged ends, where we used $\varrho_s = 2700\,\text{kg}\,\text{m}^{-3}$, $E = 7.1 \cdot 10^{10}\,\text{N}\,\text{m}^{-2}$, and $\nu = 0.34$. The problem geometry is displayed in Fig. 2.

Second order dual time stepping in combination with four level multigrid acceleration was used to obtain the flow field. As in [5] and [1], we initiate our simulations by setting the flow variables equal to the far field values throughout the fluid mesh. We take the panel to be in an undeformed state with an initial velocity that would lead to a small amplitude first mode oscillation for a linear panel in vacuum. We specify the initial velocity such that this amplitude would reach approximately 20 per cent of the plate thickness.

3.4.1 Spatial and temporal convergence results

Three panel flutter cases that have been considered previously by Bendiksen & Davis [1] were chosen to evaluate our method and to test it for spatial and temporal convergence. Three fluid grids with 49 (coarse), 97 (medium fine), and 201 (fine) equidistant grid points on the panel were used. The spacing of the first grid line parallel to the panel surface was 10^{-3} for all grids, where the panel chord length was set to unity. The far field boundaries were 20 chord lengths away from the panel. For the results discussed in the following paragraph of this section we made use of coupling scheme B, see the discussion on coupling schemes below.

The first flutter case computed is a hinged aluminium panel of thickness ratio $h/c = 0.004$ at 20000 ft above sea level in a $M = 1.1$ flow. Bendiksen & Davis [1] found limit cycle flutter for this case, see their Fig. 7, with minimum and maximum plate midpoint deflections $(u/h)_{\text{min}} = -2.91$ and $(u/h)_{\text{max}} = 2.54$, respectively, and a flutter period $T_{\text{f}} = 9.94\,t^* = 7.644\,\tau$, where $\tau = \sqrt{\gamma}Mt^*$ is the time unit used within the FLOWer code. The coarse fluid grid and a time step $\Delta t/\tau = 0.1$ were sufficient to produce a space and time converged flutter solution [11], displayed in Fig. 6. Our computed flutter period, $T_{\text{f}} = 7.645\,\tau$, agrees very well with the value found in [1], while minimum and maximum plate midpoint deflections, $(u/h)_{\text{min}} = -2.78$ and $(u/h)_{\text{max}} = 2.46$, agree within 4.7 % and 3.25 %, respectively. Both, the modal Galerkin method, with 20 modes included in the analysis, and the FE method, where we used 40 elements of equal length and took the switching parameter $\Phi = 0$, as appropriate to von Kármán plate dynamics, were used to obtain flutter solutions. The different structural methods gave results in very close agreement with each other. This is also true for the two flutter cases discussed below.

The second flutter case is the $h/c = 0.004$ aluminium panel at 20000 ft and $M = 1.2$. Bendiksen & Davis [1] found $T_{\text{f}} = 9.42\,t^* = 6.63\,\tau$, $(u/h)_{\text{min}} = -2.85$ and $(u/h)_{\text{max}} = 2.71$. For this case, the coarse fluid grid and $\Delta t/\tau = 0.05$ were needed to compute a converged flutter solution, which gave $T_{\text{f}} = 6.56\,\tau$ and $(u/h)_{\text{min}} = -2.78$, $(u/h)_{\text{max}} = 2.79$ (Fig. 7). Our flutter period and peak midpoint deflections agree within 1.0 % and 3.0 %, respectively, with the results of Bendiksen & Davis.

The third flutter case considered is a hinged aluminium panel, $h/c = 0.002$, in a $M = 0.95$ flow at sea level. Here, Bendiksen & Davis [1] found $T_f = 27.97\,t^* = 24.88\,\tau$, and $(u/h)_{\rm min} = -5.65$, $(u/h)_{\rm max} = 6.2$. The medium fine fluid grid and $\Delta t/\tau = 0.2$ gave a converged flutter solution with $T_f = 26.48\,\tau$, and $(u/h)_{\rm min} = -5.81$, $(u/h)_{\rm max} = 6.5$ (Fig. 8). For this flutter case the present method predicts a flutter period 6.4 % greater than the value given by Bendiksen & Davis for this case.

3.4.2 Comparison of coupling schemes

Results of all three coupling schemes have been obtained for the three flutter cases considered above, where we used $\Delta t/\tau = 0.1$ ($\Delta t/\tau = 0.05$, $\Delta t/\tau = 0.2$) and the coarse (coarse, medium fine) fluid grid for the $M = 1.1$ ($M = 1.2$, $M = 0.95$) case and either the modal method (20 modes) or the FE method (40 elements) on the structural side. In all three cases we found that, for the time steps given above, coupling scheme B and the tight scheme C gave results that are in very close agreement with each other. The loose coupling scheme A predicted a flutter period that is longer than the periods computed by schemes B and C for $M = 1.1, 0.95$, and shorter for $M = 1.2$. We believe that this tendency is due to the phase error that is introduced through the lagging approach of scheme A. For the $M = 1.1$ case the flutter period predicted by scheme A was 3.3 % above the value computed by schemes B and C, for the $M = 0.95$ case the difference was 24.2 % (Fig. 9), while for the $M = 1.2$ case the scheme A flutter period was 10.7 % below the scheme B/C values. Further flutter simulations with the loose scheme A were performed, where we subsequently reduced the time steps to check whether the scheme A results converge against the scheme B/C results. This was the case for the $M = 1.1$ case at $\Delta t/\tau = 0.025$. The scheme A flutter period for the $M = 0.95$ case converged towards the scheme B and C periods up to 3.1 % at $\Delta t/\tau = 0.025$. However, for the $M = 1.2$ case, we found that the time converged scheme A flutter solution, $T_f = 5.81\,\tau$, $(u/h)_{\rm min} = -3.31$, $(u/h)_{\rm max} = 3.23$ computed at $\Delta t/\tau = 0.0125$, was different from the scheme B and C solutions by 11.4 % in terms of flutter period, see Fig. 10.

Since coupling scheme B gave essentially equal results compared to the tight scheme C at a lower computing work load, scheme B was used for panel flutter results presented below.

3.4.3 Reissner-Mindlin plate results

Calculations with the RM plate model on the structural side were computed for the three flutter cases of the previous section. Again, we used 40 elements, $\Delta t/\tau = 0.1$ ($\Delta t/\tau = 0.05$, $\Delta t/\tau = 0.2$) and the coarse (coarse, medium) fluid grid for the $M = 1.1$ ($M = 1.2$, $M = 0.95$) case. In all cases, the RM solutions exactly reproduced our von Kármán results, indicating that the von Kármán model gives an adequate prediction of plate dynamics for the transonic panel flutter problem.

3.4.4 Hinged panel stability boundary

The stability boundary for a hinged aluminium plate, $h/c = 0.00454$, at sea level and hence, mass ratio $\mu = \varrho_\infty c/\varrho_s h = 0.1$, has been computed for a Mach number range 0.8 to 1.3, where we used the coarse fluid grid, nonlinear RM plate dynamics (40 elements) and $\Delta t/\tau = 0.1$. The symbols in Fig. 11 represent the results of the present method, while the solid line depicts the stability boundary computed by Davis & Bendiksen[5]. Our calculations are in fair agreement with the Davis & Bendiksen stability boundary.

3.4.5 Flutter of a linear panel

In addition to the simulations of nonlinear panels, we performed calculations of linear panels. A slightly different fluid mesh, details of which are given in [10], was used for these calculation. The absence of the nonlinear stretching term in the plate equation can lead to large plate deflections for flow conditions beyond the stability boundary. In some cases, the plate deflection became so large that fluid grid lines near the panel surface intersected and the calculation broke down. In Figs. 12-15 we present results of a simulation of a linear panel with $h/c = 0.005$ at 11 km above sea level and $M = 0.85$ and an upwards directed small initial velocity of the undeformed panel. The modal method and $\Delta t/\tau = 0.025$ were used to compute the results discussed here. While for a nonlinear panel we find stationary divergence with a relatively low amplitude ($\sim 0.5\ h$) first mode deflection for these flow parameters, in the case of a linear panel our simulation predicts oscillatory plate motion around a mean amplitude of approximately 22 h, see Figs. 12 and 13. Figures 14 and 15 show the Mach number distribution over the oscillating panel for two instants corresponding to minimal (Fig. 14, $t/\tau = 185.0$) and maximal (Fig. 15, $t/\tau = 195.0$) deflection during the panel oscillation. For $t/\tau = 185.0$, the panel deflection is such that the shock wave terminating the supersonic region on the panel surface is located near the trailing edge (at $x = 0.93$), i.e., a relatively large portion of the panel is exposed to a high speed flow and thus experiences lift that the linear panel bending stiffness cannot compensate. Consequently, the amplitude of the panel deflection grows with time. This in turn causes the shock to move towards the leading edge of the plate, whereby the total lift of the panel decreases again, until for $t/\tau = 195.0$ the shock is located at $x = 0.79$ and the motion of the panel is reversed by the structural forces. In the course of its downward motion the shock wave moves towards the trailing edge again and the cycle is repeated. Note that our Euler solution predicts the existence of a vortex near the trailing edge of the plate. We find that the size of vortex region grows as the panel moves upwards, while its size decreases during the downward motion of the panel, see Fig. 14.

For a downwards directed initial velocity with the same absolute value as before the solution becomes completely different. Depending on the magnitude of the velocity disturbance the result may be e.g. divergence with negative final deflection or flutter about a negative mean deflection in place of the results discussed before. Such dependence on initial conditions is well known from nonlinear dynamics.

4 Conclusions

Numerical simulations of 2D Euler flows over flat aluminium panels with hinged leading and trailing edges were presented. We used the DLR FLOWer code to compute the flow field, while two independent structural methods, a FE method and a modal Galerkin method, were employed to predict the plate dynamics. Both methods gave identical results for the flutter cases considered.

Three different fluid-structure coupling schemes have been used to obtain flutter results. Two of these schemes, which avoid time lagging either through extrapolation (scheme B) or a subcycling approach (scheme C), produced time converged flutter solutions in very close agreement with each other, when approximately 75 – 125 time steps per flutter period where used. The loose coupling scheme A needed significantly more time steps per cycle (at least by a factor of four) to attain a time converged flutter solution. In one case, the loose scheme computations converged against a flutter solution different from the one obtained by schemes B and C.

Reissner-Mindlin plate dynamics exactly reproduced the von Kármán results for the panel flutter cases considered in this work, indicating that the von Kármán plate model gives a sufficiently accurate prediction of plate motion for the panel flutter problem.

Fair agreement with the results given by Bendiksen & Davis [1], [5] was found for the flutter cases and stability boundary presented in this work. In the case of a simulation of a linear panel, we found panel vibrations around a high amplitude mean deflection.

References

[1] O.O. Bendiksen, G.A. Davis: Nonlinear traveling wave flutter of panels in transonic flow, AIAA Paper 95-1496, 1995

[2] G. Britten, M. Werle, M. Hesse, J. Ballmann: Analysis of an Elastic Wing in Subsonic Flow using Direct Numerical Aeroelastic Simulation, in E. Krause, W. Jäger (eds.), High Performance Computing in Science and Engineering, 2000, Springer Verlag, pp. 305-316

[3] J.R. Cebral, R. Löhner: Conservative load projection and tracking for fluid-structure problems, AIAA Journal, Vol. 35, No. 4, 1997, pp. 687-692

[4] R. Dautray, J.-L. Lions: Mathematical Analysis and Numerical Methods for Science and Technology, Vol. V, Springer-Verlag, Berlin, 1988

[5] G.A. Davis, O.O. Bendiksen: Transonic panel flutter, AIAA Paper 93-1476, 1993

[6] E.H. Dowell: Aeroelasticity of plates and shells, Noordhoff International Publishing, Leyden, 1975

[7] Foda, M.A.: Influence of shear deformation and rotary inertia on the nonlinear free vibration of a beam with pinned ends, Computers and Structues, Vol. 71, 1999, pp. 663-670

[8] R. Heinrich, K. Pahlke, H. Bleecke: A three dimensional dual-time stepping method for the Solution of the unsteady Navier-Stokes equations, Unsteady Aerodynamics Conference Preceedings, London, paper No. 5, 1996

[9] T.J.R. Hughes: The finite element method, Prentice-Hall, Englewood Cliffs, 1987

[10] J. Hurka, J. Ballmann: Preliminary results of numerical simulations of elastic panels in transonic flow, Notes on Numerical Fluid Mechanics, Vol. 77, 2002, pp. 171-178

[11] J. Hurka, J. Ballmann: Elastic panels in transonic flow, AIAA Paper 2001-2722, 2001

[12] A. Jameson: Time dependent calculations using multigrid, with applications to unsteady flows past airfoils and wings, AIAA Paper 91-1596, 1991

[13] Z. Friedman, J.B. Kosmatka: An improved two-node Timoshenko beam finite element, Computers & Structures, Vol. 47, No. 3, 1993, pp. 473-481

[14] N. Kroll, C.-C. Rossow, K. Becker, F. Thiele: The MEGAFLOW project, Aerospace Science and Technology 4, 2000, pp. 223-237; also: FLOWer installation and user handbook, Release 115, 1998

[15] J.-L. Lions: Quelques méthodes de résolution des problèmes aux limites non linéaires, Dunod, Paris, 1969

[16] R. Massjung: Uniqueness Results for Aeroelasticity – Lagrangian Approach, Technical Report No. 172, IGPM, RWTH-Aachen, 1999

[17] R.D. Mindlin: Influence of rotary inertia and shear on flexural motions of isotropic, elastic plates, Journal of Applied Mechanics 18, 1951, pp. 31-38

[18] N.F. Morozov: Nonlinear vibrations of thin plates with allowance for rotational inertia, Soviet Math. Vol.8, 1967

[19] W.H. Press, B.P. Flannery, S.A. Teukolsky, W.T. Vetterling: Numerical Recipes, Cambridge University Press, Cambridge, 1986

[20] P. Secchi, A. Valli: A free boundary problem for compressible viscous fluids, Journal für reine und angewandte Mathematik, 341, 1983, 1-31

[21] J. Serrin: On the Uniqueness of Compressible Fluid Motions Arch. Rat. Mech. Ana., 3, 1959, 271-288

[22] M. Werle: Gittergenerierung für die Strömungssimulation von elastischen Auftriebskörpern, Diplomarbeit, RWTH-Aachen, 1997

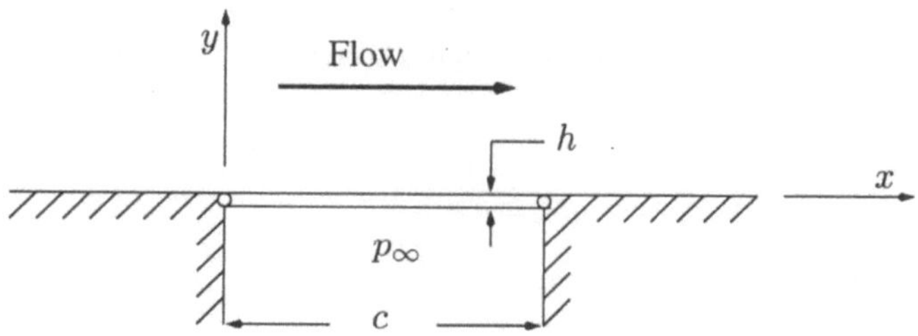

Figure 2 Sketch of problem geometry.

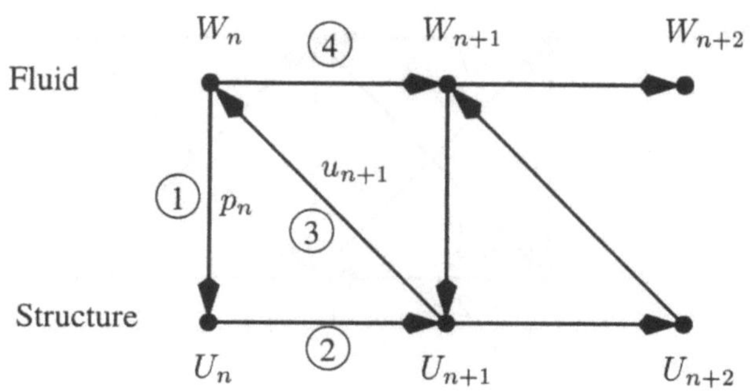

Figure 3 Coupling scheme A (loose coupling).

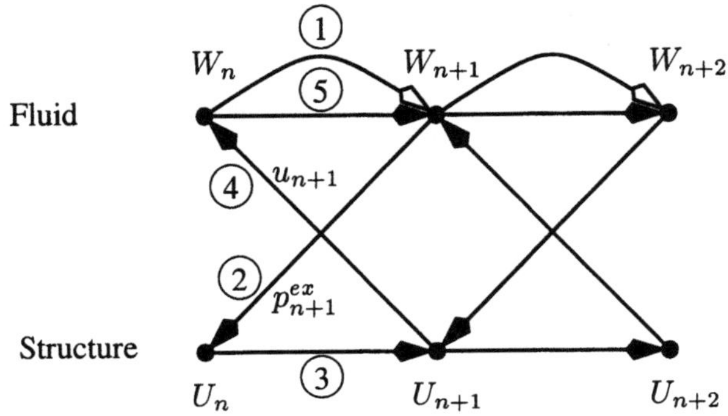

Figure 4 Coupling scheme B.

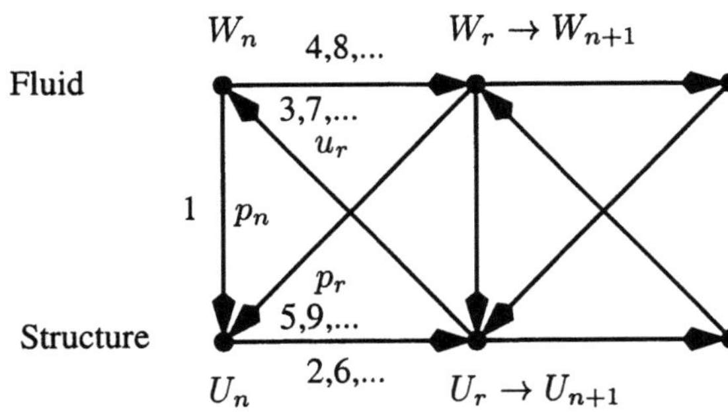

Figure 5 Coupling scheme C (tight coupling).

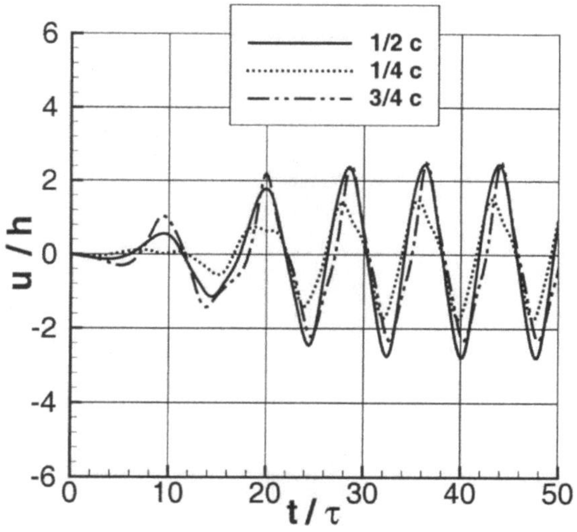

Figure 6 Transonic limit cycle flutter of a hinged aluminium plate at 20000 ft above sea level, $M = 1.1$, $h/c = 0.004$. Time history of the displacement of three points on the plate's surface: $1/4c$, $1/2c$, $3/4c$. Euler flow – von Kármán plate.

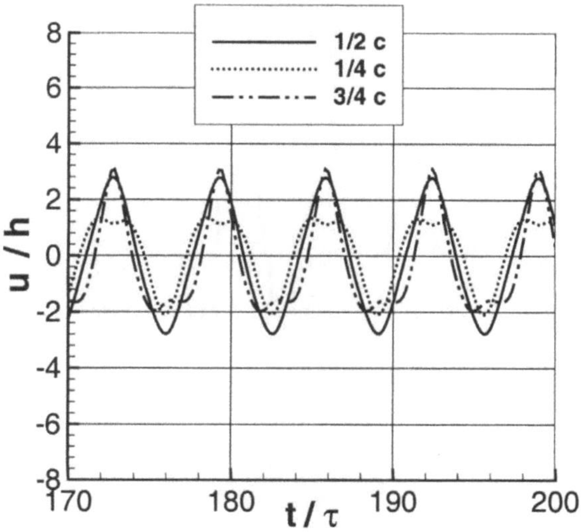

Figure 7 Limit cycle flutter of a hinged aluminium plate, $h/c = 0.004$, $M = 1.2$, at 20000 ft above sea level. Euler flow – von Kármán plate.

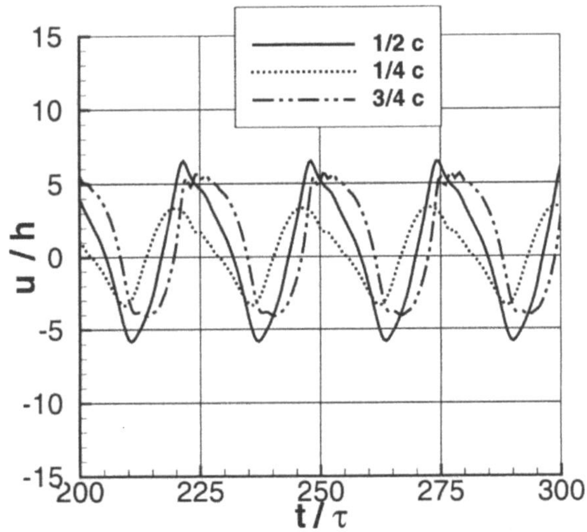

Figure 8 Transonic limit cycle flutter of a hinged aluminium plate at sea level, $M = 0.95$, $h/c = 0.002$. Time history of the displacement of three points on the plate's surface: $1/4c$, $1/2c$, $3/4c$. Euler flow – von Kármán plate.

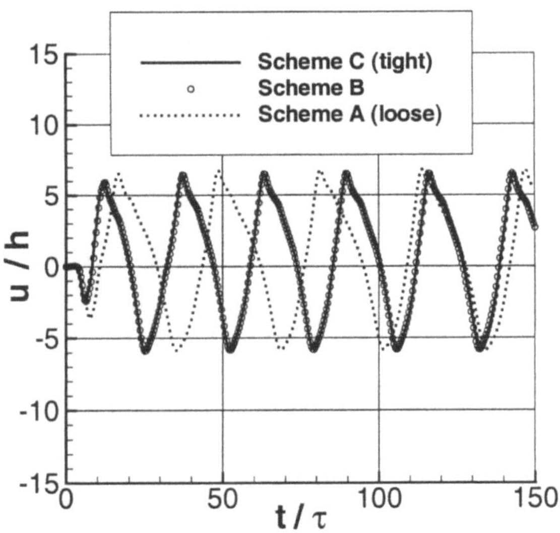

Figure 9 Comparison of coupling schemes A, B, and C at a time step $\Delta t/\tau = 0.2$. Time history of plate midpoint deflection for the transonic flutter of a hinged aluminium panel at sea level, $M = 0.95$, $h/c = 0.002$. Euler flow – von Kármán plate.

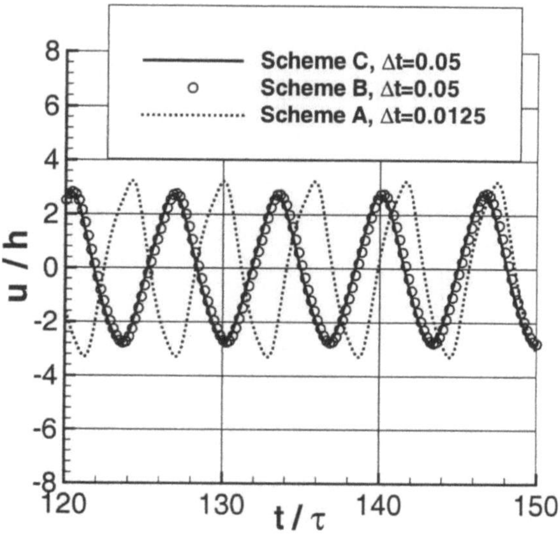

Figure 10 Comparison of time converged flutter solutions computed by schemes A, B, and C. Time history of plate midpoint deflection for the flutter of a hinged aluminium panel at 20000 ft above sea level, $M = 1.2$, $h/c = 0.004$. Euler flow – von Kármán plate.

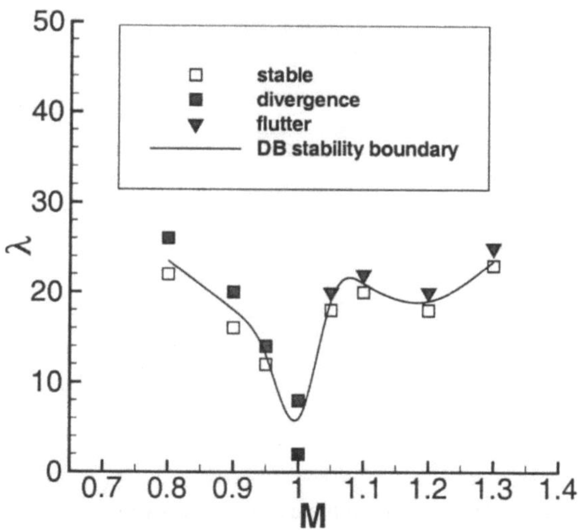

Figure 11 Stability boundary for a hinged panel at sea level, $h/c = 0.00454$, and mass ratio $\mu = 0.1$. Dimensionless dynamic pressure λ vs. Mach number.

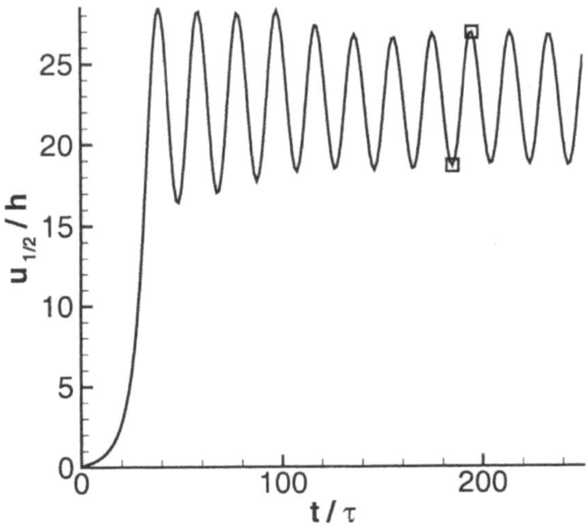

Figure 12 Simply supported aluminium plate at 11 km above seal level, $M = 0.85$, $h/c = 0.005$. Euler flow – (linear) Kirchhoff plate. Time history of the plate midpoint displacement.

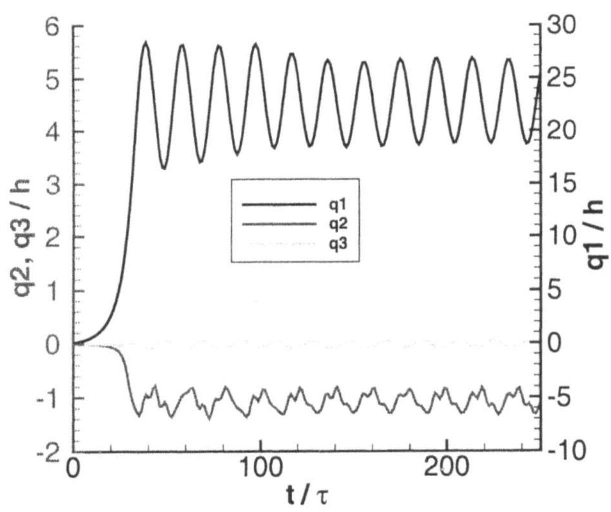

Figure 13 Simply supported aluminium plate at 11 km above seal level, $M = 0.85$, $h/c = 0.005$. Euler flow – (linear) Kirchhoff plate. Contributions of the first three modal amplitudes q_1, q_2, q_3 to the time history of plate midpoint displacement. The scale of the ordinate on the right refers to the q_1, the one on the left to q_2 and q_2.

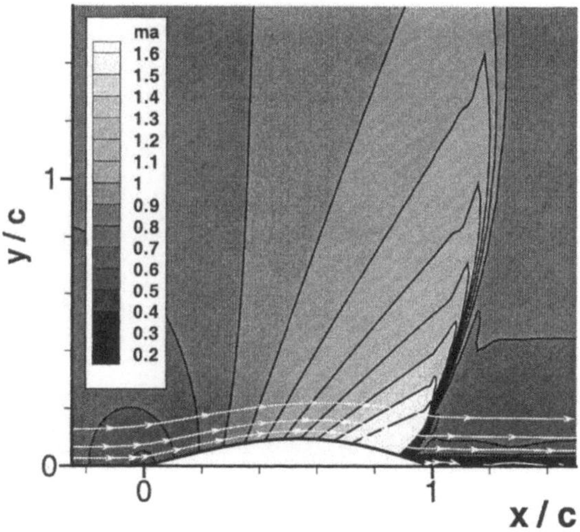

Figure 14 Mach number distribution over the oscillating panel of Fig. 12 at $t/\tau = 185.0$ (labeled in Fig. 12 by the square symbols). The white lines indicate tangents to the instantaneous velocity field near the panel surface.

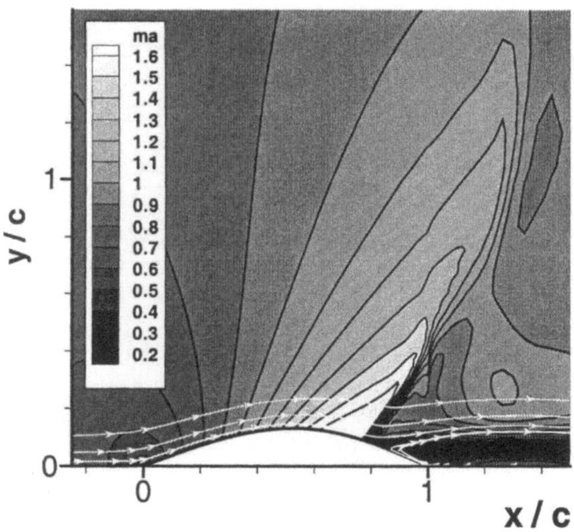

Figure 15 Mach number distribution over the oscillating panel of Fig. 12 at $t/\tau = 195.0$ (labeled in Fig. 12 by the square symbols). The white lines indicate tangents to the instantaneous velocity field near the panel surface.

Concepts for Reduced Structural Models of Airplane Wings in Aeroelasticity

W. Jung, H.-G. Reimerdes
Lehrstuhl und Institut für Leichtbau, RWTH Aachen
Wüllnerstraße 7, D-52062 Aachen, Germany

Summary

A beam model for the idealization of airplane wings for numerical studies in aeroelasticity is presented. The structural model gives a good approximation of the static and dynamic (eigenfrequencies) wing behaviour. It is to be implemented in the routines of computational aeroelasticity within the framework of this Collaborative Research Center. It extends the Timoshenko beam by incorporation of additional effects, such as taper, restrained warping and the effect of ribs and considers different material properties within the wing cross section. Examples are presented which demonstrate the influence of the additional effects on the static and dynamic behaviour of the wing. The reduced structural wing model yields results that are in good agreement with finite–element calculations and provides a suitable tool for sensitivity analyses.

Nomenclature

A_R, A_S, A_{SB}	cross-sectional area of the ribs (R) and stringers (S)
$\tilde{A}$	cross sectional area of the wing box bound by the contour midline
[**C**]	stiffness matrix
C_T, C_S	resistance to torsion– and shear related warping
d_R	distance of neighbored ribs
d_S	distance of neighbored stringers
$E_{CS}, E_S,$	Young's moduli of the unstiffened cover sheets (CS), stringers (S)
$E_R, E_{SB},$	ribs (R), spar booms (SB),
E_{Web}, E_{SS}	stiffened webs (Web) and stiffened upper and lower skins (SS)
$E_N^\star, E_y^\star,$ $E_z^\star, E_T^\star, E_S^\star$	effective Young's moduli

G	shear modulus
I_T	torsional rigidity
I_y, I_z	moments of inertia
$K_N, K_z,$ $K_y, K_{u_T^0}$	constants
$\vec{K_M}, \vec{K_T}$	node force vectors
l_y, l_z	measures of length
M_x, M_y, M_z	twisting and bending moments
m_x, m_y, m_z	twisting and bending moments per unit length
m_{WT}, m_{WS}	torsion– and shear–related warping moments per unit length
N_x	normal force
$n_{x_{CS}}, n_{\theta_{CS}}$	axial and circumferential normal stress-flow components of the skins and webs
$n_{x\theta_{CS}}$	shear stressflow of the cover sheets
p_x, p_y, p_z	normal and shear forces per unit length
$\vec{p_M}, \vec{p_T}$	distributed load vectors
Q_y, Q_z	shear forces
$Q_y^\star, Q_z^\star$	effective shear forces for the determination of the shear deformation
Q_{WT}, Q_{WS}	torsion– and shear–related warping shear forces
r_s	perpendicular distance from the shear center to the tangent at any point of the mid–surface contour
$\vec{r_M}, \vec{r_T}$	node displacement vectors of Membrane-Panel (M) and Timoshenko beam (T)
t	wall thickness
u_T, u_S	torsion– and shear related warping displacements
u_T^{00}, u_S^{00}	Saint Venant torsion– and shear related warping functions
W_T, W_S	torsion– and shear–related warping moments
$\varepsilon_x, \varepsilon_\theta$	normal and circumference strains
$\gamma_{x\theta}$	shear strain
$\varphi_x, \varphi_y, \varphi_z$	angle of twist
κ	shear correction factors
ν	Poisson's ratio
ϑ	rate of twist
σ	normal stress
ψ_T, ψ_S	measures of intensity of the torsion– and shear–related uniform warping distribution

1 Introduction

As the concept of high capacity and hence highly flexible aircraft with low-structural weight gathers more impetus, the aircraft designers are faced with a variety of challenging technical problems. Among those, aeroelastic problems are generally accepted to be of primary concern [1].

The main focus of this Collaborative Research Center is to study the aerodynamic and aeroelastic effects of large elastic wings. For this, a code solving the coupled aero-structural problem is developed [2]. The task of the work presented here is the structural idealization of a transport aircraft wing to be implemented in this code. The structural model shall consider the main physical effects with much less degrees of freedom than typical Finite–Element–descriptions of a real wing structure. The advantage is not only a reduction of necessary CPU-time needed to calculate the structural response in each time step but also a simple generation of the structural model of a wing. A fast modification of the structure and thus systematical investigations of different physical effects are possible. At last, the developed structural model shall be able to consider wing taper, different anisotropic materials in a cross-section, restrained warping, transverse shear and wing swept.

In the last decades several simplified models for the description of aircraft wings were developed. Common approaches are based on modified beam theories ([3], [4], [5]), on plate like structures ([6], [7], [8], [9]), or on shearing theory ([10], [11], [12],). In this work a modified beam theory called 'Membrane-Panel' idealization is used in order to consider transverse shear and restrained warping. In addition to the above mentioned studies wing taper is considered and the typical cross section of a load carrying wing box is idealized.

2 The Membrane–Panel structural model

2.1 Geometry of the wing box

For the static analysis of the wing an idealized single–cell wing box is chosen as load carrying structure. The idealized wing box with arbitrary unsymmetrical cross section consists of straight webs, parabolic skins and discrete spar booms as shown in figure 1. Stringers are assumed to be continuously distributed over the upper and lower skin. The external geometry of the wing box in figure 1 is described by six parameters (h_1, h_2, a_1, a_2, b and δ_z). Additionally, the equivalent thicknesses of the skins and webs as well as the area of the spar and rib booms must be given. Thus changes in geometry are easily performed. Pertinent quantities are referred to a global coordinate system xyz, where y and z denote the cross sectional coordinates and x is the spanwise coordinate.

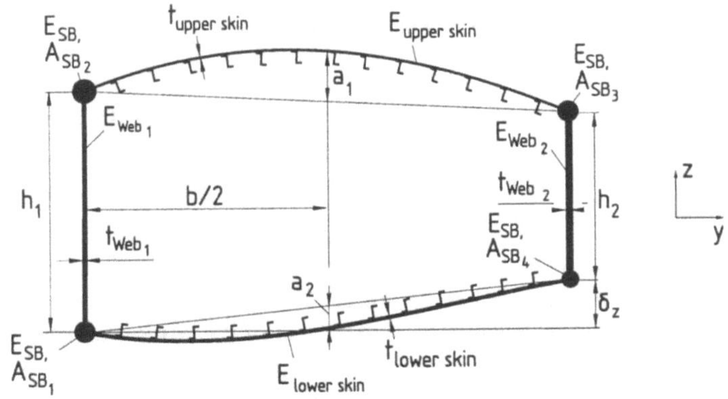

Figure 1 Structural model of the idealized wing box cross section with parabolic skins

2.2 The effect of ribs

In the Membrane–Panel structural model the skins and webs are idealized as membrane shells, which carry normal and shear forces. The following assumptions are made:

- The normal and shear stresses are constant over the skin and web thickness hence neglecting corresponding internal moments.
- Rib- and stringer elements are idealized as one-dimensional structures.
- In circumference direction of the wing box respective external moments are carried by infinitely close spaced ribs.
- The ribs of the wing are assumed to be parallel to the root section and have infinite shear stiffness in their own plane. Only the restraint of the in–plane contour deformation of the wing cross section is considered while neglecting restraints of the out–of–plane contour deformations.
- The stringers are continuously distributed along the cover sheets while discrete spar booms are assumed. The stringers, spar– and rib booms carry only normal forces and have no stiffness with respect to other forces.

The relations between stress flows and strain components for a section i of an unstiffened skin or web element is given by Hooke's law as follows:

$$\begin{pmatrix} n_{xCS} \\ n_{\theta CS} \\ n_{x\theta CS} \end{pmatrix}_i = [\mathbf{C}]_i \begin{pmatrix} \varepsilon_{xCS} \\ \varepsilon_{\theta CS} \\ \gamma_{x\theta CS} \end{pmatrix}_i . \tag{1}$$

For an isotropic skin section i the stiffness matrix $[\mathbf{C}]_i$ is

$$[\mathbf{C}]_i = \frac{E_{CS_i}\, t_{CS_i}}{1-\nu_i{}^2} \begin{pmatrix} 1 & \nu_i & 0 \\ \nu_i & 1 & 0 \\ 0 & 0 & \frac{1}{2}\,(1-\nu_i) \end{pmatrix} . \tag{2}$$

The normal stress flows of the stringers (index S) and ribs (index R) are

$$n_{x_S} = \frac{\varepsilon_{x,S}\, E_S A_S}{d_S} \tag{3}$$

$$n_{\theta_R} = \frac{\varepsilon_{\theta,R}\, E_R\, A_R}{d_R} . \tag{4}$$

The resulting normal stress flows in the stringer and rib stiffened skin section i at wing axis section ξ are equal to the sum of the normal stress flows of the unstiffened skin, stringers and ribs. Because of equal strains of the stringer and skin contact elements as well as skin and rib contact elements the following relation is derived

$$\begin{pmatrix} n_{x_i} \\ n_{\theta_i} \\ n_{x\theta_i} \end{pmatrix}_{\xi} = [\mathbf{C}_S]_{\xi,\,i} \begin{pmatrix} \varepsilon_x \\ \varepsilon_\theta \\ \gamma_{x\theta} \end{pmatrix}_{\xi} \tag{5}$$

$$\text{where } [\mathbf{C}_S]_{\xi,\,i} = \begin{pmatrix} \frac{E_{CS_i} t_i}{1-\nu_i{}^2} + \frac{E_S A_S}{d_S} & \frac{\nu_i E_{CS_i} t_i}{1-\nu_i{}^2} & 0 \\ \frac{\nu_i E_{CS_i} t_i}{1-\nu_i{}^2} & \frac{E_{CS_i} t_i}{1-\nu_i{}^2} + \frac{E_R A_R}{d_R} & 0 \\ 0 & 0 & \frac{E_{CS_i} t_i}{2\,(1+\nu_i)} \end{pmatrix} .$$

For the determination of the equivalent Young's modulus and thickness of the cover sheets the following assumptions are made:

- The resulting normal stress flow of the wing box in circumferential direction n_θ is zero.
- A non–zero normal stress flow $n_{\theta_{CS}}$ in the pure stringer stiffened skins occurs, but is equilibrated by respective reverse normal stress flows of the ribs.

The mean normal stress σ_x of the stringer stiffened skin section i is

$$\sigma_{x_i} = \frac{n_{x_i}}{t_{SS_i}} = E_{SS_i}\, \varepsilon_{x_i} \tag{6}$$

where the thickness of the stiffened skin is

$$t_{SS_i} = t_i + \frac{A_S}{d_S} .$$

With these assumptions the equivalent Young's modulus for the stringer– and rib–stiffened skin section (orthotropic stiffening) i is

$$E_{SS_i} = \frac{E_{CS_i}{}^2\, t_i^2\, d_R\, d_S + E_S\, E_R\, A_S A_R\left(1-\nu_i{}^2\right) + E_{CS_i} t_i \left(E_S A_S d_R + E_R A_R d_S\right)}{\left(t_i d_S + A_S\right)\left(E_{CS_i} t_i c_{R_i} + E_R A_R\left(1-\nu_i{}^2\right)\right)} \,. \tag{7}$$

The equivalent Young's moduli of the webs are:

$$E_{Web_i} = \frac{\left(E_{CS_i}\, t_i \left(E_{CS_i}\, t_i\, d_R + E_R A_R\right)\right)}{t_i\left(E_{CS_i} t_i c_{R_i} + E_R A_R\left(1-\nu_i{}^2\right)\right)} \,. \tag{8}$$

The equivalent Young's moduli of the skins and webs consider different (isotropic) materials around the cross section as well as the restrained in–plane contour deformation of the wing box by the ribs.

2.3 The effect of restrained torsion–related warping

The deformation of the wing box is described by stretching, bending, twisting, and warping. The torsion–related warping displacements of a closed section for restrained warping beam is approximated by

$$u_T = \psi_T(x)\; u_T^\circ(y,z) \tag{9}$$

where $u_T^\circ(y,z)$ is a cross-sectional dependent torsion–related warping function, which describes the uniform warping distribution out of the cross sectional plane. A measure of the intensity of the torsion-related warping is given by the independent generalized displacement $\psi_T(x)$ which is a function of the wing spanwise coordinate x. Like in Heilig [14] or more recently in Bhaskar [15] the Saint–Venant's torsion–related warping displacements for free warping $u_{T,f}$ are used as a starting point for the determination of the torsion–related warping displacements u_T of a beam with restrained warping. The latter ones can be expressed by the sum of the warping displacements of the shear free contour midline of an open section beam and the warping displacements caused by the constant shear flow distribution of the closed section beam because of twist only. Thus $u_{T,f}$ is defined by

$$u_{T,f} = -\vartheta \int_0^s r_s ds + \int_0^s \gamma_T ds + u_a \tag{10}$$

where r_s represents the perpendicular distance from the shear center to the tangent at any point of the mid–surface contour. Using the Bredt–Batho formula for a thin walled single–cell closed beam section

$$\gamma_T = \frac{M_T}{2\,\tilde{A}\,G\,t} \tag{11}$$

$$\vartheta = \varphi'_x = \frac{M_T}{\overline{G} I_T} \tag{12}$$

$$I_T = \frac{4\,\tilde{A}^2}{\overline{G} \oint \frac{ds}{G\,t}} \tag{13}$$

$$\text{with } \overline{G} = \frac{\oint \frac{ds}{t}}{\oint \frac{ds}{G\,t}} \quad \text{and} \quad \tilde{A} = \frac{1}{2} \oint r_s ds$$

the integral of the shear strains $\int_0^s \gamma_T ds$ is

$$\int_0^s \gamma_T ds = \vartheta \frac{2\tilde{A}\overline{G}}{\oint \frac{ds}{t}} \int_0^s \frac{1}{G\,t} ds \;. \tag{14}$$

Dividing equation (10) by the rate of twist ϑ, which is a function of the wing spanwise coordinate, yields the Saint-Venant uniform torsion–related warping function $u_T^\circ(y,z)$

$$u_T^\circ(y,z) = \underbrace{- \int_0^s r_s ds + \frac{2\tilde{A}}{\overline{G} \oint \frac{ds}{G\,t}} \int_0^s \frac{1}{t} ds}_{u_T^{00}} + u_a^0 \tag{15}$$

which considers different (orthotropic) materials along the contour of the wing box. The constant u_a^0 in equation (15) is determined in a way that the component of the normal force $N_{x,WT}$ due to the displacements of restrained torsion–related warping becomes zero. With this condition

$$N_{x,WT} = \psi'_T \int E\, u_T^\circ dA = 0 \tag{16}$$

and consideration of u_T^{00} in equation (15) the constant u_a^0 is

$$u_a^0 = \frac{u_a}{\vartheta} = \frac{-1}{E_N\,A} \oint E\, u_T^{00} dA \quad \text{where } E_N\,A = \oint E\, dA \;. \tag{17}$$

The Saint-Venant uniform torsion–related warping function $u_T^\circ(y,z)$ in equation (15) for free warping is also applied in the Membrane–Panel structural model for the uniform torsion–related warping displacements for restrained warping. In case of restrained warping the constant rate of twist ϑ is replaced by the spanwise dependent variable $\psi_T(x)$. With help of the equations (15) and (17) the uniform torsion–related warping function in equation (9) is

$$u_T^0(y,z) = -\int_0^s r_s\,ds + \frac{2\tilde{A}\overline{G}}{\oint \frac{ds}{t}} \int_0^s \frac{1}{Gt} ds$$

$$- \frac{1}{E_N A} \oint E \left(-\int_0^s r_s\,ds + \frac{2\tilde{A}\overline{G}}{\oint \frac{ds}{t}} \int_0^s \frac{1}{Gt} ds \right) dA \,. \qquad (18)$$

For a wing box with parabolic skins the distribution of the Saint-Venant uniform torsion–related warping function is given in figure 2.

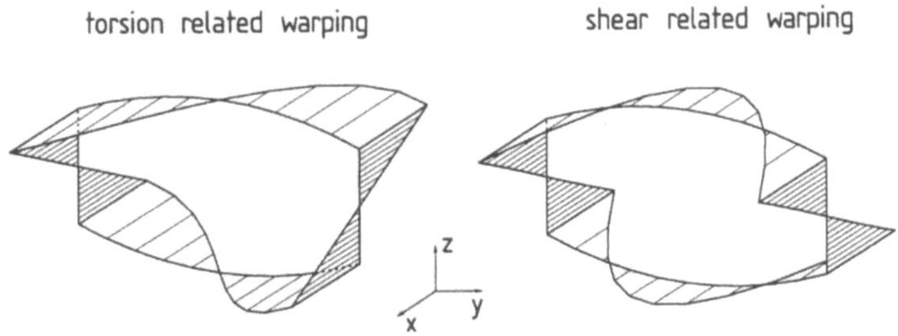

Figure 2 The distribution of the uniform warping functions for a wing box with parabolic skins

2.4 The effect of restrained shear–related warping

The shear–related warping displacement for restrained warping u_S is gained by integration of the shear strains along the wing box contour on condition that its consequential normal stress distribution forms a group of eigenforces. Furthermore, the shear–related warping normal stress distribution has to be orthogonal to the torsion–related warping normal stress distribution. Thus u_S is approximated as follows

$$u_S(x,y,z) = \int_0^s \gamma_{Q_z} ds + K_N + K_z\,z + K_y\,y + K_{u_T}\,u_T^0 \,. \qquad (19)$$

The constants K_N, K_z, K_y und K_{u_T} in equation (19) are determined on the condition in which the normal forces, bending moments and torsion–related warping moments due to u_S become zero. The shear strain γ_{Q_z} in equation (19) is

$$\gamma_{Q_z} = \frac{Q_z}{E_y I_y} \underbrace{\frac{1}{Gt} \left(\int_s^{s_e} E\,z\,dA - \frac{\oint \frac{\int E\,z\,dA}{tG} ds}{\oint \frac{ds}{Gt}} \right)}_{\gamma_{Q_z}^{00}} \qquad (20)$$

$$\text{where} \quad E_y\, I_y = \oint E\, z^2\, dA\,.$$

Apart from the shear force Q_z equation (20) only consists of cross-sectional dependent variables. Division of the shear–related warping displacements in equation (19) by the spanwise dependent term $\frac{Q_z(x)}{E_y\, I_y}$ yields the uniform shear–related warping function $u_s^0(y,z)$ as

$$u_s^0(y,z) \;=\; \underbrace{\int_0^s \gamma_{Q_z}^{00}\, ds}_{u_s^{00}} + K_N^0 + K_z^0\, z + K_y^0\, y + K_{u_T}^0\, u_T^0\,. \tag{21}$$

For a wing box with parabolic skins the distribution of the uniform shear–related warping function is given in figure 2. The following definitions for the constants K_N^0, K_z^0, K_y^0 and $K_{u_T}^0$ can be derived from the shear–related warping displacements.

$$\begin{aligned} K_N^0 &= -\frac{1}{E_N\, A} \oint E\, u_s^{00} dA \\ K_y^0 &= -\frac{1}{E_z\, I_z} \oint E y\, u_s^{00} dA \\ K_z^0 &= -\frac{1}{E_y\, I_y} \oint E z\, u_s^{00} dA \\ K_{u_T}^0 &= -\frac{1}{E_T\, C_T} \oint E\, u_T^0\, u_s^{00} dA\,. \end{aligned}$$

Analogous to equation (9) the shear–related warping displacements for restrained warping can be expressed in the following product form

$$u_S = \psi_S(x)\, u_S^0(y,z) \tag{22}$$

where $u_S^0(y,z)$ is the shear–related warping function, which is cross–sectionally dependent and describes the uniform warping distribution out of the cross sectional plane due to shear forces. A measure of the intensity of the shear–related warping is given by the independent generalized displacement $\psi_S(x)$ which is a function of the wing spanwise coordinate x.

2.5 Wing taper

In beam models taper is normally achieved by variation of the cross sections, where the geometry of each section itself is constant. The further consideration of taper in the single sections has, among others, an effect on the shear deformation. Figure 3 shows the plan view of a wing box section subjected to bending moments and shear

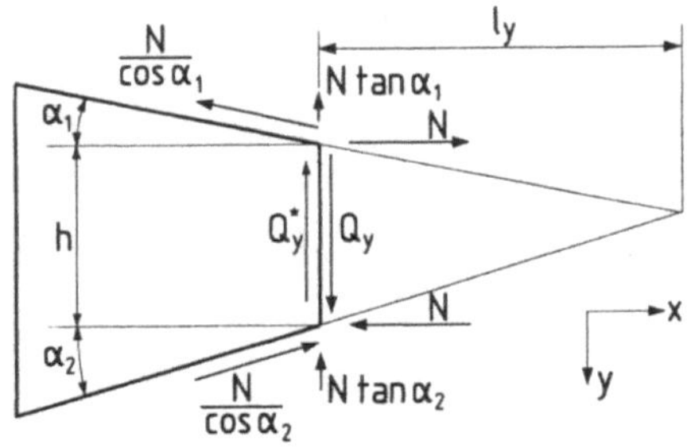

Figure 3 Effect of taper on the "effective shear force"

forces.

A part of the external shear forces is carried by the skew spars, which has an effect that the "effective shear force" $Q_y^\star$ due to shear loads is lower than the applied external load Q_y. The equilibrium of forces yields the effective shear force $Q_y^\star$ as

$$Q_y^\star = Q_y - N(\tan(\alpha_1) + \tan(\alpha_2))$$

$$\text{Considering} \quad \begin{aligned} M_z &= N * h \quad \text{and} \\ \frac{h}{l_y} &= tan(\alpha_1) + tan(\alpha_2) \end{aligned}$$

the effective shear force $Q_y^\star$ is

$$Q_y^\star = Q_y - \frac{M_z}{l_y} \tag{23}$$

In the z–direction the effective shear force is

$$Q_z^\star = Q_z + \frac{M_y}{l_z} \ . \tag{24}$$

Furthermore, the consideration of taper results in a variation of the normal stress distribution. Figure 4 shows a tapered wing section.

Extension of the dashed–lined cuboid yields a displacement of point P1 of the cuboid to P1'. The contour line of the tapered wing section through point P1' has an angle of inclination α towards the corresponding contour line of the non–tapered cuboid. If the contour line through point P1 is subjected to an infinitesimal displacement du, as shown in figure 4, then the corresponding contour line of the tapered cuboid is shifted by the absolute value Δds. Due to geometric considerations and Hooke's law follows

$$\Delta ds = du \cos\alpha \tag{25}$$

$$\varepsilon_\alpha = \frac{\Delta ds}{ds} = \frac{du}{dx}\cos^2\alpha \tag{26}$$

$$\sigma_\alpha = E\,\varepsilon_\alpha = E\,\varepsilon_x \cos^2\alpha \ . \tag{27}$$

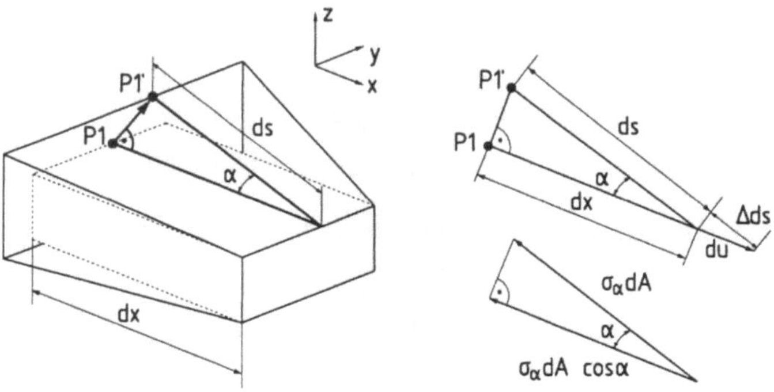

Figure 4 The effect of taper

With respect to figure 4 the bending moment M_y in the tapered wing box is

$$M_y = \oint \sigma_\alpha \cos\alpha \, z \, dA \,. \tag{28}$$

With consideration of the strain component due to the bending moment

$$\varepsilon_{x_{\varphi_y}} = \varphi_y' \, z \tag{29}$$

the equations (25), (26) and (27) yield the bending moment M_y as

$$M_y = \varphi_y' \oint E \cos^3\alpha \, z^2 dA \,. \tag{30}$$

The effect of taper can also be considered by a modified Young's modulus $E^\star$ as follows

$$M_y = \varphi_y' \, E_y^\star \, I_y \tag{31}$$

$$\text{where} \quad E_y^\star = \frac{\oint E \cos^3\alpha \, z^2 dA}{I_y} \,. \tag{32}$$

In the same way the other stress resultants can be determined.

2.6 Extension to Euler-Bernoulli hypothesis

The consideration of different (isotropic) materials along the cross section does not allow the use of stress hypotheses for the normal stress distribution along the cross section. For consistency of the strains at boundaries with different material properties an extension to Euler-Bernoulli hypothesis is applied for the determination of the displacements normal to the wing box cross section. For the consideration of warping, the linear distribution of the normal displacements in Euler-Bernoulli hypothesis is extended by incorporating additional torsion– and shear–related out of plane warping displacements as follows

$$\begin{aligned} u(x,y,z) &= u_N(x) + \varphi_y(x)\,z - \varphi_z(x)\,y \\ &\quad + \psi_T(x)\,u_T^0(y,z) + \psi_S(x)\,u_S^0(y,z)\,. \end{aligned} \tag{33}$$

With the help of Hooke's law the normal stress distribution can be derived as

$$\begin{aligned} \sigma_x(x,y,z) &= E(y,z)\left(\frac{N_x}{E_N^\star A} + \frac{M_y}{E_y^\star I_y} z - \frac{M_z}{E_z^\star I_z} y + \right. \\ &\quad \left. \frac{W_T}{E_T^\star C_T} u_T^0 + \frac{W_S}{E_S^\star C_S} u_S^0\right)\,. \end{aligned} \tag{34}$$

Considering the effect of taper (see subsection 2.5) the denominator terms in equation (34) are defined as

$$E_N^\star\,A = \oint E\,\cos^3\alpha\,dA \tag{35}$$

$$E_y^\star\,I_y = \oint E\,z^2\cos^3\alpha\,dA \tag{36}$$

$$E_z^\star\,I_z = \oint E\,y^2\cos^3\alpha\,dA \tag{37}$$

$$E_T^\star\,C_T = \oint E\,u_T^{0^2}\cos^3\alpha\,dA \tag{38}$$

$$E_S^\star\,C_S = \oint E\,u_S^{0^2}\cos^3\alpha\,dA\,. \tag{39}$$

2.7 The system of governing equations

Consistent with the equilibrium of forces and moments, the definition of the stress resultants and the principle of virtual works, the system of 16 governing equations can be derived, whose solution yields the unknown generalized forces and generalized displacements. For an arbitrary wing box section subjected to

- concentrated forces (N_x, Q_y and Q_z) and concentrated moments (M_x, M_y, M_z)
- uniform forces ((p_x), p_y, p_z) and moments per unit length (p_x), p_y, p_z, m_x, m_y, m_z, m_{WT} and m_{WS})

the system of governing equations is presented.

Normal– and shear forces, twisting–, bending– and warping moments:

$$
\begin{aligned}
N_x' &= -p_x && (40)\\
Q_y' &= -p_y && (41)\\
Q_z' &= -p_z && (42)\\
M_x' &= -m_x && (43)\\
M_y' &= -m_y + Q_z && (44)\\
M_z' &= -m_z - Q_y && (45)\\
W_T' &= -m_{WT} - \frac{M_x}{\kappa_T} + && (46)\\
& \frac{\overline{G}\, I_d}{\kappa_T}\left(\psi_T + \frac{\kappa_{zT}\, Q_z^\star}{\overline{G}\,\sqrt{I_d\, A}} + \frac{\kappa_{yT}\,, Q_y^\star}{\overline{G}\,\sqrt{I_d\, A}} - \frac{\kappa_{TS}\,(W_S' + m_{WS})}{\overline{G}\,\%sqrt C_S\, A}\right)\\
W_S' &= -m_{WS} + \frac{\overline{G}\, A\, I_y}{\kappa_S}\left(\psi_S - \frac{Q_z^\star}{E_y^\star\, I_y}\right. && (47)\\
& \left. + \frac{\kappa_{yS}\, Q_y^\star}{E_z^\star\, I_z} + \frac{\kappa_{xS}\, M_x}{\overline{G}\,\sqrt{C_S\, A}} - \frac{\kappa_{TS}\,(W_T' + m_{WT})}{\overline{G}\,\sqrt{C_S\, A}}\right).
\end{aligned}
$$

Displacements, rotations and measures of warping:

$$
\begin{aligned}
u_N' &= \frac{N_x}{E_N^\star\, A} && (48)\\
v' &= \varphi_z + \frac{\kappa_y\, Q_y^\star}{\overline{G}\, A} + \frac{\kappa_{yz}\, Q_z^\star}{\overline{G}\, A} - \frac{\kappa_{yT}\,(W_T' + m_{WT})}{\overline{G}\sqrt{I_d\, A}} && (49)\\
& - \frac{\kappa_{yS}\,(W_S' + m_{WS})}{E_z^\star\, I_z}\\
w' &= -\varphi_y + \frac{\kappa_z\, Q_z^\star}{\overline{G}\, A} + \frac{W_S' + m_{WS}}{E_y^\star\, I_y} + && (50)\\
& \frac{\kappa_{yz}\, Q_y^\star}{\overline{G}\, A} - \frac{\kappa_{zT}\,(W_T' + m_{WT})}{\overline{G}\,\sqrt{Id\, A}}\\
\varphi_x' &= \frac{M_x + W_T' + m_{WT}}{\overline{G}\, I_d} - \frac{\kappa_{xS}\,(W_S' + m_{WS})}{\overline{G}\,\sqrt{C_S\, A}} && (51)\\
\varphi_y' &= \frac{M_y}{E_y^\star\, I_y} && (52)\\
\varphi_z' &= \frac{M_z}{E_z^\star\, I_z} && (53)\\
\psi_T' &= \frac{W_T}{E_T^\star\, C_T} && (54)\\
\psi_S' &= \frac{W_S}{E_S^\star\, C_S}\ . && (55)
\end{aligned}
$$

The shear correction factors κ in equations (50) to (51) are described in detail in [13].
In case of ribs with infinite shear stiffness, as assumed in the presented Membrane–Panel structural model, the torsion– and shear–related warping forces Q_T and Q_S depend on the derivation of the respective warping moments. Thus, they can be replaced by the following relations:

$$Q_T = -W_T' - m_{WT} \quad \text{and} \tag{56}$$

$$Q_S = -W_S' - m_{WS} \; . \tag{57}$$

For the special case of a symmetric wing box cross section the system of governing equations is given in [13].

2.8 The stiffness matrix

For the calculation of displacements and stresses the wing is subdivided into several sections in the spanwise direction, where each section provides constant cross

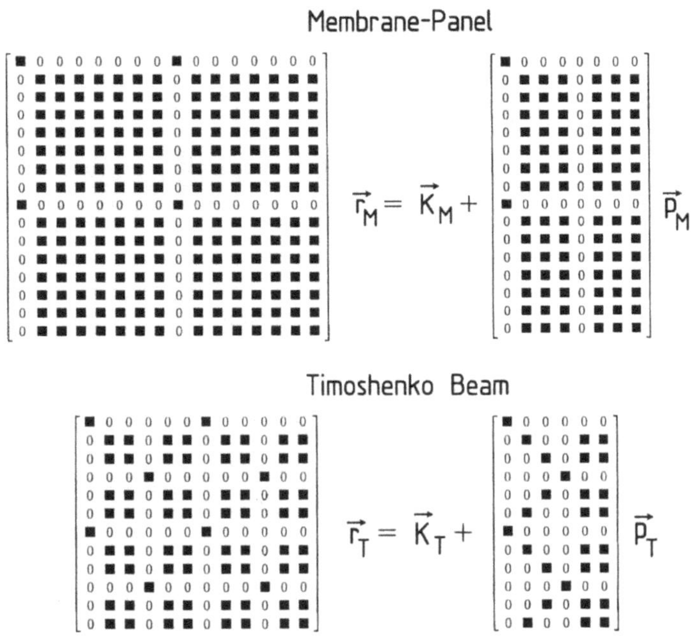

Figure 5 The sparsity patterns of the stiffness matrices of Membrane–Panel and Timoshenko beam in case of an unsymmetrical wing cross section

section geometry. Nevertheless, the effect of taper is considered in the prescribed way (see subsection 2.5). A stiffness matrix has been calculated for the sections with arbitrary section length, where eight degrees of freedom are considered for each node in the Membrane–Panel structural model. Thus in comparison to Timoshenko beam two additional warping degrees of freedom are considered for each node. Figure 5 illustrates the sparsity patterns of the stiffness matrix and the coefficient matrix of the uniform load vector $\vec{P}$ in comparison to the Timoshenko beam. In the Membrane–Panel idealization the components of the node displacement vector $\vec{r}_M$ and node force vector $\vec{K}_M$ are

$$\begin{aligned} \vec{r}_M^{\,T} &= [u_{N_i}\,,\, v_i\,,\, w_i\,,\, \psi_{T_i}\,,\, \psi_{S_i}\,,\, \varphi_{x_i}\,,\, \varphi_{y_i}\,,\, \varphi_{z_i}\,, \\ &\quad\; u_{N_{i+1}}\,,\, v_{i+1}\,,\, w_{i+1}\,,\, \psi_{T_{i+1}}\,,\, \psi_{S_{i+1}}\,,\, \varphi_{x_{i+1}}\,,\, \varphi_{y_{i+1}}\,,\, \varphi_{z_{i+1}}\,] \end{aligned} \tag{58}$$

$$\begin{aligned} \vec{K}_M^{\,T} &= [N_{x_i}\,,\, Q_{y_i}\,,\, Q_{z_i}\,,\, W_{T_i}\,,\, W_{S_i}\,,\, M_{x_i}\,,\, M_{y_i}\,,\, M_{z_i}\,, \\ &\quad\; N_{x_{i+1}}\,,\, Q_{y_{i+1}}\,,\, Q_{z_{i+1}}\,,\, W_{T_{i+1}}\,,\, W_{S_{i+1}}\,,\, M_{x_{i+1}}\,,\, M_{y_{i+1}}\,,\, M_{z_{i+1}}\,]\;. \end{aligned} \tag{59}$$

3 Numerical examples

This section presents four different examples which provide numerical results obtained with the Membrane–Panel model, detailed FE discretizations and the Timoshenko–beam. The improvement of the Membrane–Panel model compared with the Timoshenko beam is demonstrated. FE-calculations have been performed by the commercial FE-program "MARC".

3.1 Rib stiffened rectangular wing box subjected to shear load

The first example is a clamped, rib stiffened rectangular aluminium wing box, subjected to a cantilever load (F=10 kN). The vertical deflection of this box was computed. Figure 6 shows the used finite–element discretization (4640 elements) and the load introduction. For the calculations with the Timoshenko beam and the Membrane–Panel idealization the structure was subdivided in 80 elements. The achieved results are presented in figure 7.

3.2 Straight, tapered wing subjected to torsion loads

The second example is a bilinear tapered straight aluminium wing with rectangular cross section and 30 m in length subjected to a uniform load 0.5 N/mm^2 on front and rear spar. For the finite–element discretization 10272 elements compared to 120 elements in the beam idealizations were used. The wing is orthotropically stiffened by spar booms, skin stringers and ribs. Figure 9 illustrates the distribution of the angle of twist in the spanwise direction. Again, a good agreement between the beam and FE-results is obtained. Figure 10 illustrates the normal stress distribution in the upper skin at the root section in direction of the wing depth. The Membrane–Panel structural model gives a good approximation of the normal stresses due to

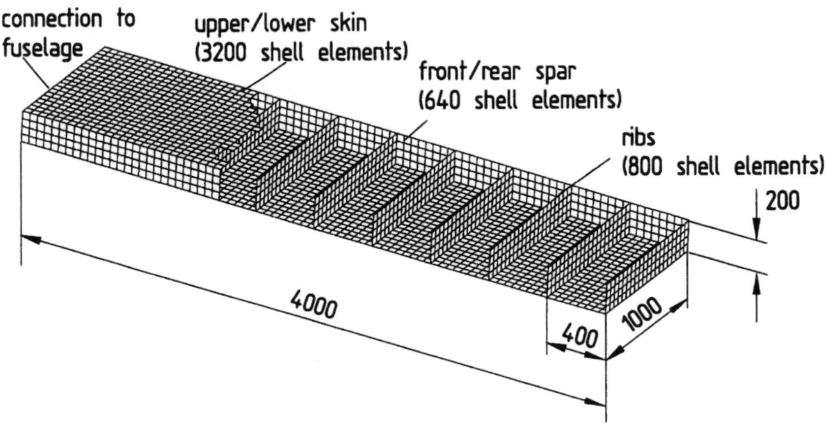

Figure 6 The finite–element model of the investigated wing box

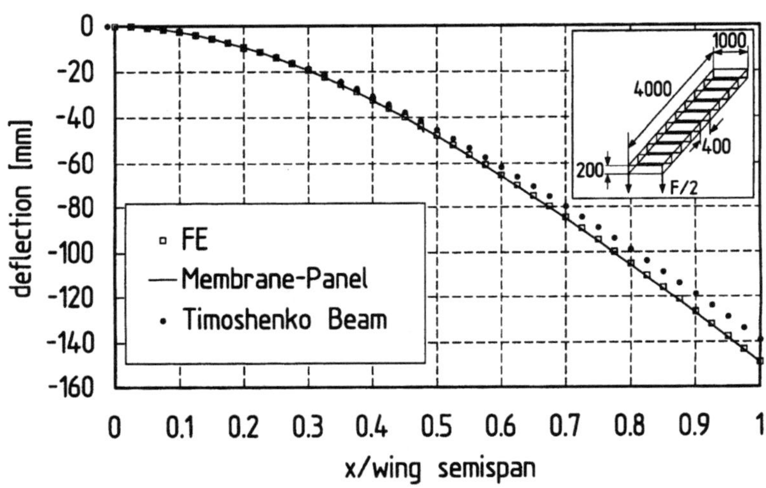

Figure 7 Distribution of the wing deflection in the spanwise direction

torsion-related warping. The Timoshenko beam gives here zero stresses as it neglects warping effects. The correct description of the stresses is especially important when dealing with anisotropic or swept wings, because the subsequent coupling of normal stress and shear stress resultants affects the magnitude of the wing displacements.

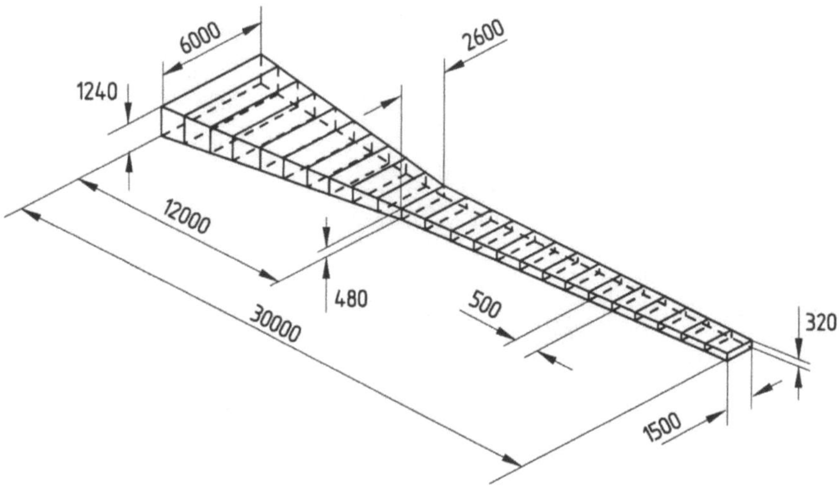

Figure 8 The straight, tapered wing box

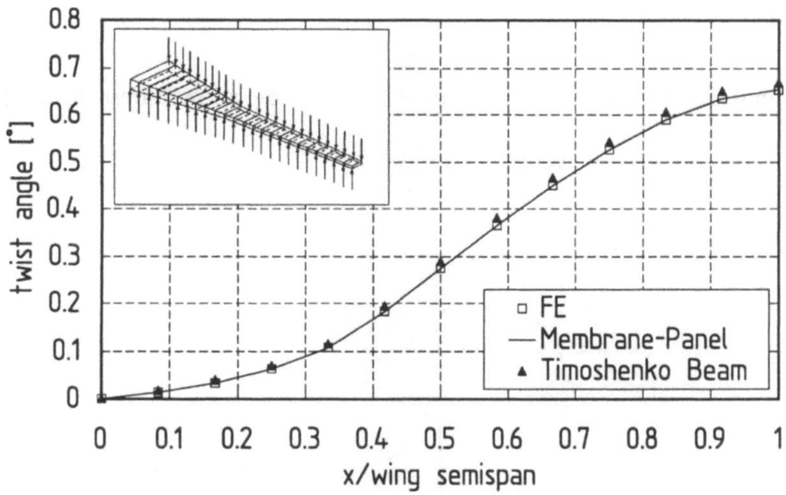

Figure 9 Distribution of the angle of twist in the spanwise direction

3.3 Straight, tapered wing subjected to distributed and concentrated loads

Next, the tapered wing was subjected to a constant shear force load per unit length at the front spar and additional single shear forces and twisting moments to investigate the effect of engine loads (figure 11). Figure 12 illustrates the normal stress distribution in the lower skin. Compared to the Timoshenko beam, the warping forces are not considered at the positions of the load introduction, the Membrane–Panel

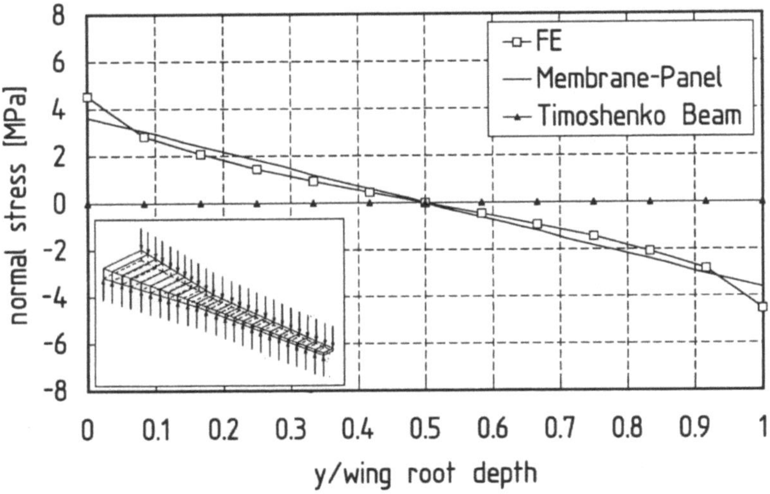

Figure 10 Normal stress distribution in the upper skin at the root section of a wing subjected to pure torsion loads

structural model yields a significant modification in the normal stress distribution.

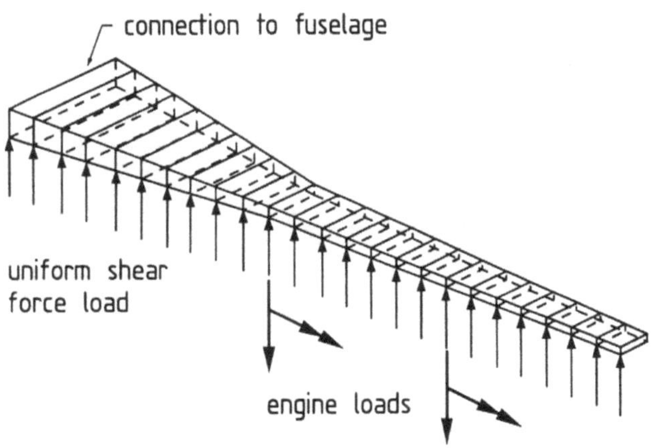

Figure 11 Tapered wing subjected to single loads and loads per unit length

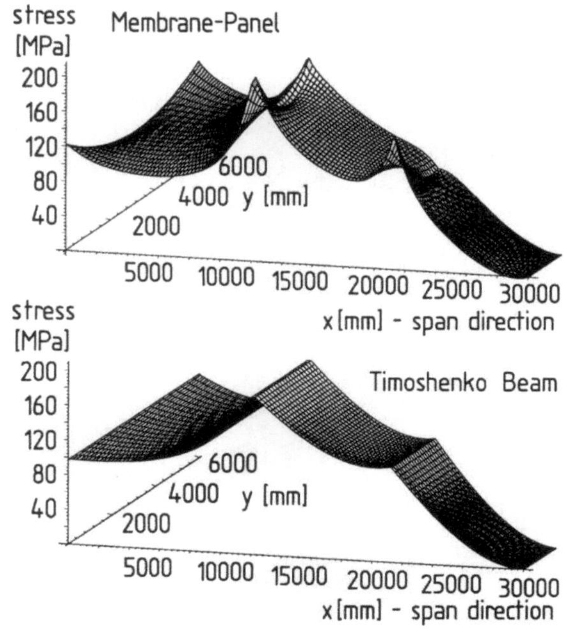

Figure 12 Distribution of the normal stresses in the lower skin

3.4 Straight wing with parabolic cover sheets subjected to distributed and single loads

Numerical investigations of the static and dynamic (eigenfrequencies) behaviour of a straight wing with parabolic skins have been performed. Figure 13 illustrates the detailed finite–element structural model with about 18000 elements, which has been used for the comparison with the Membrane-Panel (80 elements) results. For the determination of the wing box cross section geometry the following parameters ($h_1 = 400mm$, $h_2 = 330mm$, $a_1 = 70mm$, $a_2 = 30mm$, $b = 2000mm$ and $\delta_z = 30mm$) were chosen (see figure 1). A constant wall thickness of 1mm for the skins, webs and ribs was assumed. For the cross sections of the spar booms, rib booms and stringers $800mm^2$, $100mm^2$ and $100mm^2$ were considered respectively. For all structural elements isotropic material behaviour ($E = 70GPa$, $\nu = 0.3$) was assumed. The wing was subjected to uniform shear force loads of 1kN, which were introduced at the rib positions of the front spar and to an additional shear force (-9kN) and a twisting moment (6kNm) for the simulation of the effect of an engine. The figures 14, 15 and 16 illustrate the distributions of the wing deflection, the angle of twist and normal stresses in the upper skin computed with the Membrane-Panel model and the FE method.
In all cases the Membrane–Panel model yields results that are in good agreement

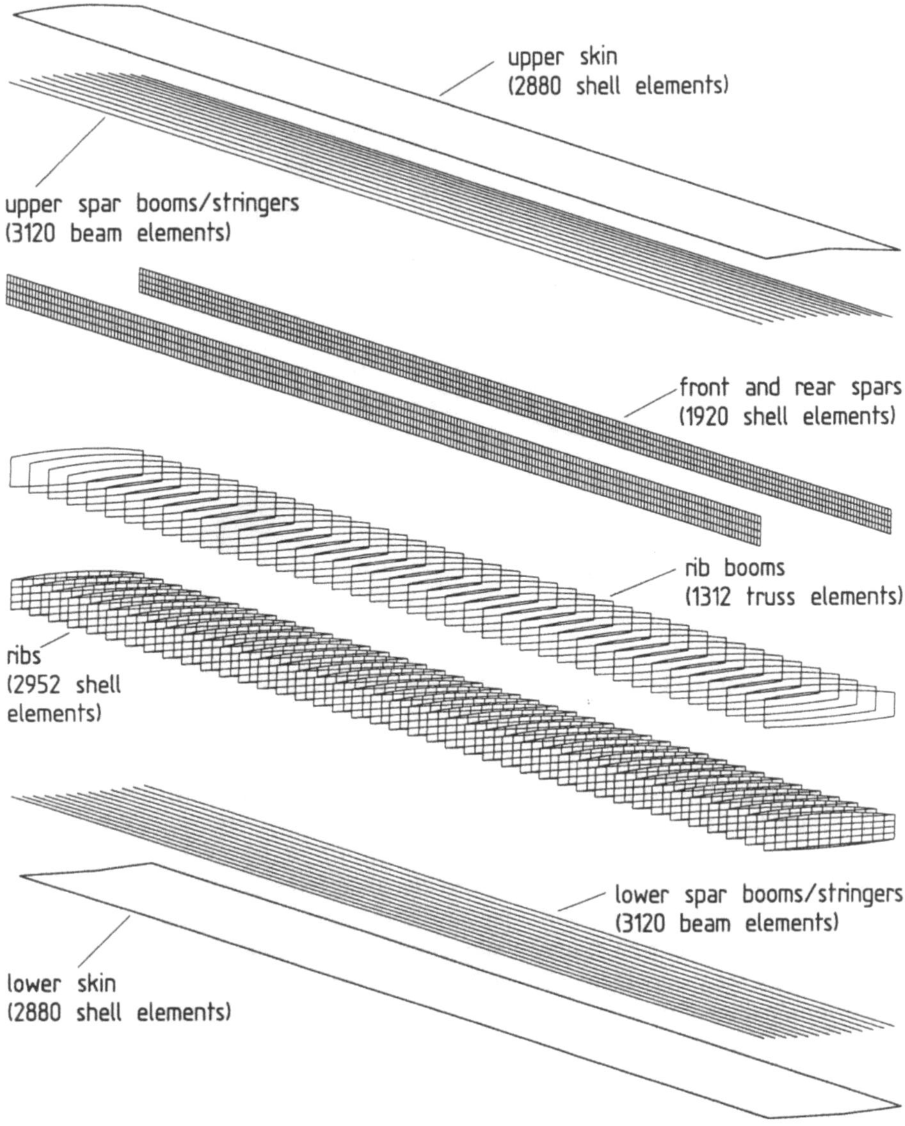

Figure 13 The finite–element model used to compare the numerical results of the Membrane-Panel structural model

with finite–element calculations.

In order to investigate the dynamic wing behaviour, its eigenfrequencies and eigenmodes were calculated. Table 1 presents the first nine eigenfrequencies of the Membrane–Panel idealization and the corresponding FE results.

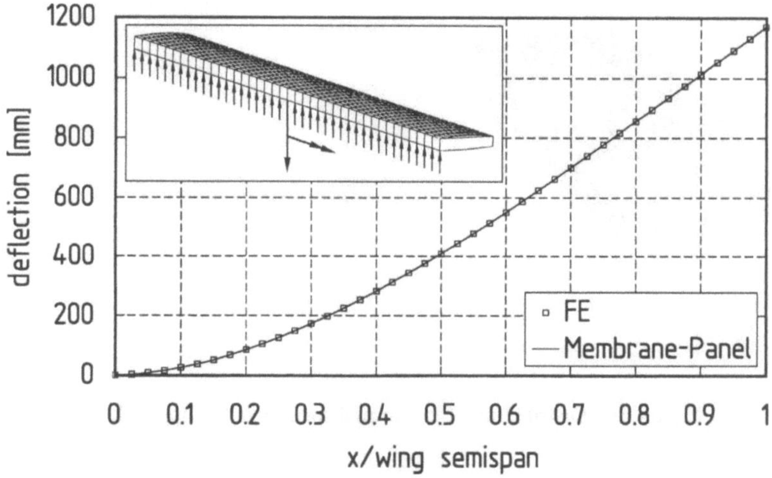

Figure 14 Distribution of the wing deflection of the bottom line of the front spar in the spanwise direction

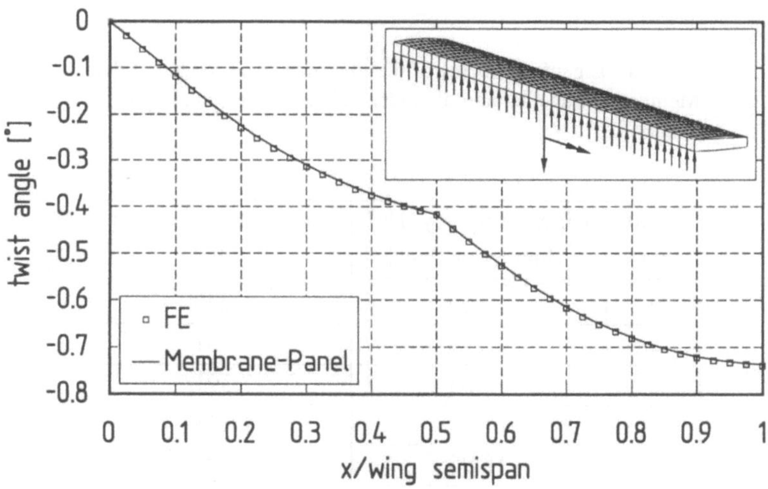

Figure 15 Distribution of the angle of twist for a straight wing with parabolic skins

Figure 17 illustrates the corresponding eigenmodes. The slight lower eigenfrequencies of the FE–results are caused by additional vibrations of the skin, web or rib elements. Especially for the sixth eigenfrequency additional vibrations can be detected in figure 17.

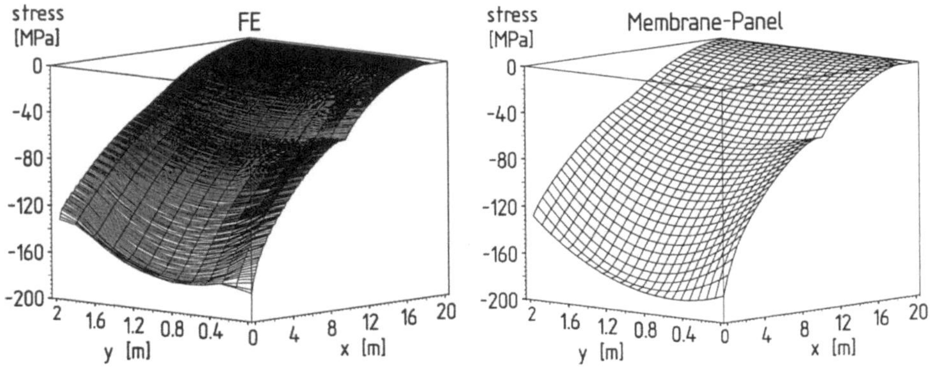

Figure 16 Distribution of the normal stresses in the upper skin

Table 1 Eigenfrequencies of the first nine wing modes of the Membrane–Panel structural model and their corresponding FE–results

Mode No.	Eigenfrequency [Hz] Membrane-Panel	Eigenfrequency [Hz] FE
1	1.266	1.260
2	4.729	4.701
3	7.403	7.391
4	11.85	11.73
5	18.95	18.94
6	26.35	25.73
7	33.45	33.36
8	35.74	35.30
9	49.78	49.28

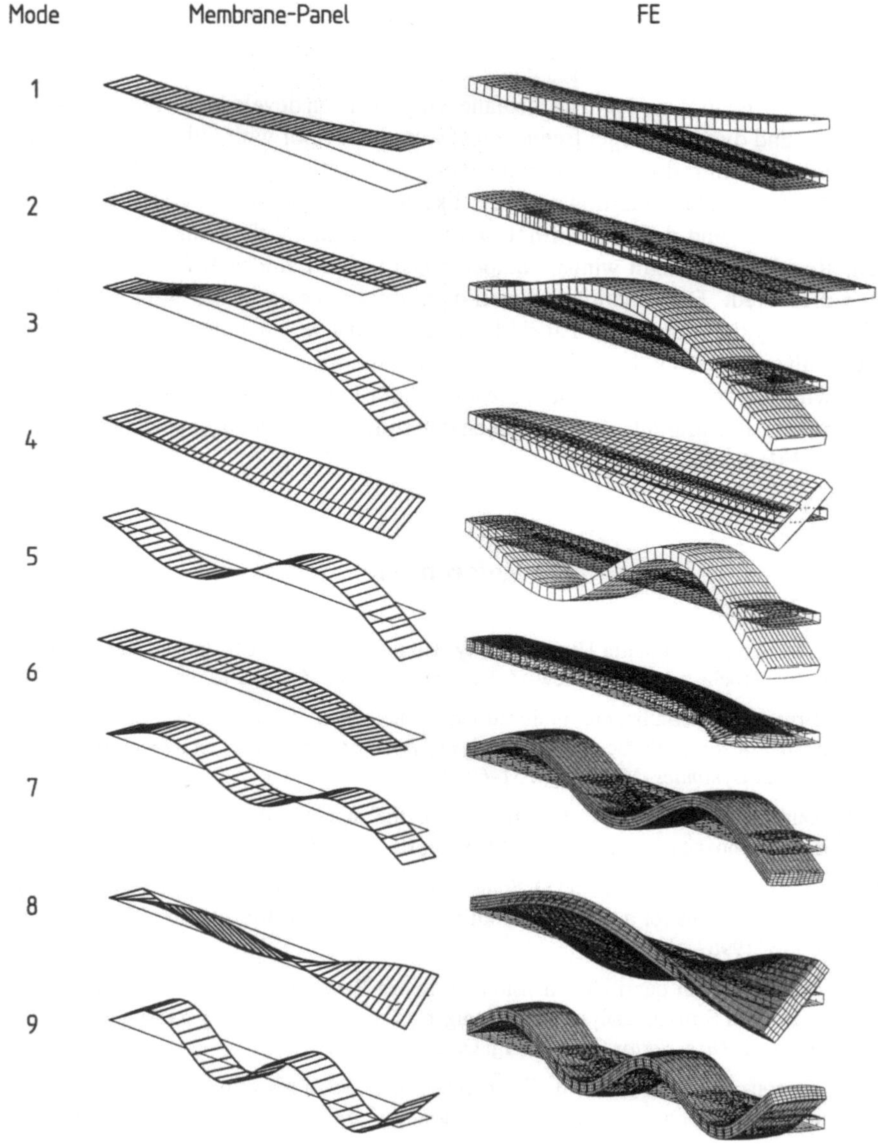

Figure 17 The first nine eigenmodes (from top to bottom) of the Membrane-Panel and their corresponding FE eigenmodes

4 Conclusion

A reduced structural model of an airplane wing has been developed, which describes the static and dynamic (eigenfrequencies) wing behaviour with only few degrees of freedom. The introduced Membrane–Panel idealization extends the Timoshenko beam by considering additional effects like torsion– and shear–related restrained warping, taper, and different (orthotropic) materials along the wing box cross section. Results for straight wings, untapered and tapered, under different loads have been presented. Deformations as well as normal stresses in the skins have been calculated by the Membrane–Panel idealization and a very detailed FE-idealization. The results of the wing displacements as well as the wing stresses are in good agreement with respective finite–element results. Additionally, the eigenfrequencies and eigenmodes of a straight wing have been determined with satisfying accuracy compared to FE–results.

References

[1] H. Försching. New Ultra High Capacity Aircraft (UHCA) - Challenges and problems from an aeroelastic point of view. *Z. Flugwiss. Weltraumforsch.*, 18, pp. 219–231, 1994.

[2] G. Britten and J. Ballmann. Fluid–structure interaction at panels and large wings with computational aeroelasticity *Proceedings of International Forum on Aeroelasticity and Structural Dynamics*, Madrid, pp. 51–62, 2001

[3] S. Duan and M. Piening. Investigation of the torsion-related warping behavior of anisotropic boxbeam structure. *DLR Bericht*, IB-131-95/33, 1995.

[4] L. Librescu, L. Meirovitch and O. Song. Integrated structural tailoring and control using adaptive materials for advanced aircraft wings. *Journal of Aircraft*, Vol. 33, No. 1, pp. 203–213, 1996.

[5] V. Matzdorf. Ein quasi-eindimensionales finites Element zur globalen statischen und dynamischen Strukturanalyse dünnwandiger anisotroper Stab- oder stabähnlicher Konstruktionen. *Dissertation*, Magdeburg, Germany, 1991.

[6] G. Karpouzian and L. Librescu. Comprehensive model of anisotropic composite aircraft wings suitable for aeroelastic analyses. *Journal of Aircraft*, Vol. 31, No. 3, pp. 703–712, 1994.

[7] R.Q. Borchert. Improvements in design oriented equivalent plate modeling of wing structures. *PhD-Thesis*, University of Washington, 1995.

[8] G.L. Giles. Equivalent plate modeling for conceptual design of aircraft wing structures. *AIAA Journal*, 95-3945, pp. 1–17, 1995.

[9] B. Langesfor. A suggested method for calculating the stresses in wings with non–rectangular plates. *Saab TN 23*, Linköping, Sweden, 1953.

[10] E. Giencke. Über die Erweiterung des Schubfeldschemas. *Jahrbuch der WGLR*, pp. 413–421, 1962.

[11] C. Brühl. Beitrag zur Berechnung schiefwinklig eingespannter Kastenträger. *DVL Bericht*, No. 18, 1960.

[12] H. Ebner. Die Beanspruchung dünwandiger Kastenträger auf Drillung bei behinderter Querschnittswölbung. *ZFM*, Vol. 24, No. 23, 1933.

[13] W. Jung and H.-G. Reimerdes. Ein Beitrag zur aeroelastischen Untersuchung mit idealisierten Tragflügeln. In *DGLR Jahrestagung*, Vol. 3, pages 1371–1378, Berlin, Germany, 1999.

[14] R. Heilig. Beitrag zur Theorie der Kastenträger beliebiger Querschnittsform. *Der Stahlbau*, 11, pp. 333–349, 1961.

[15] K. Bhaskar and L. Librescu. A geometrically non–linear theory for laminated anisotropic thin-walled beams. *Int. J. Engng Sci.*, 33, pp. 1331–1344, 1995.

Computational Aeroelasticity with Reduced Structural Models

G. Britten, C. Braun, M. Hesse, J. Ballmann
Lehr- und Forschungsgebiet fur Mechanik, RWTH Aachen
Templergraben 64, D-52062 Aachen, Germany

Summary

In order to predict the aeroelastic pattern of large span airplane wings already in the design phase, it is proposed to represent the supporting structure by a multi-axial Timoshenko beam model, while the fluid flow is modelled by the conservation laws according to Navier-Stokes or Euler. Finite element discretization of the structure and fluid-structure coupling are realized via Hamilton's principle. Finite volume discretization on a time dependent deforming structured grid is applied on the fluid side. The discretized equations are solved fully coupled and time-consistently. Preliminary results are presented for an oscillating airfoil and the deformation of a wing in steady transonic flow. More extended results are given and compared with experiments in the subsonic regime for a highly flexible rectangular wing.

1 Introduction

The design of high performance wings for large commercial aircrafts requires the inclusion of the aeroelastic quality into the aerodynamic and structural design process. Hitherto in the first design the geometry of the wing is defined as a compromise between good flight performance during take-off, landing and cruise flight on the one hand and load capacity and weight of the structure on the other hand. In an iterative fashion the aerodynamic shape, the loads, the construction of the wing assembly and the deformation are studied sequentially and more or less independently. Aerodynamic wind tunnel testing with rigid or nearly rigid reduced scale models plays a key role. But in those tests similarity with the full scale body can only be achieved in a very limited manner, primarily with respect to the aerodynamic parameter Mach number and to a certain extent with respect to Reynolds number, too. Aeroelastic similarity is usually not achieved. Based on the aerodynamic analysis, wing loads, deformations and particularly the aerodynamic twist are determined. Then wing geometry and construction of the wing assembly are modified a posteriori such that after taking into account the static aeroelastic deformation in cruise flight, sufficient lift and minimum drag are secured. (Nevertheless dynamic aeroelastic performance and efficiency of control devices remain sometimes uncertain until first flight.) The described design and construction procedure requires several iterations because in every step the aeroelastic coupling and the nonlinearity of the problem can not be captured completely. Reliable prediction of nonlinear flutter performance in the transonic regime is not possible with those methods.

Therefore, it is necessary to develop design methods, which trustworthily predict the interaction between aerodynamic, structural and inertial forces. The actual aerodynamic shape in flight, the stability of the aeroelastic equilibrium configuration and the manoeuvrability of the elastic airplane must be predicted already in the design period. Since dynamic instabilities can cause a rapid growth of vibration amplitudes of the elastic lift generating structures, it is very important to analyse the characteristics of the fluid-structure system under non-stationary conditions. For this purpose methods of Computational Aeroelastic Simulation (CAS) are under development worldwide. Surveys of recent attempts are presented in [10]. Usually codes developed originally and separately for fluid flow or for structural dynamics computations are coupled in the time domain. For that purpose the fluid code is modified such that it can operate on a dynamic grid which means a body surface fitting time-dependent grid. The discretization on the moving grid has to satisfy the so-called geometric conservation law [22]. Time integration is performed mostly sequentially in an explicit manner with a uniform time step size prescribed from the fluid side because usually the ratio of the smallest characteristic dimension of grid cells and disturbance propagation speed governing the CFL number on the fluid side is much lower than on the structure side. Often a phase delay of one time step between the fluid and the structure side is accepted.

Most of the methods make use of multi-field techniques with domain decomposition and loose coupling [24, 7, 25]. This procedure is favoured mainly because it avoids the new developement of unified coupled codes. Emphasis is put on conservative load projection [7] from the fluid side on the structure and in the opposite direction on the motion of the fluid-solid interface calculated by the structure code. In [11], for a correct energy transfer between fluid and structure, a virtual interface is introduced along the aerodynamic surface of the structure. Different interpolation methods for the transfer between fluid and solid are discussed in [15]. In [9] improved parallelized strategies are presented. Only very few methods apply strong coupling using implicit time integration. In [27] a Newton-type subiteration has been implemented for this purpose to achieve optionally first or second order time accuracy. Despite the higher effort per time step, the total computational effort for the problem may be reduced due to larger admissible time steps. In [3] a direct Euler-Lagrange-Scheme has been applied to derive a unified system of partial differential equations (PDE) for inviscid compressible fluid flow and deforming wing motion. The system of PDE's is integrated with respect to time completely simultaneously avoiding a priori any phase difference in the mutual transfer between fluid and structure.

This work focusses on the numerical method SOFIA (SOlid Fluid InterAction) for direct numerical aeroelastic simulation, being under development in the Collaborative Research Center SFB401 "Flow Modulation and Fluid Structure Interaction at Airplane Wings". The method is implicit in time and its development has been started from existing flow solvers, the Euler solver INFLEX [4] and the Navier-Stokes solver FLOWer in the version of [12].

SOFIA is used within the Center of SFB401 on one hand to support the experimental projects, while on the other hand the aeroelastic experiments are used to validate the code in every state of development [8].

The paper is organized as follows. In section 2 we describe the physical models for the fluid and the structure. Thereafter in section 3 we will comment the numerical methods used on the fluid and on the structure sides including the dynamic grid deformation tool and the strategy of coupling. In the sections following thereafter, we will present results, first for static divergence and the dynamic stability limit of the NLR7301 airfoil on elastic suspension [20]. For the LANN wing we discuss the difference between stationary solutions for the rigid and the elastic wing in transonic flow. Finally the configuration of a very compliant elastic rectangular wing with the cruise profile of the SFB reference configuration in stationary subsonic flow and its dynamic response and aerodynamic damping in non-stationary subsonic flow are examined. The profile of the reference configuration corresponds to the BAC 3/11/RES/30/21 airfoil [26].

2 Physical and mathematical modelling

2.1 Non-stationary flow in a deforming space volume

The conservation laws for unsteady viscous compressible fluid flow form the system of governing equations in the flow domain. They are solved by finite volume techniques with control volumina dependent on time in order to describe consistently the fluid flow and motion of a moving and deforming aerodynamic surface. Thereby the cell faces on the surface are taken as material elements of the structure surface throughout the motion of the structure, whereas the cell faces at the far field boundary, which is the outer boundary of the computational domain, remain fixed in a rigid body fixed coordinate system. Thus, the grid is deforming with time and the edges of the finite volumes within the flow field are moved such that a grid fitting the structure's aerodynamic surface is ensured. For a moving and deforming geometrical volume the balance laws of unsteady fluid flow read in integral form

$$\frac{\partial}{\partial t} \int\limits_{V(t)} \mathbf{U} dV + \int\limits_{\partial V(t)} \mathbf{F n} dS + \int\limits_{V(t)} \mathbf{G} dV = 0 \,, \tag{1}$$

where $\mathbf{U} = (\varrho, \varrho u, \varrho v, \varrho w, \varrho e)^T$ is the algebraic vector of conserved quantities: density, Cartesian components of momentum and total specific energy density. Contributions due to a possible rigid body motion of the grid, which may be rotation with the angular velocity vector $\omega = (\omega_x, \omega_y, \omega_z)^T$, are contained in the source term $\mathbf{G}$ (see [12]). $\mathbf{F} = \mathbf{F}^c + \mathbf{F}^d$ represents the flux function with its parts $\mathbf{F}^c$ designating the convective and in isentropic flows physically conservative terms and $\mathbf{F}^d$ denoting the diffusive part,

$$\mathbf{F^c} = \begin{bmatrix} \varrho(\mathbf{v} - \mathbf{v}_B) \\ \varrho u(\mathbf{v} - \mathbf{v}_B) + p\mathbf{e}_x \\ \varrho v(\mathbf{v} - \mathbf{v}_B) + p\mathbf{e}_y \\ \varrho w(\mathbf{v} - \mathbf{v}_B) + p\mathbf{e}_z \\ \varrho e(\mathbf{v} - \mathbf{v}_B) + p\mathbf{v} \end{bmatrix} , \tag{2}$$

$$\mathbf{F^d} = \begin{bmatrix} 0 \\ \tau_{xx}\mathbf{e}_x + \tau_{xy}\mathbf{e}_y + \tau_{xz}\mathbf{e}_z \\ \tau_{yx}\mathbf{e}_x + \tau_{yy}\mathbf{e}_y + \tau_{yz}\mathbf{e}_z \\ \tau_{zx}\mathbf{e}_x + \tau_{zy}\mathbf{e}_y + \tau_{zz}\mathbf{e}_z \\ \psi_x\mathbf{e}_x + \psi_y\mathbf{e}_y + \psi_z\mathbf{e}_z \end{bmatrix} . \tag{3}$$

The vector $\mathbf{v}_B$ denotes the velocity vector field of the time-dependent surface of the geometrical, non-material volume. A detailed description of the diffusive part including the formulation of the symmetrical stress tensor components with Stokes' hypothesis of vanishing pressure viscosity ($\mu_v = \frac{2}{3}\mu$) can be found in [12]. The viscosity is approximated in case of laminar flow by Sutherland's formula as a function of the statical temperature only,

$$\mu(T) = \mu_\infty \left(\frac{T}{T_\infty}\right)^{\frac{3}{2}} \frac{T_\infty + 110K}{T + 110K} . \tag{4}$$

The vector of heat conduction is modeled by Fourier's law ($\mathbf{q} = -k \operatorname{grad} T$) with the heat conductivity k given for a constant Prandtl number Pr

$$k = c_p \frac{\mu}{Pr} , \qquad Pr = 0.72 . \tag{5}$$

c_p denotes the specific heat at constant pressure. For turbulent flows, the viscosity μ is replaced by $\mu + \mu_t$ with the turbulent part provided by a turbulence model. In this paper only the algebraic model of Baldwin-Lomax is used. Finally, since air is assumed to behave as a perfect gas (with constant specific heats), the pressure is calculated applying the ideal gas equation of state in terms of pressure, density and internal energy

$$p = (\kappa - 1)\varrho\left(e - \tfrac{1}{2}\mathbf{v}^2\right) \tag{6}$$

with κ denoting the ratio of specific heats.

2.2 Structural dynamics models

2.2.1 Elastically suspended rigid airfoil section

For the two-dimensional case of an elastically suspended rigid airfoil of total heaving mass m, inertia moment of rotation I_φ, static mass unbalance S_φ with respect to the pitching center, the structural model of a spring/mass/-damper system with two degrees of freedom, pitch φ and heave z, is used. For the other quantities, see Figure 1. Balancing aerodynamic, structural and inertial forces and moments yields a system of two coupled non-linear second-order ordinary differential equations which is non-linear due to the aerodynamic force L and moment M,

$$m\ddot{z} - S_\varphi \ddot{\varphi} + D_z \dot{z} + K_z z = L \quad , \tag{7}$$

$$-S_\varphi \ddot{z} + I_\varphi \ddot{\varphi} + D_\varphi \dot{\varphi} + K_\varphi \varphi = M \quad . \tag{8}$$

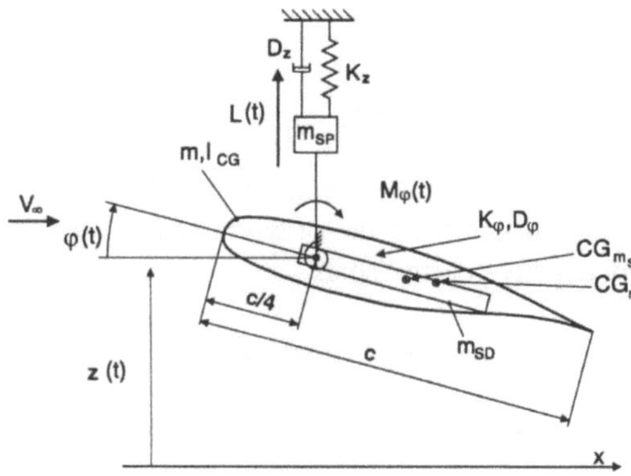

Figure 1 Sketch of the suspended airfoil and the spring/mass/damper system

2.2.2 Timoshenko Beam

Considering the unsteady deformations of large span wings, we assume a beam-type supporting structure. We include the influence of shear deformation according to Timoshenko's theory. Euler/Bernoulli-type beam models which do not account for shear deformation and rotatory inertia of the material cross-sections, yield parabolic PDEs which allow for infinite (and hence unphysical) speed of wave propagation and group velocity, the velocity with which energy is transmitted, going towards infinity with wave-length tending to zero. This phenomenon is known as anomalous dispersion in the literature [23]. Compared with that, Timoshenko-type beam models, which incorporate shear deformation and, according to Lord Rayleigh's suggestion the rotatory inertia of material cross-sections, result in hyperbolic sets of PDEs which are characterized by finite wave speeds within the correct physical limits of elastic materials. The model difference will exhibit big quantitative discrepancies not for every aeroelastic problem but for deformation propagation involving short wave length pulses, e.g. in connection with local active control of large flexible structures, it may be very important [1]. The structural material is assumed linear elastic throughout. For the mathematical formulation of the problem we choose Hamilton's principle in the extended form, which includes terms Z_{BC} and Z_{IC} for boundary and initial conditions. From this formulation the complete initial value problem for the beam vibrations can be derived.

For the purpose of practical application to real wing structures we take a multi-axial Timoshenko-type beam structure, where the centerlines of mass, bending and torsion are generally not straight and non-coinciding. By this assumption, besides the coupling via aerodynamics, all degrees of freedom may be coupled with each other mechanically, too. Six degrees of freedom are assigned to every material cross-section, the aerodynamic surface contour of which is assumed to remain undeformed. The variational equation for the complete initial boundary value problem reads

$$\delta \mathcal{I}^e(\mathbf{u}_S, \varphi) = \delta \int_{t_a}^{t_e} \int_0^l L_B \, d\xi \, Dt + \int_{t_a}^{t_e} \delta W^{aero} Dt + \delta \left(Z_{IC} + Z_{BC} \right) , \quad (9)$$

where the symbol δ designates the variation, δW^{aero} is the virtual work of the aerodynamic forces resulting from the pressure and shear stress distributions and L_B the Lagrangian density per unit span,

$$\begin{aligned} L_B \quad = \quad & \frac{1}{2} \Big(\varrho A \dot{\mathbf{u}}_S \, \dot{\mathbf{u}}_S + \dot{\boldsymbol{\varphi}} \, \boldsymbol{\Theta}_S \, \dot{\boldsymbol{\varphi}} - G \, A \, \boldsymbol{\gamma} \, \mathbf{K} \, \boldsymbol{\gamma} \\ & - E \, A \, u_{B1|1}^2 - \boldsymbol{\varphi}_{|1} \, \mathbf{C}_{BTW} \, \boldsymbol{\varphi}_{|1} + \varrho \, A \, \mathbf{u}_S \, \mathbf{g} \Big) \quad . \end{aligned} \quad (10)$$

Therein ξ is the spanwise coordinate and the subscript $|1$ denotes differentiation with respect to ξ. Notations are $\mathbf{u}$ displacement vector, φ angular deflection vector containing bending and torsion, γ shear strain vector, A cross section, $\boldsymbol{\Theta}$ mass inertia tensor of a material cross-section, G shear modulus, $\mathbf{K}$ equivalent shear distribution tensor, E Young's modulus, $\mathbf{C}_{BTW}$ stiffness tensor containing the resistance of the beam structure against bending, torsion, and warping. $\mathbf{g}$ denotes gravity, and the indices S and B represent the centers of gravity and bending, respectively. Within Timoshenko's theory the shear angles are represented as

$$\gamma_3 = u_{S2|1} - \varphi_3 - (\zeta_{SD}\varphi_1)_{|1}, \qquad -\gamma_2 = u_{S3|1} + \varphi_2 + (\eta_{SD}\varphi_1)_{|1}, \qquad (11)$$

where η_{SD} and ζ_{SD} are the cartesian coordinate differences between the center of gravity and the shear center in a cross-section. Indices 2 and 3 correspond the the η and ζ directions, respectively. A detailed description of the remaining terms can be found in [28]. Linear damping, the so-called Rayleigh damping may be included in the variational equation as a virtual work term in a similar way like the virtual work of the aerodynamic forces.

2.3 Dynamic Grid

One significant point in aeroelastic analysis is the grid generation procedure. It may be very hard to generate a structured grid about a complex aircraft configuration, so a block structured grid approach helps to simplify this problem. Nevertheless, the grid generation task still is enormous for complex configurations. In addition, this difficulty increases when the aeroelastic deformation of the aircraft is considered because the grid must match the instantaneous shape of the vehicle at every time and no intersections of grid lines must take place in the dynamic grid motion. Assume we are given an initial reference configuration of the structure and its aerodynamic surface with a multi-block structured computational grid generated by elliptic non-dimensionalized coordinates [31, 13, 32].

The most commonly used strategy for deforming grids in aeroelastic applications is transfinite interpolation (TFI). This algebraic method is used to generate a new grid in every time step, but the resulting grid only suits for small surface deformations. For moderate or large deflections, however, grid quality may suffer. In particular, mesh orthogonality at deforming surfaces which is desired for viscous flow simulation, is not guaranteed. On the other hand, the possible loss of orthogonality is alleviated by the fact that each grid deviates only slightly from the grid in the preceeding time step. As a consequence, there is the idea for a more robust and much more time saving grid deformation algorithm than generating a new grid at each iteration level [29, 27].

Our strategy to adapt a multi-block dynamic grid in three-dimensional space to the deformation of the aerodynamic configuration is to adjust the block faces first. For this purpose we use a fictitious elastic framework for the block edges.

As the effort of generating a grid solving elliptic differential equations can drastically be reduced when the given starting grid is close to the desired grid, an algebraic grid deformation method is coupled to an elliptic smoothing process [32]. This way the quality of the given starting grid can be maintained and the amount of work for grid smoothing with the help of elliptic differential equations can drastically be reduced.

Given a starting grid at time level t^n — the lines of which emanate approximately perpendicularly from the body's surface and are curvilinear in general — and a surface deformation during $\Delta t := t^{n+1} - t^n$, which is represented by a translation vector and a rotation matrix in each surface node, each grid line pointing into the flow field is rigidly moved, according to the local displacement of the surface and the rotation of its local normal vector (Figure 2, left and middle). Then the grid lines are deformed using these shifted and rotated lines for an appropriate blending procedure. In order to ensure, that the angles with which the new grid lines emanate from the surface remain approximately perpendicular to the surface, even if the surface undergoes strong deformation, the blending function is constructed in such a way that the grid lines in the deformed grid adapt to the rigidly moved lines near the deformed surface and to the lines of the original grid at the far field boundary

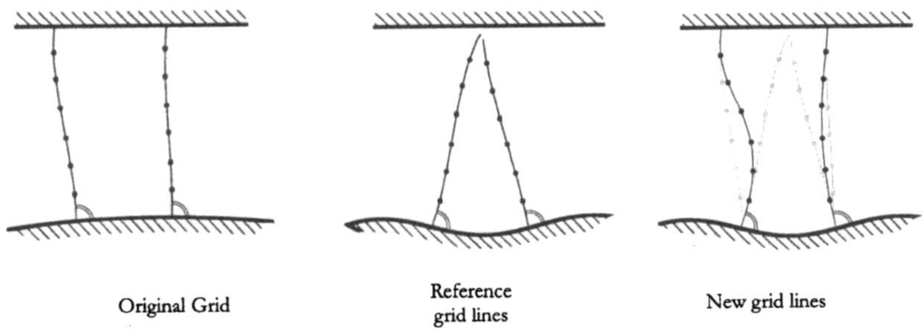

Figure 2 Algebraic deformation algorithm

(Figure 2, right picture).

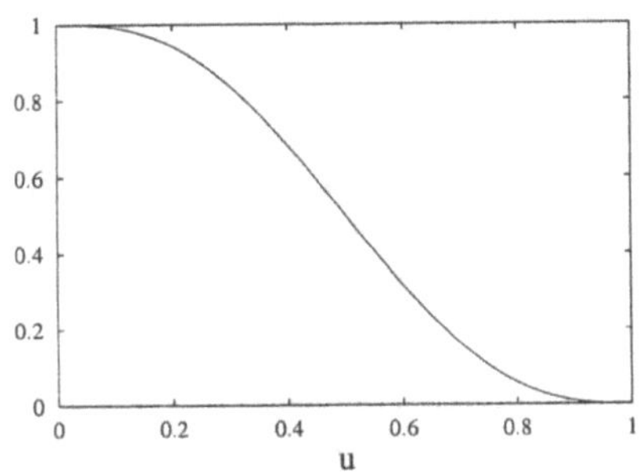

(a) Sym. blend function $f(u) = w(u)$

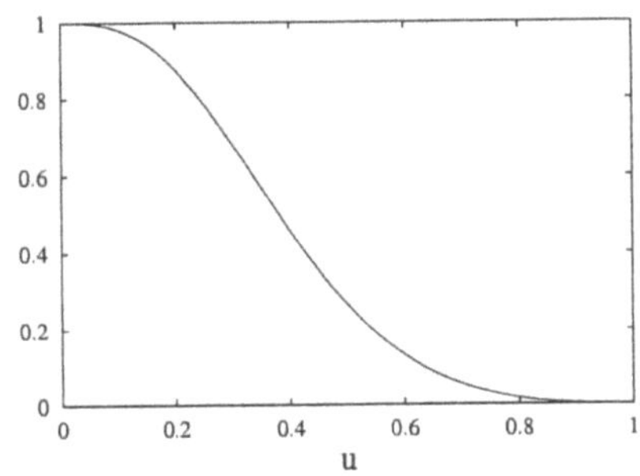

(b) Polynomial of higher degree: $f(u) = 0.25\, w\,(w^3 + 2\,w + 1)$

Figure 3 Blending functions based upon $w(u) = 1 - 10u^3 + 15u^4 - 6u^5$

Let $f(u)$ be the blending function, u being the arc-length parameter defined in the reference grid. The choice of this function can have a severe influence on the mesh quality. In our code special polynomials of degrees higher than five in u with only one inflection point in the interval $[0, 1]$ are used. All these blending functions have zero slope at the endpoints ensuring that wall orthogonality is maintained and that the grid adapts to the original grid in the far field. Figure 3 shows two common blending functions, the left one being symmetric, the right one showing a shift to the left, which means that more points in the far field are kept fixed.

If the deformation causes the grid to overlap, which is likely to occur in boundary layers if the surface undergoes large deformations, there is the need for smoothing the grid with an elliptic technique. Then, only a few iterations – the Ansatz of Hilgenstock [14] or Steger and Sorenson [30] can be used – are necessary to unfold and smooth the grid.

3 Numerical Methods

The code components used in SOFIA and the numerical methods used therein are described briefly. Details are presented in the references cited in the sequel.

3.1 Fluid

Two different codes for the calculation of the flow field have been implemented in SOFIA. In the first version the code INFLEX [4], solving the Euler equations, has been implemented in SOFIA [28]. Thereafter a version using the FLOWer code has been created, which solves optionally the Euler or the Reynolds averaged Navier-Stokes equations of three-dimensional flow. The discretization in both codes is based on structured grids with hexahedral cells.

In the INFLEX code the numerical integration of the strong conservation form of the Euler equations is performed using a cell centered discretization with upwind differencing and characteristic flux extrapolation. Time integration is implicit. The unfactored equations are solved by applying a Newton iteration method combined with a point Gauß-Seidel algorithm which results in a robust numerical scheme. Grid generation is performed applying elliptical smoothing in every time step [4]. The INFLEX code has been tested succesfully against various AGARD test cases. This validation can be found in [4].

The FLOWer code has been progressively developed in the project MEGAFLOW by different German research organisations under leadership of DLR Braunschweig [21].

In the version of the FLOWer code for unsteady fluid flow used here, cell vertex central differencing is implemented throughout for the spatial discretization. Time integration is performed by dual-time stepping [18], [12]. Within each pseudo-time step an explicit multi-stage Runge-Kutta method is employed which is accelerated by techniques of local pseudo-time stepping, implicit residual smoothing and multi-grid techniques. By coupling with an algebraic dynamic grid generator [5], [17] the code has been made applicable for aeroelastic analysis. Grid deformation is performed according to section 2.3. The velocity vectors of the nodes $\mathbf{v}_B$ are determined as the difference quotient formed with the displacement vectors of the nodes and the time step size.

The FLOWer code has been validated versus many experiments and other codes for stationary and non-stationary flow including rotating wings [12, 21]. The implementation of the Geometric Conservation Law (GCL) has been verified for a domain with a homogeneous flow field solution employing general rigid body motions of the grid and general sinusoidal time dependent harmonic deformations of the grid inside the solution domain, with fixed and with oscillating or rotating boundaries. Further validation tests for the deforming grid will be given in the results section.

3.2 Structure

To determine the generalized deflections, a second order in time system of ordinary differential equations (ODEs) is derived by applying Hamilton's principle and the method of Ritz/Kantorowitsch in a finite element approach [28]. In order to calculate faster the static equilibrium configuration of the elastic wing Rayleigh-damping is employed, such that the damping forces vanish when all nodes of the beam have come to a standstill, asymptotically. Discretization is done by isoparametric, two-noded finite elements yielding a system of ODEs for the set of unknown structural parameters $\mathbf{W}$. A reduced integration scheme avoids shear locking [16]. The system of ODEs is integrated by Newmark's method, where the resulting linear system of equations is solved directly using LU-decomposition. The external forces are assumed to vary linearly during a time-step. Alternatively, the system of ODEs is diagonalized by solving the generalized eigenvalue problem (EVP), and then the time integration is done by evaluating Duhamel's integral formulation. To validate the structural part of SOFIA, which is named ODISA (One DImensional Structural Analysis), many static and dynamic problems as well as eigenvalue problems have been considered [28].

3.3 Fluid-structure coupling scheme

Within the SOFIA code, the solution schemes for the flow field and the structure are combined in an iterative process. The central time loop works as follows: Firstly, the velocities and the displacements of the structure are calculated under the instantaneous aerodynamic loads. Then the deforming aerodynamic mesh is computed (including position and velocity of the cell vertices) such that the topology of the structured grid is preserved. Finally the flow field and the new airloads are computed optionally by the INFLEX or the FLOWer code for the current time.

Two different coupling schemes have been implemented and tested in this work. The first of these schemes, is a conventional loose coupling scheme which allows for a phase delay of one time step between fluid and structure motion. Given numerical solutions $\mathbf{U_n}$ and $\mathbf{W_n}$ at time level t_n for the flow field and for the velocity and displacement state of the structure, respectively, the fluid stresses $\mathbf{p_n}$ on the wing surface are used to advance the structure dynamics solution $\mathbf{W}$ to time level $n + 1$ (while strictly, in the implicit time integration scheme used here, the surface stress $\mathbf{p}$ at time level $n + 1$, $\mathbf{p_{n+1}}$, should be involved). The structural displacements, $\mathbf{w_{n+1}}$ thus obtained are subsequently used to update the fluid boundary and flow mesh including the grid velocities. Finally, the flow solution $\mathbf{U}$ is advanced to time level $n + 1$, see Figure 4(a). In the notations used by Farhat et al. [9] this method is called a conventional serial staggered procedure (CSS).

In the second scheme a tight coupling between the fluid and structurel solvers is achieved by subcycling, i.e. iterating the steps of the loose coupling scheme during the time step according to Figure 4(b) until a convergence criterion is satisfied in the

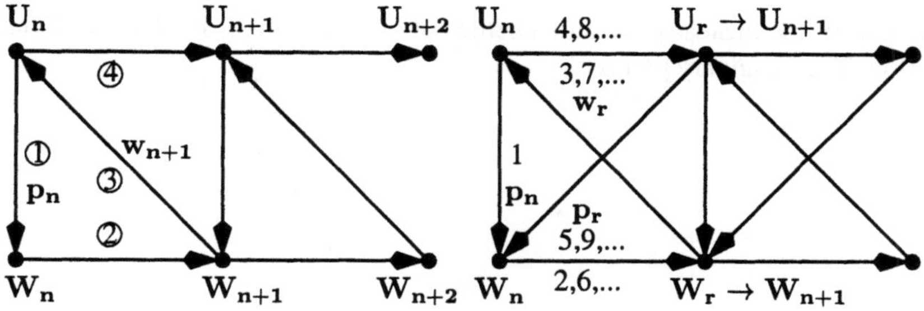

(a) loose coupling (CSS) (b) tight coupling by subcycling

Figure 4 Fluid-structure coupling, **U** Flow solution, **W** Structural dynamic solution

sense of a fixed-point iteration.

4 Results

4.1 NLR 7301 airfoil

4.1.1 Forced oscillation results

For the validation of the present method for unsteady applications on a moving grid, a numerical simulation of an oscillatory motion in pitch of the NLR 7301 airfoil has been performed corresponding with the test case AGARD CT5 [2]. In this test case, the angle of attack is updated in every time step according to the equation

$$\varphi(t) = \varphi_m + \varphi_0 \sin \omega t \quad , \tag{12}$$

In the considered test case, an oscillation with a reduced frequency of $\omega^* \equiv \omega c/2U_\infty = 0.192$ is imposed with an amplitude of $\varphi_0 = 0.5^o$ and a mean angle of attack of $\varphi_m = 2^o$. The Mach number of the oncoming flow is $M_\infty = 0.7$ and the Reynolds number $Re_\infty = 2.14 \cdot 10^6$, based on the wing chord. The rotation axis is located at 40 percent of the chord length. The time-dependence of the local pressure coefficient on the surface can be represented as

$$\begin{aligned} c_p(x,t) &= c_{pm}(x) + c_p'(x) \sin \omega t + c_p''(x) \cos \omega t + \\ &\quad c_{p1}'(x) \sin 2\omega t + c_{p1}''(x) \cos 2\omega t + \ldots \end{aligned} \tag{13}$$

Higher harmonic parts can be neglected. c_{pm} is the mean local pressure coefficient which is not necessarily equal to the local pressure coefficient c_{ps} for a stationary flow at an angle of attack of φ_m. In the general case of a non-linear dependence of the pressure on the change of the angle of attack, c_{pm} is not equal c_{ps}. c_p' is the real part of the unsteady pressure coefficient, which changes in phase with the oscillation. The imaginary part is c_p''.

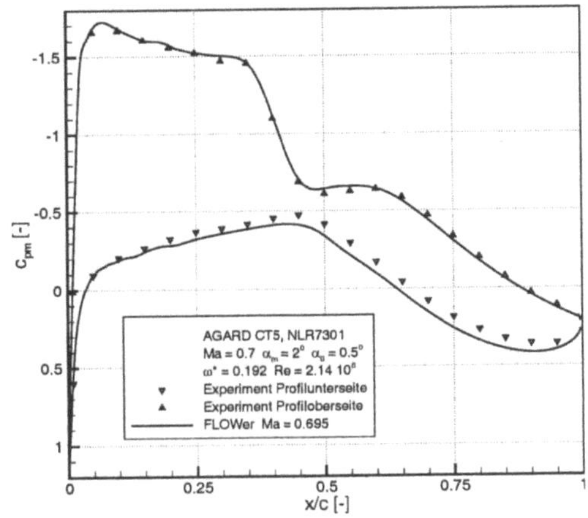

Figure 5 Mean pressure distribution (NLR7301 airfoil, forced oscillation), ▼ experimental value lower side, ▲ experimental value upper side

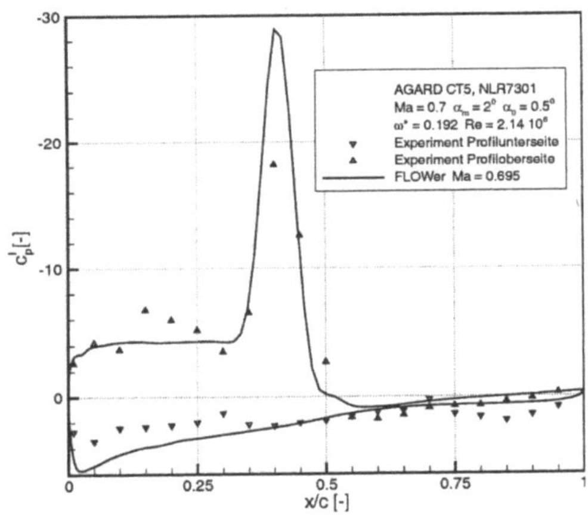

Figure 6 Real part of the unsteady pressure distribution (NLR7301 airfoil, forced oscillation), ▼ experimental value lower side, ▲ experimental value upper side

Best agreement between experimental and numerical results could be observed for a slightly corrected Mach number of $M_{\infty,c} = 0.695$. The computed c_{pm}-distribution is shown in Figure 5. The comparison with the experimental results shows a good agreement on the upper side of the airfoil. Especially the position of the shock is predicted correctly. Slight deviations can be observed on the lower side, but the agreement is still satisfactory.
The real and the imaginary part of the unsteady pressure distribution is depicted in

Figure 6 and Figure 7. Again, the agreement with experimental data is relatively good, even though higher differences can be found for the real part c_p'. Both the c_p' and the c_p'' distribution show small values on the lower surface, indicating a weak time-dependence of the local pressure. On the other hand, a large peak can be observed on the upper surface in the shock region. This corresponds to the fact, that the region of shock-boundary layer interaction is very sensitive to changes of the angle of attack, and therefore a strong time-dependence has to be expected.

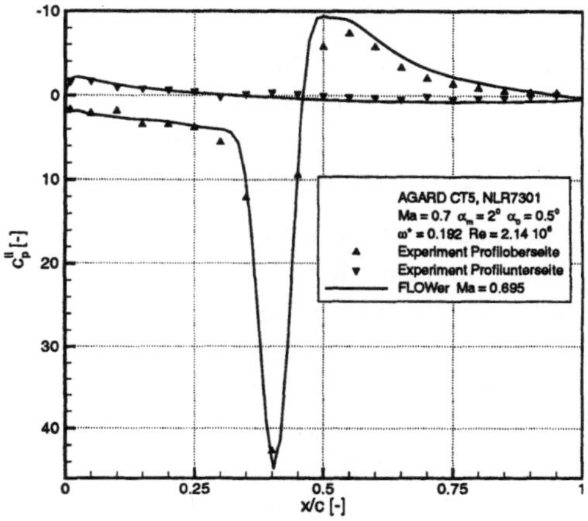

Figure 7 Imaginary part of the unsteady pressure distribution (NLR7301 airfoil, forced oscillation), ▼ experimental value lower side, ▲ experimental value upper side

4.1.2 Flutter boundary

The stability diagram of the NLR 7301 airfoil was determined by direct aeroelastic simulation in the time domain under consideration of a mean angle of attack of $\varphi_m = 0.5^o$ and free flight conditions at Mach numbers yielding transonic flow fields. For these conditions the stationary flow solutions showed acceptable approximations of the wind tunnel experiments reported in [20] for 1.2^o angle of attack. For all calculations the Baldwin-Lomax turbulence model has been used and transition was fixed at 3 percent chord. Corresponding to the experiments performed in the DLR center Göttingen, the airfoil is elastically suspended with a rotatory and a translatory degree of freedom, and the rotation axis is located at 25 percent of the chord length. The structural parameters were taken from [20]. To determine the flutter boundary a number of computations have been performed for different Mach numbers and dynamic pressure, and the dynamic behaviour has been observed. For each of these computations the procedure is as follows:

First, a steady state solution for the airfoil assumed fixed is computed for the mean angle of attack $\varphi_m = 0.5^o$, for a chosen Mach number M_∞ at a dynamic pressure q_∞. Then the deflections of the springs are determined such that the aerodynamic loads and the spring forces are statically balanced for the considered Mach number and dynamic pressure to ensure static equilibrium for $\varphi = \varphi_m = 0.5^o$.

In order to analyse the stability behaviour of the equilibrium configuration, the airfoil is disturbed from its steady-state position by prescribing a small angular velocity as initial condition which yields a small increase of the total mechanical energy of the structural system, included the elastic energy of the undisturbed system. If for $t > 0$ the total mechanical energy of the structure decreases, the system is called "asymptotically stable", if it increases the system is called "unstable", and if the total energy neither increases nor decreases, it is called "stable". The parameter combination, for which a stable behaviour is observed, is a point of the flutter boundary. In the computations we always found a consistent behaviour of the pitching and the plunging motion. When the pitch amplitude was increasing with time then also the heaving amplitude increased and vice versa.

The computed flutter boundary is shown in Figure 8. It shows the transonic dip, i.e. the minimum of the flutter boundary, for a Mach number of $M_\infty = 0.74$ and a total pressure of $p_0 = 20kPa$ which corresponds to a flutter index of 0.172 whereas in the experiment the dip was at 0.78 and a flutter index of 0.175. So, the position of the dip differs considerably from the position, that was found in experiments [20]. The discrepancy may have different reasons: i) the true experimental conditions have not been simulated. The computation was performed for plane flow, fixed transition and free flight conditions, whereas in the experiment transition was free and three dimensional effects could be present due to the side walls of the wind tunnel. ii) Baldwin-Lomax turbulence model seems not to be sufficient because of its unability to simulate separation correctly.

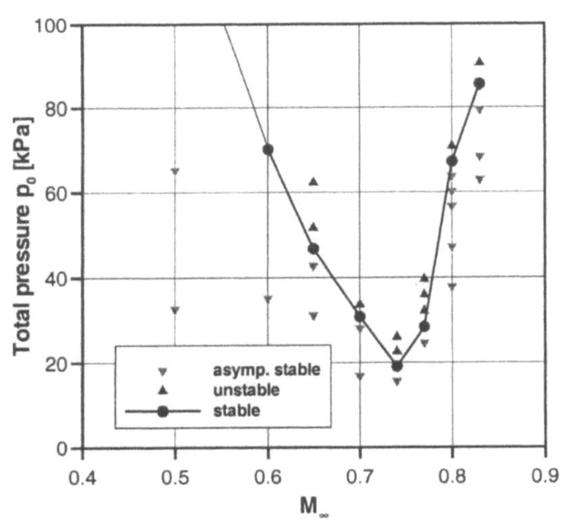

Figure 8 Stability diagram of the NLR 7301 airfoil ($\varphi_m = 0.5^o$), ▼ unstable, ▲ stable

4.2 LANN wing

4.2.1 Forced oscillations result

Another validation test for the FLOWer code extended by implementation of the algebraic grid generator is presented in Figure 9. The figure shows the total lift coefficient of the well known AGARD LANN wing configuration ([2], see Figure 10) which is oscillating about the z-axis. The origin of the coordinate system is located in the nose of the wings root cross-section, and the z-axis is perpendicular to it. The solution calculated on the dynamic grid corresponds very well with the one calculated on a grid performing rigid body motion which is a validated standard option in the FLOWer code with respect to the AGARD test case CT5, see [12].

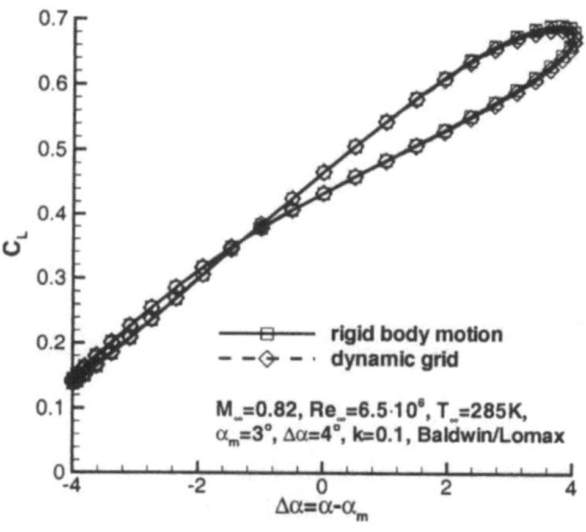

Figure 9 Total lift coefficient for an oscillating rigid LANN wing

4.2.2 Aeroelastic equilibrium configuration

We consider an example for finding the aeroelastic equilibrium configuration by coupled fluid/structure numerical analysis using the unsteady SOFIA code. The LANN wing geometry was chosen to define an elastic wing with generic elastic data and dimensions. Data is determined by extrapolating a data set provided by Daimler Benz Aerospace Airbus for a mean-size transport aircraft [28]. Figure 10 (a) shows the pressure distribution on the wing and in two cross-planes as a representative of the Navier-Stokes solution for the steady flow field about the rigid wing, i.e. the wing with the undeformed shape which we call reference configuration (jig shape). The aeroelastic computation is started in the unsteady mode of SOFIA with the aerodynamic loads upon the undeformed wing. After an oscillation period of the wing during which artificial Rayleigh damping has been applied to limit the computational effort, the steady state achieved. The pictures belonging to Figure

10 (b) show the final static equilibrium configuration and the pressure distributions corresponding to Figure 10 (a) for the deformed wing. As can be seen, remarkable differences concerning the shock position and its strength occur in the tip region.

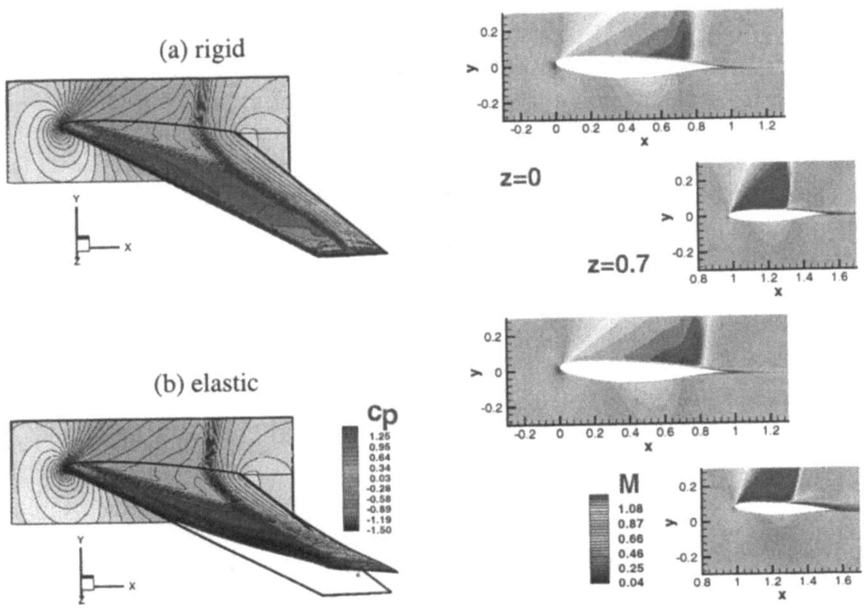

Figure 10 Pressure distribution on the wing's surface (left) and Mach number distribution in planes at 0% and 70% span (right) of a) a rigid and b) an elastic wing with the aerodynamic configuration of the LANN wing ($M_\infty = 0.82$, $\varphi_\infty = 3^o$, $Re = 6.5 \cdot 10^6$, Baldwin-Lomax)

4.3 Large span wing results

4.3.1 Wing data

The focus of this section is the application of the numerical method SOFIA for the dynamic analysis of a rectangular elastic wing in subsonic flow and the comparison with experimental results of our partners in the Collaborative Research Center SFB401, see [8] and the paper [19] in chapter 3 of this volume. The profile of the wing is presented in Figure 11. It corresponds to the cruise configuration of the airfoil BAC 3/11/RES/30/21 from [26] as mentioned above. The wing has been designed especially for aeroelastic wind tunnel experiments in low speed flow. A hybrid construction with an aluminium spar and foam defining the aerodynamic shape was chosen, see Figure 12. The total weight of the fully equipped wing amounts to approximately 2.5 kg. Average stiffnesses of the wing model are for flap, lag, and torsion 586.39 Nm^2, $2.177 \cdot 10^4$ Nm^2, and 141.329 Nm^2, respectively. The model has a set of pressure gauges at 50% semi-span. The root of the wing was fixed in a very stiff flange which is part of a piezoelectric 6-components balance measuring root forces and moments. Therefore, the elastic deformation of the support of the wing could be neglected in the computation. In the measurement campaign high precision deformation information has been made available by a reconstruction method analysing strain gauges at 16 different spanwise positions. Further details on the experimental equipement can be found in [19] as well as results for another flow speed in the wind tunnel.

Figure 11 Cruise flight reference configuration (based on the 3 element high-lift configuration BAC 3/11/RES/30/21

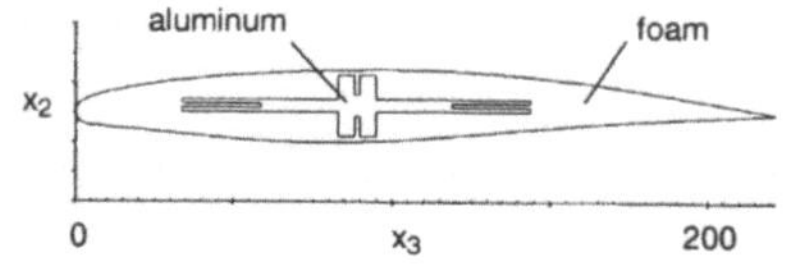

Figure 12 Cross-section of the aeroelastic wing model of the Institute of Lightweight Structures

4.3.2 Aeroelastic equilibrium configuration

The aeroelastic equilibrium configuration is determined in the time asymptotic procedure as described above for the LANN wing. First results for aeroelastic equilibrium configurations and zero lift aeroelastic configurations were given in [6].

Here, we have selected four test cases to compare the numerical results for steady states computed using the SOFIA code with the data measured by our partners of the Department of Aerospace Structures [19]:

Table 1 Test case definition (*) cases with vanishing total lift)

case	v_∞ $[\frac{m}{s}]$	ϱ_∞ $[\frac{kg}{m^3}]$	φ_R^{exp} [°]	φ_R^{calc} [°]
1	65	1.24	6.93	6.93
2	45	1.24	6.6	6.6
3*)	65	1.21	-1.38	-1.27
4*)	45	1.22	-1.87	-1.7

Cases 1 and 2 both show a relatively high angle of attack and mainly differ in the wind speed. Cases 3 and 4 are the corresponding cases in which the total lift measured at the root of the elastic wing vanishes. Since this zero lift angle of attack can be determined very precisely, we have chosen slightly different angles of attack φ_R^{calc} than the measured one (φ_R^{exp}).

Table 2 gives an overview of the total lift F_2 and the total torque M_1 at the wing's root. The maximum error in the lift reaches only about 4% (case 2) and about 7% in the torque (case 4).

Table 2 Total forces and moments at the wing's root (*) denotes cases with vanishing total lift)

	F_2 [N]		M_1 [Nm]	
case	exp	calc	exp	calc
1	486.9	498.4	9.58	9.57
2	204.3	212.7	4.04	3.71
3*)	< 1	< 1	-6.52	-7.02
4*)	< 1	< 1	-3.17	-3.41

The complete data set of case 1 is presented in detail in Figures 13 to 16.

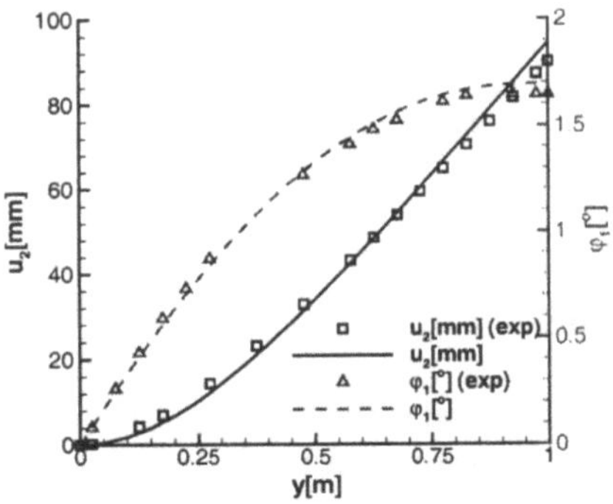

Figure 13 Comparison of measured and calculated spanwise bending and torsional deformation (case 1)

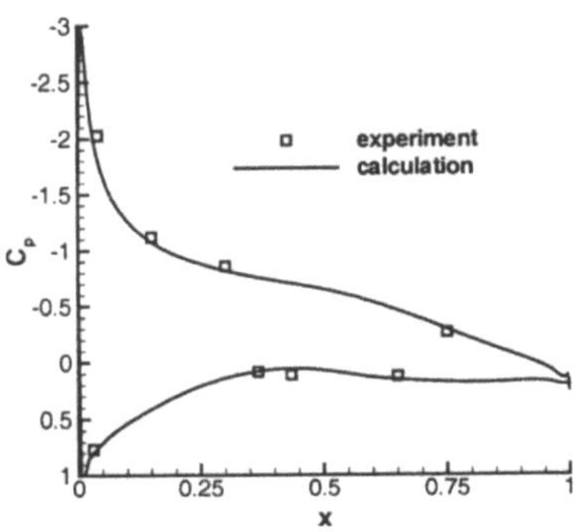

Figure 14 Comparison of measured and calculated pressure distribution at $y = 0.5$ (case 1)

The spanwise distributions of the bending and torsional deformation u_2 and φ_1 respectively are shown in Figure 13. While the torsional deformation agrees very well over the whole wing span, the bending stiffness at the inner part of the wing seems to be slightly overestimated in the calculation. In Figure 14 the pressure coefficient C_P at 50% of the semi-span is discussed. A very good agreement with the measured data is evident. Figure 15 gives information about the spanwise distribution of lift and moment coefficient C_L and C_M, respectively. In this figure the distributions for the rigid case have been added to demonstrate the big difference for this very flexible wing. Figure 16 shows the time history of the translation and the rotation of the wing tip during the unsteady simulation. Due to high artificial Rayleigh-type damping chosen here the translation reaches its asymptotically steady state relatively fast, while it takes a little more time for the rotation to converge.

Concerning case 2 we refer to the work of our partners [19], too.

The angle of attack for which the total lift of the elastic wing vanishes represents another parameter suitable for code validation. Figure 17 shows the zero lift angle of attack versus the wind speed. As can be seen in this figure the agreement

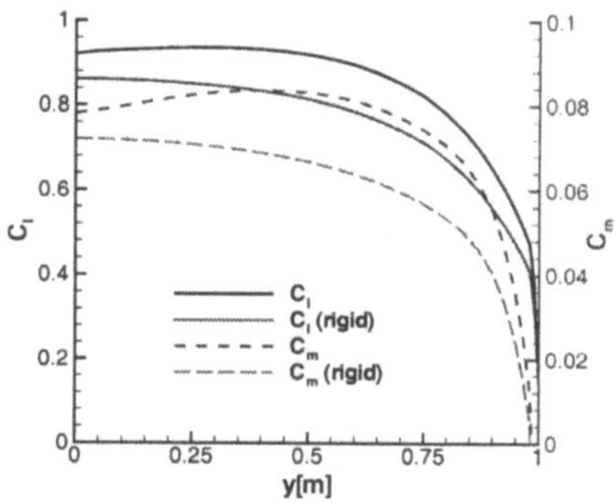

Figure 15 Spanwise lift and moment coefficient distributions (case 1)

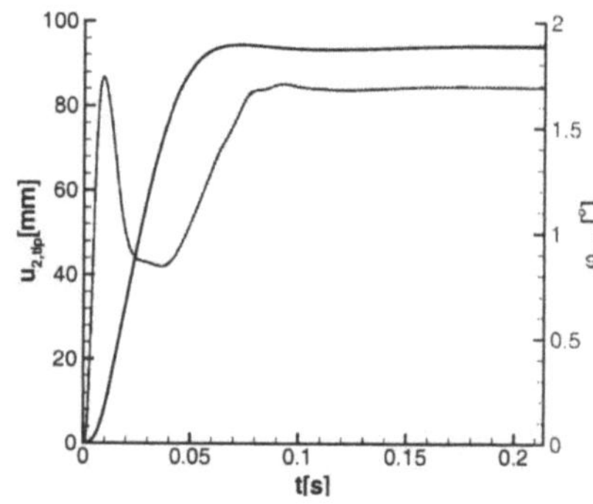

Figure 16 Time history of the torsional and bending deformation at the wing's tip (case 1)

between the computational and experimental data becomes very good with higher wind speeds. The reason for that can be found in Table 3. Herein, the total lift values for the rigid wing with root angles of attack corresponding to the zero lift cases of the elastic wing are presented. Since the total lift reaches only 3.8 N for the oncoming flow velocity 35 m/s, the deformation of the wing remains very small which is corresponding to a relative great error in both, experiment and calculation. On the upper limit at a wind speed of 85 m/s, we have a load of 125.5 N in the rigid case leading to considerable deformation and thus to a smaller error expressed in percentage.

Figure 18 and Figure 19 finally show the spanwise deformation distribution and the pressure coefficient at 50% semi span for case 3. Though the relative differences between measured data and the calculation using the SOFIA code are greater than in the cases discussed before, the results are quite satisfactory.

Table 3 Total lift vs. wind speed for the rigid wing with root angle of attack yielding zero lift in the elastic case.

v_∞ [m/s]	35	45	55	65	75	85
φ_{W0}^{calc}[°]	-1.8	-1.7	-1.5	-1.27	-1.0	-0.64
F_{W2} [N]	3.8	9.8	22.3	42.7	75.1	125.5

4.3.3 Aeroelastic response problem

Starting with the steady state aeroelastic solution for the zero lift case at 65 m/s wind speed, an additional static force of 50 N was applied by pulling a filament fixed at 2.57 percent relative choord in the wing tip cross-section which causes a deflected static configuration, as depicted in Figure 20. In the numerical calculation

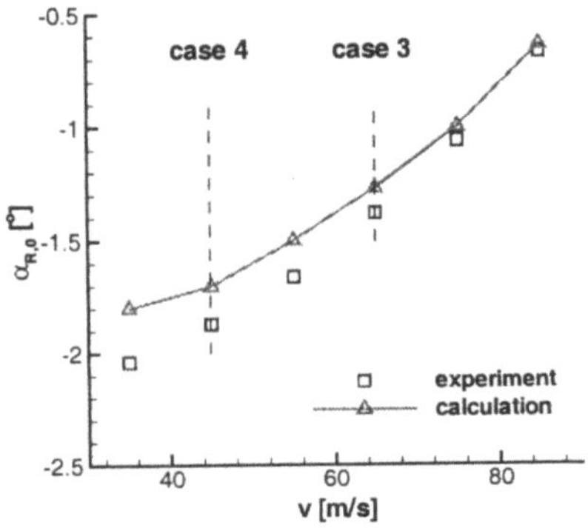

Figure 17 Comparison of measured and calculated zero-lift angles of attack at the wings root

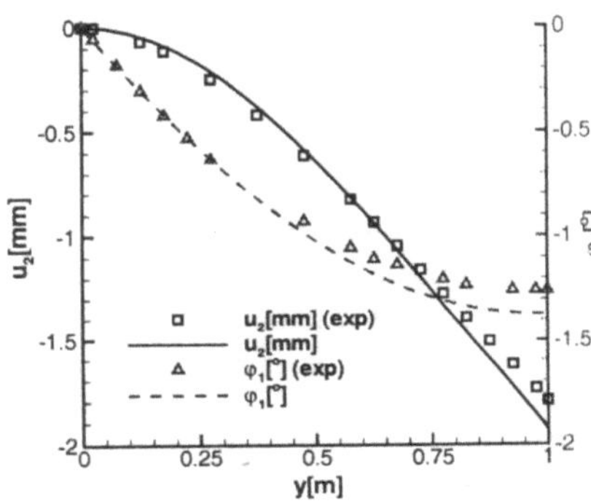

Figure 18 Comparison of measured and calculated spanwise bending and torsional deformation (case 3)

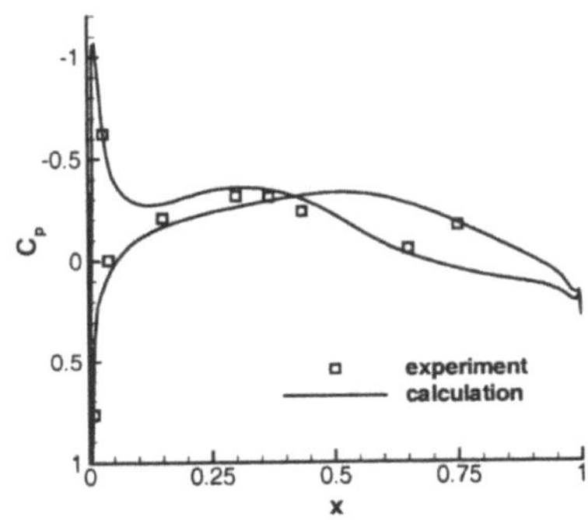

Figure 19 Comparison of measured and calculated pressure coefficient in a cross section at mid-span (case 3)

the additional static force was 48 N. At zero time, the filament is burned (i.e. the additional force is removed) and the wing starts to oscillate. In Figure 21 representative computational results for time histories of the most important deflection quantities, vertical displacement and torsion angle at the wing tip, are presented in comparison with the experimental data. The computed time histories of the torque $M_{1,\mathrm{R}}$, the bending moment $M_{3,\mathrm{R}}$ and the total lift measured and calculated, respectively, at the wing's root are depicted in Figure 22 a), b) and c). As can be seen, measurement and calculation agree very well. As is shown in Figure 22 d) by comparison of the measured and calculated time history of a local pressure value during the dynamic response period of time, the aerodynamic damping could be fully captured by the simulation even in detail. The pressure sensor is located on the upper side of the wing at half span and 3% chord.

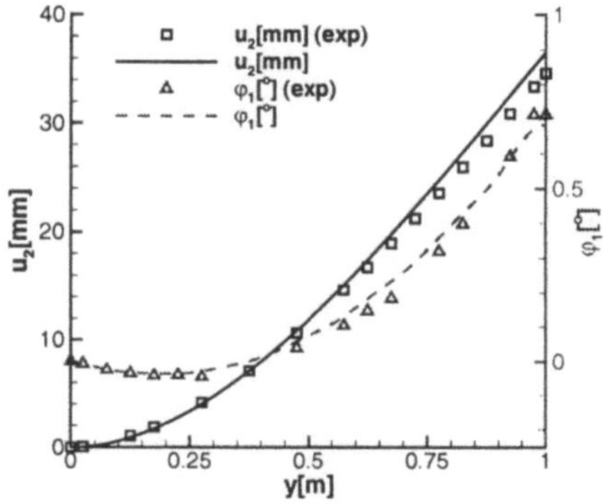

Figure 20 Spanwise vertical displacement and torsion angle distribution in the static equilibrium situation before removing the additional load of 50 N (exp.) and 48 N (calc.)

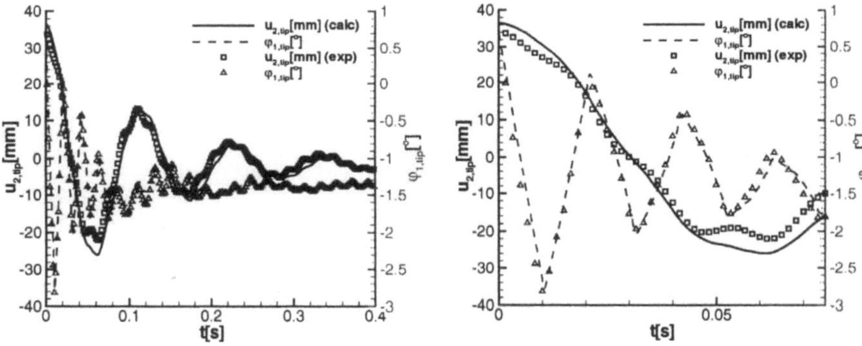

(a) time interval 0.4 s after removing the load

(b) time interval 0.075 s after removing the load

Figure 21 Time history of vertical displacement and torsion angle at the wing tip after removing the additional static load of 50 N (exp.) or 48 N (calc.)

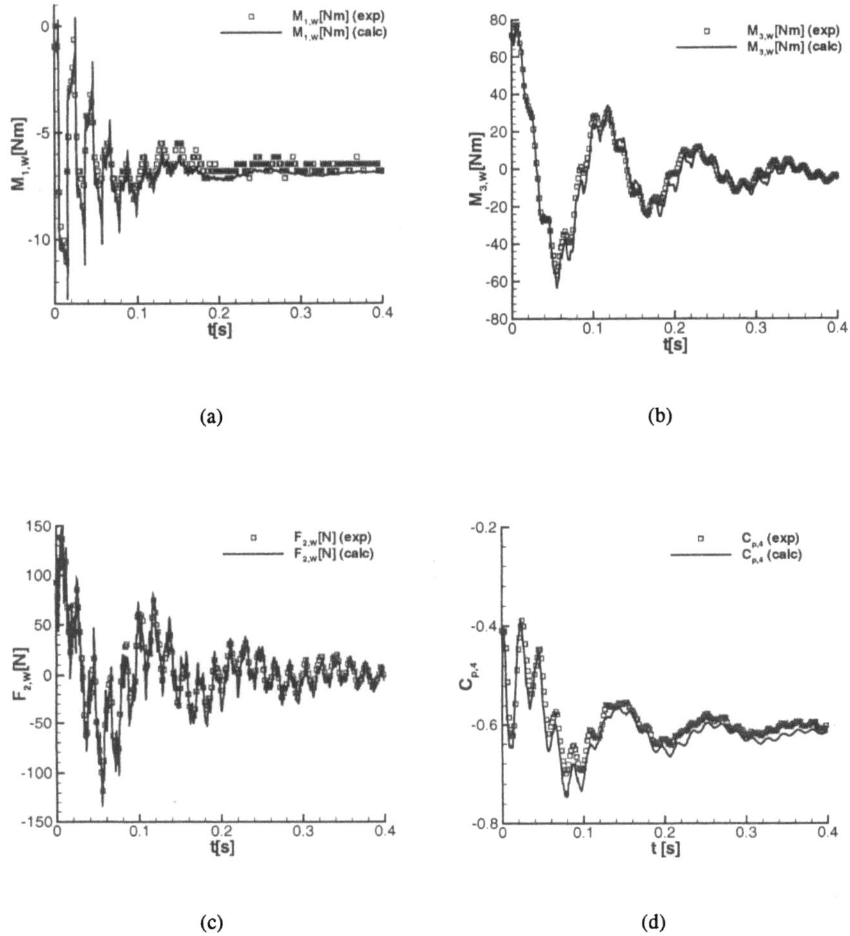

Figure 22 Time histories of measured and calculated torsional moment (a), bending moment(b), lift (c) and local local pressure (d)

5 Conclusions and outlook

In the paper the progress in the development and validation of the SOFIA code for computational aeroelasticity with reduced structural models has been reported. For a flexible rectangular wing results for aeroelastic equilibrium and for a dynamic response problem have been presented. First comparison with low speed aeroelastic experiments is very promising. Validation work will be continued with emphasis on an elastic swept wing model in subsonic flow [8, 19] and airfoils and wings in the transonic regime.

References

[1] H.N. Abramson, H.J. Plass, E.A. Ripperger, "*Stress wave propagation in Rods and Beams*", Advances in Applied Mechanics V (eds. H.L. Dryden, Th. von Kármán), 1958, pp. 111-188

[2] AGARD-R-702, "*Compendium of Unsteady Aerodynamic Measurements*", 1982, pp. 4-1 - 4-25 and 5-1 - 5-26

[3] O.O. Bendiksen, G.Y. Hwang, "*Nonlinear Flutter Calculations For Transonic Wings*", Proc. CEAS International Conference on Aeroelasticity and Structural Dynamics, Vol. II, Rome, 1997, pp. 105-114

[4] A. Brenneis, A. Eberle, "*Evaluation of an Implicit Euler Code Against Two- and Three-Dimensional Standard Configurations*", AGARD CP-507: Transonic Unsteady Aerodynamics and Aeroelasticity, Paper No. 10, 1992

[5] G. Britten, J. Ballmann, "*About the Uniqueness of Aeroelastic Solutions of Elastic Wings in Transonic Flow*", NNFM 72, 1999, pp. 77-84

[6] G. Britten, J. Ballmann, "*Navier-Stokes-Based Direct Numerical Simulation*", Proc. ICAS 2000, Paper No. ICA0483, 2000

[7] J.R. Cebral, R. Löhner, "*Conservative load projection and tracking for fluid-structure problems*", AIAA Journal, Vol. 35, No. 4, 1997, pp. 687-692

[8] A. Dafnis, M. Kämpchen, H.-G. Reimerdes "*Aero-structural investigation on highly flexible wind tunnel wing models*", Proc. International Forum on Aeroelasticity and Structural Dynamics, Madrid 2001, Paper No. 093

[9] C. Farhat, M. Lesoinne, "*Fast Staggered Algorithms for the Solution of Three-Dimensional Nonlinear Aeroelastic Problems*", AGARD-AR-822, 1998, pp. 7-1 - 7-11

[10] P.P. Friedmann, "*Renaissance of Aeroelasticity and its Future*", J. Aircraft, Vol. 36, No. 1, 1999, pp. 105 - 121

[11] G.P. Guruswamy, C. Brun, "*Fluid-Structural Interaction Using Navier-Stokes Flow Equations with Shell Finite Element Structures*", AIAA Paper 93-3087, 1993

[12] R. Heinrich, K. Pahlke, H. Bleecke, "*A Three-Dimensional Dual-Time Stepping Method for the Solution of the Unsteady Navier-Stokes Equations*", Proc. RAS 'Unsteady Aerodynamics' Conference in London, pp. 5.1-5.12, 1996

[13] M. Hesse, "*Aspekte der numerischen Gittergenerierung*", Interner Bericht, SFB 401, RWTH Aachen, 1998

[14] A. Hilgenstock, "*A Method for the Elliptic Generation of Three- Dimensional Grids with Full Boundary Control*", DFVLR-IB 221-87 A09, 1987

[15] M.H.L. Hounjet, J.J. Meijer, "*Evaluation of Elastomechanical and Aerodynamic Data Transfer Methods for Non-Planar Configurations in Computational Aeroelastic Analysis*", Proc. CEAS International Conference on Aeroelasticity and Structural Dynamics, Manchester, 1995, pp. 10.1 - 10.25

[16] T.J.R. Hughes, "*The finite element method*", Prentice-Hall, Englewood Cliffs, 1987, p. 376

[17] J. Hurka, J. Ballmann, "*Elastic panels in transonic flow*", AIAA-Paper 2001-2722, 2001

[18] A.J. Jameson, "*Time Dependent Calculation Using Multigrid, with Applications to Unsteady Flows past Airfoils and Wings*", AIAA paper 91-1596, 1991

[19] Kämpchen M., Korsch H., Dafnis A., Reimerdes H.-G., "*Design, Qualification and Experimental Investigation on Flexible Wind-Tunnel Wing Models*", This Vol.

[20] A. Knipfer, G. Schewe, V. Wendt, "*Numerische und experimentelle Untersuchungen an einem schwingenden NLR7301-Profil in transsonischer Strömung*", IB 232-98J05, DLR-Institut für Aeroelastik, Göttingen, 1998

[21] N. Kroll, C.-C. Rossow, K. Becker, F. Thiele, "*The MEGAFLOW project*", Aerospace Science and Technology 4, 2000, pp. 223-237

[22] M. Lesoinne, C. Farhat, "*Geometric Conservation Laws for Flow Problems with Moving Boundaries and Deformable Meshes, and their Impact on Aeroelastic Computations*", Comput. Methods Appl. Eng., 134, 1996, pp. 71-90

[23] H. Levine, "*Unidirectional wave motions*", North-Holland Publ. Amsterdam, New York, Oxford, 1978, p. 164

[24] R. Löhner, C. Yang, J. Cebral, J.D. Baum, H.Luo, D. Pelessone, C. Charman, "*Fluid-Structure Interaction Using a Loose Coupling Algorithm and Adaptive Unstructured Grids*", Computational Fluid Dynamics Review 1995, Wiley, New York, 1995, pp. 755 - 776

[25] J.J. Meijer, M.H.L. Hounjet, B.J.G. Eussen, B.B. Prantana, "*NLR-TU Delft Experiments in Unsteady Aerodynamic and Aeroelastic Simulation Applications*", AGARD-AR822, 1998, pp. 11.1 - 11.21

[26] I.M.R. Moir, "*Measurements on a Two-Dimensional Airfoil with High Lift Devices*", AGARD-AR-303, Vol. II, 1994, pp. 58-59

[27] S.A.Morton, R.B.Melville, M.R.Visbal, "*Accuracy and Coupling Issues of Aeroelastic Navier-Stokes Solutions on Deforming Meshes*", J.Aircraft, Vol. 35, No. 5, Sept.-Oct. 1998

[28] D. Nellessen, "*Schallnahe Strömungen um elastische Tragflügel*", Fortschrittsberichte VDI Reihe 7: Strömungstechnik, Nr. 302, 1995

[29] B.A.Robinson, J.T.Batina, H.T.Y.Yang, "*Aeroelastic Analysis of Wings Using the Euler Equations with a Deforming Mesh*", J.Aircraft, Vol. 28, No. 11, November 1991

[30] R.L. Sorenson, "*A Computer Program to Generate Two-Dimensional Grids about Airfoils and Other Shapes by the Use of Poisson's Equations*", NASA Ames Research Center, NASA TM 81198, 1980

[31] J.F. Thompson, B.K. Soni, N.P. Weatherill, "*Handbook of Grid Generation*", CRC Press, 1999 (ISBN 0-8493-2687-7)

[32] M.Werle, "*Gittergenerierung für die Strömungssimulation von elastischen Auftriebskörpern*", Diplomarbeit am Lehr- und Forschungsgebiet für Mechanik, RWTH-Aachen, 1997

Approximated Nonlinear Stability Analysis of the Dynamics of Flexible Aircraft

N. Siepenkötter, W. Kasberg, W. Alles
Lehrstuhl für Flugdynamik, RWTH Aachen
Wüllnerstraße 7, D-52062 Aachen, Germany

Summary

A method for the nonlinear analysis of dynamic systems and its application to an aircraft with a flexible wing is presented. In extension of existing averaging methods this procedure follows the linearizing eigenvalue analysis, but defines the eigenvalue and eigenvector as functions of the deflection a from the examined equilibrium. A test case to verify the robustness of the method and its application to the spiral mode and the short period mode of a rigid body aircraft with a wing of four flexible degrees-of-freedom is shown. The analysis of the unstable spiral mode shows the existence of additional stable equilibria in the vicinity of the examined equilibrium due to system nonlinearities.

Nomenclature

a:	deflection amplitude	p:	roll rate in mean axis system
C_k:	spring constants	p_i:	deformed location of the ith mass element
d_i:	local deformation of the ith mass element	q:	pitch rate in mean axis system
dm_i:	ith mass element	q_i:	generalized displacement of elastic mode i
D_k:	damper constants	$Q_{(\cdot)}$:	generalized forces
F_{C_k}:	spring force	r:	yaw rate in mean axis system
g:	gravitational parameter	s:	arclength
$\underline{I}$:	inertia tensor	s_i:	undeformed location of the ith mass element
$I_{(\cdot)}$:	components of the inertia tensor	T1:	high lift configuration
M:	total mass of aircraft	$\mathbf{u_0}$:	vector of the system parameters
M_i:	generalized mass of the ith elastic mode	u, v, w:	velocity components
N:	number of DOF	V:	flight velocity
N_e:	number of elastic modes	ζ_i:	structural damping of elastic mode i
x_k:	mass coordinate	η:	imaginary part of the eigenvectors
$\mathbf{x}$:	vector of the state variables	η_{HR}:	elevator deflection
$\mathbf{x_0}$:	vector of the reference state	Θ:	pitch angle
		κ:	eigenvector
α:	angle of attack	Φ:	roll angle
β:	angle of sideslip	Φ_{ij}:	elastic mode shapes
Γ:	error integral	φ:	phase angle
Δ:	deflection from equilibrium	μ:	eigenvalue
δ:	real part of the eigenvalue	$\boldsymbol{\nu}$:	displacement vector
ε:	vector of the errorapproach	ϱ:	real part of the eigenvector
		ω:	imaginary part of the eigenvalue
		ω_i:	frequency of elastic mode i

1 Introduction

Linearizing methods to analyse the dynamic behavior of aircraft are generally limited in their validity range to the near neighbourhood of the chosen equilibrium. These limits largely depend on the degree of nonlinearity the system holds. For flight conditions with high angles of attack (e.g. take off or landing phase) and corresponding strong aerodynamic nonlinearities or the nonlinear coupling of structure dynamics and rigid body dynamics, the stability prediction is therefore limited to a narrow range around the equilibrium examined.

Within the Collaborative Research Center SFB 401 the project B8 analyses the nonlinear dynamics of High Capacity Aircraft with flexible structure. For this purpose the nonlinear equations of motion of a rigid body reference aircraft have been extended by the first four elastic degrees-of-freedom for the motion of the wing. The dynamic analysis of this model has to predict the stability behavior under consideration of nonlinear effects, while still maintaining the typical evaluation criteria for flight dynamics: frequency and damping of the motion. Therefore the analysis is carried out under use and extension of a method for approximated stability analysis of nonlinear systems (ASTAN-method, Chapter 2.1), [1][2][3].

This averaging method is based in a generalized way on the linear eigenvalue analysis. It assumes that the equations of motion of a nonlinear system can still be met by eigenvalue solutions - even for large deflections from the equilibrium - if its parameters (eigenvalues and eigenvectors) are described as functions of the deflection itself. These functions are calculated by an approximated approach which, in extension of the BTS-method [4],[5], additionally includes the derivatives with respect to the deflection of the eigenvector components. Utilizing this approach, the method differs considerably from other modern analysis methods, e.g. bifurcation methods, where equilibria are examined with respect to varying system parameters using continuation techniques [6].

As result of the approximation the damping, the frequency, the amplitude relations and the phase relations between all the system's state variables are determined over the chosen deflection amplitude.

This approach allows the incorporation of the influence of nonlinearities into the system while maintaining an eigenvalue representation. Nonlinear effects like additional equilibria or limit cycles are therefore covered adequately.

In this paper we will present the ASTAN-method and it's application to a nonlinear spring-mass-system and to the model of a flexible aircraft. It will be shown that the nonlinear behavior of these systems at large deflections from the equilibrium can be easily examined, which would not have been possible using linear eigenvalue analysis.

2 Nonlinear stability analysis

2.1 ASTAN method

For a nonlinear autonomous system represented by the ordinary differential equations

$$\dot{\mathbf{x}}(t) = \mathbf{f}(\mathbf{x}(t), \mathbf{u_0}), \quad (1)$$

the following solution approach for an oscillating motion in eigenvalue form is used:

$$\mathbf{x}(\mathrm{t}) = \mathbf{x_0} + \mathrm{a}(\mathrm{t})\{\boldsymbol{\varrho}(\mathrm{a})\cos\varphi(\mathrm{t}) - \boldsymbol{\eta}(\mathrm{a})\sin\varphi(\mathrm{t})\} + \mathrm{a}(\mathrm{t})\boldsymbol{\nu}(\mathrm{a}). \tag{2}$$

Deviating from the classical approach, the eigenvalues $\mu_i(a) = \delta_i(a) + i\omega_i(a)$ and eigenvectors $\boldsymbol{\kappa}_i(a) = \boldsymbol{\varrho}_i(a) + i\boldsymbol{\eta}_i(a)$ in equation (2) are not considered constant any more. They are now defined as functions of the amplitude a of the deflection from the examined equilibrium. Therefore, the evolution of the amplitude a and phase φ in the time domain now contain the integral of damping and frequency:

$$\begin{aligned} a(t) &= a(0)e^{\int_0^t \delta(a(t))dt} \\ \varphi(t) &= \int_0^t \omega(a(t))dt. \end{aligned} \tag{3}$$

Additionally a vector $\boldsymbol{\nu}(a)$ is introduced in approach (2). This vector allows to include asymmetric nonlinearities. Nonlinearities that are not symmetric with respect to the equilibrium cause a deformation of the state progression in the time domain that can be considered as a displacement of the oscillation center. The vector $\boldsymbol{\nu}(a)$ is therefore called the displacement vector.

With the relations

$$\begin{aligned} &\dot{a} = a\delta, \;\; \dot{\varphi} = \omega \;\text{ and} \\ \{\dot{\boldsymbol{\varrho}}; \dot{\boldsymbol{\eta}}; \dot{\boldsymbol{\nu}}\} &= \frac{d}{da}\{\boldsymbol{\varrho}; \boldsymbol{\eta}; \boldsymbol{\nu}\} \cdot \frac{da}{dt} \\ &= \{\boldsymbol{\varrho}'; \boldsymbol{\eta}'; \boldsymbol{\nu}'\} \cdot a\delta \quad \text{with} \quad ()' = \frac{d}{da}, \end{aligned}$$

taking the time derivative of equation (2) gives

$$\dot{\mathbf{x}} = \mathrm{a}\{(\delta\boldsymbol{\varrho} + \mathrm{a}\delta\boldsymbol{\varrho}' - \omega\boldsymbol{\eta})\cos\varphi - (\delta\boldsymbol{\eta} + \mathrm{a}\delta\boldsymbol{\eta}' + \omega\boldsymbol{\varrho})\sin\varphi\} + \mathrm{a}(\delta\boldsymbol{\nu} + \mathrm{a}\delta\boldsymbol{\nu}'). \tag{4}$$

Here, the time derivative of the solution approach (2) is expressed by the deflection dependent parameters (eigenvalue, eigenvector and displacement vector) and their derivatives with respect to the deflection, as well as the amplitude deflection and the phase themselves.

In order to represent the system, equation (2) and its time derivative (4) must meet the given differential equations (1) of the system. This relation is used to determine the shape of the functions $\mu_i(a) = \delta_i(a) + i\omega_i(a)$, $\boldsymbol{\kappa}_i(a) = \boldsymbol{\varrho}_i(a) + i\boldsymbol{\eta}_i(a)$ and $\boldsymbol{\nu}(a)$. To do this, the following error function is defined:

$$\boldsymbol{\varepsilon}(t) = \dot{\mathbf{x}}(\mathrm{t}) - \mathbf{f}(\mathbf{x}(\mathrm{t}), \mathbf{u_0}) \;,$$

wherein $\mathbf{x}, \dot{\mathbf{x}}$ and $\mathbf{f}$ are given by equations (2), (4) and (1) respectively. The parameters have to be determined in such a way, that the error $\boldsymbol{\varepsilon}(t) = \boldsymbol{\varepsilon}\ (a, \delta, \varphi, \omega, \boldsymbol{\varrho}, \boldsymbol{\varrho}',$

η, η', ν, ν') minimizes over a given time interval T:

$$\Gamma = \int_{\tau}^{\tau+T} \varepsilon^2 dt \quad \Rightarrow \quad MIN. \tag{5}$$

For an oscillating motion the time dependency of the error integral (5) can be replaced by

$$dt = \frac{1}{\omega} d\varphi. \tag{6}$$

If the time interval is chosen as one period of the oscillation, the error integral (5) transforms to

$$\Gamma = \int_{0}^{2\pi} \varepsilon^2 \frac{1}{\omega} d\varphi \quad \Rightarrow \quad MIN. \tag{7}$$

Under the assumption (averaging approach) that the parameters $\delta, \omega, \varrho, \eta, \nu$, as well as their derivatives ϱ', η' and ν' change only 'slightly' within a single oscillation, they can be approximated by means of a least square approximation. The normal equations arising from equation (7) are yet still dependent on the oscillation amplitude. They can only be solved, if the amplitude a is approximated by its mean value over the oscillation too. Using the mean value has the effect, that a can be considered constant regarding the integration in equation (7) and thus, due to the arbitrariness of the chosen oscillation in time, that a can be chosen arbitrary too. As the result of the approximation, the parameters of the system, that is the eigenvalue, eigenvector and displacement vector, are determined for a given deflection amplitude a.

The number of normal equations arising from equation (7) is not sufficient to determine all the parameters of the system. Additional conditions have to be found. These conditions are introduced by estimating the derivative of the eigenvector approximation ϱ', η' with respect to the deflection and the derivative of the displacement vector ν' with respect to the deflection. In extension of the BTS method ([4],[5]) this is done as follows (Fig. 1):

The parameters of the linearized system are used as a starting point. This matches an oscillation amplitude of $a = 0$. Proceeding stepwise for greater values of a, the progression of $\delta(a), \omega(a), \varrho(a), \eta(a)$ and $\nu(a)$ is developed via a Newton iteration process, while using divided differences with the previously developed values to estimate the derivatives $\varrho'(a), \eta'(a)$ and $\nu'(a)$ [1],[2],[3].

In case of an aperiodic motion, the solution approach is simplified to

$$\mathbf{x}(\mathrm{t}) = \mathbf{x}_0 + \mathrm{a}(\mathrm{t})\varrho(\mathrm{a}), \tag{8}$$

and its time derivative becomes

$$\dot{\mathbf{x}}(\mathrm{t}) = \mathrm{a}\delta \left\{ \varrho + \mathrm{a}\varrho' \right\}.$$

To obtain the parameters the same error approach is made like in the oscillatory case (5), but here the time dependency is replaced by

$$dt = \frac{1}{a\delta} da,$$

leading to

$$\Gamma = \int_{a_\tau}^{a_\tau + T} \varepsilon^2(a, \delta, \varrho, \varrho') \frac{1}{a\delta} da \quad \Rightarrow \quad MIN. \tag{9}$$

Similar to the oscillating case the parameters are obtained by solving the normal equations arising from (9), while simultaneously estimating their derivatives with respect to the deflection by divided differences with the previously determined parameters following Fig. 1.

2.2 N degree-of-freedom system

Following [5], a mechanical system of masses connected by springs and dampers (Fig. 2) is used as a test case for the analysis method. This test case allows an easy modification of the systems nonlinearities and number of degrees-of-freedom. The motion of each mass is limited to one degree-of-freedom, giving a system of N degrees-of-freedom for a system with N masses. To simplify matters $m = m_1 = m_2 = \cdots = m_N$ is assumed. The spring and damper coefficients C_k and D_k are given per mass unit and their units are dropped for the purpose of simplicity. All springs are unloaded if $x_k = 0$ $(k = 0, \cdots, N)$.

This model was used with different nonlinearities, numbers of degrees-of-freedom and spring and damper coefficients. The results for an $N = 30$ system are presented here. For this system, it is assumed that all springs other than those connected to the walls hold nonlinear characteristics, defined by

$$F_{C_k} = C_k \left((x_{k-1} - x_k) + (x_{k-1} - x_k)^3 \right) \quad \text{for } k = 2, 4, 6, \cdots \quad (k \le 30)$$

$$\text{and} \quad F_{C_k} = C_k sin(x_{k-1} - x_k) \quad \text{for } k = 3, 5, 7, \cdots \quad (k \le 29).$$

Here, the systems dampers hold linear characteristics. For the factors C_k and D_k uniformly distributed random numbers are used, for C_k in the range from 5 to 25 and for D_k in the range from 0.5 to 1.5 (Fig. 1). The systems equations of motion consequently evolve to:

$$
\begin{aligned}
\ddot{x}_1 &= -C_1 x_1 - D_1 \dot{x}_1 - C_2\left((x_1 - x_2) + (x_1 - x_2)^3\right) - D_2(\dot{x}_1 - \dot{x}_2) \\
\ddot{x}_2 &= C_2\left((x_1 - x_2) + (x_1 - x_2)^3\right) + D_2(\dot{x}_1 - \dot{x}_2) - C_3 sin(x_2 - x_3) \\
&\quad -D_3(\dot{x}_1 - \dot{x}_2) \\
\ddot{x}_3 &= C_3 sin(x_2 - x_3) + D_3(\dot{x}_1 - \dot{x}_2) - C_4\left((x_3 - x_4) + (x_3 - x_4)^3\right. \\
&\quad -D_4(\dot{x}_3 - \dot{x}_4) \\
&\quad \vdots \\
\ddot{x}_{28} &= C_{28} sin(x_{27} - x_{28}) + D_{28}(\dot{x}_{27} - \dot{x}_{28}) - C_{29}\left((x_{28} - x_{29})\right) \\
&\quad \left.+(x_{28} - x_{29})^3\right) - D_{29}(\dot{x}_{28} - \dot{x}_{29}) \\
\ddot{x}_{29} &= C_{29}\left((x_{28} - x_{29}) + (x_{28} - x_{29})^3\right) + D_{29}(\dot{x}_{28} - \dot{x}_{29}) \\
&\quad -C_{30} sin(x_{29} - x_{30}) - D_{30}(\dot{x}_{29} - \dot{x}_{30}) \\
\ddot{x}_{30} &= C_{30} sin(x_{29} - x_{30}) + D_{30}(\dot{x}_{29} - \dot{x}_{30}) - C_{31} x_{30} - D_{30} \dot{x}_{30}
\end{aligned}
$$

A linear analysis (amplitude $a = 0$) of the system shows 30 oscillating eigenforms of motion, with the corresponding 30 complex conjugates (Table 1). For the first eigenform ($-1.7820 \pm 7.9405i$) the nonlinear analysis with regard to its governing state variable x_{10} leads to the eigenvalue and eigenvector behaviour shown in Figure 3.

Both, the eigenvalue and the components of the normalized eigenvector show a strong nonlinear progression. The value of the damping δ decreases after reaching a local maximum at $\Delta x_{10} \approx 0.75$, but finally increases towards zero at $\Delta x_{10} \approx 1.465$. Given $\dot{a} = a\delta$ it is obvious that $\delta = 0$ indicates a new equilibrium, that would have to be examined by a new linearization at this point. The frequency ω increases monotonously for greater values of the deflection amplitude, but with a very unsteady gradient. The progressions of the eigenvector components normalized with $\varrho_{10} + i\eta_{10}$ also show varying gradients, some even with a changing sign.

The damping and frequency progressions in Fig. 3 show very steep gradients at the end of the examined deflection interval. Under some circumstance this can cause convergence problems in the stepwise evolution of the parameters. In such a case, changing the parameterization of the process can avoid these problems. Instead of the deflection amplitude a another parameter, e.g. the arclength s of the damping curve can be chosen. Therefore, the normal equations arising from equation (7) are extended by a parameterizing equation

$$0 = p_s(\delta, a, s) = ((\delta - \delta(s_j))^2 + (a - a(s_j))^2 - (s - s_j)^2,$$

and the deflection amplitude a is included in the set of unknowns that have to be evaluated. $\delta(s_j)$ and $a(s_j)$ indicate the solutions previously calculated at arclength s_j during the stepwise evolution of the parameters. This strategy allows the calculation of the parameters with respect to the arclength s and thus independently of the local gradients.

Table 1 Spring and damper coefficients and eigenvalues of the 30 N system

Spring coefficients $C_1 \cdots C_{31}$	Damper coefficients $D_1 \cdots D_{31}$	Eigenvalues $\delta_n \pm \omega_n i \ (n = 1, \cdots, 30)$
7.7304	0.9966	$-1.7820 \pm 7.9405i$
5.2351	1.3998	$-2.2117 \pm 7.8489i$
22.8780	1.3216	$-1.9558 \pm 7.7997i$
8.9828	1.1449	$-1.9705 \pm 7.4564i$
10.9745	1.3180	$-2.3095 \pm 6.9676i$
18.2289	1.1602	$-2.0215 \pm 6.8595i$
10.6882	0.8420	$-1.9646 \pm 6.7917i$
14.3845	0.7897	$-1.8233 \pm 6.7434i$
6.2956	0.8412	$-1.7503 \pm 6.2316i$
24.7667	1.0341	$-1.5195 \pm 6.2041i$
16.6558	1.2271	$-1.7072 \pm 6.0010i$
13.4699	0.8093	$-1.3442 \pm 5.9176i$
15.3102	1.3385	$-1.3252 \pm 5.5616i$
11.6790	1.0681	$-1.1064 \pm 5.2407i$
13.6581	0.8704	$-0.8903 \pm 4.9873i$
9.5190	1.2027	$-1.2297 \pm 4.6142i$
16.5961	1.0466	$-0.8160 \pm 4.2773i$
20.2073	0.9449	$-0.6054 \pm 3.8237i$
15.5965	1.1946	$-1.3806 \pm 3.6327i$
17.8105	1.1213	$-0.5575 \pm 3.3282i$
9.1814	1.2948	$-0.4472 \pm 3.1010i$
12.5964	1.4568	$-0.3654 \pm 2.6677i$
20.6666	1.0226	$-0.2842 \pm 2.3461i$
18.6169	1.3801	$-0.2296 \pm 2.0156i$
14.2219	0.6730	$-0.1007 \pm 1.3579i$
16.3566	1.4797	$-0.1637 \pm 1.7750i$
20.8842	0.7714	$-0.0599 \pm 1.0145i$
6.1837	0.7523	$-0.0066 \pm 0.3238i$
17.0574	1.3757	$-0.0278 \pm 0.6656i$
6.0054	1.2373	$-1.2944 \pm 3.8148i$
7.9720	1.1046	

An evaluation of the results can be performed in the time domain. For a given initial deflection, the time history of the state variables is reconstructed from the approximation results and compared with the results of a direct Runge-Kutta-Fehlberg integration (RKF) of the systems equations of motion. Therefore, using equations (3) and the now known functions $\delta(a)$ and $\omega(a)$ the time history of the amplitude and phase angle are calculated (Fig 4) and in a next step with the solution approach (2) or (8) the time history of the state variables.
The time histories of the governing state variables calculated by this procedure, are compared with those of the RKF integration in Fig. 5 for an initial deflection of $\Delta x_{10} = 1.465$. The good match of the curves proves, that the approximation approach (2), though based on a linear approach, is representing the system with its strong nonlinearities very well. Of course this can be attributed to the fact, that the nonlinearities are included in this approach by the deflection dependent parameters.

2.3 Rigid body aircraft with a flexible wing

As a basis for the nonlinear dynamic stability analysis, a reference aircraft was defined. This aircraft is equipped with the reference wing used throughout the Collaborative Research Center with a wing span of 80m [9]. The dimensions of the fuselage, engines and control surfaces are defined according to existing High Capacity Aircraft and flight dynamic necessities [7],[8]. The dynamic parameters were then adapted to emphasise coupling effects if required. Fuselage and control surfaces are considered rigid. Fig. 6 summarizes the main data of the configuration.

The dynamic model of the aircraft is based on the nonlinear rigid body Newton-Euler equations of motion extended to flexible aircraft. References [10],[11] and [12] describe in detail how flexible deflection and rigid body movement can be combined in a coupled set of nonlinear equations of motion in a mechanically correct way. Therefore only the basic principles will be repeated here.

The structural deformation is assumed sufficiently small, such that linear elastic theory is valid. Furthermore, a set of vibration modes for the deformation with frequencies and mode shapes is assumed to be available. Then, considering the aircraft as a large set of lumped masses dm_i with constant density, the local deformation e_i can be written as a linear combination of the mode shapes Φ_{ij} and the corresponding generalized displacements q_i, where the actual position p_i of each mass element consists of the undeformed location s_i and its deformed translation e_i. (Figure 7)

$$e_i = \sum_{j=1}^{N_e} \Phi_{ij} q_j \ . \tag{10}$$

By choice of an appropiate body-reference system, the *mean axis system* [10], the inertial coupling between rigid and elastic dynamics can be minimized. This non-inertial reference system, which moves with the body but is not fixed to a material point in the body, is defined such, that the relative linear and angular momenta due to elastic deformation are zero at every instant. These exact constrains are often difficult to satisfy, but with the assumptions of small deformations and co-linear deformation rates it is sufficient when the *practical* mean axis constrains are satisfied [11]. A reference system with its origin always in the center of gravity and with the axes in the direction of the rigid modes satisfies these practical mean axis constrains if the elastic vibration modes have been obtained for the unconstrained body. With the final assumption, that the inertia tensor is constant (although it will vary slightly with the elastic deformation) the equations of motion can be obtained:

$$M \left(\begin{bmatrix} \dot{u} \\ \dot{v} \\ \dot{w} \end{bmatrix} + \begin{bmatrix} q\,w - r\,v \\ r\,u - p\,w \\ p\,v - q\,u \end{bmatrix} - \begin{bmatrix} -sin\Theta \\ sin\Phi\, cos\Theta \\ cos\Phi\, cos\Theta \end{bmatrix} g \right) = \begin{bmatrix} Q_X \\ Q_Y \\ Q_Z \end{bmatrix} \tag{11}$$

$$\underline{I} \begin{bmatrix} \dot{p} \\ \dot{q} \\ \dot{r} \end{bmatrix} + \begin{bmatrix} q\,r(I_{zz} - I_{yy}) - p\,q\,I_{xz} \\ r\,p\,(I_{xx} - I_{zz}) + (p^2 - r^2)\,I_{xz} \\ p\,q(I_{yy} - I_{xx}) + q\,r\,I_{xz} \end{bmatrix} = \begin{bmatrix} Q_\varphi \\ Q_\theta \\ Q_\psi \end{bmatrix} \tag{12}$$

$$M_i \left(\ddot{q}_i + 2\,\zeta_i \omega_i \dot{q}_i + \omega_i^2 q_i \right) = Q_{q_i} \qquad i = 1, 2, 3, ...n \;. \tag{13}$$

In these nonlinear equations the rigid and flexible motion is ultimately coupled via the aerodynamic and propulsive forces $Q_{(\cdot)}$ which can be determined from the Principle of Virtual Work [11].

The aerodynamic forces have been modeled using simple strip theory. This was considered sufficient for our focus on flight mechanical aspects of the interaction of rigid body and flexible structure. However, the above model can be readily extended to unsteady aerodynamics if for example flutter behavior is to be investigated.

2.4 Results

In this chapter results of a nonlinear analysis are presented for the coupled longitudinal, lateral and wing motion of the reference aircraft with four flexible degrees-of-freedom. The included flexible modes are the symmetrical vibration modes with the lowest frequency. These are the first three bending modes $q_{1,2,3}$ and the first torsional mode q_4. The examined equilibrium represents a trimmed horizontal flight attitude with a velocity of 100 m/s. The other state variables and the eigenvalues at the equilibrium with zero deflection are given in Table 2. These are identical to those of the linearized motion.

The eigenvalues describe 9 characteristic forms of motion, which can be identified analogous to the classic rigid body forms of motion by their dominant state variable. They are the short period mode, phugoid, dutch-roll, roll and spiral mode and the new additional three bending and one torsional motion of the wing. Table 2 additionally lists the eigenvalues of a rigid model at the same reference state where the elasticity was comprised only by means of equivalent rigid body derivatives at the chosen reference point. The influence of the elastic degrees-of-freedom on the dynamics of the rigid body is most significant at the short period mode, whose frequency is the closest to the elastic modes. The damping value differs significantly

Table 2 Reference state and eigenvalues at the equilibrium

α	$2.8148\,^\circ$
V	$100\,m/s$
q	$0\,^\circ/s$
Θ	$2.8148\,^\circ$
β	$0\,^\circ$
r	$0\,^\circ/s$
Φ	$0\,^\circ$
p	$0\,^\circ$
q_1	$-0.6852\,m$
q_2	$-0.0845\,m$
q_3	$-0.0403\,m$
q_4	$0.0409\,rad$
$\dot{q}_1$	$0\,m/s$
$\dot{q}_2$	$0\,m/s$
$\dot{q}_3$	$0\,m/s$
$\dot{q}_4$	$0\,rad/s$

	rigid	flexible
short period mode	-2.0409 $\pm 3.9548\,i$	-2.5113 $\pm 3.9316\,i$
phugoid	-0.0117 $\pm 0.1341\,i$	-0.0116 $\pm 0.1349\,i$
dutch-roll	-0.0164 $\pm 0.2588\,i$	-0.0165 $\pm 0.2583\,i$
roll mode	-1.3626	-1.3624
spiral mode	$+0.0042$	$+0.0042$
1. bending motion of the wing	–	-1.1266 $\pm 9.6358i$
2. bending motion of the wing	–	-2.4217 $\pm 16.4702i$
3. bending motion of the wing	–	-6.9259 $\pm 43.8911i$
1. torsional motion of the wing	–	-1.7784 $\pm 24.8806i$

between rigid and elastic modeling. For the other two oscillating rigid body modes, phugoid and dutch-roll, the influence is less visible due to the wide separation of their frequencies from the elastic modes.

The spiral mode, as an aperiodic and unstable motion with an eigenvalue of $+0.00442$ stands out in this list. Its eigenvalue and eigenvector functions over the deflection amplitude are presented in Fig 8. Here, the eigenvector was normalized with the dominant roll angle component ϱ_Φ, and the deflection amplitude was evolved up to a roll angle of $\Delta\Phi = 40°$. With increasing deflection from the equilibrium, a decrease of the eigenvalue can be observed, finally leading to $\delta = 0$ at $\Delta\Phi = 34.5°$. This zero point represents a new equilibrium, different from the starting point of the analysis at $\Delta\Phi = 0°$. This is due to the fact that according to equation (3), the amplitude a ($\Delta\Phi$ in this case) does not change any more in the time history if the damping δ vanishes. The change in stability behavior from an unstable to a stable motion happens of course due to the nonlinear character of the system at large deflection from the equilibrium. Accordingly this behavior would be difficult to observe using linear analysis techniques.

To illustrate the ASTAN results of the spiral mode from a different perspective, Fig. 9 shows the branching diagram of the roll angle Φ. In this figure trimmed flight states (i.e. equilibria) are plotted, with the elevator deflection η_{HR} as the free parameter. Negative values of η_{HR} describe a pulled elevator. Therefore, in Fig. 9 the flight velocity increases from left to right. At low flight velocities a single solu-

tion exists describing a straight flight state with $\Phi = 0$ which is stable with respect to the spiral mode. Beginning with a critical bifurcation point at $\eta_{HR} = -0.46°$ ($V = 70.11 m/s$) the solution splits into three branches, one unstable with $\Phi = 0°$ and two stable with $\Phi \neq 0°$.
The equilibrium examined in Fig. 8 is found in the branching diagram on the unstable branch at $\eta_{HR} = 0.58°$. A variation of the deflection amplitude $\Delta\Phi$ according Fig. 8 finds its equivalence in the branching diagram in an analysis along a line $\eta_{HR} = const$ i.e. parallel to the Φ axis. Therefore, the new equilibrium found in Fig. 8 at $\delta = 0$ is located on the upper stable branch in the branching diagram. If the new equilibrium found is chosen as the starting point of an approximated analysis, the eigenvalue progression as shown in Fig. 9 on the right is the result. Regarding this equilibrium, the motion of the system is stable. It stays stable, until the deflection reaches a value beyond the unstable branch ($\Delta\Phi < -34.5°$). Then the eigenvalue becomes positive, resulting in a divergent motion towards the lower stable branch. Thus, with the help of the ASTAN method it is possible to analyse the stability behavior in the non-equilibrium between the solution branches with respect to a selected equilibrium.

The time history of the state variables at the spiral mode according Fig. 8 is shown in Fig. 10 for an initial deflection of $\Delta\Phi = 5°$. The initial deflections of the other state variables are given by the solution approach (2). As expected, in the beginning the deflections (deviations from the equilibrium) increase exponentially, due to the unstable character of the form of motion. Later though the time histories show a trend towards constant values, i.e. the new equilibrium. Within the scope of plotting accuracy the time histories reconstructed from the ASTAN results can not be distinguished from those of the direct Runge-Kutta integration.

As an example for an oscillating form of motion, Fig. 11 shows the damping and frequency of the short period mode with respect to the α-amplitude $\Delta\alpha$. Although, as shown in Table 2, the short period mode is influenced significantly by the flexibility of the aircraft the analysis reveals no strong nonlinear effects even at large deflection from the reference point. Both, damping and frequency are nearly constant over a wide $\Delta\alpha$ range, indicating a linear system behavior mainly due to the linear modeling of the flexibility. The comparison of the state variables in the time domain shows a good match of the ASTAN results and the results of the direct RKF integration too (Fig. 12).

3 Conclusions

With the method of approximated stability analysis of nonlinear systems (ASTAN method) a procedure was presented that allows the incorporation of the nonlinear character of a system into the eigenvalue analysis. The representation of eigenvalue and eigenvector as functions of the deflection amplitude a and introduction of a displacement vector $\nu(a)$ enables the method to identify nonlinear effects like new equilibria or limit cycles even far away from the examined equilibrium. By means of a generic test case the robustness of the method concerning the number of degrees-of-freedom and type of nonlinearity was shown. For a rigid body aircraft with a flexible wing of four degrees-of-freedom the nonlinear analysis of the spiral mode showed the presence of additional stable equilibria in the vicinity of the examined unstable equilibrium that would have not been obvious using linear analysis methods. The short period mode of the same model turns out to be influenced considerably by elasticity, while the nonlinear analysis shows that at least by using linear elastic modeling no nonlinear effects are introduced into this mode. The evaluation of the results in the time domain showed a very good agreement with the results of a direct Runge-Kutta integration for both modes.

4 References

[1] Reitenbacher, S.: Ein Verfahren zur approximativen Stabilitätsanalyse von nichtlinearen, dynamischen Systemen und seine Anwendung auf flugmechanische Problemstellungen. Verlag Mainz, Wissenschaftsverlag, Aachen, ISBN 3-89653-099-2, 1996, Dissertation, Aachen

[2] Kasberg, W.: An approximation method for stability analysis of nonlinear systems, GAMM-Meeting, Bremen, 1998

[3] Siepenkötter, N.: Approximative Stabilitätsanalyse der nichtlinearen Dynamik von Flugzeugen, Proceedings of the DGLR anual meeting Vol. I, Leipzig, DGLR-JT2000-058, September 2000

[4] Simpson, A.: An Algorithm for Autonomous Non-linear Dynamical Equations, Aeronautical Quarterly, August 1977

[5] Christopher, P.A.T., Thorne, R.: Some examples of the application of the Beecham-Titchener-Simpson (BTS) method to autonomous systems, Paper No. 1344, Aeronautical Journal, December 1985

[6] Seydel, R.: Practical Bifurcation and Stability Analysis. From Equilibrium to Chaos. Second Edition. Springer Interdisciplinary Applied Mathematics, Volume 5. Springer, New York 1994

[7] Majunke, C.: Simulation eines Großflugzeuges unter MATLAB/SIMULINK, student research project, Lehrstuhl für Flugdynamik, RWTH Aachen, 1999

[8] Keil, P.E.L.: Entwurf des mathematischen Modells eines flexiblen Flügels für ein Großflugzeug und Implementierung unter MATLAB/SIMULINK, diploma thesis, Lehrstuhl für Flugdynamik, RWTH Aachen, April 2000

[9] Pieper, C., Coors, D.: Definition der Referenzprofile, internal report, Collaborative Research Center SFB 401, April 1997

[10] Milne, R. D., Dynamics of the Deformable Aeroplane, Parts I and II, Her Majestys Stationery Office, Reports and Memoranda No. 3345, Sep. 1962

[11] Waszak, M. R., Schmidt,D. K., Flight Dynamics of Aeroelastic Vehicles, Journal of Aircraft, Vol. 25, No. 6, pp.563-571, 1988

[12] Buttrill, C.S., Zeiler, T.A., Arbuckle, P.D.: Nonlinear Simulation of a Flexible Aircraft in Maneuvering Flight, Proceedings of the AIAA Flight Simulation Technologies Conference, Monterey, CA, AIAA-87-2501, 1987

5 Tables & Figures

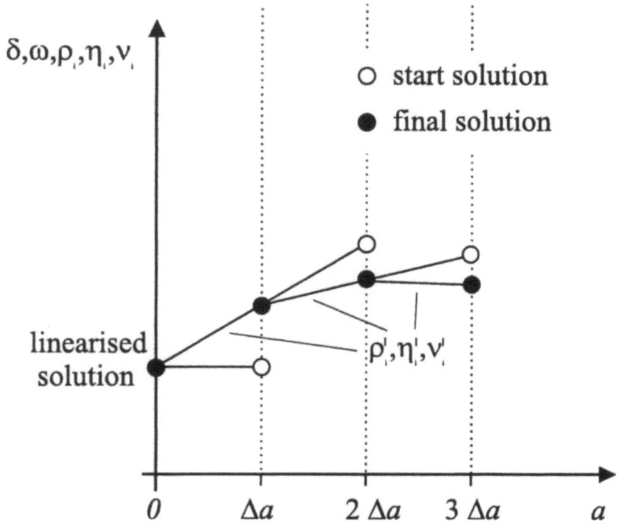

Figure 1 Evolution of the parameters

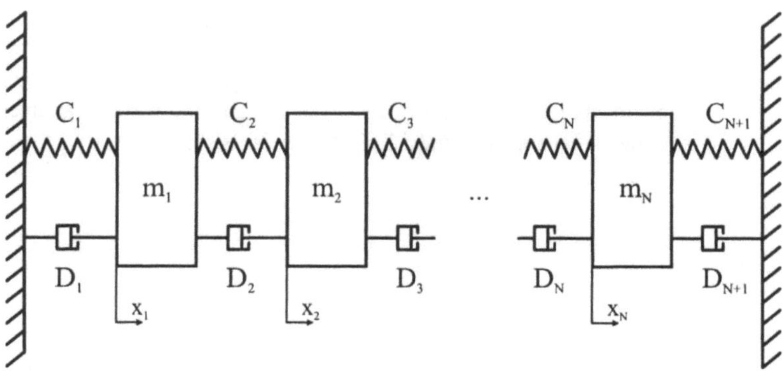

Figure 2 Horizontal system of N masses

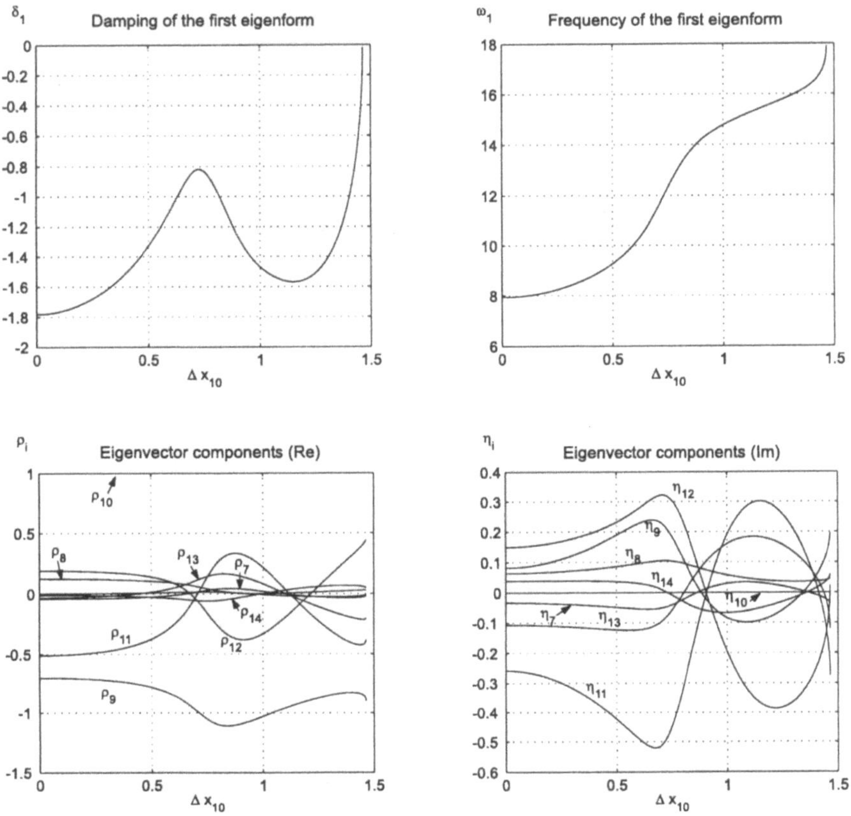

Figure 3 Progression of the eigenvalue- ($\delta(\Delta x_{10}) + i\omega(\Delta x_{10})$) and eigenvector components ($\boldsymbol{\varrho}(\Delta x_{10}) + i\boldsymbol{\eta}(\Delta x_{10})$) of the governing state variables over the deflection amplitude Δx_{10} for the 30 N system

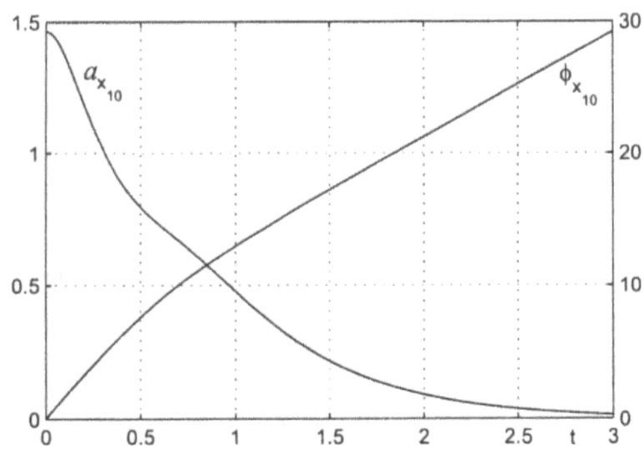

Figure 4 Time history of the amplitude $a_{x_{10}}$ and phase angle $\varphi_{x_{10}}$ of the first eigenform of the 30 N system

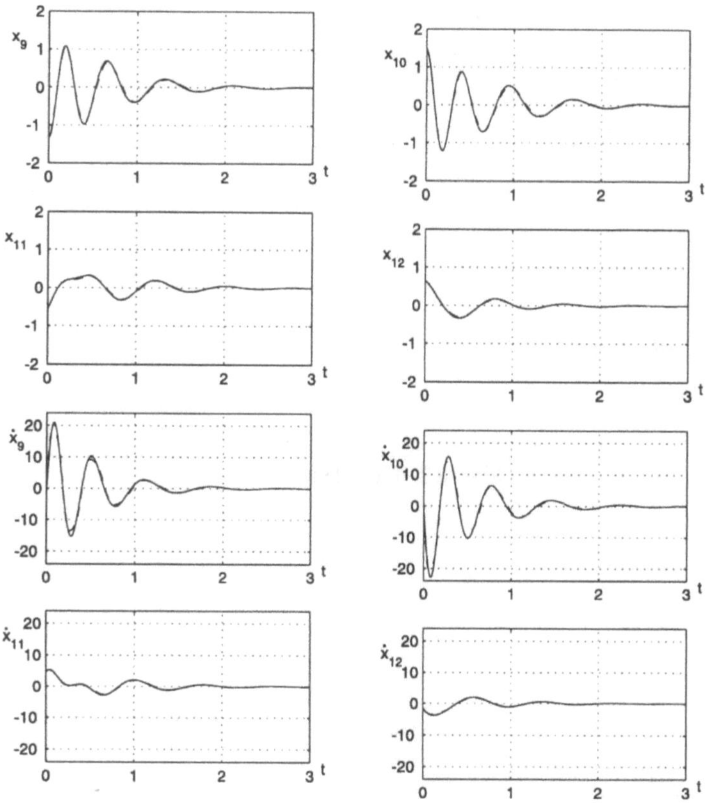

Figure 5 Time history of the first eigenform's governing state variables for the 30 N system (initial deflection $\Delta x_{10} = 1.465$) (ASTAN: —, RKF: - -)

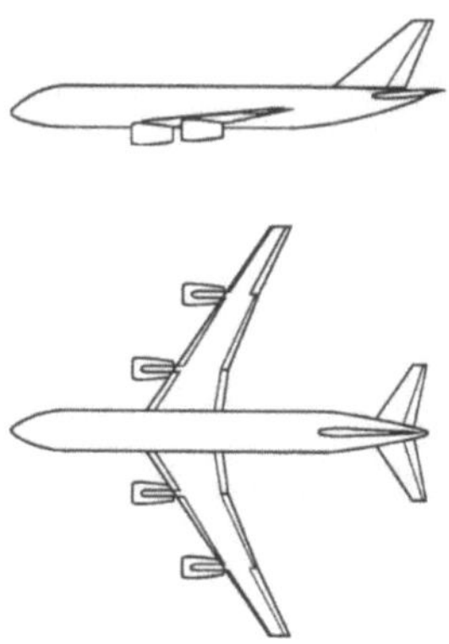

Reference aircraft

Wing / fuselage

Wing span:	80 m
Wing area:	715,4 m^2
Fuselage length:	81,5 m
Fuselage- ϕ *:*	8 m

Engines: **4 x 580 kN**

Mass and moment of inertia

MTOW:	**782 to**
Ix:	**107 10^6 kg m^2**
Iy:	**146 10^6 kg m^2**
Iz:	**261 10^6 kg m^2**

Horizontal tail

H. tail span:	**26,51 m**
H. tail area:	**156,23 m^2**

Vertical tail

Height:	**12,25 m**
V. tail area:	**107,25 m^2**

Figure 6 Reference aircraft

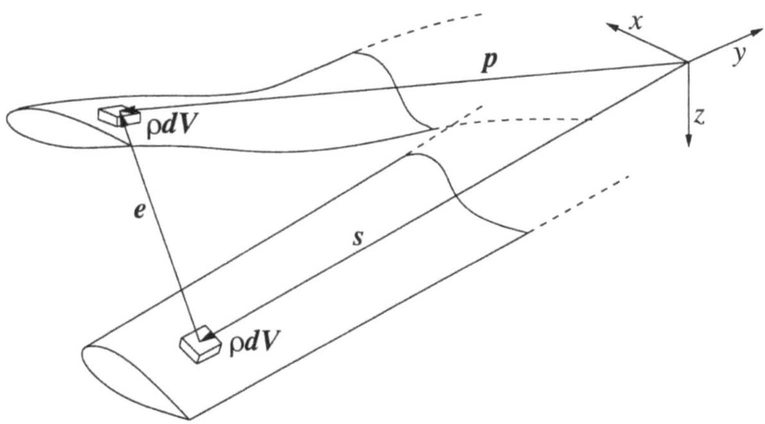

Figure 7 Flexible aircraft

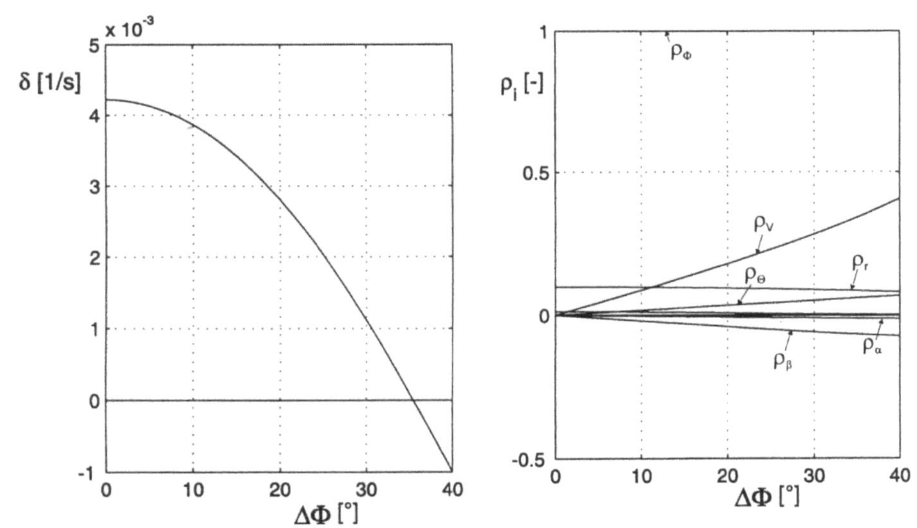

Figure 8 Damping and eigenvector of the spiral mode over the roll angle deflection $\Delta\Phi$

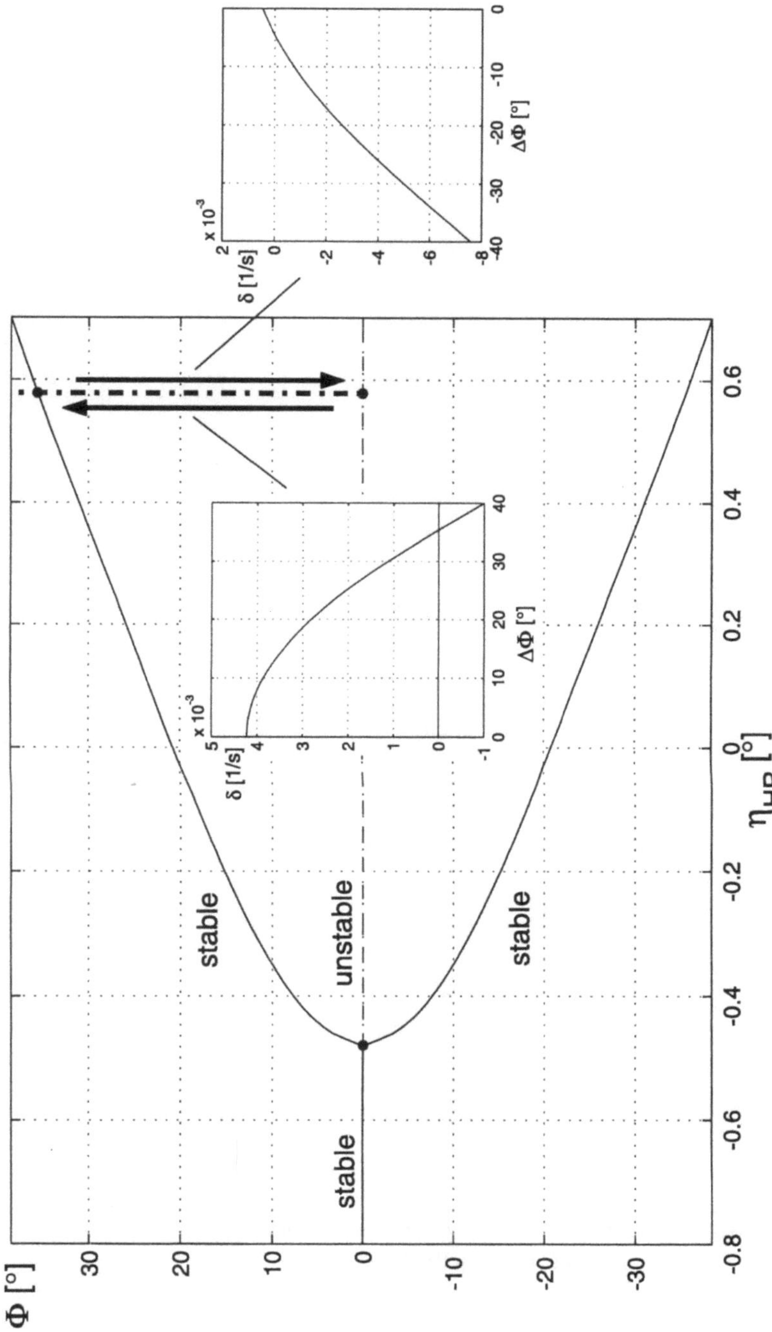

Figure 9 Branching diagram of the roll angle with respect to the elevator deflection η_{HR} ('supercritical pitchfork bifurcation')

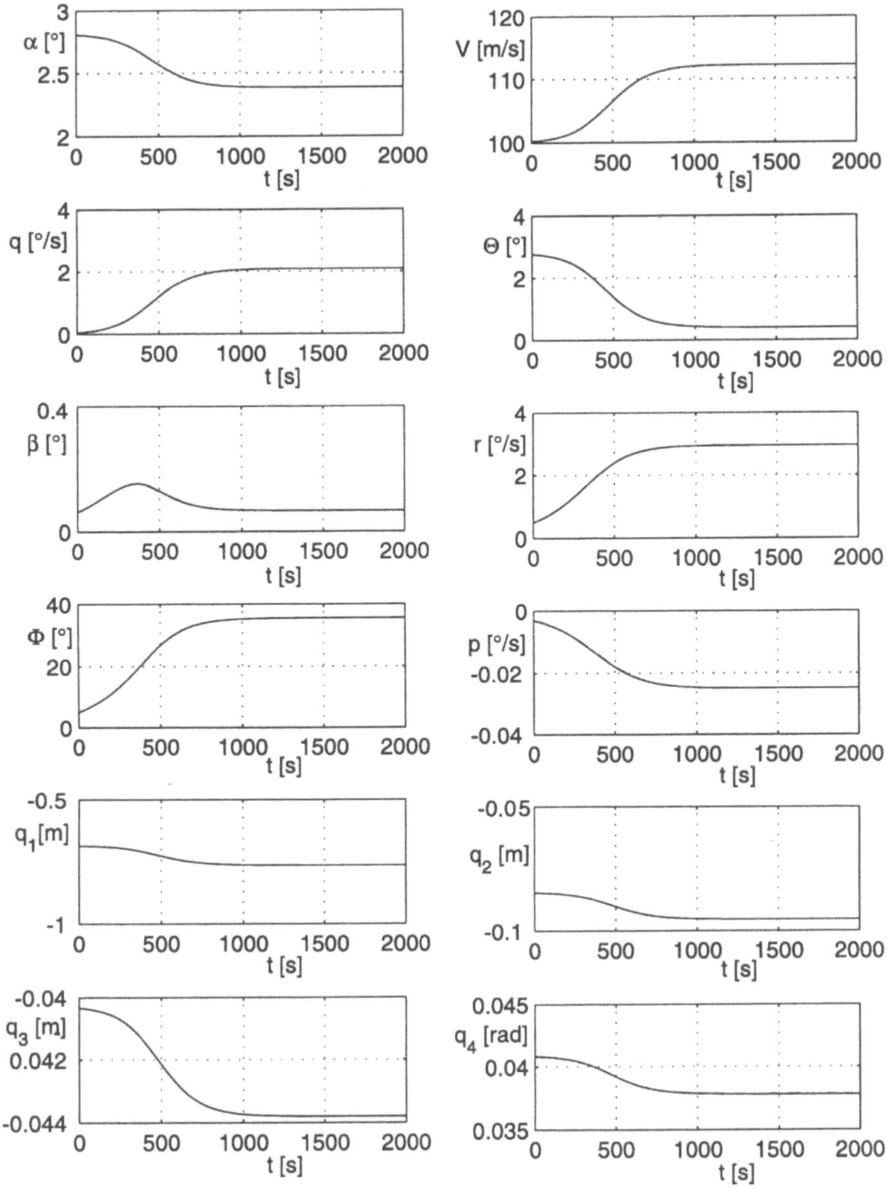

Figure 10 Time history of the state variables at the spiral mode (initial deflection: $\Delta\Phi = 5°$) (ASTAN: —, RKF: – –)

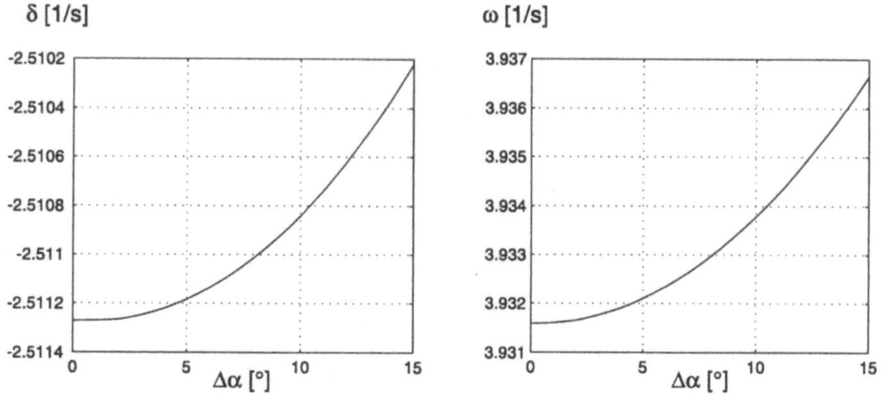

Figure 11 Damping and frequency of the short period mode with respect to the deflection amplitude $\Delta\alpha$

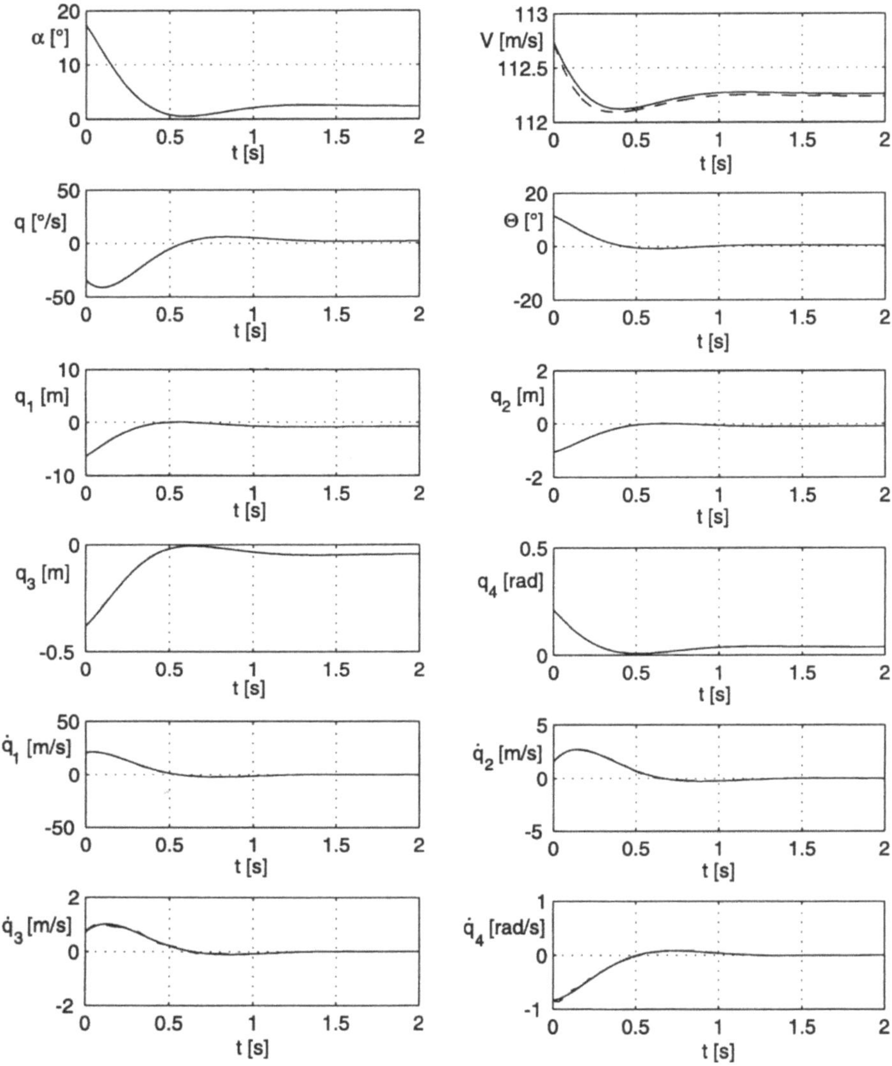

Figure 12 Time history of the governing state variables at the short period mode (initial deflection: $\Delta\alpha = 5°$) (ASTAN: —, RKF: – –)

Analysis of Unsteady Airfoils at Low Speeds

X. Bertrán[◇], H. Olivier[◇], S. Turek [‡]
[◇] Stoßwellenlabor, RWTH Aachen
Schurzelterstraße 35, D-52074 Aachen, Germany
[‡] Lehrstuhl Mathematik III, Universität Dortmund
Vogelpothsweg 87, D-44227 Dortmund, Germany

Summary

In this study exemplary results on the influence of two fundamental unsteady parameters, the reduced frequency k and the non-dimensional pitching rate α_+, on the transition from a steady to an unsteady flow field around the reference airfoil BAC3-11/RES/30/21 are presented using experimental and numerical techniques. The tools used are described and their results verified with each other. On the experimental side, water tunnel experiments are conducted using particle-image-velocimetry (PIV) and strain-gauge-balance (SGB) measurement techniques. The experimental results are enhanced by two-dimensional laminar unsteady numerical Navier-Stokes simulations of the experiments with a state of the art code from the Featflow package. The numerical simulations are found to agree well with experimental results, despite three-dimensional effects and a relatively high Reynolds number of $Re_c = 16000$. An influence of the reduced frequency on the lift coefficient at a local non-dimensional pitching rate value of $\alpha_+ = 0$ for the relevant range is not detected, while for the drag and the pitching moment a decrease and an increase respectively is observed. Nevertheless a phase difference for the lift with increasing reduced frequency was calculated, which appears at least above $k = 0.14$ and indicates unsteady effects. The same was observed for the pitching moment. The effects of the airfoil camber are clearly seen during the phase of negative angle of attack, inducing leading and trailing edge separation. The flow separation regions are discussed and the formation of a triple vortex system at the trailing edge in the pitching up motion of the airfoil is identified in the experimental and numerical data.

1 Introduction

Aeroelasticity is of major importance for the success of an airplane design since the beginning of aviation [1]. In the past two decades, the aspect ratio of modern civil aircrafts was increased significantly, leading to enhanced aeroelastic phenomena which became a fundamental role in the development of new aircraft and new tools have been designed to speed up the development process [2, 3]. While reliable tools for on-design conditions exist, there is a lack of tools for off-design conditions, where in most cases unsteady flow fields are encountered. Even though there have been numerous studies in this field for special cases such as dynamic stall, still many flow field conditions have not been explored. Especially the transition process from a quasi-steady to an unsteady flow field has only seldom been analyzed. An attempt is to be made in this study using experimental and numerical techniques available.

By definition a time varying flow field is unsteady, but there are situations where a time dependent flow field can be considered steady, thus quasi-steady. For example in the case of flutter, the initial situation may be a steady flow field, which due to increasing unsteady effects and airfoil motion, becomes unsteady. It therefore goes through a transition process from a steady to an unsteady flow field. To study this process, the model problem - the oscillating thin airfoil - is used, which has been thoroughly analyzed in the past [4]. In general the unsteady motion of oscillating thin airfoils is characterized by the reduced frequency k, which represents the ratio between the time a particle takes to pass the chord length and the time of an oscillation cycle. As the reduced frequency decreases the flow field becomes increasingly steady. The aim is to determine the influence of the reduced frequency and other parameters in the transition process from a steady to unsteady flow.

2 The Facility and Measurement Techniques

2.1 Water-Ludwieg-Tunnel

The Water-Ludwieg-Tunnel is a short duration facility for the study of unsteady hydrodynamic phenomena, which was proposed by Akamatsu in 1978 [5]. The principle is shown schematically in Fig. 1. The Water-Ludwieg-Tunnel is divided in two parts, a high and a low pressure section, thereby resembling a shock tube set-up. Both are separated by a pneumatically driven fast acting valve with a very short opening time ($\approx 9\,ms$). The high pressure section is divided into three parts, the top being the storage tank followed by the nozzle and the test section. The test section is $150\,mm$ wide and $75\,mm$ deep and has a length of about $820\,mm$. The test section ends in the fast acting valve, consisting of a perforated flat plate connected to the rod of a pneumatic actuator. The valve opening ratio Kv can be adjusted to give the desired free stream velocity u_∞ in the test section. The low pressure section consists of the retainment tank and a water pump, which is connected to the storage tank and a water supply. The facility can either be controlled manually or by a PC. Models are mounted onesided and $0.280\ m$ away from the nozzle exit in the test section. The model investigated here is the airfoil BAC3-11/RES/30/21 specified in the AGARD-AR-303 report [6] with a chord length of $54\ mm$. It is chosen as reference geometry of the collaborative research center SFB 401 "Flow Control

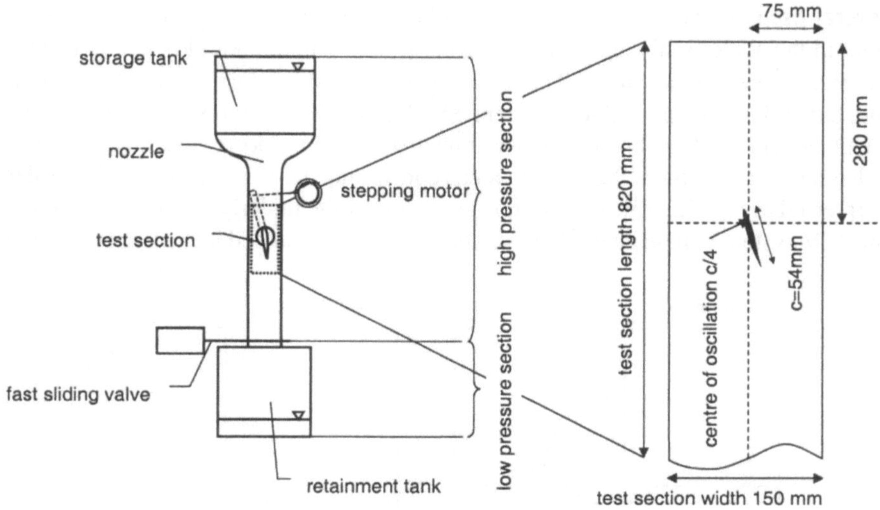

Figure 1 Schematic view of the Water-Ludwieg-Tunnel

and Fluid-Structure Interaction at Airplane Wings". The measurement techniques applied are:

- digital flow visualization via particle tracking,
- digital Particle-Image-Velocimetry (PIV),
- and the measurement of forces and the pitching moment with a three-component strain gauge balance.

2.2 Particle-Image-Velocimetry (PIV) and Flow Visualization

2.2.1 PIV-system

In the past the PIV-method has been proven to be a very effective tool studying dynamic airfoil flow phenomena [7, 8]. Here the PIV-system was designed to examine oscillating airfoils in an unsteady water flow. In the past, problems accounted to phase averaging methods used to analyze unsteady flow fields have been reported (see [9]). At the same time Wernert et al. [10] have demonstrated, that vector plots from several oscillations go aside with having non-reproducible results concerning the unsteady flow field of an airfoil in deep stall. Therefore the main aim for this PIV-system was to extract 10 vector plots from one single oscillation, in order to attain as much information as needed to examine the unsteady behaviour of the flow without having to use phase averaging methods or to assemble PIV-recordings from several oscillations. The number of oscillations is restricted by the short testing time in the Water-Ludwieg-Tunnel, making it impossible to employ phase averaging techniques with accuracy. The recording technique was kept as simple as possible, but at the same time able to generate high resolution pictures by using a digital CCD-camera. A digital recording was favoured to reduce postprocessing time. For a detailed description of the fundamentals concerning the PIV-technique, the reader is referred to [11, 12, 13]. Here a short overview of the underlying methods used in this PIV-system will be given. In PIV, a laser beam is expanded into a light sheet and

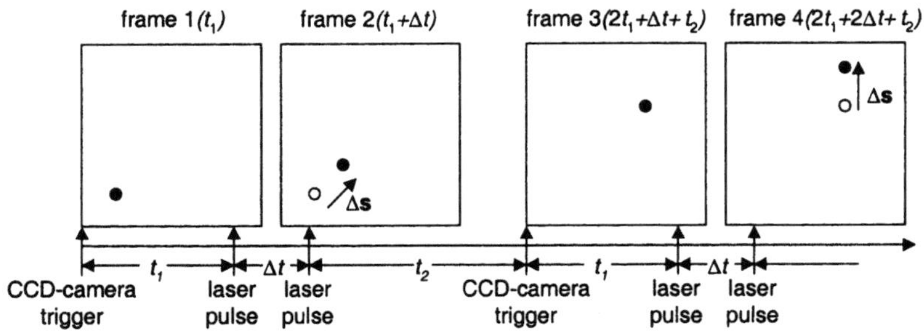

Figure 2 Single pulse multi-frame mode operation of the PIV-system

is introduced into the flow seeded with particles (see Fig. 3). The light is scattered by the particles and can be recorded perpendicular to the plane of the light sheet. In this case a CCD-camera is used to record two sequential frames a Δt apart of the flow field. The frames contain images of particles suspended in the flow, which are chosen in a way so that they follow the local flow as accurately as possible. The displacement can therefore be calculated by:

$$\mathbf{u} = \frac{\Delta \mathbf{s}}{M \Delta t}. \tag{1}$$

Where M is the magnification. For these studies highly unsteady flow fields were expected, therefore only a recording of multiple single exposure frames was considered which also has the advantage of removing directional ambiguity encountered using multiple exposure frames, illustrated in Fig. 2. In this mode the analysis is performed using cross-correlation [13, 14]. For this purpose the frames are divided into subspaces called interrogation regions 32×32 or $64 \times 64\, Pixels$ wide. The spacing between the interrogation regions is varied from 50% to 75% overlap respectively. Cross-correlation requires a sufficient number of particles $N > 10 - 20$ in the interrogation region. The laser pulse length is $200\mu s$ and laser pulse separation is $1000\mu s$.

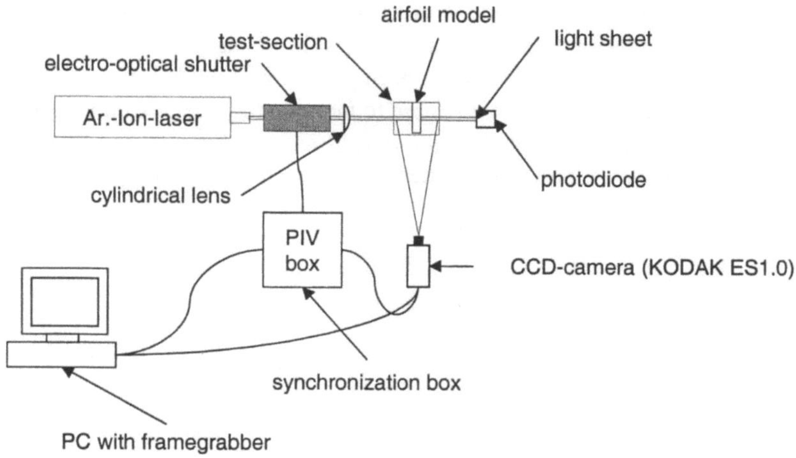

Figure 3 A view of the PIV system

Illumination

Illumination is achieved by using a continuous wave Argon-ion laser model Spectra-Physics 2000-05 with a maximum power output of 5 Watts in multiline mode. The laser power during experiments was adjusted to be typically between $2.5 - 4.2\,W$. The laser aperture was opened at its maximum and was directed straight at the test section to have its full power available. The laser beam is aligned with an electro-optical shutter of type Dantec/Invent 80x41, a cylindrical lens and a vertical aperture to create a light sheet with a thickness of $2\,mm$.

Particles

Vestamid 1111 particles are used which have a diameter of about $100\,\mu m$. Seeding is done manually by measuring the volume of the particles about to be dispensed in the water ($\sim 150\,ml$). The density is a little lower than that of water. It is therefore inevitable to run the test water several times through the facility in order to have a homogeneous seeding. The size of the particles is chosen to give an approximate particle image on the CCD-chip of $1 - 3\,pixels$, depending on the size of the flow region recorded.

Data processing and evaluation

The image acquisition system consists of a Kodak Megaplus ES1.0DC with 1024 x 1024 pixels and a ITI IC-P-4M framegrabber with maximum image acquisition speed of 30 Hz in single and 15 Hz in double frame mode. The CCD-camera is operated in double frame mode with an acquisition speed of 10 Hz. For data processing the commercial software package Visiflow is used. The PIV-system was evaluated by measuring the velocity in the empty test-section and comparing the results with Laser-doppler-anemometry used to measure the free-stream velocity. The mean error is roughly 2%, which is the overall error to be expected from such a system. Figure 4 shows the values plotted against each other and a good agreement can be found.

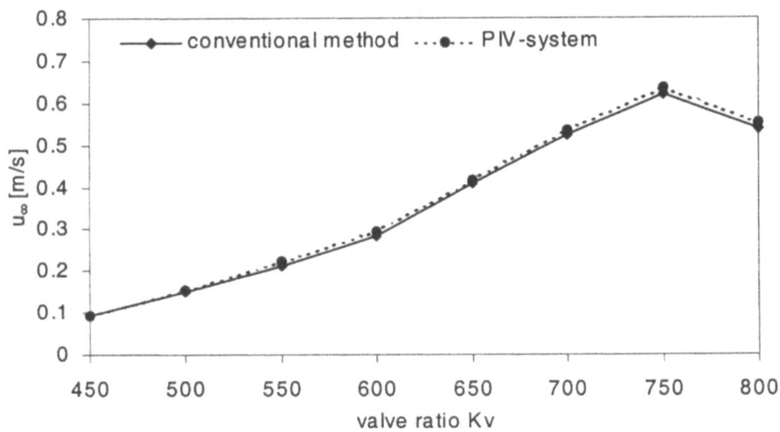

Figure 4 Comparison of the free-stream velocity measured by the PIV-system with the conventional method (LDA)

2.2.2 Flow Visualization

Digital visualization is achieved by recording the particles illuminated by a light sheet in the mid-section of the test section with a CCD-camera. The same device is a component of the PIV-system installed. Exposure times are 15 ms to generate relatively long particle lines in order to compare them well with numerical data visualization. Total frame time is 32 ms allowing for a good contrast and the frame rate is 10 Hz.

2.3 Strain-Gauge-Balance (SGB)

The three-component strain-gauge-balance is constructed out of commercially available load cells. Figure 5 shows the setup of the strain-gauge-balance. The load cell section is sealed by a rubber bellow reaching from the model mount to the back flange. The influence of the rubber bellow in constant pressure conditions is negligible, while for varying pressure conditions the measured values have to be corrected. It measures the normal and axial forces and the pitching moment in strain-gauge-balance coordinates (SGB), which are rotated relatively to the model coordinate system (AM) in order to optimize the actual force components acting in the load cell direction. The pitching moment is equal in the considered coordinate systems. The aerodynamic forces in the aerodynamic coordinate system are given by:

$$\begin{aligned} L &= N_{SGB}\cos(\varepsilon-\alpha) + T_{SGB}\sin(\varepsilon-\alpha) \\ D &= -N_{SGB}\sin(\varepsilon-\alpha) + T_{SGB}\cos(\varepsilon-\alpha), \end{aligned} \quad (2)$$

where ε is equal to 45° and α is the angle of attack. During the post-processing the signals are transformed and filtered to erase noise and vibrational influences occurring during the measurement. The forces acting on the SGB are very small and in the order of $\sim 10^{-1}\ N$ and the pitching moment even smaller in the order of $\sim 10^{-4}\ Nm$. Therefore force and moment measurements have to be taken with care as they are easily influenced by external laboratory vibrations. The strain

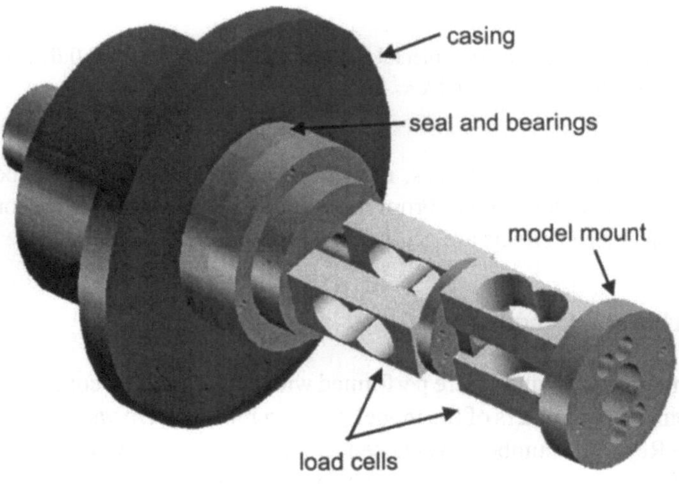

Figure 5 Schematic view of the strain-gauge-balance

gauge balance has been calibrated statically, where combinations of normal, tangential force and the pitching moment were applied to determine the complete range of interferences and the accuracy of the balance. The mean error of the calibration factors amounts to 0.1 to 0.2 % for the normal and tangential force and 1 % for the pitching moment. The interference effects between the components were negligible. This is mainly due to the double beam elements used in the load cells. The natural frequency of the balance model system installed in the tunnel was found to be 10 Hz for the force as well as the pitching moment component which is at least one order of magnitude larger than the frequency of the model motion. Figure 6 shows typical time histories of the lift and pitching moment coefficient deduced from the balance signals. The natural frequency of the system is found from the low amplitude, high frequency oscillation superimposed on the low frequency model motion. To deter-

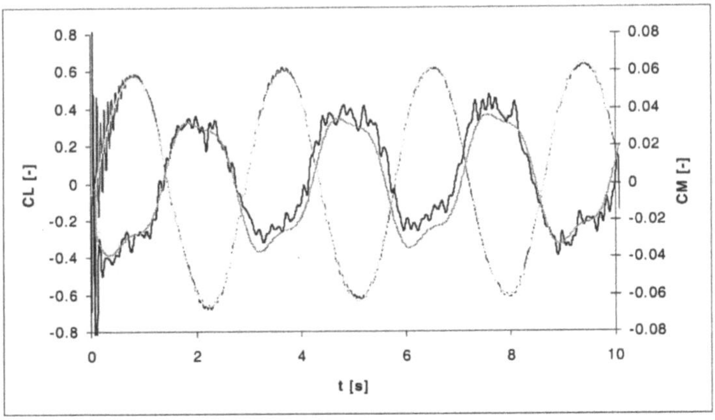

Figure 6 Typical example of experimental lift (smooth curve, $c_{Lmax} = 0.6$) and pitching moment coefficient for $\alpha_M = 0°$ and $k = 0.18$

mine the influence of inertia forces, measurements of the oscillating model have been performed in water at rest. From these, it was found that for the considered flow conditions inertia effects are negligible.

2.4 Experimental Conditions

The experiments in this study were performed with the test section conditions in Tab. 1. The kinematic parameters of the model pitching motion were varied according to Tab. 2. The Reynolds number was chosen to suppress the generation of turbulence while at the same time generating enough force and moment for the strain-gauge-balance to measure properly. Lower reduced frequencies are not possible in the experiments, as the minimum number of oscillations was set to 2 in the available time. The typical run time is 10 seconds. After opening the fast sliding valve of the tunnel, the flow starts accelerating until it reaches its nominal free stream value. This flow starting time is given in Tab. 1 as characteristic acceleration time. After this, the flow very slightly starts to decelerate due to the decreasing water level in the storage tank. This is characterized by the deceleration in Tab. 1, which is a quite small value.

Table 1 Experimental conditions in the test section

Description	Variable	Value
Free-stream velocity	u_∞	$0.3\,m/s$
Characteristic acceleration time	τ	$\sim 17\,ms$
Deceleration	a	$0.0032\,m/s^2$
Model Reynolds number	Re_c	16000

Table 2 Kinematic parameters of the model motion

Description	Variable	Value
Mean angle of attack	α_M	$0°$
Amplitude angle of attack	α_A	3 and 6°
Frequency	f	$0.25, 0.3, 0.35, 0.4, 0.45$ and $0.5\,Hz$
Reduced frequency	k	$0.14 - 0.28$

3 Numerical Techniques

3.1 The Navier-Stokes Solver

The modified flow code PP2DMOV solves the unsteady two-dimensional Navier-Stokes equations in an Arbitrary Lagrangian-Eulerian (ALE) formulation on a region Ω. It essentially is based on the discrete projection type solver PP2D from the FEATFLOW package, which was modified to simulate incompressible two-dimensional flows with moving boundaries in an efficient way. Here only a short description will be given of the PP2D solver. For a detailed description of the numerical background the reader is referred to [15, 16]. The incompressible Navier-Stokes equations can be given by:

$$\begin{aligned}
\frac{\partial u}{\partial x} + \frac{\partial v}{\partial y} &= 0 \\
\varrho\frac{\partial u}{\partial t} + \varrho\left(u\frac{\partial u}{\partial x} + v\frac{\partial u}{\partial y}\right) &= \varrho\nu\left(\frac{\partial^2 u}{\partial x^2} + \frac{\partial^2 u}{\partial y^2}\right) - \frac{\partial P}{\partial x} + \varrho f_x \qquad (3)\\
\varrho\frac{\partial v}{\partial t} + \varrho\left(u\frac{\partial v}{\partial x} + v\frac{\partial v}{\partial y}\right) &= \varrho\nu\left(\frac{\partial^2 v}{\partial x^2} + \frac{\partial^2 v}{\partial y^2}\right) - \frac{\partial P}{\partial y} + \varrho f_y .
\end{aligned}$$

Using $p = \frac{P}{\varrho}$, they can be written in a more compact form:

$$\mathbf{u}_t - \nu\Delta\mathbf{u} + \mathbf{u}\cdot\nabla\mathbf{u} + \nabla p = \mathbf{f}\,,\ \nabla\cdot\mathbf{u} = 0\,,\ \text{in}\ \Omega\times[0,T]\,. \qquad (4)$$

Equation 4 is solved by first introducing a time discretization, in this case the robust Fractional-step-θ-scheme which is of second order accuracy. In general form in each time step the following Quasi-Navier-Stokes problems are solved:

Given $\mathbf{u}^n$ and the time step $\Delta t = t_{n+1} - t_n$, then solve for $\mathbf{u} = \mathbf{u}^{n+1}$ and $p = p^{n+1}$

$$\frac{\mathbf{u} - \mathbf{u}^n}{\Delta t} + \theta\left[-\nu\Delta\mathbf{u} + \mathbf{u}\cdot\nabla\mathbf{u}\right] + \nabla p = \mathbf{g}^{n+1} \quad , \quad \nabla\cdot\mathbf{u} = 0, \tag{5}$$

with the right hand side

$$\mathbf{g}^{n+1} = \theta\mathbf{f}^{n+1} + (1-\theta)\,\mathbf{f}^n - (1-\theta)\left[-\nu\Delta\mathbf{u}^n + \mathbf{u}^n\cdot\nabla\mathbf{u}^n\right]. \tag{6}$$

The parameter θ is chosen according to the time step scheme used. Discretization in space is achieved by using finite elements. Starting from a variational formulation, the region Ω is divided by a finite mesh of quadrilaterals, where trial functions for velocity and pressure are defined. The quadrilateral used in PP2DMOV uses non-parametric piecewise rotated bilinear shape functions for the velocities spanned by $\langle x^2 - y^2, x, y, 1\rangle$ and piecewise constant pressure approximations. This gives us an element where the velocities are defined as mean nodal values on the element edges with a cell-centered nodal value for the pressure. In finite element terms this element pair $\tilde{Q}1/Q0$ is non-conforming. The non-linear 1-step projection solution process is accelerated with a multigrid technique. PP2D has successfully taken part in the DFG-Benchmark Computation within the DFG priority Research Program "Flow Simulation on High Performance Computers" [17].

To provide a little insight to the solution process commonly used in PP2D and hence in PP2DMOV, the nonlinear 1-step projection scheme will be presented in the following section. Again, for detailed information on the algorithms and approaches used, the reader is referred to [16].

Nonlinear projection scheme:

Given: Iterates $\mathbf{u}^{n-1} = \mathbf{u}(t_{n-1})$ and $p^{n-1} = p(t_{n-1})$ from time level $n-1$

Perform: 3 steps to obtain $\mathbf{u}^n$ and p^n on the new time level n

[1] Solve for an intermediate velocity $\tilde{\mathbf{u}}^n$ the corresponding nonlinear momentum equation. The right hand side has the load vector of time level t_n and the last pressure p^n,

[2] Solve a pressure-poisson problem for a right hand side f_p with the divergence from $\tilde{\mathbf{u}}$,

[3] Finally update the new velocity $\mathbf{u}^n$ and the new pressure p^n to satisfy the continuity equation.

PP2DMOV is basically just an extension of PP2D by introducing the mesh velocity in the convective term of the incompressible two-dimensional Navier-Stokes equations (see Eq. 7). In the literature this is referred to as an Arbitrary Lagrangian-Eulerian (ALE) formulation. Note that mesh velocities $\mathbf{u}_G$ do not appear in the continuity equation, as a pressure-poisson equation is solved to satisfy the continuity equation in an outer loop. Care has to be taken to satisfy the geometric conservation law (GCL), where the mesh velocity $\mathbf{u}_G$ must be equal to the movement of the mesh $\Delta\mathbf{x}$ during the time step k:

$$\mathbf{u}_t - \nu\Delta\mathbf{u} + (\mathbf{u} - \mathbf{u}_G) \cdot \nabla\mathbf{u} + \nabla p = \mathbf{f}\,,\ \nabla \cdot \mathbf{u} = 0\,,\ \text{in}\,\Omega \times [0,T]\,. \tag{7}$$

The following modifications were performed on PP2D:

- auxiliary routines to update the matrices in every sub time step according to the new position of mesh nodes,
- handling of meshes, such as interpolation and a local refinement process in the vicinity of boundaries,
- and changes in the upwinding scheme to satisfy the GCL.

No changes were made in algorithmic solution process. An update of the corresponding matrices according to the new mesh position is performed, allowing a very efficient time dependent handling of the mesh and the corresponding boundary conditions.

3.2 Moving Mesh

3.2.1 Mesh Interpolation

In order not to change the algorithmic approach of PP2D, the mesh is refined for the multigrid driver in every sub-step of the time scheme used. Therefore changes to the mesh are applied at coarse grid level only, before all other levels are created. This procedure is certainly not optimal in respect to the computational cost, but leaves many possibilities for moving the mesh and allows the direct integration of multigrid solvers. The method used in this study is very simple and uses blending functions to interpolate a mesh between two fixed positions, the original and the rigidly transformed mesh, in this case a rotation about the $1/4$ chord point of the airfoil for a given angle of attack. The blending functions are constructed as polynomials and use the closest distance to the boundaries as input. The method is derived from [18], where it has been successfully used for structured grids. Figures 7 and 8 show the original and the resulting rotated mesh for an angle of attack of 6°. Mesh deformations without cross-overs up to large angles of attack of 40° are possible using this method. The elements near to the airfoil are moved nearly rigidly with the airfoil in order to preserve the quality of the mesh in these regions.

3.2.2 Boundary Mesh Refinement

Normally, the mesh in the FEATFLOW solvers is refined for the multigrid driver with a common $h/2$ refinement, which means that a quadrilateral is refined to into

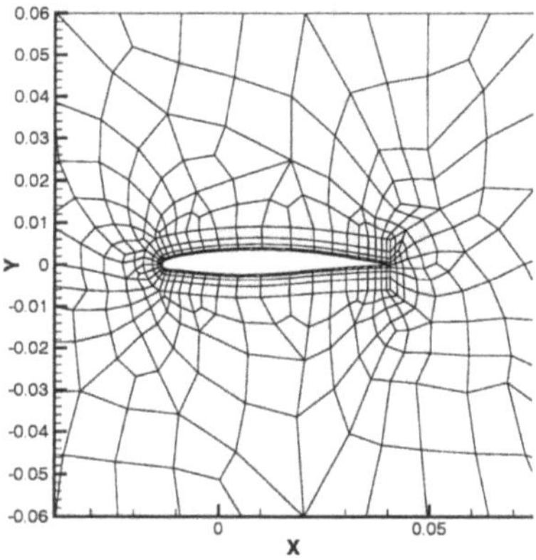

Figure 7 Coarse mesh at $\alpha = 0°$

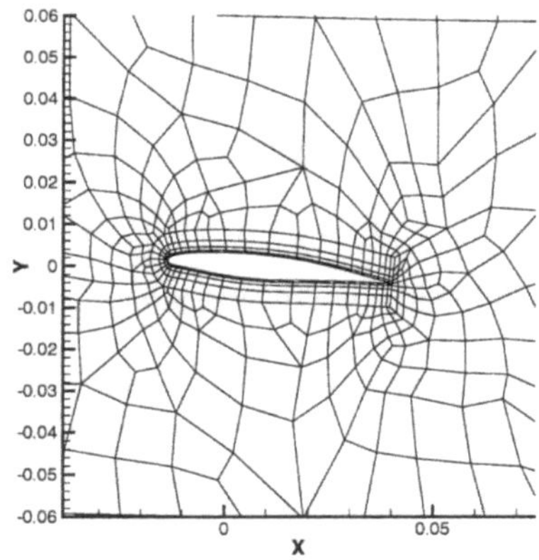

Figure 8 Coarse mesh at $\alpha = 6°$

four new elements using the midpoints of the edges and the cell-centered node. Using such a refinement with the interpolation method would result in a poor spatial resolution of the boundary layers. Therefore the refinement strategy was modified to resolve this problem. A coefficient, which can be chosen by the user, determines the position of the new points on the edges attached to boundaries creating the new four elements. Figure 9 illustrates the process. The spacing Δs between the node on the boundary and the first node in the flow can be given by:

$$\Delta s_{N_L} = c_B^{(N_L - N_{L_{MIN}})} \cdot \Delta s_{N_{L_{MIN}}}, \tag{8}$$

where c_B is a constant factor smaller than 0.5. Equation 8 can also be used to incorporate an adaptive refinement for each edge, where the distance Δs_{N_L} is given and c_B is calculated accordingly.

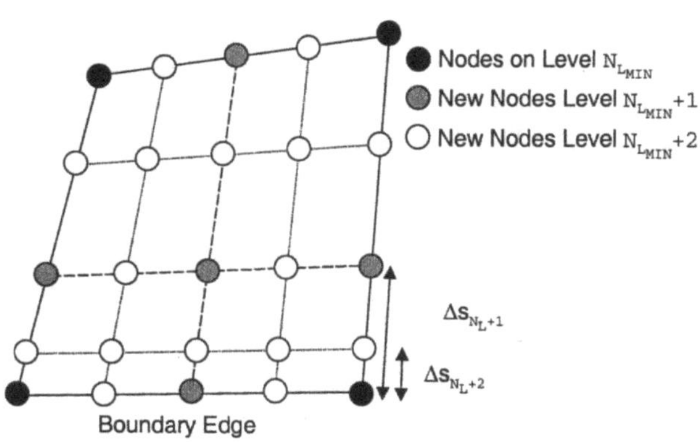

Figure 9 Boundary mesh refinement strategy to resolve boundary layers

3.2.3 GCL and the Upwind Scheme

Weighted Samarski upwinding is used for PP2DMOV, where the mesh velocities $\mathbf{u}_G$ are subtracted and calculated according to the nodal movement from the previous time step, thus satisfying the geometric conservation law automatically [19]:

$$\mathbf{u}_G = \frac{1}{\Delta t} \begin{pmatrix} x^{n+1} - x^n \\ y^{n+1} - y^n \end{pmatrix}. \tag{9}$$

Here Δt is the time step size and n denotes the time step number. The correct implementation was tested on a classic driven cavity problem with a moving mesh.

3.3 Boundary Conditions

The following boundary conditions are prescribed for the simulations concerning the Water-Ludwieg-Tunnel. At the inflow section, Dirichlet boundary conditions are applied, setting the velocity as a function of time:

$$u(t) = u_\infty \tanh\left(\frac{t}{\tau}\right) + at \ , \ v(t) = 0. \tag{10}$$

a is the deceleration factor, when the water level starts to drop in the storage tank of the Water-Ludwieg-Tunnel. There are no slip boundary conditions at the upper and lower tunnel wall (see Fig. 10). On the model boundary, the model rotation

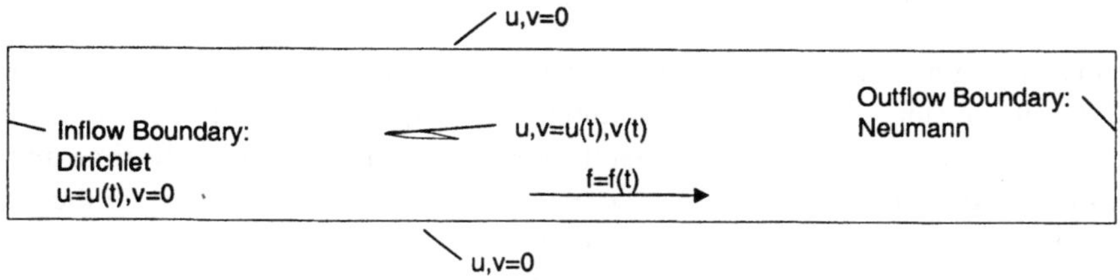

Figure 10 Boundary conditions applied to numerical simulations of Water-Ludwieg-Tunnel experiments

about a fixed point (x_M, y_M) is differentiated to give the velocities for the Dirichlet boundary condition given in Eq. 11. Thus if no motion is prescribed (i.e. fixed angle of attack), no slip boundary conditions apply to the model.

$$\begin{aligned} \alpha(t) &= \alpha_M + \alpha_A \sin(2\pi f(t - t_{osc})), \\ \dot{\alpha}(t) &= 2\pi f \alpha_A \cos\left(2\pi f(t - t_{osc})\right), \\ u(t) &= \dot{\alpha}(y(t) - y_M), \\ v(t) &= -\dot{\alpha}(x(t) - x_M). \end{aligned} \tag{11}$$

In order to reduce the computational effort, only a small part of the test section is covered by the computational domain (see Fig. 10). Therefore, the boundary conditions and external forces acting on the fluid are chosen in such a way that the real flow behaviour is simulated as it occurs during the experiment. To numerically simulate the time dependent flow velocity according to Eq. 10, a scaled force has to act on the fluid in the computational domain (see Eq. 7), which is simply the time derivative of the flow velocity given in Eq. 10.

$$f_x = \frac{u_\infty}{\tau}\left[1 - \left(\tanh\left(\frac{t}{\tau}\right)\right)^2\right] + a \ , \ f_y = 0. \tag{12}$$

3.4 Verification of the Numerical Techniques

The correct simulation of experiments with PP2DMOV is verified with experimental data. It must be emphasized that the numerical simulation is performed in a two-dimensional space whereas the experiments are carried out in three dimensions. Therefore the numerical simulation does not account for three-dimensional effects such as side wall interference in form of a horse-shoe vortex, corner vortices and many others. The small aspect ratio of the model ($\sim$ 1.4) certainly does not improve the situation. Despite these set-backs a good agreement between numerical and experimental data is found, sufficient to enhance experimental results and give a greater insight into the occurring flow phenomena.

3.4.1 BAC3-11 at $\alpha_M = 6°$, $\alpha_A = 0°$ and $f = 0\ Hz$

In this case of a static airfoil the time dependent flow phenomena are initialized by the impulsive start of the flow in the tunnel (see Eq. 10). In Figs. 11 and 12 the experimental and numerical aerodynamic coefficients are plotted versus time and a reasonable agreement can be observed, except for the dynamic effects in the first second. The trailing edge separation bubble leads to a break in the development of the aerodynamic lift. The breaking point in the form of a peak at about 0.5 s cannot be clearly detected in the experimental values. From one second onwards a good agreement is found, where three-dimensional effects can be clearly detected in the experimental lift signal as trailing edge vortex shedding is disturbed in comparison with numerical data. The mean steady aerodynamic lift and drag values agree reasonably. The mean pitching moment coefficient is slightly smaller, but as mentioned earlier pitching moment results at these conditions have to be taken with care as the order of magnitude is very small. Nevertheless the frequency of the vortex shedding process is equal in both cases. A comparison of instantaneous streamlines and flow visualization reveals a good overall agreement. Figures 13 and 14 show a good agreement of the flow phenomena encountered exactly at $t = 0.6\ s$. Especially the primary trailing edge vortex can be identified and is also situated in the same position relative to the airfoil. The boundary layer separation point on the upper side also coincides. In Figs. 15 and 16 a triple vortex structure can be observed

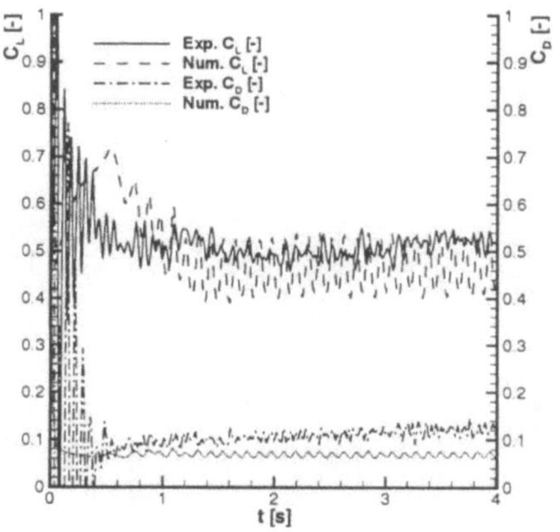

Figure 11 Lift and drag coefficients from experimental and numerical data for $\alpha_M = 6°$, $\alpha_A = 0°$ and $k = 0$

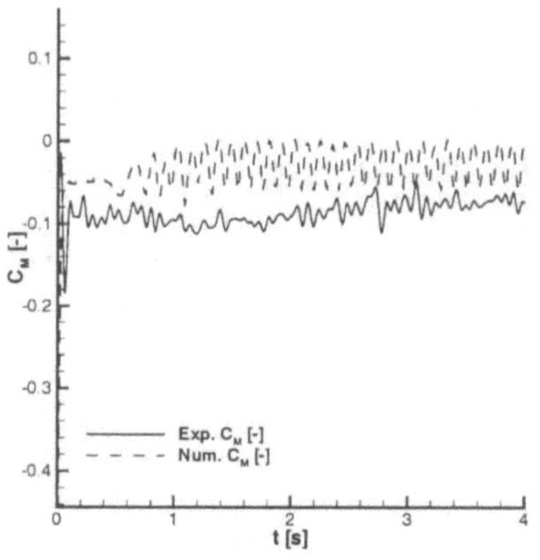

Figure 12 Pitching moment coefficient from experimental and numerical data for $\alpha_M = 6°$, $\alpha_A = 0°$ and $k = 0$

in both cases, revealing a very good agreement between experiment and numerical simulation. The laminar separation point has moved upstream in comparison with the flow at $t = 0.6\ s$. Comparing numerical results with PIV data also shows a good agreement (Figs. 17 and 18). Two vortices can be observed in the PIV and numerical data at nearly the same relative position to the airfoil.

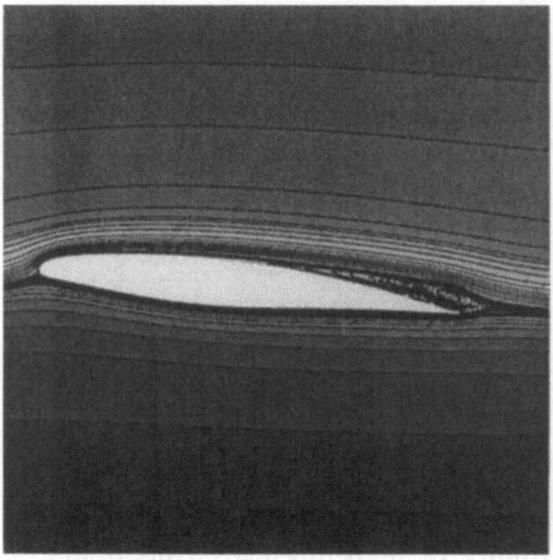

Figure 13 Numerical instantaneous streamlines at $\alpha = 6°$ and $t = 0.601\ s$ for $\alpha = 6°$ and $k = 0$

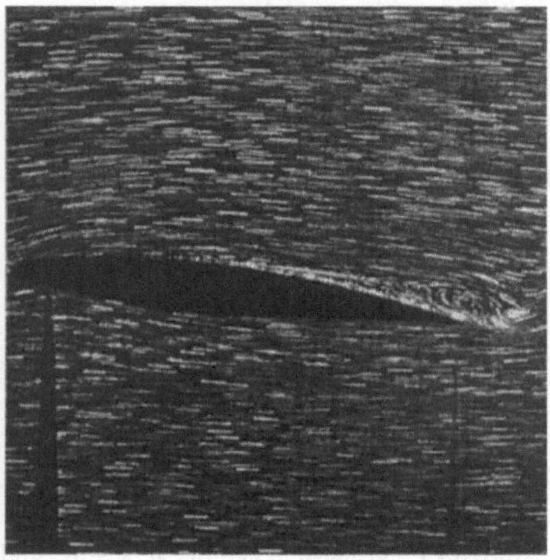

Figure 14 Experimental particle visualization at $\alpha = 6°$ and $t = 0.6\ s$ for $\alpha_M = 6°$, $\alpha_A = 0°$ and $k = 0$

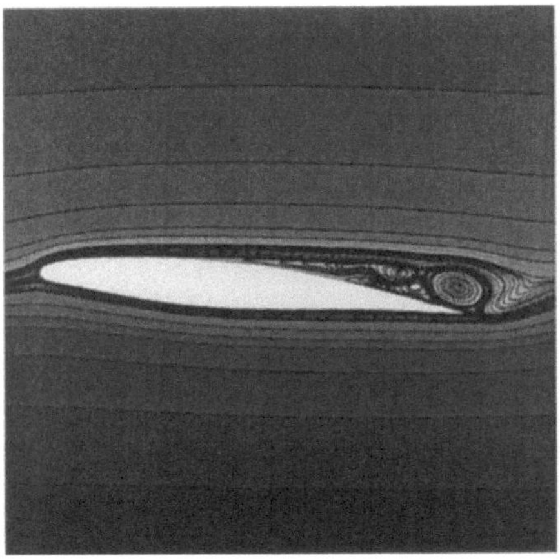

Figure 15 Numerical instantaneous streamlines at $\alpha = 6°$ and $t = 1.204$ s for $\alpha_M = 6°$, $\alpha_A = 0°$ and $k = 0$

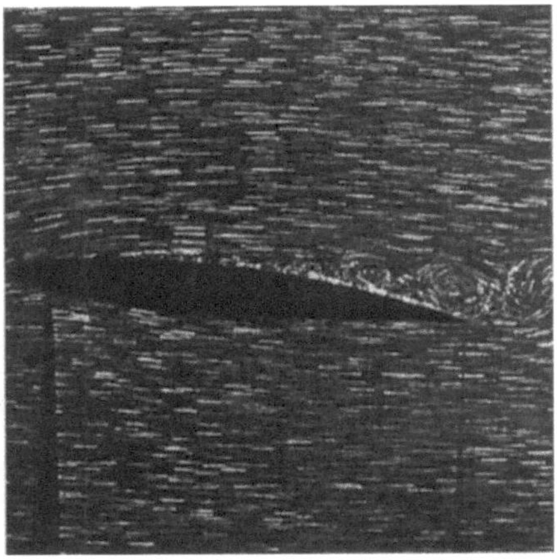

Figure 16 Experimental particle visualization at $\alpha = 6°$ and $t = 1.2$ s for $\alpha_M = 6°$, $\alpha_A = 0°$ and $k = 0$

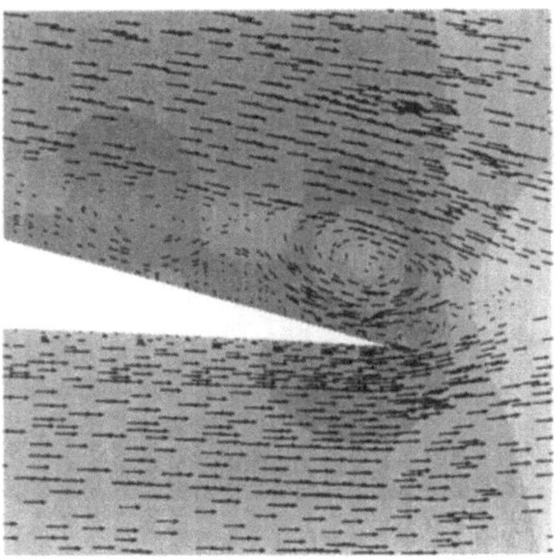

Figure 17 Numerical vectors at $\alpha = 6°$ and $t = 1.204$ s for $\alpha_M = 6°$, $\alpha_A = 0°$ and $k = 0$

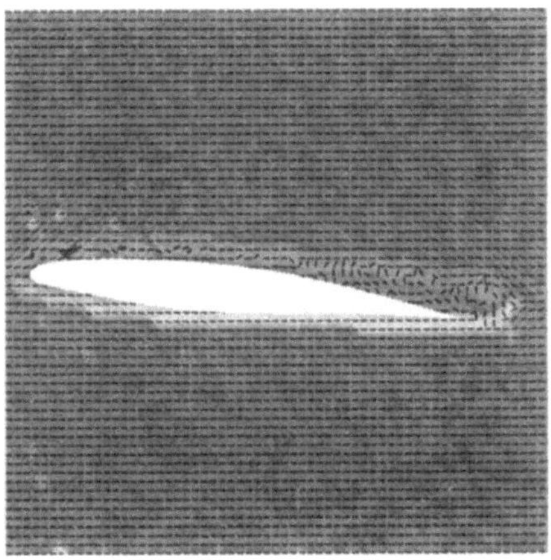

Figure 18 Experimental vectors from PIV at $\alpha = 6°$ and $t = 1.2$ s for $\alpha_M = 6°$, $\alpha_A = 0°$ and $k = 0$

3.4.2 BAC3-11 at $\alpha_M = 0°$, $\alpha_A = 6°$ and $f = 0.25\ Hz$

After discovering a good agreement at a fixed angle of attack, it is necessary to verify the agreement at a varying angle of attack. For this purpose the experiment with the lowest reduced frequency was chosen and it must be noted that the same agreement can be found for higher reduced frequencies. In Figs. 19 and 20 both numerical and experimental aerodynamic coefficients are plotted against time, revealing a reasonable agreement. The pitching moment coincides well, except for some peaks. The lift coefficient agrees, except for negative peak values of the angle of attack, where the vortices on the lower side (see Fig. 23) seem to create an extra peak in the negative lift and pitching moment. This is probably due to an inhomogeneous vortex field in spanwise direction for which the three-dimensional integration of the forces and the pitching moment by the strain-gauge-balance damps out the effect so that they are not detected in the experimental signal. The mean drag coefficient falls short of the experimental values, but has the same appearance in time. Figures 21 and 22 show a similar situation as described for the fixed angle of attack.

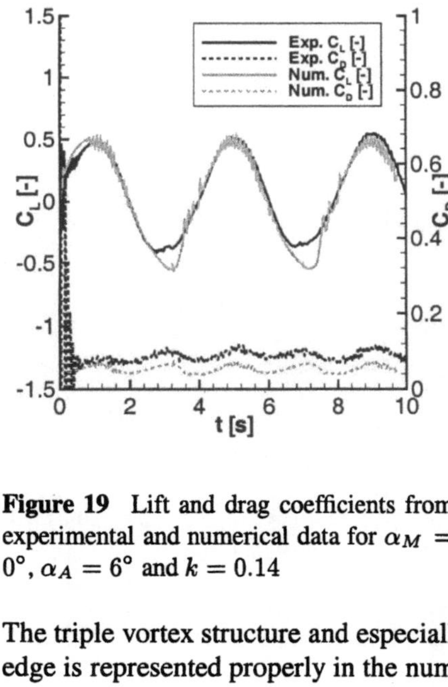

Figure 19 Lift and drag coefficients from experimental and numerical data for $\alpha_M = 0°$, $\alpha_A = 6°$ and $k = 0.14$

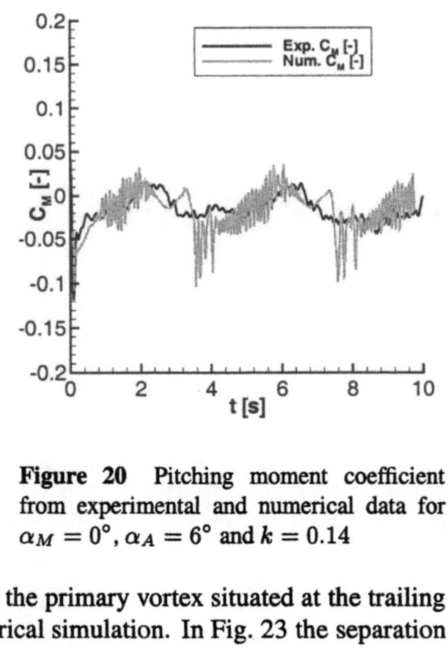

Figure 20 Pitching moment coefficient from experimental and numerical data for $\alpha_M = 0°$, $\alpha_A = 6°$ and $k = 0.14$

The triple vortex structure and especially the primary vortex situated at the trailing edge is represented properly in the numerical simulation. In Fig. 23 the separation bubbles at the leading and trailing edge can be identified, but their structure is not clearly visible in Fig. 24 due to the optical overlapping of the airfoil front edge. Nevertheless it is clear that separation occurs.

Figure 21 Numerical instantaneous streamlines at $\alpha = 5.9° \downarrow$ and $t = 5.208\ s$ for $\alpha_M = 0°$, $\alpha_A = 6°$ and $k = 0.14$

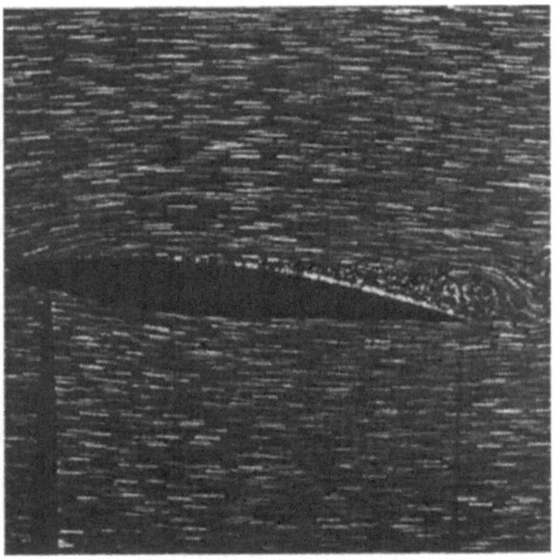

Figure 22 Experimental particle visualization at $\alpha = 5.9° \downarrow$ and $t = 5.201\ s$ for $\alpha_M = 0°$, $\alpha_A = 6°$ and $k = 0.14$

Figure 23 Numerical instantaneous streamlines at $\alpha = -5.8° \downarrow$ and $t = 6.911\ s$ for $\alpha_M = 0°$, $\alpha_A = 6°$ and $k = 0.14$

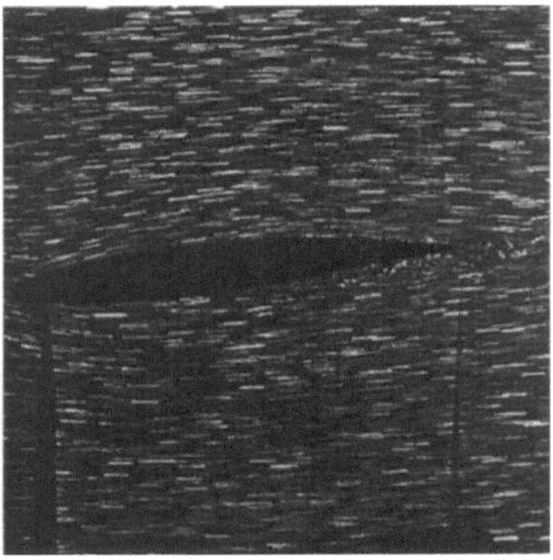

Figure 24 Experimental particle visualization at $\alpha = -5.8° \downarrow$ and $t = 6.901\ s$ for $\alpha_M = 0°$, $\alpha_A = 6°$ and $k = 0.14$

4 Results

4.1 Influence of the Reduced Frequency on the Aerodynamic Coefficients

As shown in the literature [22, 4, 21, 20] both the reduced frequency:

$$k = \frac{\pi f c}{u_\infty}, \tag{13}$$

and the non-dimensional pitching rate:

$$\alpha_+ = \frac{\dot{\alpha} c}{u_\infty} = \frac{2\pi f c \alpha_A}{u_\infty} \cos\left(2\pi f (t - t_{osc})\right) = 2k\alpha_A \cos\left(2\pi f (t - t_{osc})\right), \tag{14}$$

are the dominant parameters, which influence the unsteady aerodynamic coefficients of an oscillating airfoil. These influences are to be discussed in the reduced frequency range of 0 to 0.28.

The following figures show the influence of the reduced frequency on the aerodynamic coefficients at an instantaneous pitching rate of $\alpha_+ = 0$, which corresponds to the peak values at $\alpha_A = \alpha = 6°$. In Fig. 25 no significant influence of the reduced frequency can be detected for the range of reduced frequencies investigated. The lift stays constant at approximately the static value. This behaviour can be observed for both experimental series, as well as for the numerical data, which coincides well. The variation in the static lift coefficient for $\alpha_A = 6°$ is of experimental nature and is attributed to three-dimensional unsteady effects in the test section. The linear thin airfoil theory according to Theodorsen [22] predicts a decrease of the lift coefficient with increasing reduced frequeny. This is contrary to the experimental and numerical results shown in Fig. 25. This different behaviour is due to the influence of viscous effects, which also influence the trailing edge separation observed for this airfoil. Due to the relatively low Reynolds number viscous effects play an important role which are not considered in the thin airfoil theory. The numerical and experimental absolute pitching moment for $\alpha_A = 3°$ shows at first an increase and then slight decrease with greater reduced frequencies (see Fig. 26). For $\alpha_A = 6°$ the absolute experimental values show a gradual increase with frequency from $k = 0.14$ onwards, while the numerical data seems to reach a constant value. Experimental and numerical data is equal in this case for $k = 0.28$. The tendency, that for higher reduced frequency the agreement between experiment and numerical simluation becomes better, is caused by the fact that with increasing frequency the extent of separation regions becomes smaller because there is less time to establish a full separation. In this case numerical simulation yields more accurate results, also, due to the smaller separation, three-dimensional effects in the test section are less dominant. This is also the reason why in general for $\alpha_A = 3°$ a better agreement is found as for $\alpha_A = 6°$. It must be emphasized that the pitching moment is very sensitive and is strongly influenced by the reduced frequency, far more than the lift or drag coefficients.

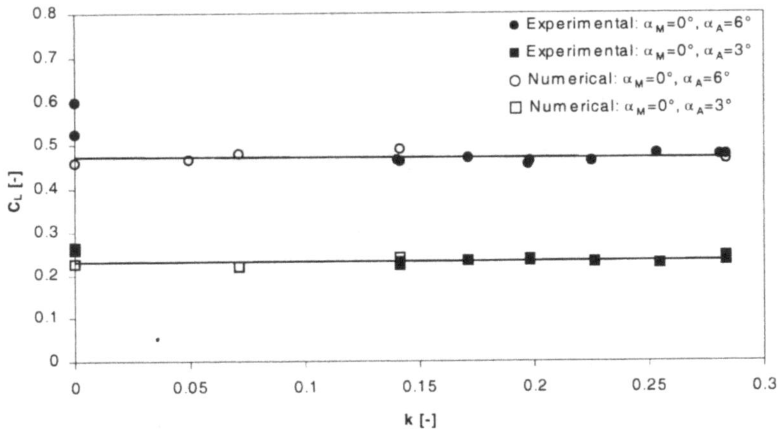

Figure 25 Numerical and experimental lift coefficient at $\alpha_A = 3°$ and $6°$ as a function of the reduced frequency

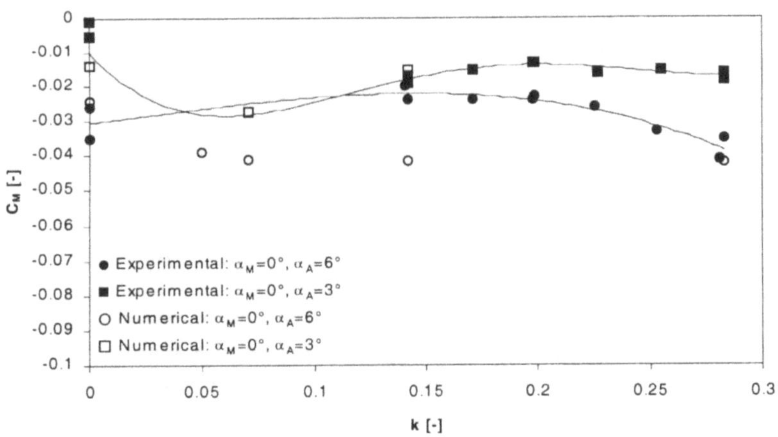

Figure 26 Numerical and experimental pitching moment coefficient at $\alpha_A = 3°$ and $6°$ as a function of the reduced frequency

4.2 Influence of the Pitching Rate on the Aerodynamic Coefficients

Figures 27 and 28 show the influence of the instantaneous positive pitching rate at $\alpha = 3°$ for $\alpha_A = 6°$ in comparison with the peak values for the same angle of attack, hence $\alpha_A = 3°$ and $\alpha_+ = 0$. For the lift and the pitching moment coefficient, dynamic effects are apparent. The lift is increased in comparison to the $\alpha_+ = 0$ values. Here the experimental data shows a greater gain in lift than the numerical data. The tendency with increasing pitching rate is equal in both data sets. For the pitching moment the results are not quite clear. For greater reduced frequencies $k > 0.14$ the absolute value is slightly increased, while the numerical data for $k = 0.07$ is smaller than the corresponding value at $\alpha_A = 3°$. Overall a positive non-dimensional pitching rate increases the lift, which corresponds to results presented by other researchers [20, 21] for large angles of attacks.

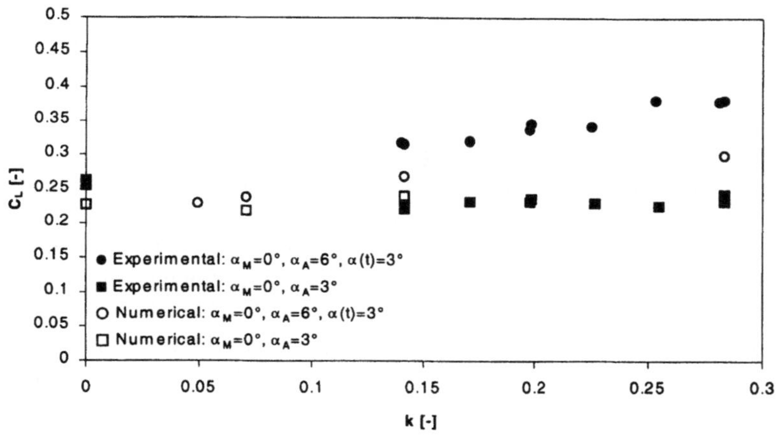

Figure 27 Influence of the non-dimensional pitching rate at $\alpha_+ = \sqrt{3}k\alpha_A$ for $\alpha_A = 6°$ and $\alpha_+ = 0$ for $\alpha_A = 3°$ on the lift coefficient

4.3 Hysteresis of the Lift and Pitching Moment Coefficient Hysteresis Loops and Phase Angles

Figures 30 to 33 show the lift and pitching moment coefficients plotted against the instantaneous angle of attack. With increasing reduced frequency the phase difference between the aerodynamic coefficients and the airfoil motion becomes apparent. In Fig. 30 the lift values show no hysteresis for large positive angles of attack. The lift travels along a more or less straight line, even though trailing edge separation occurs. For a negative angle of attack a strong influence of the leading and trailing edge separation occurring on the lower side of the airfoil can be seen (see Fig. 24), creating large fluctuations in the lift coefficient. The separation grows once the air-

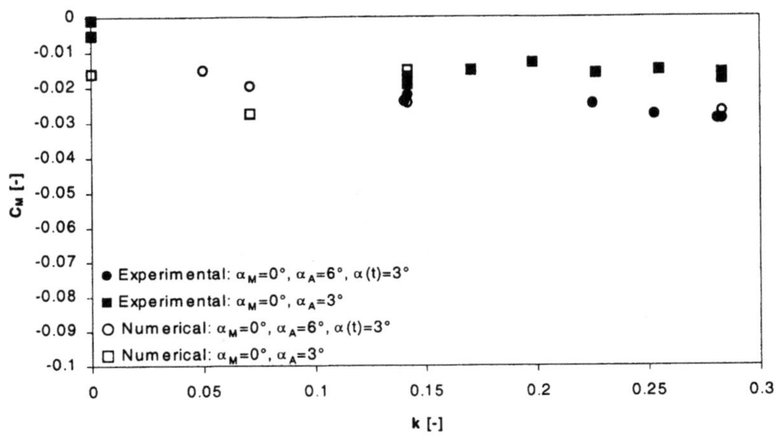

Figure 28 Influence of the non-dimensional pitching rate at $\alpha_+ = \sqrt{3}k\alpha_A$ for $\alpha_A = 6°$ and $\alpha_+ = 0$ for $\alpha_A = 3°$ on the pitching moment coefficient

foil pitches up and vanishes when a positive angle of attack of about $\alpha = 3°$ is reached. In contrast in Fig. 32 for higher reduced frequencies a hysteresis can be observed from $k = 0.14$ onwards. In general, hysteresis effects increased with increasing frequency. The effect is also evident in the pitching moment in Figs. 31 and 33. For an angle of attack between $\alpha = 0°$ and $-6°$ fluctuations are observed for $k < 0.28$ and a peak value in the range of $\alpha = 0°$ to $-2°$ for $k = 0.14$ and 0.28. The fluctuations disappear for $k = 0.28$, where the pitching moment decreases gradually from the value at $\alpha = -6°$ to a peak at $-2°$. Vortex shedding on the lower side of the airfoil is suppressed in this case. At this reduced frequency $k = 0.28$, steady separation phenomena are strongly influenced by the motion of the airfoil, in the sense that for the oscillating airfoil the separation is smaller than for the airfoil at rest, because for the oscillating case the characteristic flow time is much shorter than the time needed for generating a fully established separation. Caused by the highly dynamical behaviour of the flow obvious in Figs. 30 to 32 it is hard to determine phase angles from the numerical as well as experimental data. Nevertheless, the data has been analysed by a technqiue which determines a fitting curve for the hysteresis loops [23] allowing to determine phase angles. The result of this is shown in Fig. 29. It has to be mentioned that due to the fitting process of the hysteresis loops the quantitative values of the phase angles are not absolutely correct, nevertheless the most dominant tendencies are correctly described. The phase angle for the lift is positive and increases with increasing frequency, whereas the phase angle of the pitching moment is negative, indicating a stable behaviour of the airfoil. This is also found in the hysteresis loops which for the pitching moment do not show significant parts of clockwise sense of direction.

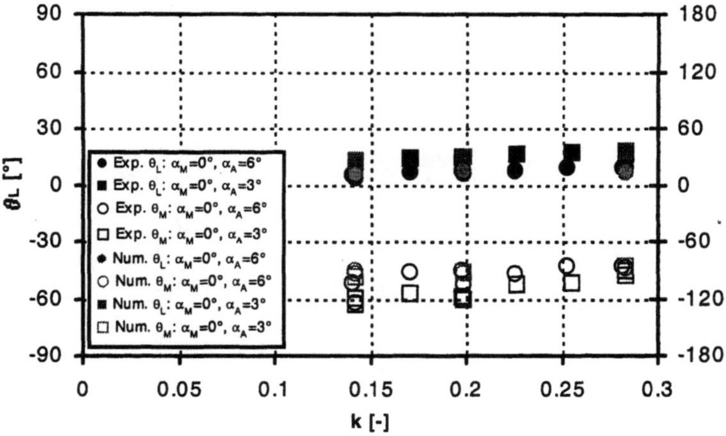

Figure 29 Experimental and numerical phase angle for lift Θ_L and pitching moment Θ_M versus reduced frequency k

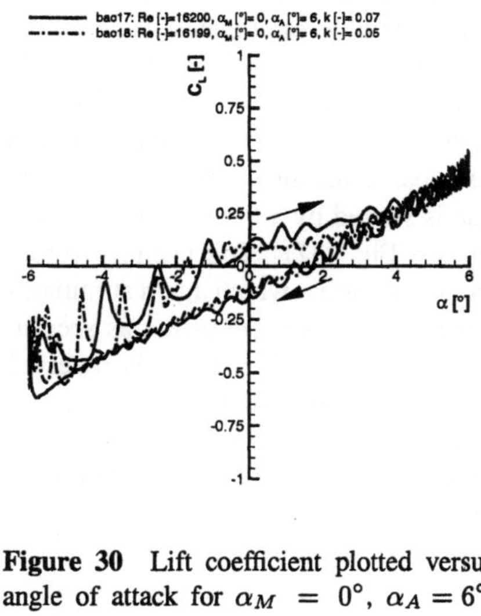

Figure 30 Lift coefficient plotted versus angle of attack for $\alpha_M = 0°$, $\alpha_A = 6°$, $k = 0.05$ and $k = 0.07$

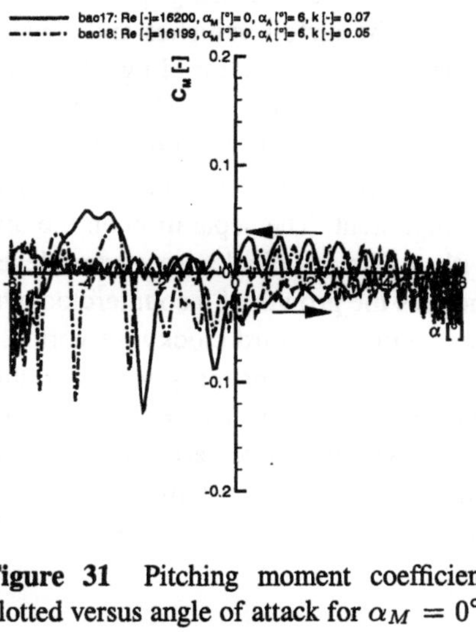

Figure 31 Pitching moment coefficient plotted versus angle of attack for $\alpha_M = 0°$, $\alpha_A = 6°$, $k = 0.05$ and $k = 0.07$

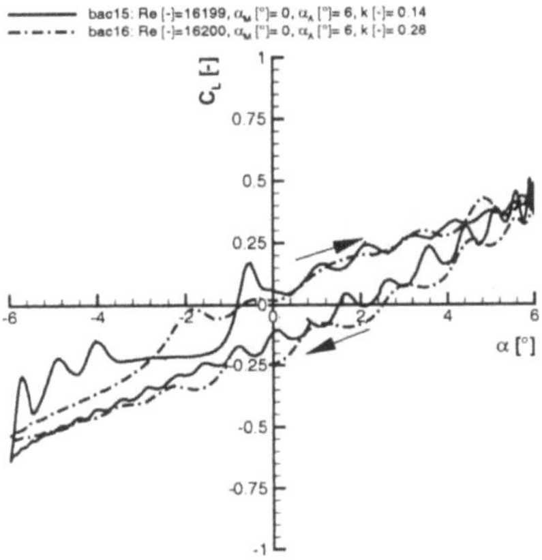

Figure 32 Lift coefficient plotted versus angle of attack for $\alpha_M = 0°$, $\alpha_A = 6°$, $k = 0.14$ and $k = 0.28$

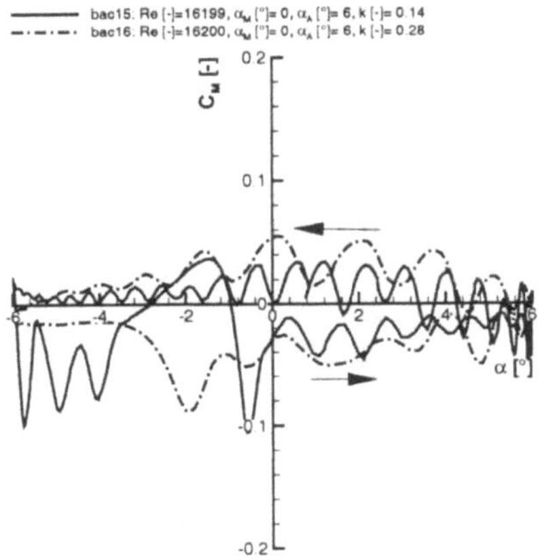

Figure 33 Pitching moment coefficient plotted versus angle of attack for $\alpha_M = 0°$, $\alpha_A = 6°$, $k = 0.14$ and $k = 0.28$

4.4 Trailing Edge Separation on the Pitching BAC3-11/RES/30/21 Airfoil

The Figs. 34 to 39 present the evolution of the flow in the pitching up phase of one oscillation cycle. During the pitching up manoeuver, the separation on the lower side of the airfoil vanishes and a new separation bubble establishes itself on the upper side near to the trailing edge. The effects of the lower side separation bubble were shown in Sec. 4.3. Unfortunately the experimental data in Fig. 34 does not show the lower side separation bubble seen in the numerical data. The reason is that the front profile edges of the airfoil cover areas near to the airfoil due to the optical arrangement. The separation on the upper side is started by a flow reversal region, with a primary clockwise vortex developing (see Fig. 37). The primary vortex induces the growth of a counterclockwise secondary vortex, which again stimulates the creation of third clockwise vortex. This triple vortex system grows as seen in Fig. 39 until the primary vortex is shed into the wake. This process is repeated periodically. It must be emphasized that the same process takes place at a static angle, where vortices are shed at about a frequency of 10 Hz into the wake (see Fig. 16). For the oscillating airfoil the separation is usually smaller. This becomes obvious comparing Figs. 16 and 38 or 15 and 39.

5 Conclusions

In the context of this project both experimental and numerical tools have been developed, which allow the study of flow fields around oscillating airfoils for various

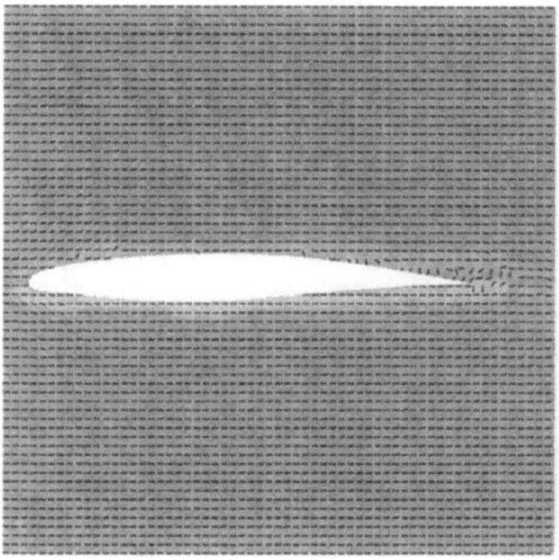

Figure 34 Experimental velocity vectors and vorticity for $\alpha_M = 0°$, $\alpha_A = 6°$ and $k = 0.14$ at $\alpha = 0°$ and $t = 4.1\ s$

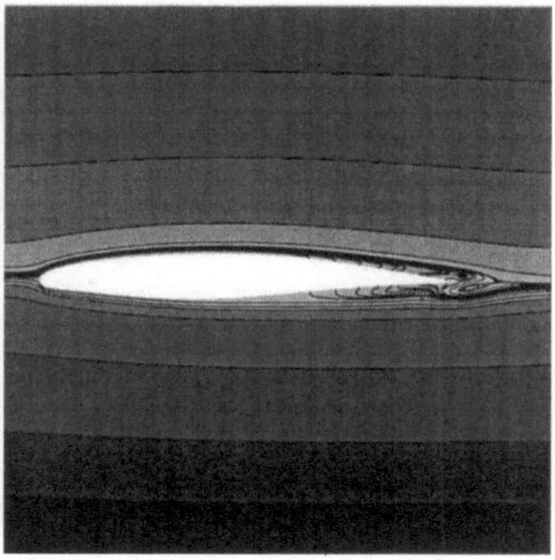

Figure 35 Numerical streamlines for $\alpha_M = 0°$, $\alpha_A = 6°$ and $k = 0.14$ at $\alpha = 0.2°$ and $t = 4.108\ s$

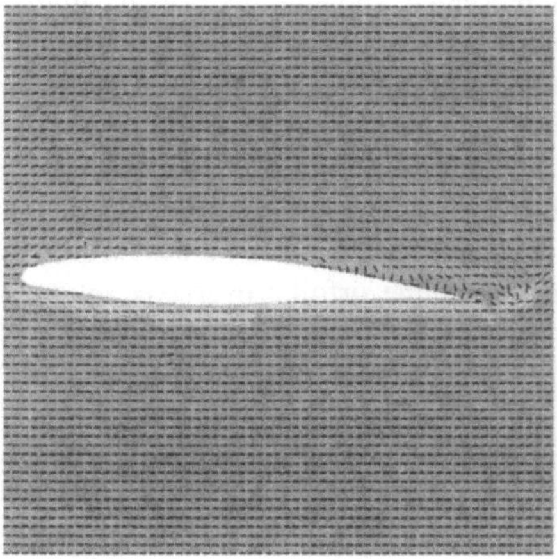

Figure 36 Experimental velocity vectors and vorticity for $\alpha_M = 0°$, $\alpha_A = 6°$ and $k = 0.14$ at $\alpha = 2.7°$ and $t = 4.4\ s$

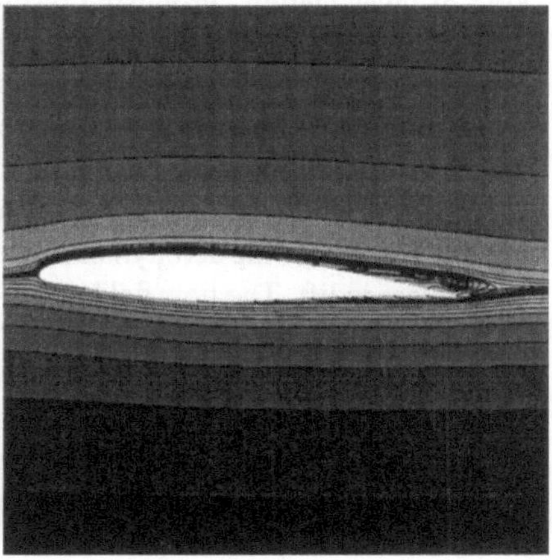

Figure 37 Numerical streamlines for $\alpha_M = 0°$, $\alpha_A = 6°$ and $k = 0.14$ at $\alpha = 2.9°$ and $t = 4.408\ s$

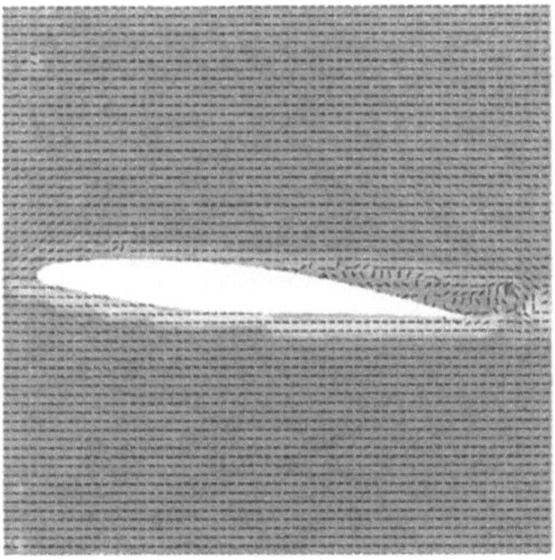

Figure 38 Experimental velocity vectors and vorticity for $\alpha_M = 0°$, $\alpha_A = 6°$ and $k = 0.14$ at $\alpha = 5.9°$ and $t = 5\ s$

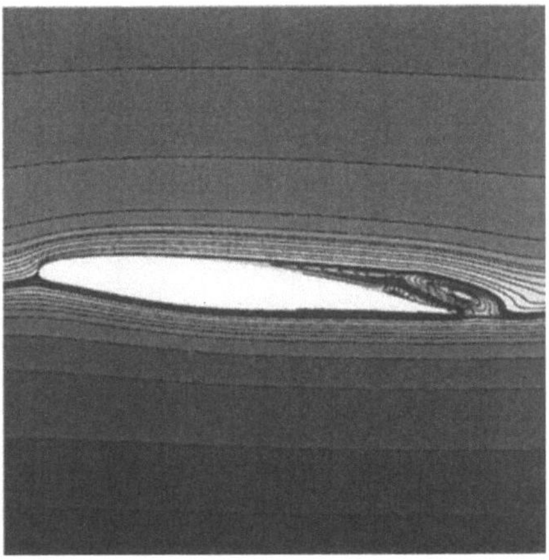

Figure 39 Numerical streamlines for $\alpha_M = 0°$, $\alpha_A = 6°$ and $k = 0.14$ at $\alpha = 6°$ and $t = 5\ s$

kinematic and geometric parameters in an acceptable amount of time. The numerical tool has been successfully verified with experiments and allows a time efficient computation of cases which cannot be run in the experimental facility and thus completes the experimental data very well. This combination is very promising for this kind of investigation. Promising results are presented, showing the primary influence of two well known parameters, the reduced frequency and the non-dimensional pitching rate, for two amplitude angles of attack of $\alpha_A = 3°, 6°$ and for a relevant range of reduced frequencies for the reference airfoil BAC3-11/RES/30/21. An increasing reduced frequency has little effect on the peak values of the aerodynamic coefficients, while the flow field is gradually affected. The non-dimensional pitching rate influences the aerodynamic coefficients, which for positive values results in an increase in lift. The flow field due to the motion of the airfoil is overlayed by separation regions of the type encountered at a static angle of attack and are influenced by the unsteady motion. Further investigation is required in order to complete the study for many variations of the kinematic and geometric parameters. Viscous effects can have a strong influence on the dynamic behaviour, which is not described by the classical thin airfoil theory based on potential flow. Under certain conditions, these viscous effects in combination with separation can be destabilizing for an oscillating airfoil.

References

[1] BUBLITZ, P.: *Geschichte der Entwicklung der Aeroelastik in Deutschland von den Anfängen bis 1945.* DFVLR-Mitteilung 86-15, (1986).

[2] BENNET, R. M.; EDWARDS, J. W.: *An Overview of Recent Developments in Computational Aeroelasticity.* AIAA Paper 1998-2421, 29th Fluid Dynamics Conference, Albuquerque, NM (1998).

[3] KUMAR, A.; HEFNER, J.: *Future Challenges and Opportunities in Aerodynamics.* ICAS 2000-0.2.1, ICAS 2000 Congress (2000).

[4] MCCROSKEY, W. J.: *Unsteady Airfoils.* Ann. Rev. Fluid Mechanics, Vol. 14, pp. 285-311, (1982).

[5] AKAMATSU, T.: *Application of shock tube technology to studies of hydrodynamics.* Shock Tube and Shock Wave Research (ed. Ahlborn, Hertzberg, Russel), University of Washington Press, Seattle and London, (1978).

[6] MOIR, I.R.: *Measurements on a Two-Dimensional Airfoil with High-Lift Devices.* AGARD-AR-303, A2.1-2.12, (1994).

[7] WERNERT, P.; GEISSLER, W.; RAFFEL, M.; KOMPENHANS, J.: *Experimental and numerical investigations of dynamic stall on a pitching airfoil.* AIAA Journal, Vol. 34, No. 5, p. 982-989, (1996).

[8] RAFFEL, M.; WILLERT, C.; KOMPENHANS, J.: *Particle Image Velocimetry: A practical guide.* Springer-Verlag, (1998).

[9] WERNERT, P.; FAVIER, D.: *Considerations about the phase averaging method with application to ELDV and PIV measurements over pitching airfoils.* Experiments in Fluids 27, p. 473-483, Springer-Verlag (1999).

[10] WERNERT, P.; KOERBER, G.; WIETRICH, F. ET AL.: *Demonstration by PIV of the Non-Reproduceability of the Flow Field Around an Airfoil Pitching Under Deep Dynamic Stall Conditions and Consequences Thereof.* Aerospace Science and Technology, Vol. 2, p. 125-135, (1997).

[11] ADRIAN, A. J.: *Particle-Image Techniques for Experimental Fluid Mechanics.* Annual Review of Fluid Mechnics, Vol. 23, p. 261-304(1991).

[12] WEIGAND, A.: *Review of some Novel Measurement Techniques for Fluid Dynamics Studies.* DGLR Jahrestagung, Göttingen, Germany (1993).

[13] WILLERT, C. E.; GHARIB, M.: *Digital particle image velocimetry.* Experiment in Fluids 10, p. 181-193, (1991).

[14] SHAND, A. M.: *The Investigation, Development and Optimisation of Global Laser Diagnostics for Combustion and Related Flow Applications.* PhD Thesis, University of Warwick, (1996).

[15] TUREK, S.; BECKER, C.: *FEATFLOW. Finite element software for the incompressible Navier-Stokes equations. User Manual. Release 1.1.* Technical Report, University of Heidelberg, Germany (1998).

[16] TUREK, S.: *Efficient solvers for incompressible flow problems: An algorithmic approach in view of computational aspects.* LNCSE 6, Springer Verlag (1999).

[17] SCHÄFER, M; TUREK, S.: *Benchmark Computations of Laminar Flow Around a Cylinder.* In: Flow Simulation with High-Performance Computers II, Ed. Hirschel, E. H., NNFM Vol. 52, Vieweg Verlag (1996).

[18] HESSE, M.; BRITTEN, G.; BALLMANN, J.: *A Multi-Block Grid Deformation Algorithm for Aeroelastic Analysis.* B.K. Soni, J. Häuser, J.F. Thomson, P. Eiseman (eds.), Numerical Grid Generation in Computational Field Simulations, Proc. 7th Int. Conf., pp. 161-170 (2000).

[19] WALL, W. A.: *Fluid structure interaction with stabilized finite elements.* Institute of Structural Mechanics, University of Stuttgart, PhD Thesis (1999).

[20] OHMI, K.; COUTANCEAU, M.; LOC, T. ET AL.: *Vortex formation around an oscillating and translating airfoil at large incidences.* J. Fluid Mechanics, Vol. 211, pp. 37-60 (1990).

[21] WALKER, J. M.; HELIN, H. E.; STRICKLAND, J. H.: *An Experimental Investigation of an Airfoil Undergoing Large-Amplitude Pitching Motions.* AIAA Journal, Vol. 23, No. 8, pp. 1141-1142 (1985).

[22] THEODORSEN, T.: *General theory of aerodynamic instability and the mechanism of flutter.* NACA Report 496 (1935).

[23] BERTRAN, X.: *Separation Effects on the Unsteady Aerodynamics of Oscillating Airfoils at Low Speeds.* Doctoral Thesis, Stoßwellenlabor, Aachen University (2002).

Experiments on Transonic Aerodynamics about Elastically Suspended Airfoils

C. Hillenherms, W. Schröder, W. Limberg
Aerodynamisches Institut der RWTH Aachen
Wüllnerstraße zw. 5 und 7, D-52062 Aachen, Germany

Summary

This paper describes the experimental investigation of the interaction of aerodynamic and structural forces on a 2–D rectangular wing section oscillating in pitch at transonic flow conditions. The setup consists of a symmetric elastic suspension on both sides of the rigid wing providing the pitching motion as self-sustained or decaying oscillations with different initial conditions. Unsteady forces and surface pressures are determined simultaneously on the wing model oscillating about mean incidence angles of zero to two degrees at Mach numbers 0.50 to 0.85, corresponding to Reynolds numbers of 1.5 to 2.2 million. The wing has a supercritical BAC 3-11/RES/30/21 airfoil with a relative thickness of 11% [1]. Besides the free-stream conditions, structural dynamic parameters like the stiffness of the elastic suspension are varied to change the eigenfrequency of the mechanical system and its influence on the aerodynamic derivatives. Preliminary investigations include measurements of the steady pressure distribution and flow visualizations at transonic Mach numbers for a fixed mounted wing model with the same airfoil. The experimental results of both, steady and unsteady investigations, serve to validate numerical methods such as the fully-coupled direct simulation of the fluid-structure interaction [2].

1 Introduction

Since airplane payloads continue to rise the design requires an extended use of lightweight structures. For this reason the reliable prediction of stability boundaries is one of the most essential tasks in the aircraft design process. The investigation of fluid–structure interaction therefore is highly important, since phenomena like flutter and limit cycle oscillations of airplane wings are affecting the stability of an aircraft escpecially in the transonic flow regime, where non-linear effects become predominant. As a consequence, efficient design tools have to be developed that are based on numerical simulations of the fluid–structure interaction. The experiments that are carried out in the frame of this project will help to understand more of the physics and to build a data base for numerical validation.

The fluid–structure interaction is investigated with a rigid wing section whose suspension provides the elastical properties for the torsional degree of freedom while the airfoil model represents the aerodynamic similarity. This is current practice in experimental investigations of aeroelastic phenomena which are generated by the interaction of aerodynamic, elastic and inertial forces [3, 4, 5, 6, 7, 8, 9].

2 Experimental setup

The mechanical system of the oscillating wing section with the interacting aerodynamic, elastic, and inertia forces and moments and one degree of freedom is sketched in Figure 1. Figure 2 shows schematic and real views of the setup from one side of the wind tunnel test section. A two–component force balance is mounted on both sides of the airfoil model to measure the unsteady forces in the direction of the airfoil chord and perpendicular to it. The torsional degree of freedom is provided by torsion springs on both sides of the airfoil model, hence defining the position of the elastic axis, which is located at quarter chord. Only the airfoil itself is subject to the flow while all parts of the suspension are mounted outside of the test section. For the experiments, these outer parts are sealed with a cover on each side because the test section side panels provide air gaps for the airfoil oscillations. For a fast pressure compensation inside the covers during the test two big bypass tubes connect the covers to the wind tunnel upper ceiling. For decaying oscillations, the airfoil model is deflected against the torsion springs to an initial angle of attack and released when the tunnel flow is steady.

2.1 Wind tunnel

The wind tunnel tests are performed in a trisonic intermittent suction tunnel with a test section of 0.4 m × 0.4 m, and a Mach number range from subsonic to supersonic (Figure 3). Prior to a tunnel run the four vacuum tanks are evacuated by a compressor down to approximately 100 hPa. The air is pumped into the reservoir balloon at the tunnel intake where it is held at ambient pressure. The flow is started by opening a fast acting valve between the tunnel and the vacuum vessels. The volume of the reservoir allows test periods of 2 to 5 seconds. Subsonic Mach numbers are regulated by changing the cross section of the convergent–divergent diffusor downstream of the test section. The Laval nozzle is adjusted to control supersonic Mach numbers, and the diffusor can then be used to regain static pressure. Due to its operational concept, the total pressure and total temperature are always equal to the ambient conditions in the laboratory. As a consequence, the chord Reynolds number varies between 1.5 and 2.2 millions for a profile of 150 mm chord length at Mach numbers of 0.50 to 0.85.
For transonic testing a second test section with twodimensional adaptive walls can

be used to minimize wall interferences [10]. The top and bottom walls are flexible and either one can be vertically adjusted by 24 stepping motors. The adaptation method is based on the linearised theory for compressible flows. It is assumed that the flow near the walls slightly differs from a parallel flow.
During a test run the wall pressure distributions are measured by 25 pressure taps along the centerlines of each of the two walls. The disturbance velocities arising from the model–wall–interactions can be simply evaluated from the measured wall pressures and the wall angles which are derived from the wall displacement coordinates [11]. The wall interference is computed using Cauchy's integral formula which allows the determination of the wall induced interference velocities within the test section as well as on the walls. The integral is taken along the path C, i.e., along the walls and across the intake and outlet sections. To be more precise, the wall interference velocity reads

$$w_{int}(z) = \frac{1}{2}w(z) + \frac{1}{2\pi i}\oint_C \frac{w(\zeta)}{z-\zeta}d\zeta \qquad (1)$$

with the disturbance velocity on the wall

$$w(\zeta) = \beta\, U(\zeta,\eta) - i\, V(\zeta,\eta) \qquad (2)$$

and the interference velocity at any point within the integration boundary

$$w(z) = \beta\, U_{int}(x,y) - i\, V_{int}(x,y) \qquad (3)$$

and the Prandtl–Glauert transformation

$$\beta = \sqrt{1-M_\infty^2}\,. \qquad (4)$$

The necessary change of the wall angle $\delta(\theta)$ to obtain interference–free flow then is equal to the opposite wall induced interference velocity V_{int} and hence the wall displacement $\delta(y)$ can be determined in one step.

$$\delta(y) = \int \theta dx = \int \frac{(V - V_{int})}{U_\infty} dx\,. \qquad (5)$$

It was shown that steady adaptation makes sense even for unsteady experiments [12]. For the assessment of unsteady wall pressures, 2×13 miniature pressure sensors are mounted next to the centerlines of the top and bottom walls.

2.2 Flow visualisation

The Schlieren technique is based on the dependency of the refractive index of a transparent medium on its density. This leads to a local deflection of a light beam traversing the medium from its undisturbed path. By cutting off the refracted part of the light it is possible to produce an image of the disturbances due to the change of the refractive index. This method is used particularly in compressible flows to visualise shock and expansion waves as well as flow separations and vortices. Figure 4 shows the basic arrangement for the color Schlieren technique. This method was applied to study the appearance and the location of shocks and eventual separation regions. The use of a circular colored filter mask as light source furnishes additional information –qualitative in nature– on the strength and the direction of the observed density gradients [13].

2.3 Airfoil models

In the present investigation two wind tunnel models are applied for steady and unsteady measurements, respectively. Both are rectangular wing sections with BAC 3-11 airfoil geometry and have a 150 mm chord and a 400 mm span. The first model consists of an aluminium core with 48 integrated pressure tubes to measure the surface pressures on both sides of the model. It is cast using Araldite to reach the final shape. The pressure taps are located on two diagonal lines in the mid section of the upper and lower side.

In a first test series large sets of pressure distributions and schlieren photographs were recorded with this model at Mach numbers of 0.50 up to 0.78 and angles of attack from -2° to 5°. Those steady results form a basis for the design of the setup for unsteady measurements and the numerical validation.

The second model is made of carbon fiber reinforced plastics in order to keep its mass and hence its moment of inertia as low as possible. This is important to determine the aerodynamic forces and moments because only the total amount of aerodynamic and inertial forces and moments can be measured during oscillation and the latter have to be subtracted afterwards. They have to be measured in a discrete wind-off, still-air test but under the same remaining conditions. The upper and lower side of the model have been manufactured separately, equipped with 11 miniaturized in-situ pressure sensors as well as two glass fiber reinforced spars and the flanges, and then glued together (Figures 5 and 6).

2.4 Pressure measurement technique

For the preliminary investigations where steady pressure distributions with the first wind tunnel model are taken a data acquisition system for up to 64 pressure and 16 analog inputs is used (HyScan®2000). It consists of pressure scanning modules supplied with differential pressure sensors whose signals are multiplexed and read by the A/D–coupling device of the system. The 48 surface pressures and the static pressure in the test section are sampled with 5 kHz. The Mach number is determined from the static pressure in the test section and the ambient pressure. Afterwards a period of steady flow is chosen out of the static pressure time record for further evaluation of all signals of the same run.

As mentioned above, the wind tunnel model for the unsteady measurements is provided with piezo-resistive pressure sensors in order to prevent damping and a low frequency response of the oscillatory pressure signals due to long tubings. The cylindrical sensors are mounted on the inner side of the upper and lower model shells (Figure 6). The fixation and the sealing of the sensors is achieved using special fittings, with a very small cavity in front of the sensor head and a connection to the surface pressure taps of 0.3 mm diameter. Only the sensors at the leading edge and at the most aft position had to be connected to the surface pressure taps by longer tubes of about 20 mm, because of the restricted space in the model.

2.5 Force measurement technique

The unsteady forces are measured by a strain gauge balance made out of maraging steel which is particularly suited for the construction of strain gauge balances due to its extremely high yield strength [14]. It consists of two pairs of circular panels mounted on both sides of the airfoil model (Figure 7). One panel of each pair is to measure the forces in the chordwise direction, whereas the other panel measures the forces perpendicular to the model chord. The panels have four bending beams equipped with strain gauges to register the beam deformation caused by an acting force. The principle of the force measurement is sketched in Figure 8, showing the theoretical balance reactions to the different stress modes. On every beam the strain gauges are assigned to a full Wheatstone bridge, so that the relationship between the strain gauge signals $\varepsilon_{1\ldots4}$ and the bridge signal U_A reads

$$\frac{U_A}{U_B} = \frac{k}{4}(\varepsilon_1 - \varepsilon_2 + \varepsilon_3 - \varepsilon_4)$$

with U_B denoting the bridge supply voltage. The first case represents the design mode yielding a maximum signal:

$$\varepsilon = \varepsilon_1 = \varepsilon_3 = -\varepsilon_2 = -\varepsilon_4 \quad \Rightarrow \quad \frac{U_A}{U_B} = k\,\varepsilon\,.$$

The second case produces only tension (l) or compression (r) in the beams so that the resulting bridge signal is zero:

$$\begin{aligned} \varepsilon &= \varepsilon_{1,l} = \varepsilon_{2,l} = \varepsilon_{3,l} = \varepsilon_{4,l} \\ &= -\varepsilon_{1,r} = -\varepsilon_{2,r} = -\varepsilon_{3,r} = -\varepsilon_{4,r} \quad \Rightarrow \quad \frac{U_A}{U_B} = 0\,. \end{aligned}$$

The last case shows the torsion mode which also generates a zero bridge signal:

$$\sum_{i=1,A}^{4} \varepsilon = \sum_{i=1,C}^{4} \varepsilon < 0; \quad \sum_{i=1,B}^{4} \varepsilon = \sum_{i=1,D}^{4} \varepsilon > 0 \quad \Rightarrow \quad \frac{U_A}{U_B} \simeq 0\,.$$

The calibration has confirmed that the balance design provides minimum interference of the different stress modes and minimum sensitivity to torsion.

2.6 Elastic suspension

The pitching motion of the airfoil model is enabled by two torsional springs on both sides of the airfoil. It is connected to the outer balance panel and fixed mounted on the other end. It is a two-part design with a central circular rod embedded in a cross spring, that allows a variation in stiffness just by changing the cylindrical rod. The contribution of the four cross spring leafs to the torsional stiffness is much less than that of the central rod, but they bear a relatively high bending stiffness instead. Figure 9 shows the deformations resulting from a finite element simulation of the spring under torsional stress. The relatively strong deflection of the four leaf springs illustrates their softness against torsion whereas the central torsion rod shows only a little deformation due to its higher twisting resistance.

2.7 Measurement of the angle of attack

A rocker is fixed to the airfoil outside of the test section between the balance and the torsion spring to transmit the pitch oscillation to two Laser triangulation sensors (Figure 2). The Laser beams are directed on the tails of the rocker so that their upward and downward displacements, respectively, are registered as changes in the distance that can be easily translated to the angle of attack. This optical method is applied because it does not produce any friction. Furthermore, this arrangement also allows the measurement of a heave motion. Oscillations in this degree of freedom will be studied in a next step of this project.

2.8 Data acquisition

The data is sampled simultaneously by a system of five data acquisition boards installed on the passive backplane of a personal computer. The synchronized boards allow maximum sampling of all 40 channels at 1.25 MHz with 12 bit resolution. The pressure signals are low-pass filtered and amplified by DC amplifiers, whereas the strain gauge signals are processed by a carrier frequency amplifier.

3 Definitions and parameters

Table 1 gives an overview of the significant aerodynamic parameters and coefficients while Table 2 lists the mechanical properties of the setup with the notation of the variables indicated in Figure 1. It should be noted that the values for the center of gravity x_{CG} and the static unbalance S_α are partly calculated. All other values are measured.

The still air natural frequencies and damping for the 10 mm torsion rod were kindly determined by the *Institut f"ur Leichtbau*. The first two pitch and plunge modes

Table 1 Aerodynamic parameters

Mach number	M
Reynolds number	$Re = \frac{u_\infty c}{\nu}$
Freestream velocity	u_∞
Density, undisturbed fluid	ϱ_∞
Dynamic pressure	$q = \varrho_\infty / 2\, u_\infty^2$
Pressure coefficient	$c_p = \frac{p - p_\infty}{q}$
Pitch angular frequency	ω_α
Reduced frequency, based on chord length c	$\omega^* = \frac{\omega_\alpha c}{u_\infty}$
Pitching angle (+ nose up)	α
Lift (+ upward)	F_L
Pitching moment (+ nose up)	M_α
Lift coefficient	$c_l = \frac{1}{q\, c\, b} F_L$
Moment coefficient	$c_m = \frac{1}{q\, c^2\, b} M_\alpha$
Mass ratio airfoil/fluid	$\mu = \frac{m}{\pi\, \varrho_\infty\, b\, c^2}$

Table 2 Mechanical parameters

Airfoil chord	c	150	mm
Airfoil span	b	400	mm
Airfoil mass	m_M	0.935	kg
Total mass	m	6.5	kg
Elastic axis location from c/4-point	x_{el}	0	mm
Center of gravity from c/4-point	x_{CG}	4.3	mm
Static unbalance about elastic axis	S_α	0.025	$kg\, m$
Torsion spring stiffness	$K_{\alpha,10mm}$	970	$\frac{Nm}{rad}$
	$K_{\alpha,8mm}$	570	$\frac{Nm}{rad}$
Pitching frequency	$f_{\alpha,10mm,1}$	55.66	Hz
	$f_{\alpha,10mm,2}$	716.79	Hz
	$f_{\alpha,8mm,1}$	42.80	Hz
Damping	$D_{\alpha,10mm}$	0.684	%
	$D_{\alpha,8mm}$	0.35	%
Plunge frequency (due to model bending)	$f_{h,10mm,1}$	129.88	Hz
	$f_{h,10mm,2}$	268.55	Hz

could be identified by this test. The amplitudes of the plunge oscillations were about one order of magnitude smaller than those of the pitching motion.

4 Results and discussion

4.1 Steady measurements

Figure 10 gives an overview of the steady data base. Pressure distributions were measured for angles of attack of -2° to 5° (7°) and Mach numbers of 0.5 to 0.9 in the two different test sections and with free and forced transition, respectively. Schlieren photographs and Shadowgraphs were taken at the same time for the transonic speed range.
The black (blue, red with dashed line) symbols show measurements with free transition while the grey (green, red with solid lines) symbols pertain to measurements with forced transition. The two vertical lines connect the lift maxima for the different angles of attack and thus evidence the decreasing lift divergence Mach numbers with rising angles.

In the test section with parallel walls the Mach number could not be enhanced above 0.78 due to tunnel blockage, which is why some curves (blue and green lines) do not extend beyond that Mach number. Using the adaptive test section, however, it was possible to adjust Mach numbers up to 0.9. Besides the extended Mach number envelope it can be stated that the maximum lift coefficients and the corresponding Mach numbers are shifted towards higher values for all measured angles of attack. For example, at 5% fixed transition and an angle of attack of 2° a lift maximum of 0.56 is found at M = 0.71 with parallel walls, whereas a maximum lift coefficient of 0.60 is found at M = 0.75 with adapted walls. For an angle of attack of 5° a direct comparison is basically impossible because of the lack of data for fixed transition and parallel walls. Additionally, the wall adaptation in this case is not able to totally eliminate wall interference effects due to the expanded supersonic region on the upper side of the model.

After a first test series with free transition it was decided to carry out a second series with forced transition. The fact that the boundary layer development is very sensitive to the Reynolds number, which is at least one order of magnitude lower in the experiments than for a full scale wing, leads to a wider area of laminar flow on the wind tunnel model compared to free-flight. In the experiments with free transition the change from laminar to turbulent flow mostly occurs in the region of the shock. Since the boundary layer state is of great importance for shock–boundary layer interactions and the resulting pressure distribution, a forcing of the transition should allow to study these phenomena in a physically similar manner to the full scale situation. Furthermore, a specification of the transition location often is required or at least desirable for the comparison with numerical results.

Two positions of the forced transition have been tested, 12% and 5% airfoil chord on the upper and lower side. A zig-zag aluminium tape of 90 μm thickness was used as transition strip. The 5% position seemed to be more favourable because the disturbance of the pressure distribution was obviously weaker.

It can be seen in Figure 10 that the lift coefficient decreases for the tripped airfoil, i.e. with an increasing area of turbulent flow on the airfoil. For a better understanding of the effects of the transition fixing pressure distributions and Schlieren photographs for free resp. fixed transition at a Mach number of 0.74 and an angle of attack of 0° are contrasted in Figure 11 (NOTE: *The dark round spot in the shock region is not a flow phenomenon but a flaw in the glass*). The drop of the lift coefficient in the case of forced transition can be explained with the upstream shift of the shock and the associated rise in pressure as can be seen in Figure 11. A reason for this shift can be found in the larger turbulent displacement thickness δ_1 which causes the supersonic region to move upstream.

Figure 11 also clearly shows how the interaction of a shock wave with a laminar boundary layer differs from the interaction with a turbulent boundary layer. The upper pressure distribution (free transition) exhibits a flattening upstream of the shock indicating a separated boundary layer. The Schlieren picture also shows the flow separation ahead of the shock and a typical λ-shock. The shear layers in the wake of the airfoil appear to be quite thick and indicate a trailing edge separation on the upper side. At fixed transition, however, the pressure distribution displays a sharp peak at the position of the shock caused by the smaller width of the boundary-layer interaction region. A strong suction peak on the lower side of the airfoil can be recognised in both pressure distributions. This is due to a change of curvature in the airfoil geometry at this position and leads to a local separation bubble in most cases.

Figure 12 depicts the lift- and moment-curve slopes for Mach numbers of 0.72, 0.75, and 0.80 with fixed transition (5 %), determined from steady pressure distributions measured in the adaptive test section. The lift slopes for Ma = 0.72 and 0.75 show a relatively constant rise up to $\alpha = 5°$ whereas the lift slope for Ma = 0.80 is already nonlinear. This corresponds to the trend of the lift curves in Figure 10 where the Mach numbers corresponding to the lift maxima are decreasing at a larger angle of attack. The behavior of the lift slope for Ma = 0.80 can be explained by the increasing flow separation on the upper side (s. also Figures 14, 15, and 16). The moment coefficient about the leading edge of the airfoil $c_{m,0}$ is similar for Ma = 0.72 and Ma = 0.80, but shows a lower value for Ma = 0.75 at $\alpha = 5°$. This is also true for the moment coefficient related to the c/4-point $c_{m,1/4}$ where a minimum value is reached for Ma = 0.75 at $\alpha = 5°$. In other words, the distribution of the moment coefficient shows that at Ma = 0.72 and Ma = 0.80 the c/4-point can be considered the aerodynamic centre. The moment slopes for Ma = 0.72 and Ma = 0.80 show slight differences only for higher angles of attack ($\alpha = 4°$ and 5°). The angle of attack for zero lift is found at about –1.55° (Ma = 0.72) and –1.30° (Ma = 0.80), respectively, with an incidental moment coefficient c_m^0 of –0.06 and –0.08, respectively.

The symbols in Figure 13 show two experimental results both at a Mach number 0.70 and an angle of attack of 0° in the non-adapted test section. The black findings are obtained with transition fixing and are compared with a fully turbulent FLOWer calculation (*by Gerd Britten, Lehr- und Forschungsgebiet für Mechanik der RWTH Aachen*) that takes into account the wind tunnel walls, whereas the grey symbols are results with free transition compared to a TFS Navier-Stokes solution (*by Ehab Fares, AIA*), that takes into account the wind tunnel walls and the transition location. TFS [15, 16] is a method of solution developed and successfully used at the AIA for predicting a wide variety of internal and external flow problems. The dashed line is a TFS Navier-Stokes calculation at free-flight conditions. The differences between the three cases not only underline the necessity of wall adaptation but also the necessity of transition fixing for wind tunnel experiments.

Figures 14, 15, and 16 show pressure distributions at three different Mach numbers, Ma = 0.75, Ma = 0.77, and Ma = 0.80, and angles of attack, $\alpha = 0°$, 1°, and 2°. Besides the expanding supersonic area on the upper surface and the increasing shock strength with rising Mach numbers and angles of attack, an attached flow in the vicinity of the trailing edge can be observed at Ma = 0.75 and Ma = 0.77, $\alpha < 2°$. At Ma = 0.77 and $\alpha = 2°$, however, a divergence of the pressure distribution in the rear part of the airfoil appears for the first time indicating a separated flow on the upper side. This leads to the assumption that the flow around this profile is of type B2 according to the classification of Pearcey [17], i.e., there exists a separation at the trailing edge besides the shock-induced separation bubble, and both types of separation more or less interact with each other depending on Mach number and incidence.

A further steady test series was conducted with the experimental setup described in section 2 but without the torsion springs to test the measurement technique and to determine the wall contours of the adaptive wind tunnel walls under steady conditions.

Figure 17 shows results of the lift coefficients for angles of attack of 0° and 2° measured with the balance in comparison with the results from the pressure measurements for $\alpha = 0°$, 1°, and 2°. The lift forces measured by the force balance are up to 50 % lower than the integrated values from the pressure measurements. The curve slopes for $\alpha = 0°$ are nearly parallel until the maximum is reached, then the balance yields a more shallow decrease of the lift coefficient. For $\alpha = 2°$ the discrepancy between both results is rising with the Mach number until the maximum is reached, then the negative gradient of the lift coefficient for Mach numbers above the maximum is also much weaker for the balance measurements than for the pressure measurements.

These remarkable differences can be partly explained by the fact that the lift coefficients calculated from pressure distributions are based on pointwise measurements in the mid-section of the airfoil and therefore represent a local information in a region where one can assume a two-dimensional flow field. In contrast to this, the force balance measures the reactions produced by the whole airfoil, thus also including all three-dimensional effects arising from the side walls. Due to the relatively unfavourable span to chord ratio of only 2.67 these effects have a strong influence on the global reaction forces measured by the balance. The extent of the three-dimensional wind tunnel side wall interference area on the airfoil model can be estimated by the oilflow pattern in Figure 18.

The root mean square (RMS) values of the lift coefficient are displayed in Figure 19 for the same Mach numbers and angles of attack. They are indicating increasing fluctuations up to Mach numbers of 0.72 for an angle of attack of 0° and 0.69 for 2°, respectively. At these Mach numbers, the shock is developing on the suction side of the airfoil, generating shock–boundary layer interaction such as shock induced boundary layer separation and thus, higher fluctuation levels. Further downstream, the RMS values possess another local maximum at Mach numbers of 0.76 and 0.75, respectively, which is followed by a sharp decline for both angles of attack.

4.2 Unsteady measurements

Unsteady forces and surface pressures have been measured simultaneously on the wing model oscillating about mean incidence angles of -0.5 to 2.3 degrees at Mach numbers between 0.50 and 0.85. The torsional stiffness has been varied by changing the diameter of the central torsion rods from 10 mm to 8 mm, representing torsional stiffnesses of 970 Nm/rad, and 570 Nm/rad, respectively (s. Table 2). Transition was always fixed at 5% chord length on the upper and lower side.

For any reading point, the airloads are obtained by subtracting the measured forces and moments of a wind-off test under the same conditions to eliminate the inertia forces. The measured time series are evaluated to determine the unsteady pressure, lift and moment coefficients.

The initial amplitudes of the angle of attack decreased in every measured case, i.e., the pitch oscillations were always damped and the system remained stable. In some cases the amplitudes reached higher values after the decaying process. This may also be caused by a yet unknown non-linear behaviour of the mechanical system.

Figure 20 shows time series of the pressure fluctuations on the upper and lower side of the airfoil during the decaying oscillation for a Mach number of 0.74. The static angle of attack (the angle of attack in still air) was 2.0°, resulting in a mean angle of attack of 1.64 after an initial deflection of about +1.0°. The reduced frequency ω^* was 0.217. From the analysis of the steady state flow presented in Figures 14, 15, and 16 it can be concluded that under these circumstances the flow is attached.

While the pressure coefficients at the first three positions on the upper surface are fluctuating mainly with the first harmonic frequency of the pitch oscillation $f_\alpha = 56.15$ Hz, when the pitch amplitudes are higher, the other signals show higher frequency portions (Figure 21). A spectral analysis of the signals by means of the discrete Fourier transform yields a dominant frequency of 56.15 Hz for the first sensor positions which are oscillating out of phase with the angle of attack. The sensors in the rear part of the upper side show the largest power spectral densities at frequencies of 238 Hz but have also higher magnitudes at the pitch frequency. This is also true for the signals from the sensors on the lower side that are oscillating in phase with the angle of attack. The power spectra of the lift and the moment coefficients also indicate a higher magnitude at 56.15 Hz.

Time series of the fluctuations of the angle of attack $\Delta\alpha = \alpha(t) - \bar{\alpha}$, and the lift and moment coefficients $\Delta c_{l,m} = c_{l,m}(t) - \bar{c}_{l,m}$ are plotted in Figure 22. The phase angle between the lift coefficient and the angle of attack is nearly zero (Figure 23) for the pitch frequency, whereas the phase angle between the moment coefficient and the angle of attack is 79°, i.e., the pitch motion follows the moment by nearly a quarter of a period.

The torsional stiffness was reduced for the next test series by exchanging the central cylindrical rods. For a comparison, the same procedures were repeated for two static angles of attack of 0.0° and 2.5°, and for Mach numbers between 0.50 and 0.80. Due to the lower torsional stiffness, the still air pitching frequency decreased from 55.66 Hz to 42.80 Hz.

Figure 24 shows time series of the pressure fluctuations on the upper and lower side of the airfoil during the decaying oscillation for a Mach number of 0.74. The static angle of attack was 2.5°, resulting in a mean angle of attack of 1.89 after an initial deflection of about +1.0°. The reduced frequency ω^* was 0.165.

Time series of the fluctuations of the angle of attack $\Delta\alpha = \alpha(t) - \bar{\alpha}$, the lift and moment coefficients $\Delta c_{l,m} = c_{l,m}(t) - \bar{c}_{l,m}$ are plotted in Figure 25. The moment coefficient is again ahead of the angle of attack. The cross spectral analysis (not displayed) yields 97° in this case which is larger than with the thicker spring.

Figure 26 shows the change in the pitch frequency with the Mach number for the 10 mm torsion spring. The errorbars indicate the frequency selectivity as the ratio of the sampling rate and the number of evaluated points resulting from the discrete Fourier transform. It can be seen, that the frequency shows no evident trend for a static angle of attack of 0°, whereas it is increasing for the static angles of attack of 2.0° and 2.5°. (One has to keep in mind that the mean angle of attack of the pitch oscillation varies with the Mach number and is always lower than the static angle of attack.) The relationship is similar for the 8 mm torsion spring, where the pitch frequency for the static angle of attack of 2.5° is growing for Mach numbers greater than 0.77 (figure 28).

The damping coefficient, plotted in figure 27 with respect to the Mach number for the 10 mm torsion spring, seems to scatter with a slight tendency to increase for static angles of attack of 0° and 2.0°, but is rising for Mach numbers up to 0.80 and then shows a tendency to decrease for the static angle of attack of 2.5°. A more definite behaviour can be observed in Figure 29, where the damping coefficient reaches maximum values at a Mach number of 0.75 for both static angles of attack, and is decreasing for higher Mach numbers. This leads to the assumption of an unstable region when the Mach number is further increased.

5 Conclusion

The experimental setup for the measurement of unsteady pressures and forces on a rectangular wing section oscillating in pitch was described. The results of a thorough study of the steady state flow including pressure distributions and flow visualisation for transonic Mach numbers and a range of moderate angles of attack were presented and discussed. Free and forced transition were compared with emphasis on the different shock–boundary layer interaction effects. Based on the understanding of the steady data a detailed analysis of the unsteady flow field was conducted. Decaying pitch oscillations with different static angles of attack at transonic Mach numbers and a variation of mechanical parameters were investigated and presented with emphasis on the aerodynamic stability. It was found that all of the measured points have been stable or near a limit cycle pitch oscillation. The influence of mechanical non-linearities on the stability of the system remains to be studied. The gained experimental results are a data base for the validation of numerical methods. Numerical simulations were used to isolate effects that are only present in experimental testing. Besides the results presented here, the experimental data will undergo further evaluation, e.g., to study local effects like shock–boundary layer interaction and separation.

6 Remarks

Thanks to the appreciable cooperation with DLR Göttingen, Institut für Aeroelastik. Lots of technical questions could be solved during the design process of the described experimental setup. As a consequence there exist certain similarities to their setup [5, 19].

References

[1] Moir, I., Measurements on a two-dimensional aerofoil with high-lift devices. AGARD-AR-303, DRA, Farnborough (1994).

[2] Britten, G. and Ballmann, J., Navier-Stokes-based direct Numerical Aeroelastic Simulation. In: *Proc. 22nd Congress of the International Council of the Aeronautical Sciences*, Harrogate (UK), 27 Aug. - 1 Sept. 2000. ICAS-2000-4.8.3.

[3] Abel, I., *Research and Applications in Aeroelasticity and Structural Dynamics at the NASA LAngley Research Center*. NASA TM-112852 (1997).

[4] Farmer, M. G., A Two-Degree-Of-Freedom Flutter Mount System with Low Damping for Testing Rigid Wings at Different Angles of Attack. Report nr., NASA TM 83302 (1982).

[5] G. Schewe and H. Deyhle. Experiments on Transonic Flutter of a Two-Dimensional Supercritical Wing with Emphasis on the Non-Linear Effects. In *Proceedings of the Royal Aeronautical Society Conference on "Unsteady Aerodynamics"*, 17-18 July 1996.

[6] Rivera, J. A., Dansberry, B. E., Durham, M. H., Bennett, R. M. and Silva, W. A., Pressure measurements on a rectangular wing with a NACA0012 airfoil during conventional flutter. Tech. rep., NASA TM 104211 (1992).

[7] Tichy, L., *Transsonische Strömungen an einem schwingenden Profil und deren Einflußauf die Flattergrenze*. Dissertation, TU München (1992), DLR-FB 92-08.

[8] Tijdeman, H., *Investigation on the Transonic Flow around Oscillating Airfoils*. Ph.D. thesis, TU Delft (1977), NLR TR 77090 U.

[9] Triebstein, H., Schewe, G., Zingel, H. and Vogel, S., Measurement of unsteady airloads on an oscillating engine and a wing/engine combination. *J. Aircraft*, **31(1)**, 97–102 (1994).

[10] Romberg, H.-J., *Experimentelle Untersuchung der schallnahen Umströmung eines superkritischen Tragflügelprofils unter besonderer Berücksichtigung von Windkanalinterferenzen*. Dissertation, RWTH Aachen (1990).

[11] Amecke, J., Direkte Berechnung von Wandinterferenzen und Wandadaption bei zweidimensionaler Strömung in Windkanälen mit geschlossenen Wänden. Forschungsbericht, DFVLR-FB 85-62 (1985).

[12] Voß, R., *AGARDograph 336: Wind Tunnel Wall Corrections*, chap. Wall Correction Methods for Dynamic Tests, pp. 9–1 – 9–29. NATO AGARD (1998).

[13] Kleine, H. and Grönig:, H., Color Schlieren Methods in Shock Wave Research. *Shock Waves*, **(1)**, 51–63 (1991).

[14] Hufnagel, K., *Entwicklung und Optimierung von Sechs-Komponenten-DMS-Windkanalwaagen zum Einsatz unter kryogenen Bedingungen*. Dissertation, TH Darmstadt (1995).

[15] Meinke, M. and Krause, E., Simulation of Incompressible and Compressible Flow on Vector-Par allel Computers. In: C. A. Lin, A. Ecer, N. Satofuka, P. Fox and J. Periaux (eds.), *Parallel Computational Fluid Dynamics*, pp. 25–34. North Holland (1999).

[16] Fares, E., Meinke, M., and Schröder, W., Numerical Simulation of the Interaction of Wingtip Vortices and Engine jets in the Near Field. AIAA paper 2000-2222 (2000).

[17] H.H. Pearcey, J. Osborne, and A.B. Haines. The interaction between local effects at the shock and rear separation - a source of significant scale effects in wind-tunnel tests on aerofoils and wings. In *AGARD-CP-35*, pages 11–1 – 11–23, 1968.

[18] Fung, Y., *An Introduction to the Theory of Aeroelasticity*. Dover (1993).

[19] G. Schewe, A. Knipfer, H. Mai, and G. Dietz. Experimental and numerical investigation of nonlinear effects in transonic flutter. DLR IB 232-2002 J 01, 2002.

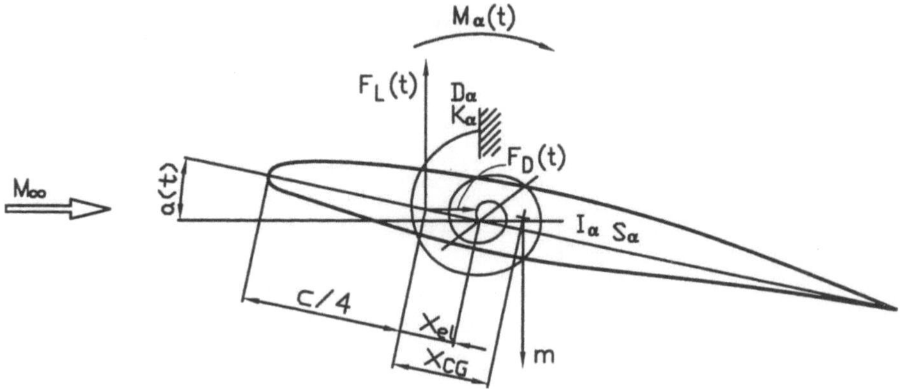

Figure 1 The airfoil oscillating in pitch

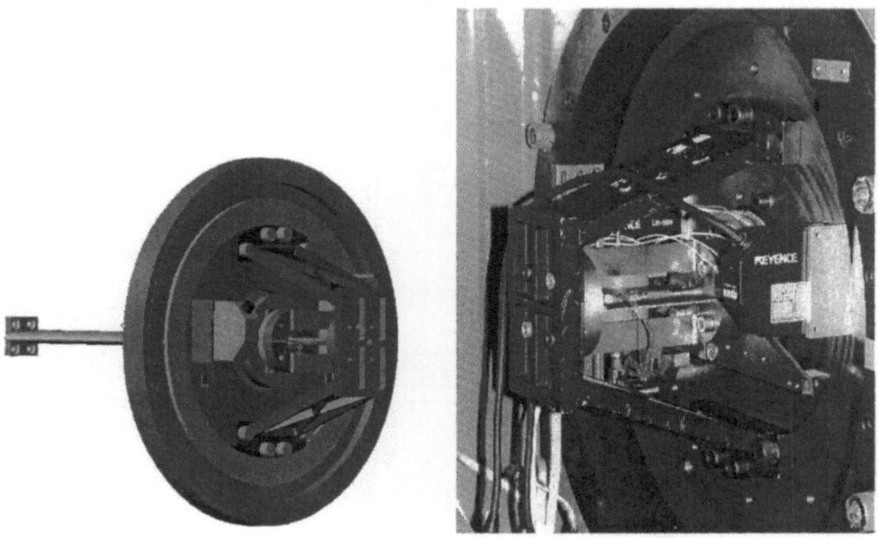

Figure 2 Design and installation of the model suspension.

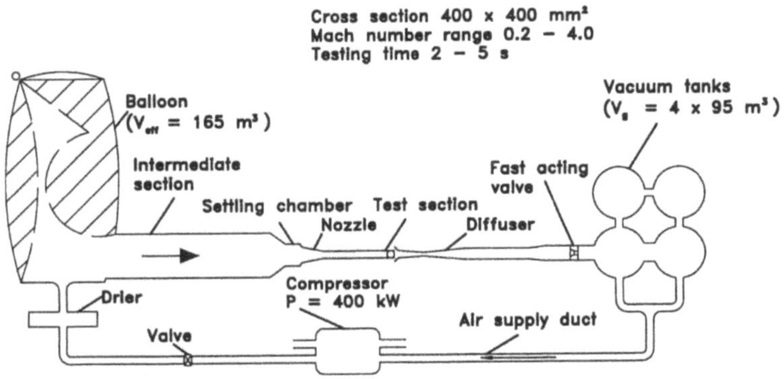

Figure 3 Trisonic wind tunnel

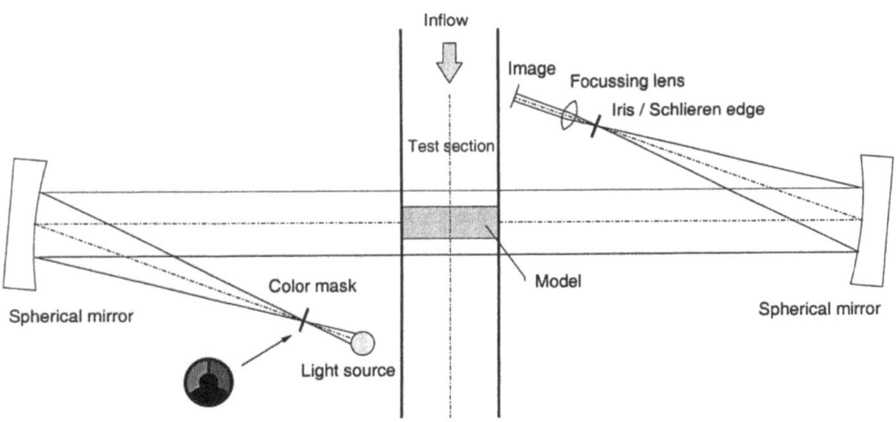

Figure 4 Principal arrangement for color Schlieren technique

Figure 5 Carbon fiber reinforced wind tunnel model...

Figure 6 ...with in situ pressure sensors

Figure 7 Two panels of the strain gauge balance. Front view of the panel for tangential reactions.

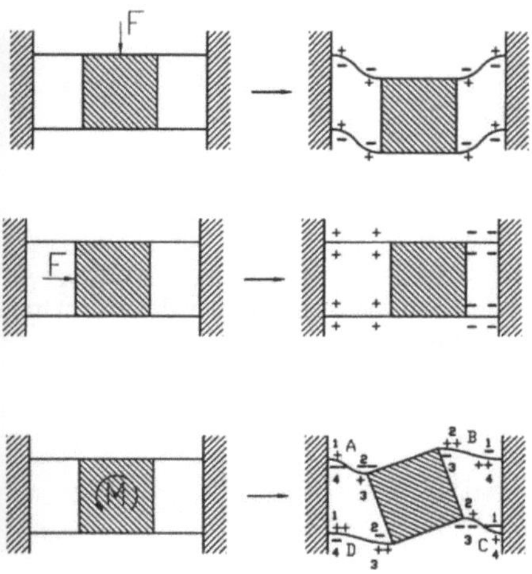

Figure 8 Principle of force measurement

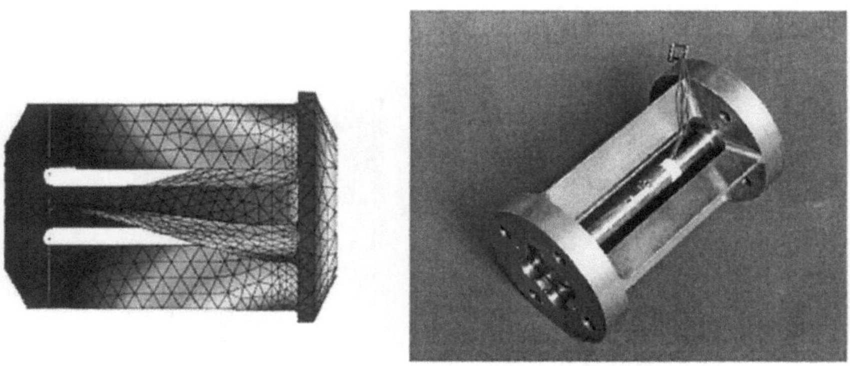

Figure 9 Finite element simulation of the torsion spring

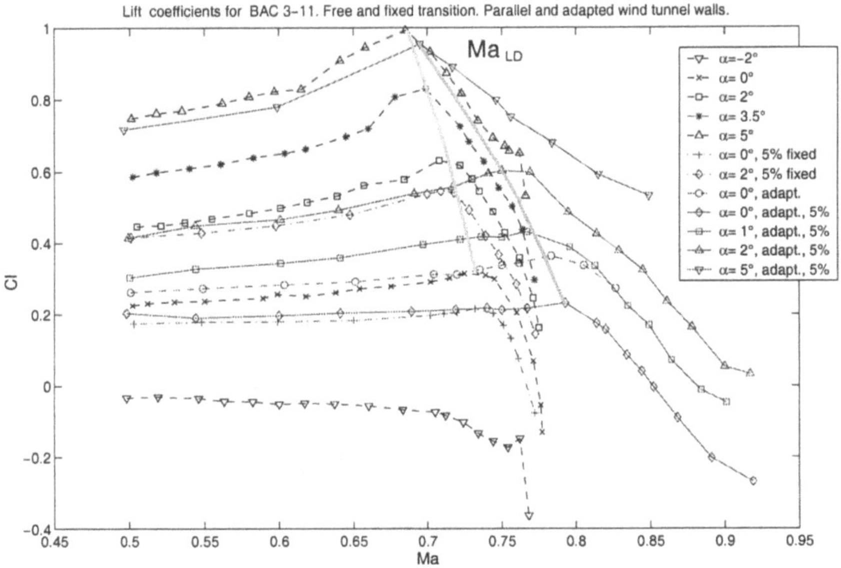

Figure 10 Measured lift coefficients at transonic Mach numbers for different angles of attack and different flow conditions.

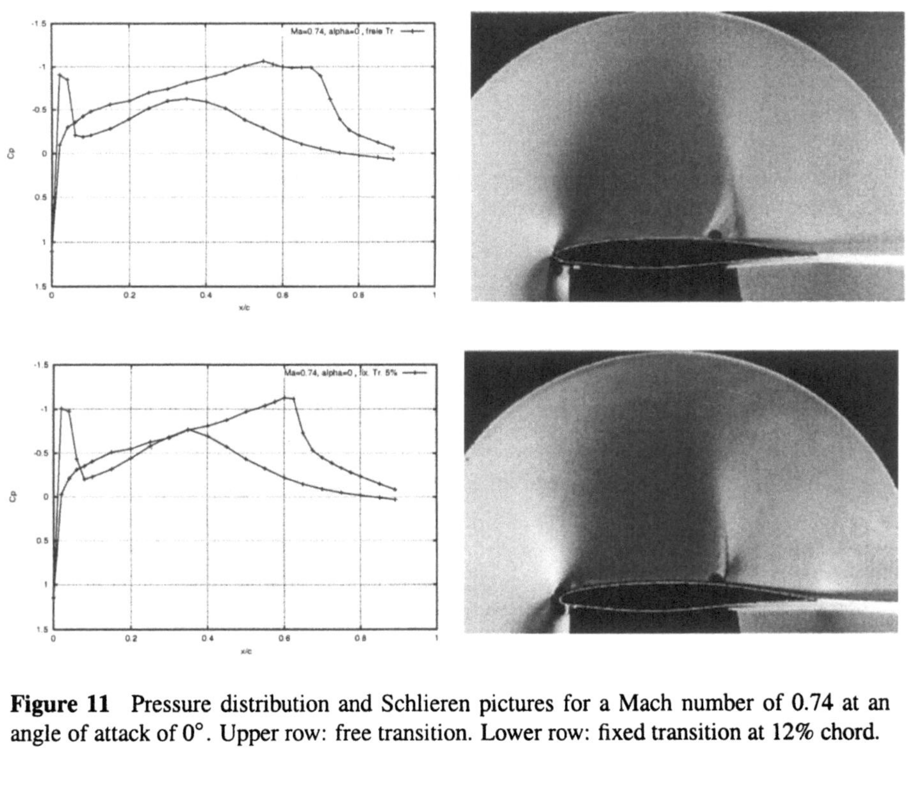

Figure 11 Pressure distribution and Schlieren pictures for a Mach number of 0.74 at an angle of attack of 0°. Upper row: free transition. Lower row: fixed transition at 12% chord.

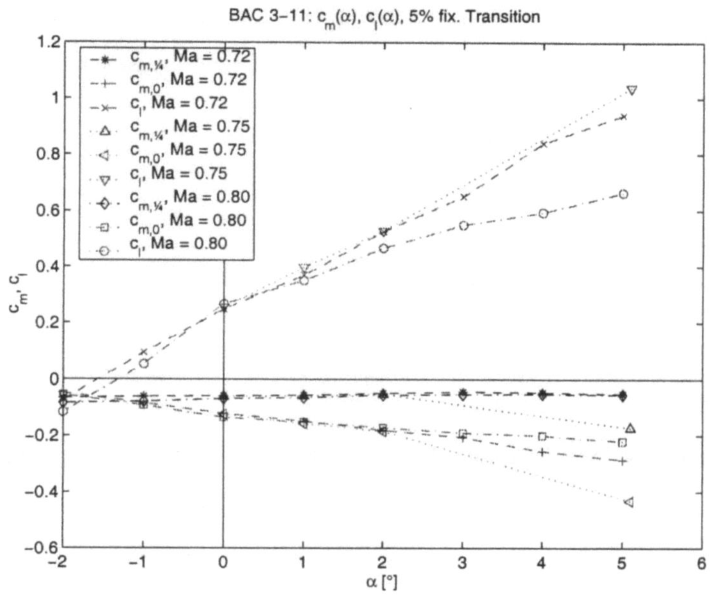

Figure 12 Lift and moment coefficients as a function of the angle of attack for Mach number of 0.72, 0.75, and 0.80, fixed transition (5%).

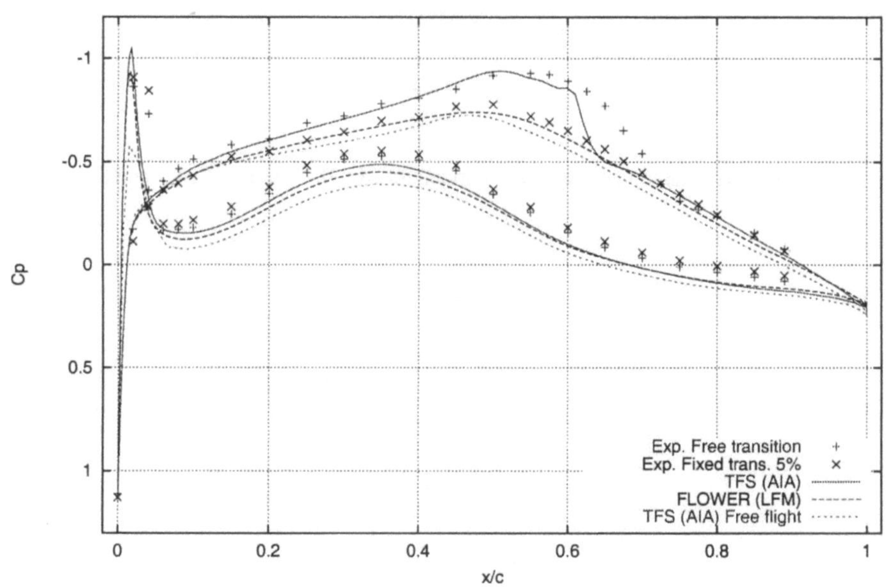

Figure 13 Comparison experiment - computation: Wind tunnel wall effect at M = 0.70, $\alpha = 0°$.

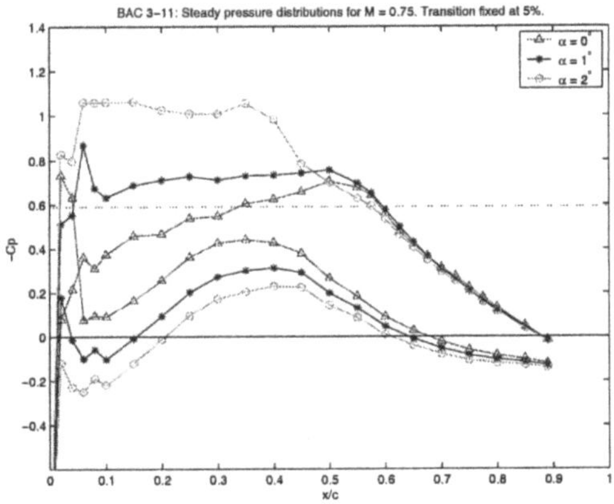

Figure 14 Steady pressure distributions for M = 0.75 at $\alpha = 0°$, 1°, and 2°.

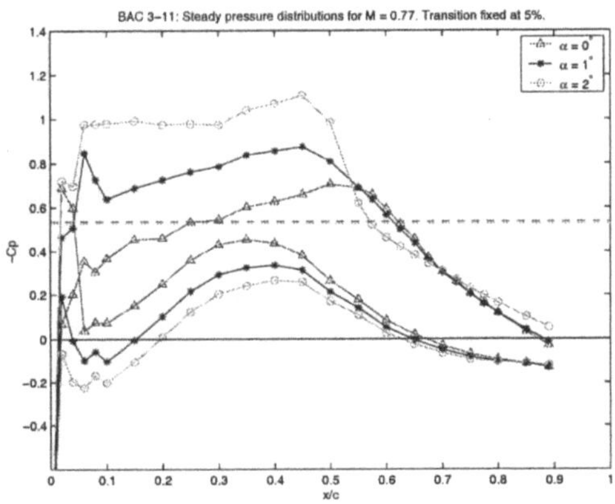

Figure 15 Steady pressure distributions for M = 0.77 at $\alpha = 0°$, 1°, and 2°.

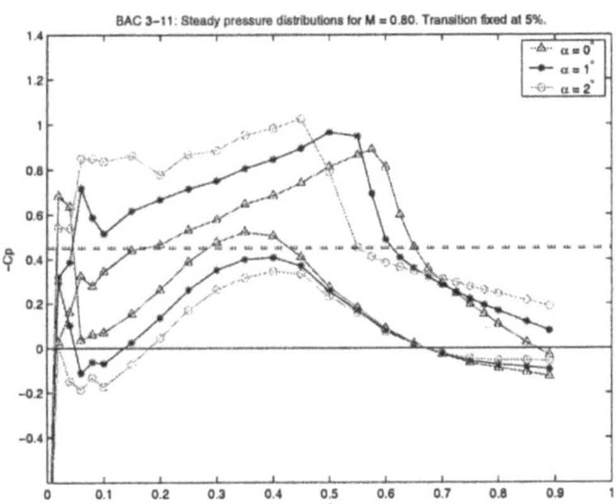

Figure 16 Steady pressure distributions for M = 0.80 at $\alpha = 0°$, 1°, and 2°.

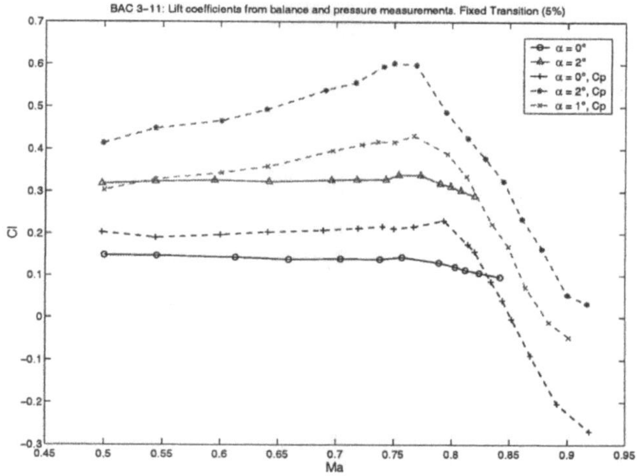

Figure 17 Lift coefficients for $\alpha = 0°$ to $2°$. Results from steady test series.

Figure 18 Oilflow pattern of the wind tunnel side wall influence.

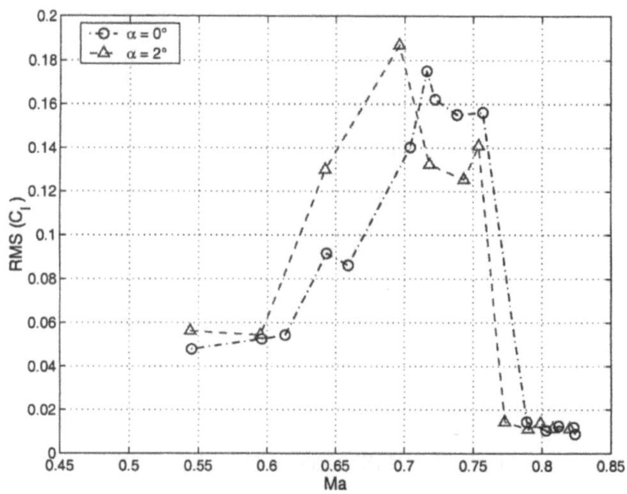

Figure 19 RMS(C_l) values for $\alpha = 0°$ and $2°$. Results from steady test series.

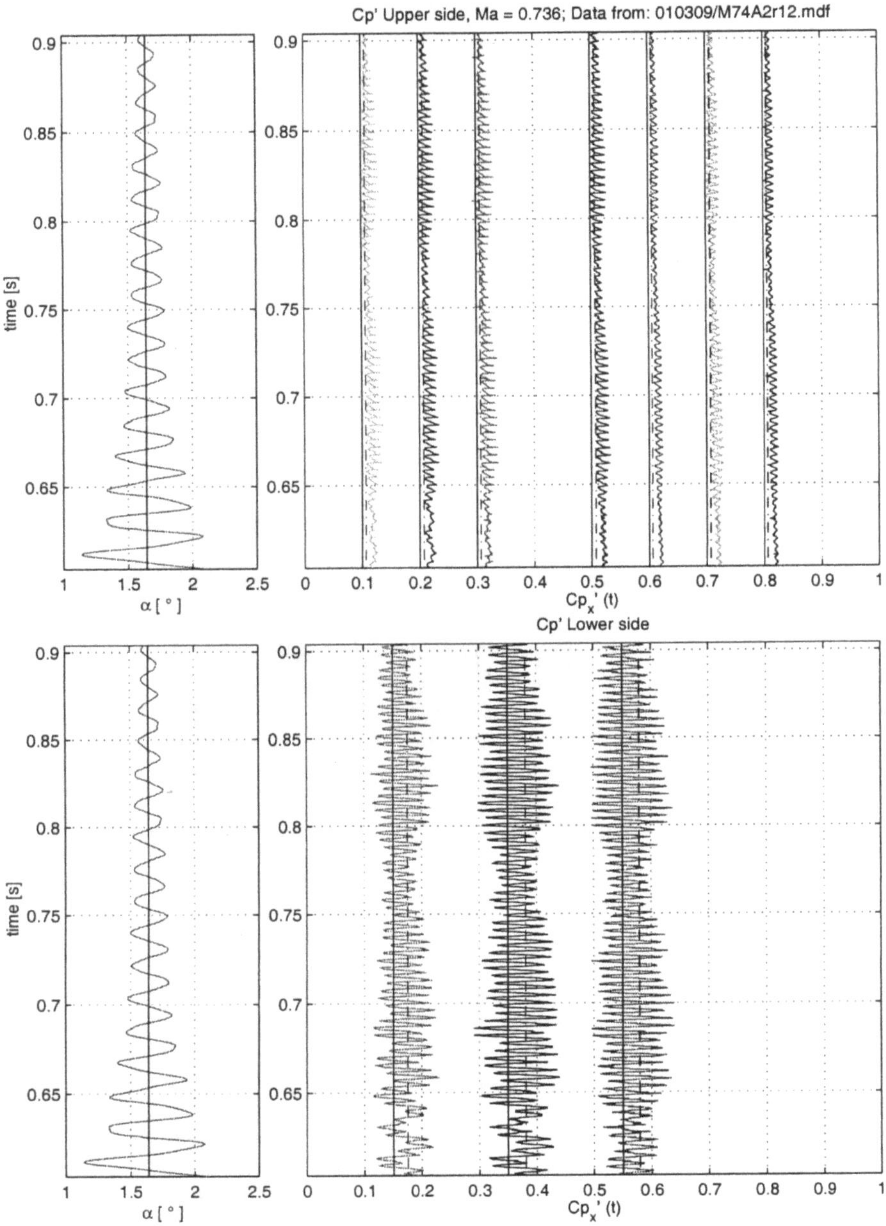

Figure 20 Fluctuations and RMS values of C_p at Mach = 0.74 and $\bar{\alpha}$ = 1.64°.

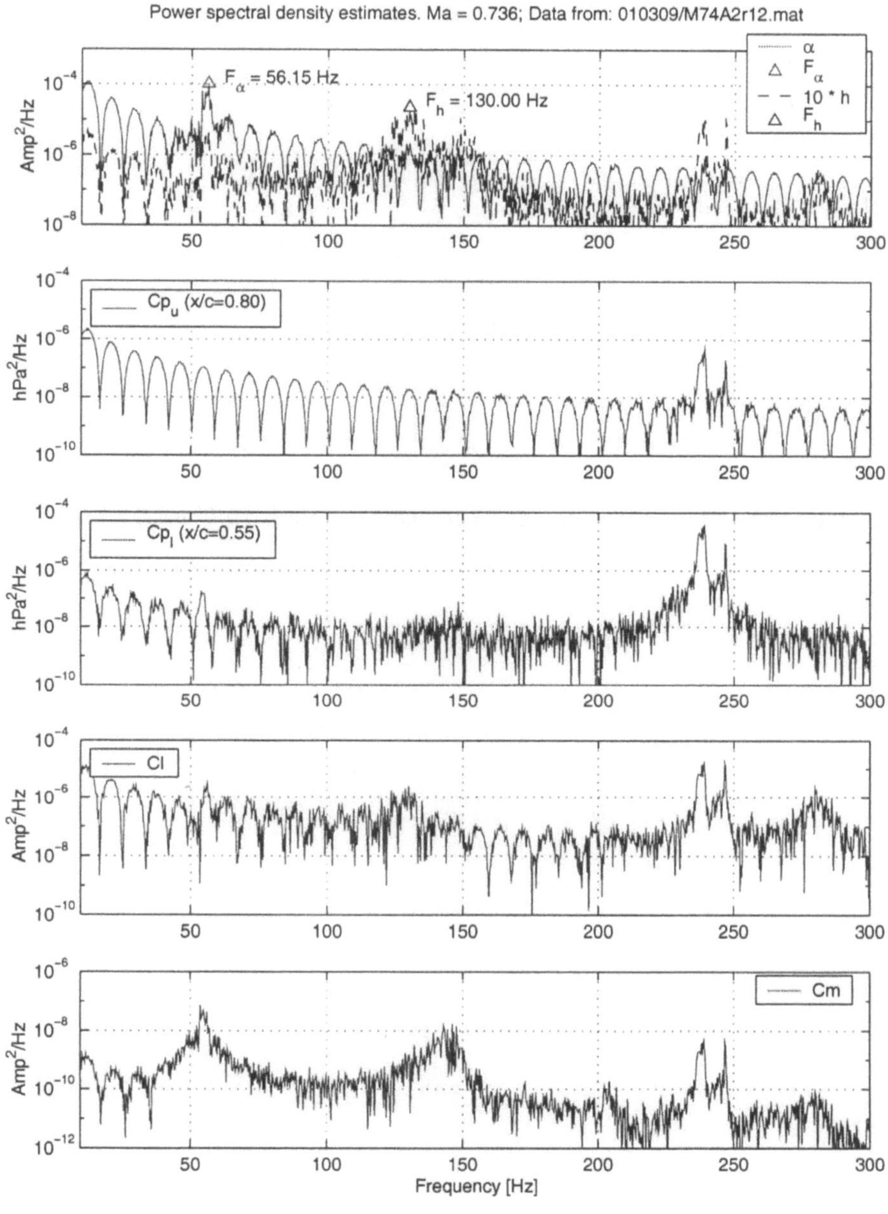

Figure 21 Power spectra of $\alpha(t)$, pressure sensors No. 8 (upper side, x/c=0.80) and 11 (lower side, x/c=0.55), c_l and c_m at Mach = 0.74.

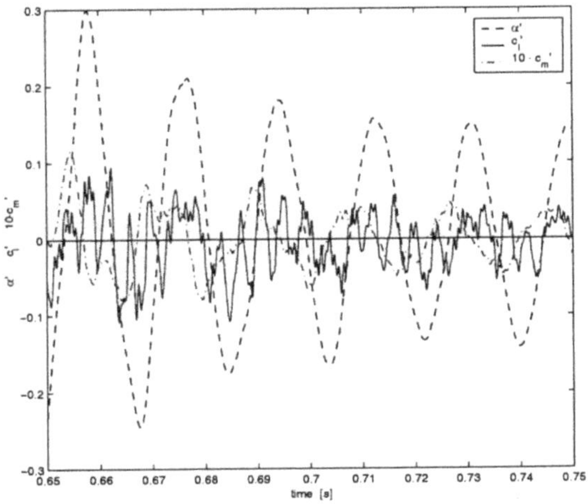

Figure 22 Time series of $\Delta\alpha$, $\Delta c_m * 10$, and Δc_l at Mach = 0.74 and $\bar{\alpha} = 1.64°$.

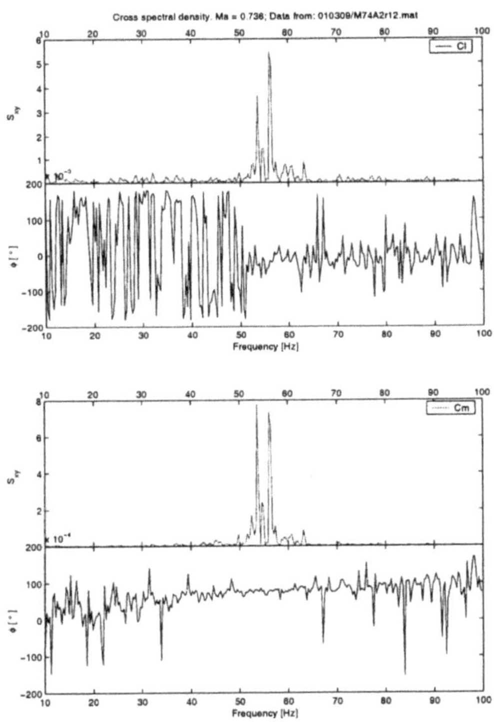

Figure 23 Phase angles φ between the aerodynamic coefficients $c_{m,l}$ and α at Mach = 0.74.

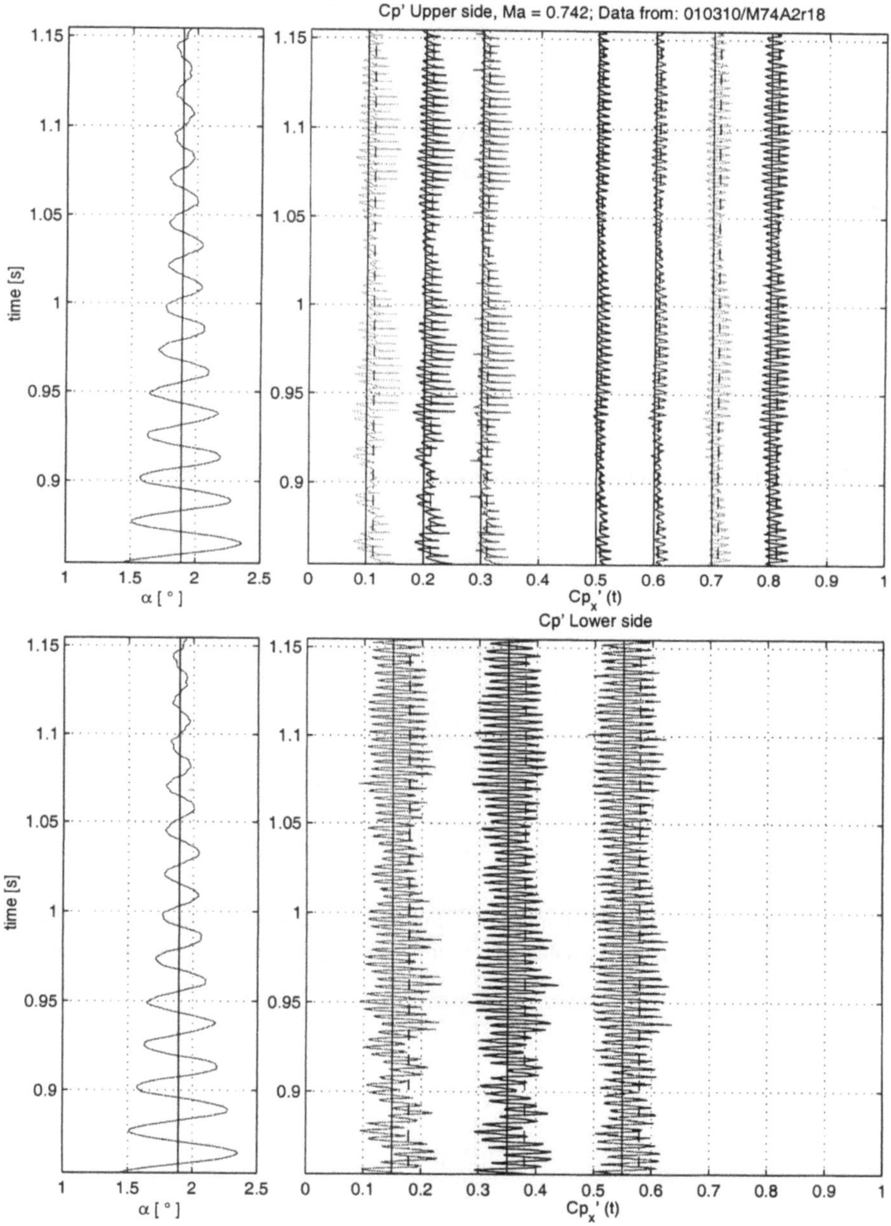

Figure 24 Fluctuations and RMS values of C_p at Mach = 0.74 and $\bar{\alpha} = 1.89°$.

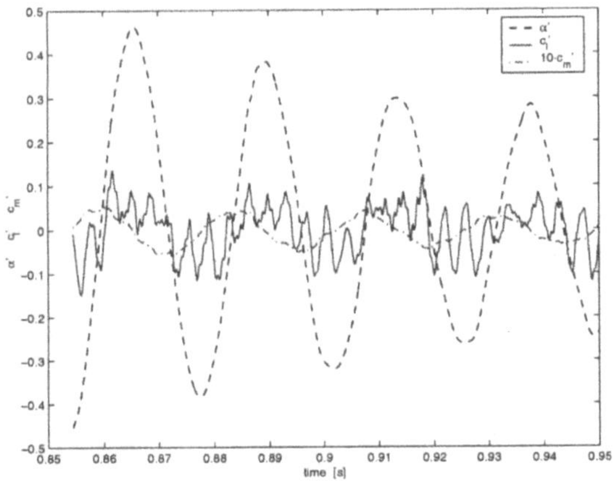

Figure 25 Time series of $\Delta\alpha$, $\Delta c_m * 10$, and Δc_l at Mach = 0.74 and $\bar{\alpha} = 1.89°$.

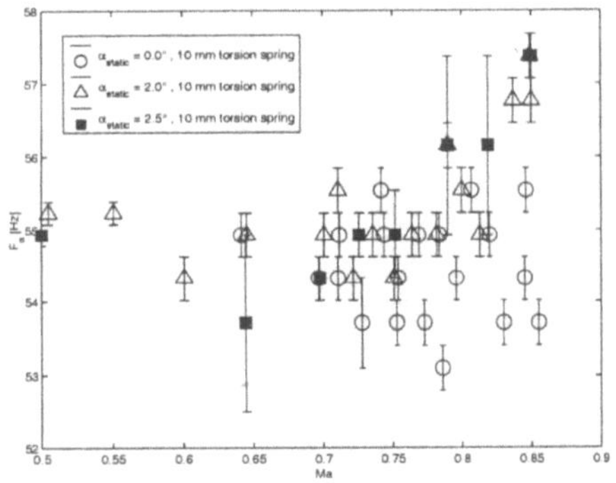

Figure 26 Pitch frequency for 10 mm torsion spring depending on Mach number

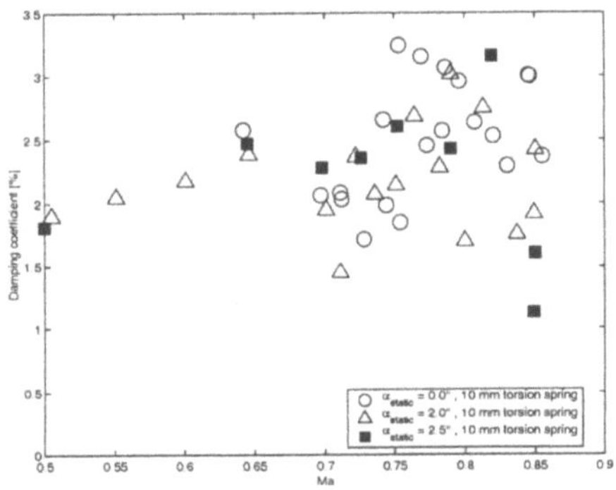

Figure 27 Damping coefficient for 10 mm torsion spring depending on Mach number

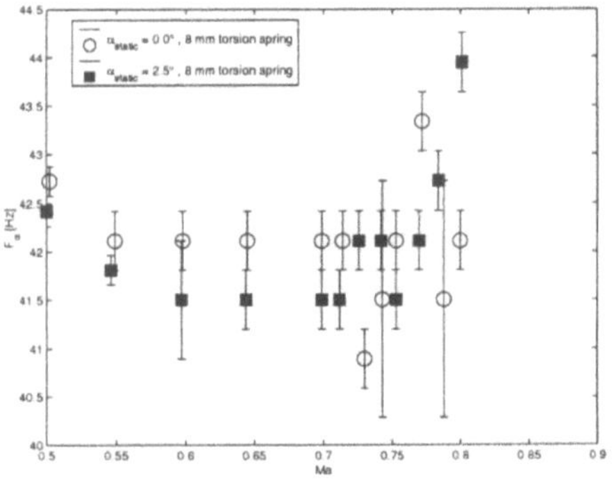

Figure 28 Pitch frequency for 8 mm torsion spring depending on Mach number

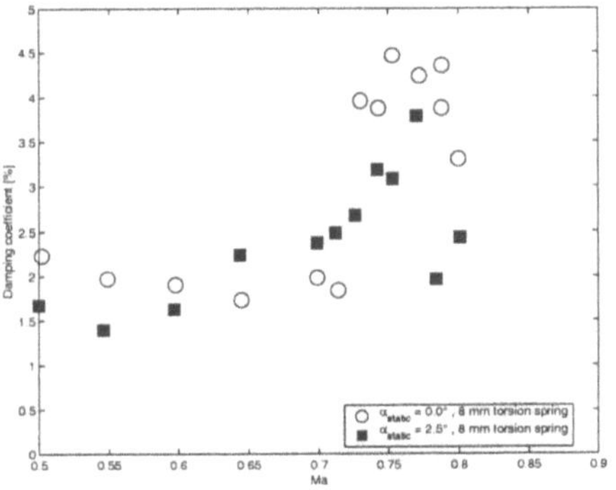

Figure 29 Damping coefficient for 8 mm torsion spring depending on Mach number

Design, Qualification and Experimental Investigation on Flexible Wind Tunnel Wing Models

M. Kämpchen, H. Korsch, A. Dafnis, H.-G. Reimerdes
Lehrstuhl und Institut für Leichtbau, RWTH Aachen
Wüllnerstraße 7, D-52062 Aachen, Germany

Summary

The purpose of this work on highly flexible wing models is the validation of the numerical direct simulation, developed in the Collaborative Research Center SFB 401. The first step in the validation process is the investigation of a flexible straight wing model in the subsonic region. The design and construction of this wing model will be described under consideration of aerodynamic and structural constraints. A short introduction in the measurement techniques and the tests under wind-off and wind-on conditions will be given. Different data reduction methods to obtain the frequency and the damping factor during wind tunnel tests will be introduced. Some experiments under steady state and transient flow conditions will be presented for validation of the numerical calculations.

1 Introduction

Aeroelasticity and aero-structural phenomena become increasingly more important in the design process of large civil aircraft. Since the transport capacity raises with concurrently reduced structural weight, the deformation of the wing induces a significant influence on the aerodynamics and has to be taken into account in the early design process. For the time being a numerical code is developed within the Collaborative Research Center (SFB 401) at the Aachen University to investigate the fluid-structure interaction at airplane wings. For the validation of this numerical code experimental results from wind tunnel tests are required.

In the recent years a couple of flexible wing models have been developed suitable for aero-structural and aeroelastic investigations in the subsonic and the transsonic region. Well-known aeroelastic wing models are the ARW-2 [1] and the I-Wing 445.6 [2]. The latter was selected as the first AGARD Standard Aeroelastic Configuration. It is often used for validation of numerical simulation methods [3, 4, 5, 6]. Recently another flexible wing model named MAVRIC-I [7] was investigated at NASA Langley Research Center for validation purposes in the transonic region.

The flexible wing model used in the SFB 401 is a rectangular wing with a simple, beam-like structure. The purpose was to obtain experiments under steady state and transient flow conditions. For the respective numerical simulation a Timoshenko-beam model for the wings structure was a good representation for the behaviour of the real wing model. Due to the very low structural damping of the real wing model the damping factor could be set to zero for the simulation. This was essential for the first step in the validation process of the numerical direct simulation, particularly for the verification of the fluid-structure coupling as well as the grid generator.

In this paper, a methodology for the specification, qualification and validation of highly flexible wing models is presented and experimental results are given for the rectangular wing model. The validation process is shown in the following scheme (Figure 1). Dotted boxes indicate items whose investigation is not yet completed.

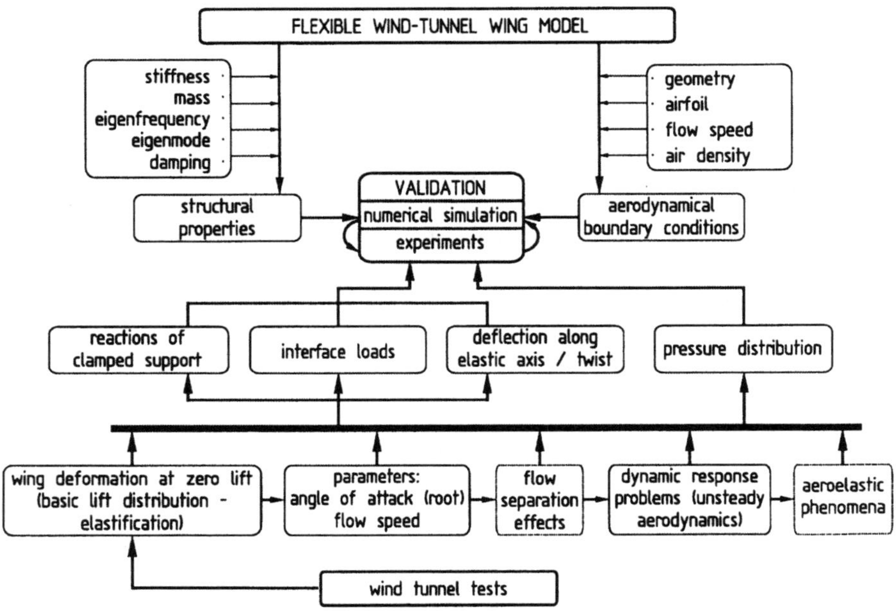

Figure 1 Scheme of the validation process

On the upper side of Figure 1 is the flexible wind tunnel wing model which determines the experiment as well as the structural and geometric input data for the numerical simulation. On the lower side there are the validation experiments which have to be compared to results of the numerical simulation.

2 Wing Model Design, Manufacturing and Instrumentation

The flexible wing model presented in this paper was designed for investigation of static and dynamic aero-structural response problems in a low speed wind-tunnel at the Aachen University (1.5 m diameter, max. 70 m/s). No aeroelastic phenomena during the first segment of research should be involved in the approach. For this reason the center line of mass and the elastic axis coincide approximately at 40 % chord length.

2.1 Wing Model Design

The design method of highly flexible wing models differs in some aspects from that of rigid wing models, normally used for wind-tunnel tests. The equilibrium state between aerodynamics and structural elasticity is affected by large deformations, in particular by twisting. Consequently the aerodynamic loads cannot be determined on the basis of airfoil/shape-theory assuming rigid structures. It is evident that critical wing characteristics, e.g. zero lift (basic lift distribution) and flow separation behaviour of flexible wings, are different from those of rigid wings. For these reasons flexible wing models have to be designed according to their mechanical behaviour being set by both stiffness and mass distribution. The highly flexible wing model used in this investigation had to fulfill the following requirements:

- flexible load sustainable structure with large deflections of about 10 % of half of wingspan
- predefined bending and torsion stiffness without stiffness coupling (see Table 2)
- relatively low eigenfrequencies, reduced frequency of 1st, 2nd bending and 1st torsion $\omega^* < 1$
- low structural damping
- no aeroelastic instabilities within the relevant speed range $V_\infty = [0 \ldots 85]$ m/s
- sufficient accessibility for the application of sensors (strain gauges, pressure transducers, accelerometers)

At the Department of Aerospace Structures an analysis tool was developed enabling the prediction of the structural behaviour of straight and swept wing models. It provides an iterative approximation for determining the equilibrium state of the flexible wing under given flow conditions. For the structural description of the wing model during the design process, arbitrary finite elements can be applied, e.g. beam, shell and/or volume elements. The aerodynamic loads are calculated by the doublet–lattice method.

The rectangular wing model used in this investigation has 1 m half-span and a chord length of 222 mm. It has the airfoil of BAC3-11/RES/30/21 and the aspect ratio $\Lambda = 9$, according to the reference configuration of the SFB 401 [8]. It meets no other criteria of similarity.

For the spar of the flexible wing model, only metallic materials were evaluated, because of their relatively low structural damping (hysteresis). This way the aerodynamic damping influence became evident.

2.2 Wing Model Construction

To fulfill the requirements mentioned in section 2.1 a hybrid concept was preferred against the compact design introduced in [9] and the construction of a shaped plate introduced in [10]. It is assembled by the following structural components:

- wing spar
- ribs
- aerodynamic encasement

in which the appropriate measurement instrumentation has been integrated. The specific requirements of the static and dynamic properties described above were met using a strength/stiffness/mass iterative analysis.

The cross shaped wing spar as shown in Figure 2 was selected, to realize the required flexibility of the wing. In order to achieve a close plunge/twist coupling of the wing movement, the needed bending/torsion stiffness has been assessed by variation of its cross section globally and locally (see vertical and horizontal notches of cross section). The exact geometry is given in Table 1.

Table 1 Geometry of the wing spar cross-section [mm]

vertical part			horizontal part		
width	height	notch	width	height	notch
12	21	1.5 x 7	110	4	1.5 x 25

A low torsion stiffness was specially required to initiate both high static elastic deformations and close coupled eigenmodes regarding to response problems due to unsteady aerodynamics.

Different material properties (steel, titanium, aluminium) have been looked at in the design process with respect to their influence on the static and dynamic deformation behaviour [11]. The spar of the straight wing model was build of aluminum (FORTAL 7075-T6) to obtain low structural damping and large deflections within the elastic region of the material.

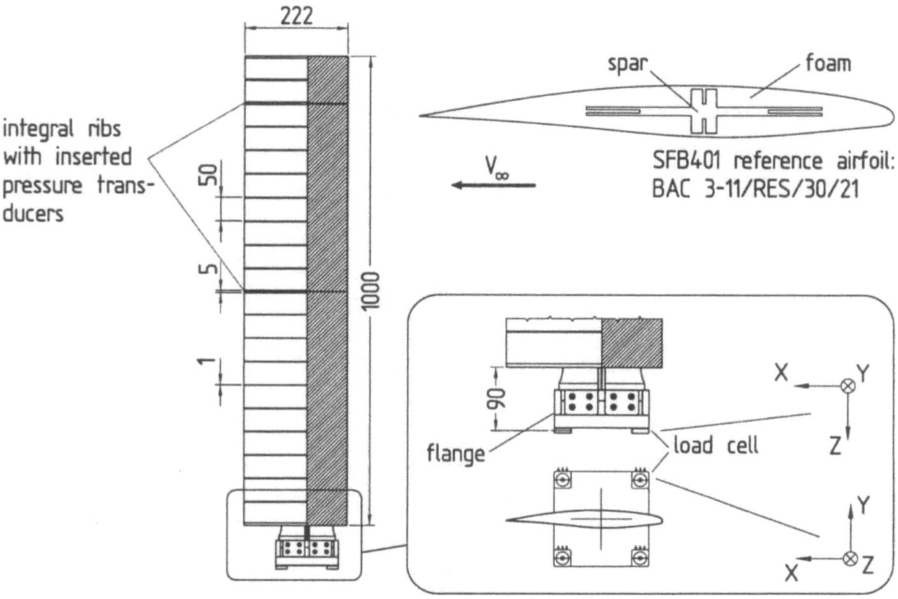

Figure 2 Shape of the straight wind-tunnel wing model

The wing spar thickens at the root and is fitted to a rigid flange. The transfer of the aerodynamic loads to the wing spar is obtained by ribs and polyurethane-integral-foam segments which fill the space between the ribs. A number of 21 aluminium-ribs are applied in order to keep a rigid airfoil geometry of the wing (inflexible airfoil camber). Two of them were prepared to receive in-situ pressure transducers. A 50 μm plastics foil provided a smooth surface.

2.3 Measurement Instrumentation

The flexible wing model is bolted to a rigid flange which is part of a piezoelectric 6-components load balance suitable to record the static and dynamic reaction forces and consequently moments. At 16 spanwise positions (see Figure 3) strain gauges were set to record bending and torsion strains. The cross-coupling of the signals was negligible because of the specific cross shape of the wing-spar. The application of strain gauges enables the determination of wing deflection under wind-on condition by means of Load Calibration Curves which have been yielded during laboratory specification tests (indirect method). It also enables the monitoring of the material strain during the tests.

A number of 9 miniaturised in-situ pressure transducers were inserted into the integral structure of one rib, being able to receive transient pressure signals directly. In addition a number of 10 very light weight accelerometers arranged in pairs were set

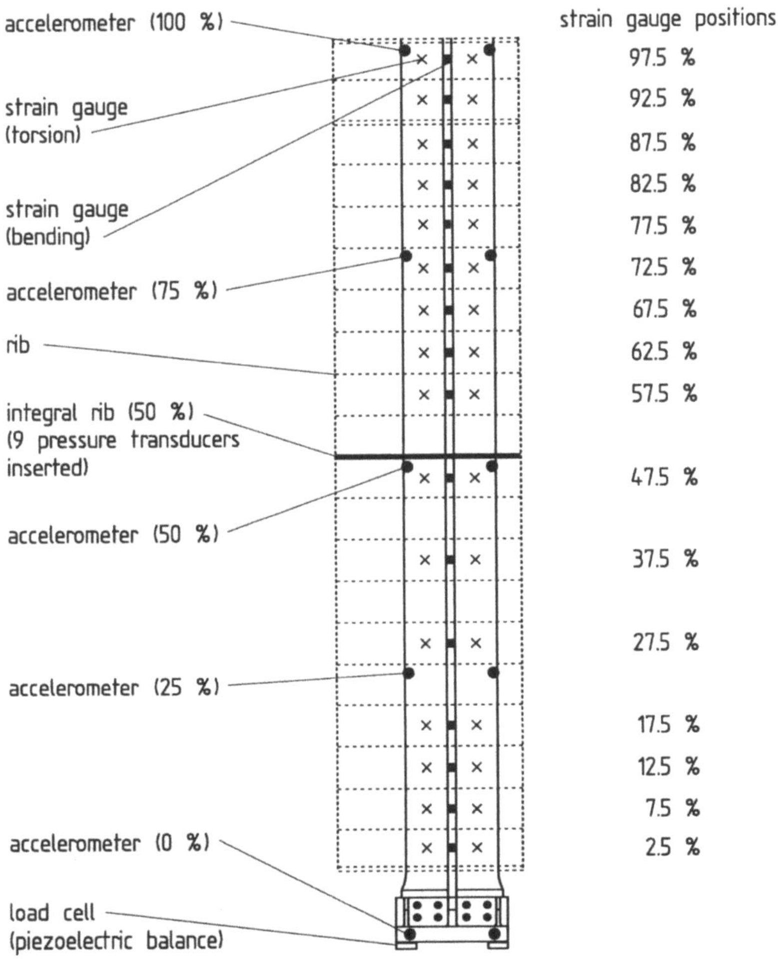

Figure 3 Sensor positions relative to half of wingspan

at 5 positions along the wingspan (see Figure 3) supporting investigation of dynamic response due to transient aerodynamics. For the time being an optical method, based on Moiré-Interferometry is engaged, being suited to record coupled bending/twist spatial deflections directly.

All components of the measurement system were calibrated and checked thoroughly by careful and proper laboratory procedures in order to eliminate systematic errors. Illegitimate and random errors were reduced by repetition of the measurements and by long time acquisition estimating possibly occuring errors, using statistical analysis. In this way the maximum expected errors of the applied measurement chains were restricted to 4 %.

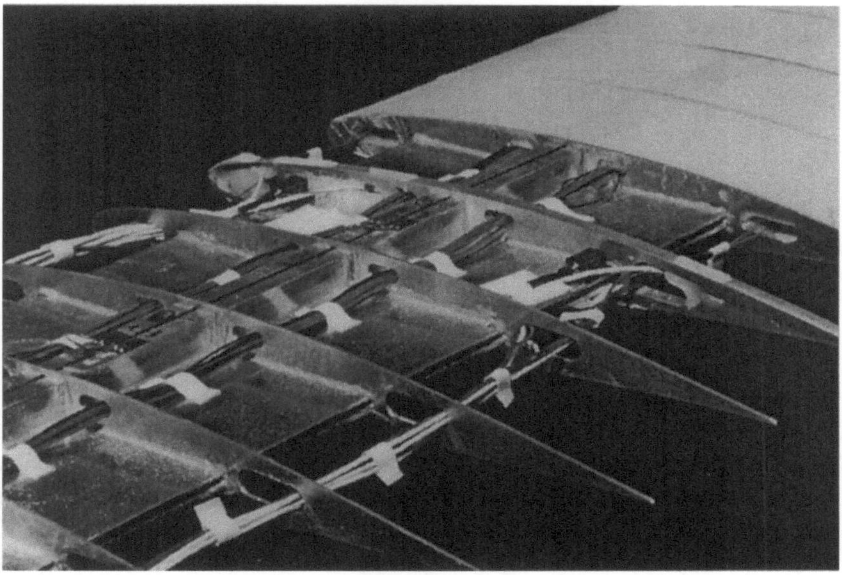

Figure 4 Measurement instrumentation of the rectangular wing model

3 Experimental Investigation

3.1 Wind-off Qualification

For the static tests a specific arrangement has been designed capable to introduce shear forces for bending and pure torsion moments into the wings structure. The relevant displacements were recorded by a three-dimensional distance measurement, consisting of two draw wire sensors and one compact laser-optical CCD displacement sensor (laser-triangulation).

First of all a laboratory identification of the structural properties of the wing has been carried out, yielding following information:

- bending/torsion stiffness distribution determining the influence numbers matrix
- determination of the elastic axis
- Load Calibration Curves which enable the reconstruction of the spatial deflection of the elastic axis of the wing under wind-on conditions (indirect method)

The stiffness of the wing model averaged along the wingspan is shown in Table 2.

In addition to the elastic properties of the wing model, the determination of the dynamic behaviour was also essential as input data for the numerical direct simulation:

Table 2 Measured/calculated stiffness of the flexible wing model [Nmm^2]

Flap Bending	Lag Bending *)	Torsion
$5.864 \cdot 10^8$	$2.178 \cdot 10^{10}$	$1.4133 \cdot 10^8$

*) calculated stiffness

- mass distribution, superimposing discrete masses
- eigenfrequency, eigenform, damping factor (parametric identification)

3.2 Wind Tunnel Specification and Test Schedule

The flexible wing model was designed for a subsonic wind-tunnel available at RWTH Aachen. It is a closed loop atmospheric wind-tunnel with an open test section. During the first test series continuous vibration disturbances occurred. They were caused by the relative high turbulence level in the test section of this wind-tunnel type, so that recorded signals had to be averaged. This led to unreliable results, particularly for the dynamic tests, and thus ineligible for validation purposes.

On this account new tests took place at the DNW-LST of the NLR in Emmeloord, a closed loop atmospheric wind-tunnel with a closed wall test section. Thanks to its proven low turbulence level the problem of the wing vibrations was eliminated. There appeared no more observable movement of the flexible wing model so that the measuring signals were characterized by a high signal-noise ratio.

Following items were investigated during the wind tunnel tests:

- determination of the angle of attack of zero-lift dependig on the dynamic pressure
- tests under steady state flow conditions ($V_\infty = [25 \ldots 65]$ m/s) and different angles of attack ($\Delta\alpha_w = [-2 \ldots + 8]°$)
- pluck tests at the respective flow speeds to determine the damping ratio depending on
  - the flow speed
  - the initial deflection
  - the angle of attack
- tests at separated flow conditions

The signals of the bending and torsion strain gauges, the reactions of clamped support, the pressures in one airfoil section and the accelerations were recorded at all tests with a sample rate of 600 Hz.

3.3 Data Reduction Methods

Different procedures have been applied to identify the dynamic behaviour of the wing model under wind-off and wind-on conditions. Three methods have been used in order to determine frequency, damping and mode shapes.

- **Fast Fourier Transformation (FFT) / Logarithmic Decrement Technique (LDT)**
  Time histories for each eigenfrequency have been estimated using Fast Fourier Transformation (FFT) techniques and bandpass filters. Damping ratios D_j have been determined in these decoupled time series by the Logarithmic Decrement Technique. The method is generally applicable to free responses.

- **Ibrahim Time Domain Method (ITD)**
  An alternative method to analyze free response time series is ITD [16]. It assumes a linear and time-invariant elasto-mechanical system with viscous damping. The dynamic behaviour in case of free responses can be described by the following equation:

  $$[M]\ddot{\underline{x}} + [D]\dot{\underline{x}} + [K]\underline{x} = \underline{0}. \tag{1}$$

  Here $[M]$, $[D]$ and $[K]$ are the n x n - matrices of mass, damping and stiffness whereas $\underline{x}$, $\dot{\underline{x}}$ and $\ddot{\underline{x}}$ are vectors of displacement, velocity and acceleration. The equation can be solved assuming free decay functions:

  $$\underline{x}(t) = \sum_{j=1}^{2n} \underline{\Phi}_j \, e^{\lambda_j t} \tag{2}$$

  $\underline{\Phi}_j$ are the eigenvectors and λ_j the eigenvalues of the problem which occurs in complex conjugate pairs in case of underdamped systems:

  $$\lambda_j = a_j \pm i b_j \tag{3}$$

  b_j defines the damped natural frequency ω_j for each mode j and the damping ratio D_j can be calculated by:

  $$D_j = \frac{a_j}{\sqrt{a_j^2 + b_j^2}} \quad . \tag{4}$$

  ITD deals with the problem as follows. Equation (1) will be solved by two special matrices $[U_1]$ and $[U_2]$ which can be set up by test data. $[U_1]$ and $[U_2]$ enable the determination of the eigenvectors $\underline{\Phi}_j$ and the eigenvalues λ_j by an eigenvalue problem [16]. Instead of displacement signals the usage of velocity or acceleration signals is possible as well. The application of ITD is restricted to free responses (Figure 5).

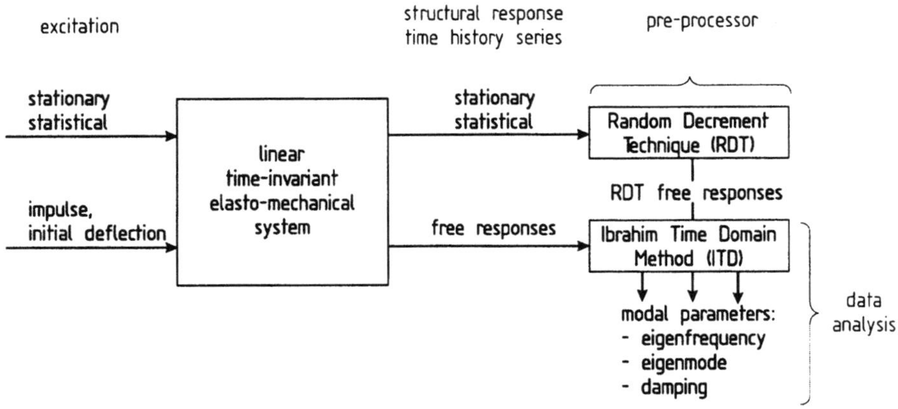

Figure 5 Combination of Random Decrement Technique and Ibrahim Time Domain Method

- **Random Decrement Technique (RDT)**
  The RDT is used as a pre-processor to determine free response time series called Random Decrement function's $y(\tau)$ which can be analyzed afterwards with regard to the structural vibration parameters frequency, damping and mode shapes. It assumes a stationary statistical random excitation so that the structural response is also stationary statistical.

  The Random Decrement function $y(\tau)$ can be calculated by ([14]):

$$y(\tau) = \frac{1}{n_E} \sum_{i=1}^{n_E} x(t_i + \tau) \quad , \; x(t_i) = x_E \; . \tag{5}$$

  The signal $x(t)$ is split into n_E time segments which start with the triggering value x_E which is a predefined amplitude or gradient. The arithmetic average of these segments yields the Random Decrement function dependent on the time parameter τ.

  In case of using multiple sensor signals which have been recorded simultaneously a so-called leading station has to be selected. The time step in which the signal of the leading station crosses the triggering value x_E determines the start-time for the other signals [15].

  For in-flight or windtunnel tests the natural turbulence can be assumed as a stationary statistical excitation and no additonal excitation is necessary. Here the RDT function is used to provide free response time series for the ITD (Figure 5).

4 Numerical Simulation

For the numerical direct simulation the SOFIA code is used in which higher order solutions for the aerodynamics and the structure dynamics are coupled. For the flow analysis of the here presented results the implicit Euler-method INFLEX was used. Alternatively, the utilisation of the Navier-Stokes-method FLOWer would also be possible. For the structural analysis a reduced one-dimensional FE-beam-model (Timoshenko-beam) with arbitrary bending and torsional stiffness distribution as well as mass inertia is applied, assuming a rigid airfoil cross-section. Within the code SOFIA the fluid-structure interaction was realized iterative in such a way that the new velocity and deformation state of the wing is calculated, based on the results of the previous time step. Thus a new airload condition can be determined. This procedure will be repeated for each time step until convergence is achieved. A detailed description of this method is given in item 4.3 of this issue of NNFM and in [12, 13].

5 Results

5.1 Angle of Attack of Zero-Lift

The first experimental step, using highly flexible wing models is the accurate determination of the angle of attack, with reference to the root chord line, in which the resulted lift becomes equal to zero. This investigation was carried out on the basis of the recorded signals of the piezoelectric balance. The pitching moment of the used non-symmetric BAC 3-11 airfoil twists the wing model proportionally to the dynamic pressure and it still takes effect if zero-lift is achieved. In order to keep the total lift equal to zero, stepping up the flow speed, the angle of attack of the wing root has to be raised (Figure 6). In this case the flexible wing model produces positive lift at the inner section which is balanced by the negative lift in the outer section of the wing. This phenomenon becomes obvious both in the experiment and the numerical simulation.

Figure 7 shows the "elastification" of the wing model at the given experimental boundary condition powering up and down the wind-tunnel continuously. This phenomenon is typical for highly flexible wing models during static aerodynamic investigations and can be explained as follows: the overall lift increases at the beginning due to the stepping up of velocity. At a later time the twist of the wing model becomes relevant so that the resulted lift approaches to zero continuously. The small lift force of about 1 N remaining (Figure 7) is caused by the limited adjustment capability of the angle of attack of the test rig.

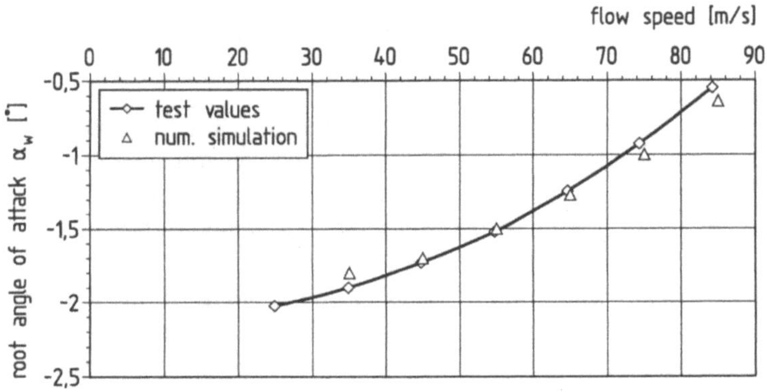

Figure 6 Root chord line angle of attack of zero-lift vs. flow-speed

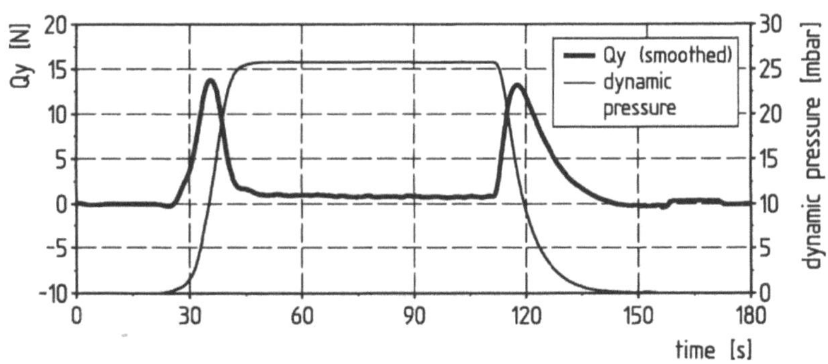

Figure 7 Zero-lift at V_∞=65 m/s, α_w=-1.33°

5.2 Steady State Flow Conditions

In the following step both the reactions of clamped support measured by means of the piezoelectric balance and the spatial elastic deflections were experimentally determined and compared with the respective numerical simulations. At this point the deflections with reference to the elastic axis were extracted indirectly from the calibrated signals of the bending-strain gauges. Experimental results, with regard to the tests at the DNW-LST in Emmeloord (V_∞=45 m/s, α_w=6.6°) are exemplary presented. Table 3 presents the measured and calculated lift and torque at the root of the wing. The graphs in Figure 8 and Figure 9 show that the deviation between the numerical simulation and the experiment is acceptable.

To verify the indirect evaluation of the deformations several twice-exposed photographs were analyzed, on which the flexible wing model is pictured with and without incoming flow (Figure 10). In future time this verification will be substi-

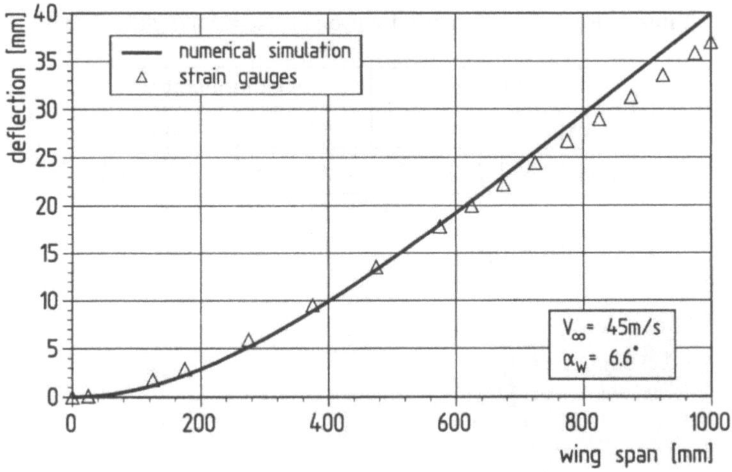

Figure 8 Deflection line of the wing model

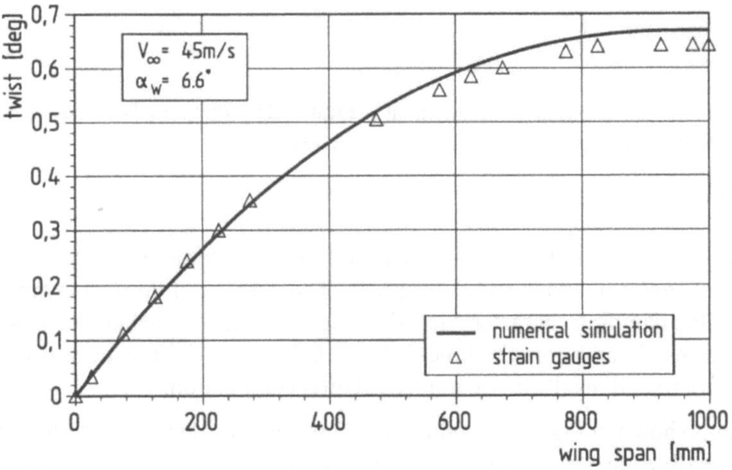

Figure 9 Twist distribution of the wing model

tuted by the Moiré-Interferometry.

The good results of the measured and calculated wing deflection, are also verified comparing the recorded and calculated pressure coefficients in the airfoil section at 50 % wingspan as shown in Figure 11.

Table 3 Responses at clamped support at V_∞=45 m/s and α_w=6.6°

lift force		twist moment	
exp.	num.	exp.	num.
204.30 N	212.97 N	-4.03 Nm	-3.72 Nm

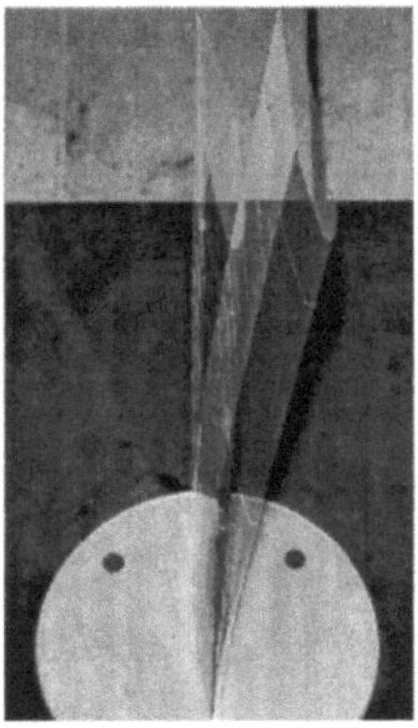

Figure 10 Twice exposed photograph at V_∞=0 / 65 m/s, α_w=6.93°

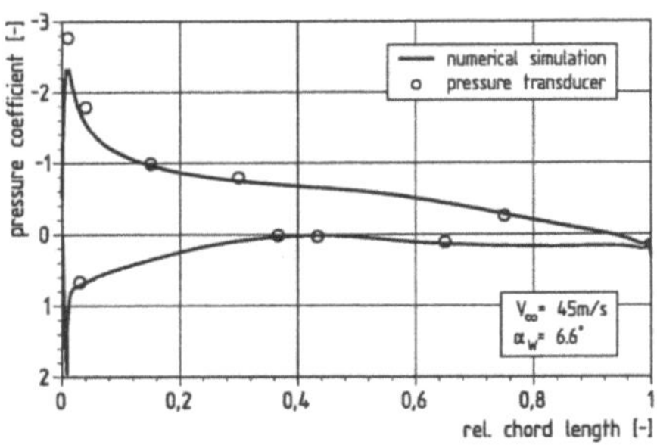

Figure 11 Pressure Distribution at 50 % wingspan

5.3 Pluck Tests

Apart from the tests under steady state flow conditions some pluck tests were done at different angles of attack and flow speeds. The purpose of this test campaign was to extract the damping characteristics of the flexible wind-tunnel model under wind-on conditions. In addition, the time-decayed behaviour of the wing was compared with the computed results of the numerical direct simulation.

The graphs shown in Figure 12 and Figure 13 present the decaying oscillation of the straight flexible wing model during a flow condition of V_∞=45 m/s and the zero-lift angle of attack (α_w=-1.87°). For the pluck tests the wing was pulled by a thin fiber fixed at the wing tip close to the leading edge. Thereby both a bending and a torsion moment could be generated.

Cutting the fiber at the time point t=0 s decaying plunge and twist oscillations of the wing model could be initiated simultaneously. The numerical simulation and the experiment show a good compliance. One of the targets of the design process was to obtain a simple load carrying structure with negligible structural damping. This wing structure could be idealized as a Timoshenko-beam with minor error. Obviously it was not necessary to consider structural damping for the numerical simulation, there was only aerodynamic damping in effect regarding Figure 12 and Figure 13.

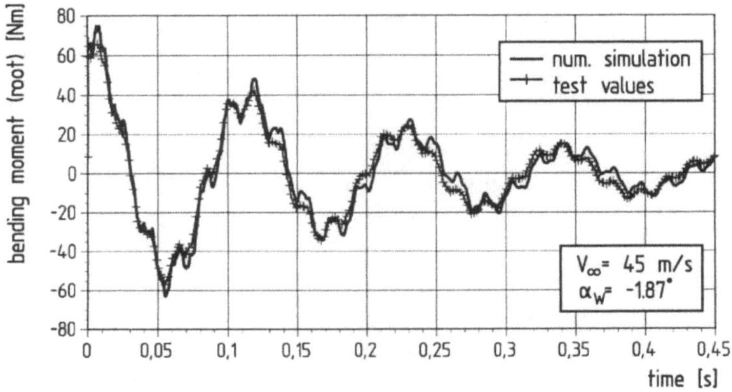

Figure 12 1st bending oscillation under wind-on conditions (pluck test)

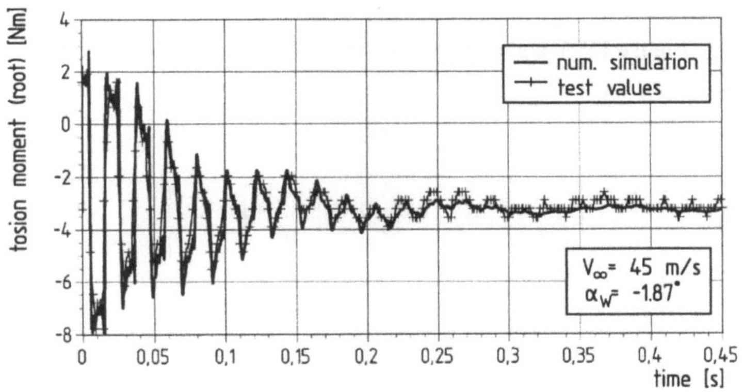

Figure 13 1st torsion oscillation under wind-on conditions (pluck test)

5.4 Flow Separation Tests

Additionally some flow separation tests have been undertaken. Because the respective numerical simulations are still under development, no comparison of the results of the flow separation tests to numerical simulation has been possible.

During the test the root angle of attack was increased stepwise (1°, 0.5°) starting with $\alpha_w = 5°$. First flow separation effects have been detected at 10.8° root angle of attack. In the following graphs two states of the flexible wing model will be shown:

[1] entirely clean flow without separation at $\alpha_w = 10°$

[2] highest measured angle of attack ($\alpha_w = 15°$) including flow separation effects

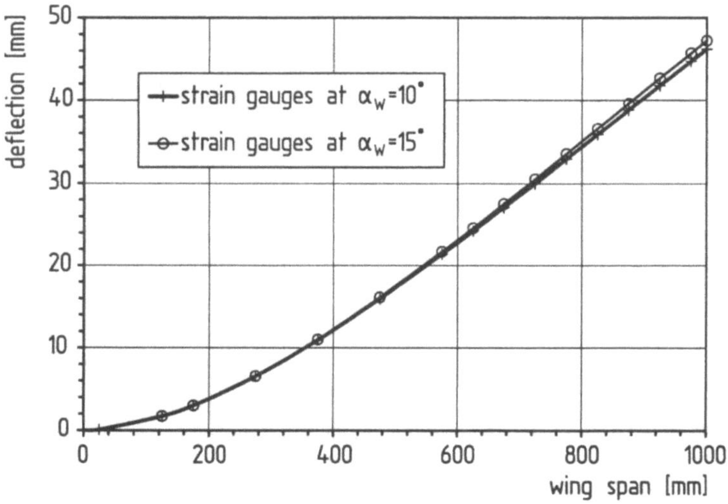

Figure 14 Wing model deflection at V_∞=45 m/s, α_w=10° / 15°

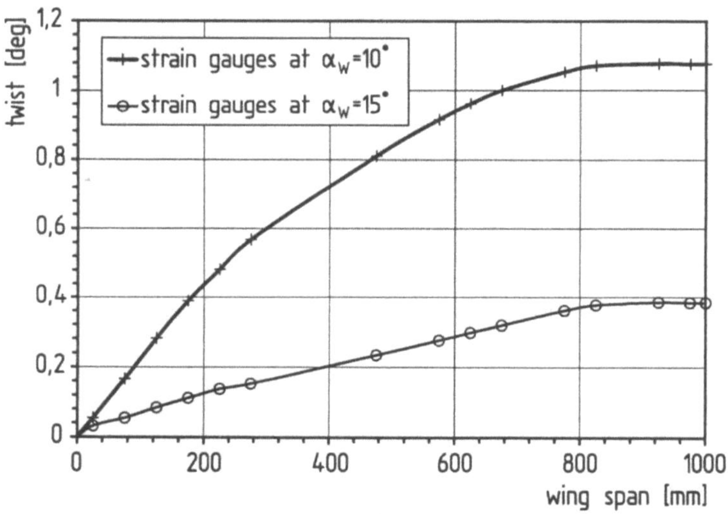

Figure 15 Twist of the flexible wing model at V_∞=45 m/s, α_w=10° / 15°

After the occurence of flow separation, the total lift, root bending moment and the deflection at the wing tip remained approximately constant while the root angle of attack was being increased (see Table 4 and Figure 14). This behaviour was particularly caused by the flexibility of the wing model, which was less twisted under the conditions of separated flow (Figure 15) causing lower local angles of

Table 4 Responses at clamped support at V_∞=45 m/s and α_w=10° / 15°

α_w	lift force	bending moment	twist moment
10°	254.5 N	146.4 Nm	-5.68 Nm
15°	243.50 N	143.6 Nm	-1.75 Nm

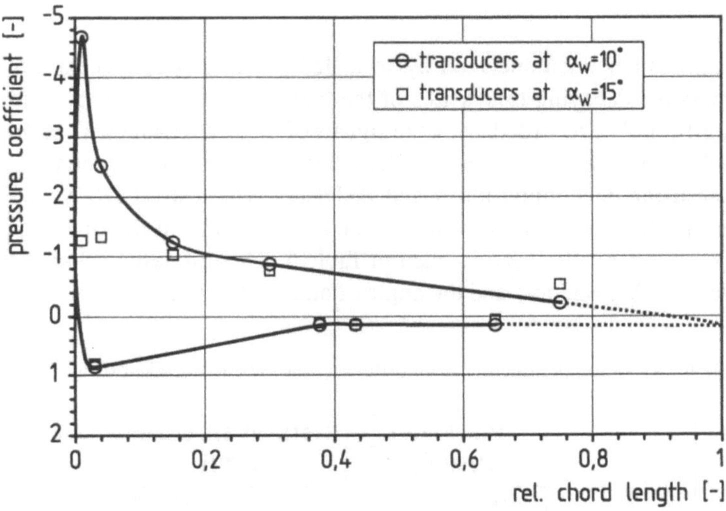

Figure 16 Pressure distribution at 50 % halfspan, α_w=10° / 15°

attack. Regarding the respective pressure distribution (Figure 16), measured at 50 % halfspan, representative for the area with separated flow, the reason for the reduced twisting moment becomes obvious.

5.5 Dynamic Analysis

The wind-off tests of the wing model have been carried out in the laboratory. In this case the model has been stimulated by an impulse at the wing tip and time history responses have been recorded by accelerometers. The free decay time series have been analyzed by Fast Fourier Transformation/Logarithmic Decrement Technique (FFT/LDT). The results are presented in Table 5.

The wind-on testing has been carried out in two subsonic wind-tunnels, a closed loop atmospheric wind-tunnel with an open test section at Aachen University (RWTH Aachen) and the DNW-LST of the NLR in Emmeloord (NL) with a closed wall test section (see section 3.2) resulting in a low turbulence level.

The dynamic behaviour of the wing model under wind-on conditions (unsteady

Table 5 Analyses of free decay time responses under wind-off conditions

	$f\,[Hz]$	$D\,[\%]$
1st plunge	8.79	0.23
1st torsion	51.27	0.50
2nd plunge	54.20	0.11

aerodynamics) has been analyzed by so called pluck tests (see section 5.3). Besides others the decaying time series of the first three modes (1st and 2nd plunge, 1st torsion) could be recorded using the signals of the strain gauges. Post-processing analysis determining frequencies as well as damping of the excited modes have been performed using decoupling filters and FFT/LDT. The tests at DNW-LST lead to better results because of the low turbulence level and with it the high signal-noise ratio. Exemplary results are presented in Table 6. The test case is characterized by a flow velocity V_∞=35 m/s and the angle of attack of zero-lift.

Table 6 Vibration parameters from different analyses at V_∞=35 m/s and zero-lift

	Pluck Test DNW-LST		**Natural Turbulence Test RWTH Aachen**		
	FFT & LDT		**FFT**	**RDT & ITD**	
	$f\,[Hz]$	$D\,[\%]$	$f\,[Hz]$	$f\,[Hz]$	$D\,[\%]$
1st plunge	8.84	6.21	9.17	9.23	5.97
1st torsion	49.88	2.93	–	48.59	2.16
2nd plunge	54.08	0.76	55.17	55.02	0.53

This procedure of the external excited wing model in order to determine the dynamic structural parameters (frequency, damping and normal modes) under wind-on conditions is a time consuming and complex task in view of the experimental setup. Looking to full scale wings the measurement of the dynamic behaviour requires an experimental setup which enables the excitation of all significant modes without influencing both the mass and stiffness distribution of the structure as well as the aerodynamics. The determination of excitation forces to calculate the transfer functions which yield the structural vibration parameters represents an additional problem.

Looking for an alternative method to circumvent the above mentioned problems in future the application of two well known methods in combination (see Figure 5), the Random Decrement Technique (RDT) [14] and Ibrahim Time Domain Method (ITD) [16], [15] is used simultanously. Tests in the subsonic wind-tunnel at RWTH Aachen submitted the application of this method to determine the dynamic structural parameters. The tests called "natural turbulence tests" had shown a dynamic response under constant wind-on conditions (V_∞ = const.; α = const.) and a wind-tunnel turbulence level of 1.5-2%. Consequently a stationary statistical excitation of the wing model could be assumed in these tests and the RDT and the ITD could be applied. In this case the structural responses were determined by accelerometers. Tests under similar conditions at DNW-LST effected no dynamic response of the wing model because of the very low turbulence level of 0.03%.

Table 6 includes the results of this analyzing method as well. The "natural turbulence tests" under the respective flow conditions described above submit the application of the combination of RDT and ITD (Figure 5). Additionally, an FFT has been used to estimate the frequencies during these tests.

6 Conclusion

The dimensioning of highly flexible wind-tunnel wing models as well as the approach of experimental investigations for the purpose of validation work of numerical calculations were presented. The measurement instrumentation and techniques for the determination of the wing model's deformation state, the aerodynamic loading and local pressure coefficients were introduced. The accurate wind-off qualification process was leading to reliable input data for the numerical simulation. Different tests in a subsonic atmospheric wind-tunnel under steady and transient flow conditions were conducted with a flexible straight wing model. The derived results corresponded well with these of the numerical direct simulations.

References

[1] C.V. Eckstrom, D.A. Seidel and M.C. Sandford. Measurements of Unsteady Pressure and Structural Response for an Elastic Supercritical Wing. *NASA Technical Paper 3443*, 1994

[2] E.C. Yates. AGARD Standard Aerolastic Configurations for Dynamic Response – I.-Wing 445.6. *AGARD Report No.765*, 1998

[3] F. Liu, J. Cai, Y. Zhu, A.S.W. Wong and H.M. Tsai. Calculation of Wing Flutter by a Coupled CFD-CSD Method. AIAA-2000-0907, *38th Aerospace Sciences Meeting & Exhibit*, Reno, 2000

[4] W.A. Silva and D.E. Raveh. Development of Unsteady Aerodynamic State-Space Models From CFD-Based Pulse Responses. AIAA 2001-1213,*42nd AIAA/ASME/ASCE/AHS/ASC Structures, Structural Dynamics and Materials Conference*, Seattle, 2001

[5] J.J. Meijer and B.B. Prananta. Static Aeroelastic Simulation of Military Aircraft Configuration in Transsonic Flow. *IFASD International Forum on Aeroelasticity and Structural Dynamics*, Madrid, pp. 119-132, 2001

[6] R. Palacios, H. Climent, A. Karlsson and B. Winzell. Assessment of Strategies for Correcting Linear Unsteady Aerodynamics Using CFD or Test Results. *IFASD International Forum on Aeroelasticity and Structural Dynamics*, Madrid, pp. 195-210, 2001

[7] J.W. Edwards, D.M. Schuster, C.V. Spain, D.F. Keller and R.W. Moses. MAVRIC Flutter Model Transsonic Limit Cycle Oscillation Test. *NASA TM-2001-210877*, 2001

[8] E. Özger, I. Schell and D. Jacob. On the Structure and Attenuation of an Aircraft Wake. *Journal of Aircraft*, Vol. 38, No. 6, 2001

[9] E. Lee-Rausch and J. Batina. Wing flutter boundary prediction using unsteady euler aerodynamic method. *Journal of Aircraft*, Vol. 32, No 1, pp 416-422, 1995

[10] O. Bendiksen and G.-Y. Hwang. Nonlinear flutter calculations for transsonic wings. *CEAS Forum on Aeroelasticity and Structural Dynamics*, Rome, 1997

[11] A. Dafnis, W. Jung and H.-G. Reimerdes. Auslegung elastischer Windkanal-Flügelmodelle zur Validierung aeroelastischer Berechnungen. *Aeroelastik-Tagung der DGLR*, pages 261–316, Göttingen, Germany, 1998.

[12] R. Heinrich, K. Pahlke and H. Bleecke. A Three Dimensional Dual-Time Stepping Method for the Solution of the Unsteady Navier-Stokes Equations. *RAS 'Unsteady Aerodynamics' Conference.*, London, pp. 5.1-5.12, 1996.

[13] D. Nellessen and J. Ballmann. Schallnahe Strömungen um elastische Tragflügel. *Fortschrittsberichte VDI.*, Reihe 7: Strömungstechnik, Nr. 302, 1995.

[14] H. A. Jr. Cole. On-Line Failure Detection and Damping Measurement of Aerospace Structures by Random Decrement Signatures. *NASA Contractor Report 2205*, Washington D.C., 1973.

[15] S. R. Ibrahim. Random Decrement Technique for Modal Identification of Structures. *Journal of Spacecraft*, Vol. 14, No. 11, 1977

[16] S. R. Ibrahim and E. C. Mikulcik. A Method for the Direct Identification of Vibration Parameters from the Free Response. *Shock and Vibration Bulletin*, No. 47, 1977

Correction to: Automated Gradient Calculation

Christian H. Bischof, H. Martin Bücker, Bruno Lang, Arno Rasch

Correction to: Chapter 8 in: Christian H. Bischof, H. Martin Bücker, Bruno Lang and Arno Rasch, *Automated Gradient Calculation*, DOI: 10.1007/978-3-540-44866-2_8

Author name in this chapter was incorrectly captured as H. Martin Bicker. The author name should read as H. Martin Bücker. This has been corrected.

The updated online version of the original chapter can be found at https://doi.org/10.1007/978-3-540-44866-2_8.

J. Ballmann, *Flow Modulation and Fluid—Structure Interaction at Airplane Wings*,
DOI 10.1007/978-3-540-44866-2_16

Notes on Numerical Fluid Mechanics and Multidisciplinary Design

Available Volumes

Volume 83: L. Davidson, D. Cokljat, J. Fröhlich, M. A. Leschziner, C. Mellen, W. Rodi (eds.): LESFOIL: Large Eddy Simulation of Flow Around a High Lift Airfoil - Results of the Project LESFOIL supported by the European Union 1998 - 2001. ISBN 3-540-00533-1

Volume 82: E. H. Hirschel (ed.): Numerical Flow Simulation III - CNRS-DFG Collaborative Research Programme, Results 2000-2002. ISBN 3-540-44130-1

Volume 81: W. Haase, V. Selmin, B. Winzell (eds.): Progress in Computational Flow-Structure Interaction - Results of the Project UNSI, supported by the European Union 1998-2000. ISBN 3-540-43902-1

Volume 80: E. Stanewsky, J. Délery, J. Fulker, P. de Matteis (eds.): Drag Reduction by Shock and Boundary Layer Control - Results of the Project EUROSHOCK II, supported by the European Union 1996-1999. ISBN 3-540-43317-1

Volume 79: B. Schulte-Werning, R. Grégoire, A. Malfatti, G. Matschke (eds.): TRANSAERO - A European Initiative on Transient Aerodynamics for Railway System Optimisation. ISBN 3-540-43316-3

Volume 78: M. Hafez, K. Morinishi, J. Periaux (eds.): Computational Fluid Dynamics for the 21st Century. Proceedings of a Symposium Honoring Prof. Satofuka on the Occasion of his 60th Birthday, Kyoto, Japan, 15-17 July 2000. ISBN 3-540-42053-3

Volume 77: S. Wagner, U. Rist, H.-J. Heinemann, R. Hilbig (eds.): New Results in Numerical and Experimental Fluid Mechanics III. Contributions to the 12th STAB/DGLR Symposium Stuttgart, Germany 2000. ISBN 3-540-42696-5

Volume 76: P. Thiede (ed.): Aerodynamic Drag Reduction Technologies. Proceedings of the CEAS/DragNet European Drag Reduction Conference, 19-21 June 2000, Potsdam, Germany. ISBN 3-540-41911-x

Volume 75: E.H. Hirschel (ed.): Numerical Flow Simulation II. CNRS-DFG Collaborative Research Programme, Results 1998-2000. ISBN 3-540-41608-0

Volume 74: D. Henry, A. Bergeon (eds.): Continuation Methods in Fluid Dynamics. Contributions to the ERCOFTAC/EUROMECH Colloquium 383, Aussois, France, 6.-9. September 1998. ISBN 3-540-41556-4

Volume 73: M. Griebel, S. Margenov, P. Yalamov (eds.): Large-Scale Scientific Computations of Engineering and Environmental Problems II. Proceedings of the Second Workshop on „Large-Scale Scientific Computations", Sozopol, Bulgaria, June 2-6, 1999. ISBN 3-540-41551-3

Volume 72: W. Nitsche, H.-J. Heinemann, R. Hilbig (eds.): New Results in Numerical and Experimental Fluid Mechanics II. Contributions to the 11th AG STAB/DGLR Symposium Berlin, Germany 1998. ISBN 3-540-41552-1

Volume 70: W. Hackbusch, G. Wittum (eds.): Numerical Treatment of Multi-Scale Problems. Proceedings of the 13th GAMM-Seminar, Kiel, January 24-26, 1997. ISBN 3-540-41562-9

Volume 69: D. P. Hills, M. R. Morris, M. J. Marchant, P. Guillen (eds.): Computational Mesh Adaptation. ECARP - European Computational Aerodynamics Research Project. ISBN 3-540-41548-3

Volume 67: C. H. Sieverding, G. Cicatelli, J. M. Desse, M. Meinke, P. Zunino: Experimental and Numerical Investigation of Time Varying Wakes behind Turbine Blades. Results of the Industrial and Material Technology Aeronautics Research Project AER2-CT92-0048, 1992-1996. ISBN 3-540-41546-7

Volume 66: E. H. Hirschel (ed.): Numerical Flow Simulation I. CNRS-DFG Collaborative Research Programme. Results 1996-1998. ISBN 3-540-41540-8

Volume 64: R. Friedrich, P. Bontoux (eds.): Computation and Visualization of Three-Dimensional Vortical and Turbulent Flows. Proceedings of the Fifth CNRS-DFG Workshop on Numerical Flow Simulation, München, Germany, December 6 and 7, 1996. ISBN 3-540-41557-2

Volume 63: M. Fiebig, N. K. Mitra (eds.): Vortices and Heat Transfer. Results of a DFG-Supported Research Group. ISBN 3-540-41558-0

Volume 62: M. Griebel, O. Iliev, S. D. Margenov, P. S. Vassilevski (eds.): Large-Scale Scientific Computations of Engineering and Environmental Problems. Proceedings of the First Workshop on „Large-Scale Scientific Computations" Varna, Bulgaria, June 7-11, 1997. ISBN 3-540-41560-2

Volume 57: H. Deconinck, B. Koren (eds.): Euler and Navier-Stokes Solvers Using Multi-Dimensional Upwind Schemes and Multigrid Acceleration. Results of the BRITE/EURAM Projects AERO-CT89-0003 and AER2-CT92-00040, 1989-1995. ISBN 3-540-41535-1

MIX
Papier aus verantwortungsvollen Quellen
Paper from responsible sources
FSC® C105338

If you have any concerns about our products,
you can contact us on
ProductSafety@springernature.com

In case Publisher is established outside the EU,
the EU authorized representative is:
Springer Nature Customer Service Center GmbH
Europaplatz 3, 69115 Heidelberg, Germany

Printed by Libri Plureos GmbH
in Hamburg, Germany